Vol. 31. **Photometric Organic Analysis** (*in two parts*). By Eugene Sawicki

Vol. 32. **Determination of Organic Compounds: Methods and Procedures.** By Frederick T. Weiss

Vol. 33. **Masking and Demasking of Chemical Reactions.** By D. D. Perrin

Vol. 34. **Neutron Activation Analysis.** By D. De Soete, R. Gijbels, and J. Hoste

Vol. 35. **Laser Raman Spectroscopy.** By Marvin C. Tobin

Vol. 36. **Emission Spectrochemical Analysis.** By Morris Slavin

Vol. 37. **Analytical Chemistry of Phosphorus Compounds.** Edited by M. Halmann

Vol. 38. **Luminescence Spectrometry in Analytical Chemistry.** By J. D. Winefordner, S. G. Schulman and T. C. O'Haver

Vol. 39. **Activation Analysis with Neutron Generators.** By Sam S. Nargolwalla and Edwin P. Przybylowicz

Vol. 40. **Determination of Gaseous Elements in Metals.** Edited by Lynn L. Lewis, Laben M. Melnick, and Ben D. Holt

Vol. 41. **Analysis of Silicones.** Edited by A. Lee Smith

Vol. 42. **Foundations of Ultracentrifugal Analysis.** By H. Fujita

Vol. 43. **Chemical Infrared Fourier Transform Spectroscopy.** By Peter R. Griffiths

Vol. 44. **Microscale Manipulations in Chemistry.** By T. S. Ma and V. Horak

Vol. 45. **Thermometric Titrations.** By J. Barthel

Vol. 46. **Trace Analysis: Spectroscopic Methods for Elements.** Edited by J. D. Winefordner

Vol. 47. **Contamination Control in Trace Element Analysis.** By Morris Zief and James W. Mitchell

Vol. 48. **Analytical Applications of NMR.** By D. E. Leyden and R. H. Cox

Vol. 49. **Measurement of Dissolved Oxygen.** By Michael L. Hitchman

Vol. 50. **Analytical Laser Spectroscopy.** Edited by Nicolo Omenetto

Vol. 51. **Trace Element Analysis of Geological Materials.** By Roger D. Reeves and Robert R. Brooks

Vol. 52. **Chemical Analysis by Microwave Rotational Spectroscopy.** By Ravi Varma and Lawrence W. Hrubesh

Vol. 53. **Information Theory As Applied to Chemical Analysis.** By Karel Eckschlager and Vladimír Štěpánek

Vol. 54. **Applied Infrared Spectroscopy: Fundamentals, Techniques, and Analytical Problem-solving.** By A. Lee Smith

Vol. 55. **Archaeological Chemistry.** By Zvi Goffer

Vol. 56. **Immobilized Enzymes in Analytical and Clinical Chemistry.** By P. W. Carr and L. D. Bowers

Vol. 57. **Photoacoustics and Photoacoustic Spectroscopy.** By Allan Rosencwaig

Vol. 58. **Analysis of Pesticide Residues.** Edited by H. Anson Moye

Vol. 59. **Affinity Chromatography.** By William H. Scouten

Vol. 60. **Quality Control in Analytical Chemistry.** By G. Kateman and F. W. Pijpers

Vol. 61. **Direct Characterization of Fineparticles.** By Brian H. Kaye

Vol. 62. **Flow Injection Analysis.** *Second Edition*. By J. Ruzicka and E. H. Hansen

Vol. 63. **Applied Electron Spectroscopy for Chemical Analysis.** Edited by Hassan Windawi and Floyd Ho

Vol. 64. **Analytical Aspects of Environmental Chemistry.** Edited by David F. S. Natusch and Philip K. Hopke

Vol. 65. **The Interpretation of Analytical Chemical Data by the Use of Cluster Analysis.** By D. Luc Massart and Leonard Kaufman

(*continued on back*)

Laser Remote Chemical Analysis

CHEMICAL ANALYSIS

A SERIES OF MONOGRAPHS ON
ANALYTICAL CHEMISTRY AND ITS APPLICATIONS

VOLUME 94

A WILEY-INTERSCIENCE PUBLICATION

JOHN WILEY & SONS

New York / Chichester / Brisbane / Toronto / Singapore

Laser Remote Chemical Analysis

Edited by

RAYMOND M. MEASURES
Professor of Applied Science and Engineering
University of Toronto
Institute for Aerospace Studies
Toronto, Ontario

A WILEY-INTERSCIENCE PUBLICATION

JOHN WILEY & SONS

New York / Chichester / Brisbane / Toronto / Singapore

Library of Congress Cataloging in Publication Data:

Laser remote chemical analysis.

(Chemical analysis, ISSN 0069-2883; v. 94)
"A Wiley-Interscience publication."
1. Atmosphere—Remote sensing. 2. Atmosphere—Laser observations. 3. Atmospheric chemistry—Remote sensing. 4. Optical radar. I. Measures, Raymond M. II. Series.

QC871.L37 1988 551.5′028 87-13380
ISBN 0-471-81640-X

Printed in the United States of America

10 9 8 7 6 5 4 3 2 1

CONTRIBUTORS

Reinhard Beer, Jet Propulsion Laboratory, Pasadena, California

R. Stephen Brown, Chemistry Department, University of Toronto, Toronto, Ontario, Canada

Kent A. Fredriksson, National Swedish Environment Protection Board, Lund Institute of Technology, Lund, Sweden

E. David Hinkley*, Jet Propulsion Laboratory, California Institute of Technology, Pasadena, California

Frank E. Hoge, NASA Goddard Space Flight Center, Wallops Island, Virginia

Ulrich J. Krull, Chemistry Department, University of Toronto, Toronto, Ontario, Canada

Raymond M. Measures, University of Toronto Institute for Aerospace Studies, Toronto, Ontario, Canada

G. Mégie, Service d'Aéronomie, Centre Nationale des Recherches Scientifiques, Paris, France

Robert T. Menzies, Jet Propulsion Laboratory, California Institute of Technology, Pasadena, California

Christopher R. Webster, Jet Propulsion Laboratory, California Institute of Technology, Pasadena, California

*Present affiliation: Lockheed Advanced Aeronautics Company, Valencia, California

To my family
for their love and understanding,
and in particular to my wife
for her assistance in the preparation of this book

PREFACE

This book is concerned with "chemical analysis at a distance." Astrophysicists have been involved in such matters for decades as the extreme environment of the objects they study emit radiation that makes this possible. The benign environment of our planet makes such an undertaking practical only if we excite the atoms and molecules of interest. Although searchlights were initially used for this purpose, optical probing of the atmosphere (and later the hydrosphere) was significantly advanced by the invention of the laser due to its inherent capability of generating well-collimated beams of high irradiance and spectral purity.

Today lasers are having a dramatic impact in many diverse areas, ranging from medicine to material processing. They are also making it possible for photons to increasingly compete with electrons in many arenas, such as communications, signal processing, computing and mass storage of information. Indeed, the laser has been referred to as the most versatile tool ever invented by man. Its ability to perform chemical analysis at a distance should have a profound impact on remote sensing.

Within the past two decades we have grown acutely aware of the finite nature of the Earth and the fragile balance of its ecosystems. Environmental issues such as the influence of fluorocarbons and nitric oxide on the Earth's protective shield of ozone, the effect of carbon dioxide and volcanic dust on the climate, the formation of photochemical smog, oil pollution, and acid rain have drawn to our attention the ease with which the biosphere can be perturbed. This book reveals how lasers can be employed to undertake remote chemical analysis of our environment and should be of value to scientists, researchers, and students concerned with such matters.

Chapter 1 provides the fundamentals of laser remote sensing, while the subsequent chapters present comprehensive overviews of the different facets of the subject, written by contributers who are highly respected in each field. In a book of this nature it is not practical, nor desirable, to avoid some overlap of material since this permits each chapter to be reasonably self-contained.

Chapter 2 presents one of the most detailed studies of atmospheric transmission to be found anywhere and has been written in a manner that makes it useful to most people engaged in laser remote sensing.

Chapter 3 represents a comprehensive overview of environmental sensing based on infrared absorption of laser radiation, including the foundations of the subject and several examples of its application.

Chapter 4 serves to illustrate how lasers can be used to map specific atmospheric pollutants using the powerful **di**fferential **a**bsorption **l**idar, DIAL, technique.

Chapter 5 demonstrates how lasers can measure trace constituents in the atmosphere. This includes the use of resonance scattering to evaluate the distribution of trace metals in the upper atmosphere and DIAL to observe the ozone profile in the middle atmosphere.

Chapter 6 presents a comprehensive overview of how Earth-pointing lidar (using laser-induced fluorescence in many instances) can undertake airborne oceanic and terrestrial measurements that range from mapping of chlorophyll in natural bodies of water, to oil slick detection, thickness evaluation, and identification.

Chapter 7 offers a possible glimpse of the future where remote chemical and biochemical analysis might be undertaken with light guided through optical fibers to extremely sensitive and chemically specific sensors.

RAYMOND M. MEASURES

Thornhill, Ontario, Canada
October 1987

CONTENTS

Laser Remote Chemical Analysis

CHAPTER

1

FUNDAMENTALS OF LASER REMOTE SENSING

RAYMOND M. MEASURES

University of Toronto Institute For Aerospace Studies
Toronto, Ontario, Canada

The first laser studies of the atmosphere were undertaken by Fiocco and Smullin (1963), who recorded laser echoes from the upper regions of the atmosphere, and by Ligda (1963), who probed the troposphere. In the period that followed this pioneering work great strides were made, based on the development of *Q*-switching by McClung and Hellwarth (1962). This led to the generation of very short, single pulses of laser energy and made possible range-resolved measurements, in a manner somewhat analogous to radar. From this we derive the acronym lidar, standing for light detection and ranging.

The energy compression afforded by *Q*-switching also gave rise to laser power densities of such magnitude that remote measurements based on inelastic scattering from specific molecules was plausible. Additional developments in laser technology, such as second-harmonic generation and the invention of the nitrogen and the tunable dye laser, led to new prospects for laser remote sensing based on absorption and fluorescence. In the case of these techniques lidar might more appropriately stand for light identification, detection, and ranging. This important advance involved recognition of the fact that the radiation detected at wavelengths different from that of the laser's output contained highly specific information that could be used to determine the composition of the target region. The ability of lidar systems to perform an effective spectral analysis at a distance has added a new dimension to remote sensing and has made possible an extraordinary variety of observations that can be made from the ground or from mobile platforms such as helicopters, aircraft, or eventually satellites.

1.1 LASER REMOTE SENSORS

1.1.1 Laser Remote Sensing Techniques

Today the range of processes amenable to laser remote sensing includes Rayleigh scattering, Mie scattering, Raman scattering, resonance scattering, fluorescence, absorption, and differential absorption and scattering. A brief description of each is provided in Table 1.1. Although resonance scattering, sometimes referred to as laser-induced fluorescence, also has an inherently large cross section, collision quenching with the more abundant atmospheric species generally ensures that the detected signal is small; consequently, this technique is used to best effect in studies of the trace constituents in the upper atmosphere. This will be demonstrated in Chapter 5.

Raman scattering is an inelastic scattering process wherein laser radiation raises the molecule to a virtual level from which it immediately decays in a very short time ($< 10^{14}$ s), with the subsequent emission of radiation having a different wavelength. The difference in energy between the incident and emitted photons is a characteristic of the irradiated molecule and usually corresponds to a change of one vibrational quantum. Unfortunately, the Raman cross sec-

Table 1.1. Optical Interactions of Relevance to Laser Environmental Sensing

Technique	Physical description
Rayleigh scattering	laser radiation elastically sca'tered from atoms or molecules is observed with no change of frequency  VIRTUAL LEVEL hν hν GROUND LEVEL
Mie scattering	laser radiation elastically scattered from small particulates or aerosols (of size comparable to wavelength of radiation) is observed with no change in frequency 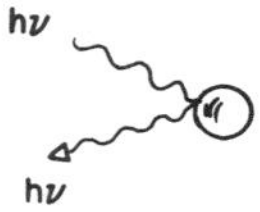 hν hν
Raman scattering	laser radiation inelastically scattered from molecules is observed with a frequency shift characteristic of the molecule ($h\nu - h\nu^* = E$) 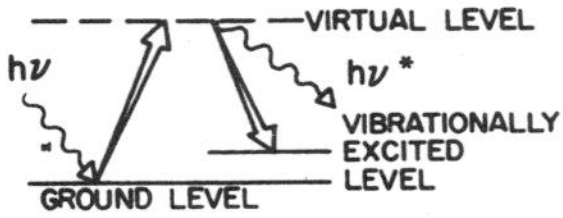 VIRTUAL LEVEL hν hν * VIBRATIONALLY EXCITED LEVEL GROUND LEVEL
Resonance scattering	laser radiation matched in frequency to that of a specific atomic transition is scattered by a large cross section and observed with no change in frequency 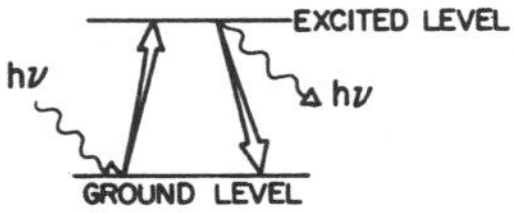 EXCITED LEVEL hν hν GROUND LEVEL
Fluorescence	laser radiation matched to a specific electronic transition of atom or molecule suffers absorption and subsequent emission at lower frequency; collision quenching can reduce effective cross section of this process; broadband emission is observed with molecules 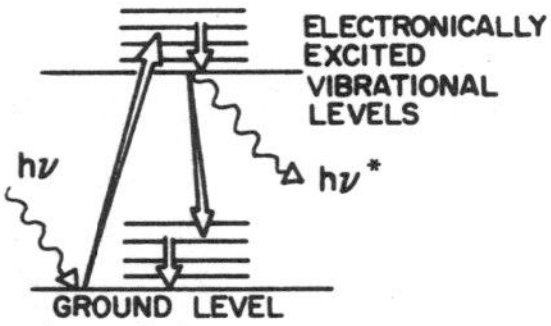 ELECTRONICALLY EXCITED VIBRATIONAL LEVELS hν hν * GROUND LEVEL

Table 1.1. (*Continued*)

Technique	Physical description
Absorption	observe attenuation of laser beam when frequency matched to the absorption band of given molecule $h\nu$ EXCITED LEVEL GROUND LEVEL
Differential absorption and scattering (DAS)	the differential attenuation of two laser beams is evaluated from their backscattered signals when the frequency of one beam is closely matched to a given molecular transition while the other's frequency is somewhat detuned from the transition $h\nu_1$ EXCITED LEVEL $h\nu_2$ GROUND LEVEL $h\nu_2$

tions are so small that the range and sensitivity of this technique are limited, and consequently, it is most likely to be employed for remote monitoring of effluent plumes, where the concentrations can be quite high (hundreds of ppm) (Melfi et al., 1973) or the range limited to about 1 km, (Houston et al., 1986).

The cross section for absorption of radiation is in general much greater than either the effective (quenched) fluorescence cross section or the cross section for Raman scattering. Consequently, the attenuation of a beam of suitably tuned laser radiation can be a sensitive method of evaluating the mean density of a given constituent. In order to separate absorption by the molecule of interest from other causes of attenuation, a differential approach is usually adopted. In this instance two frequencies are employed, one centered on a strong line within the absorption band of interest, the other detuned into the wing of the line. With a few notable exceptions, most of the absorption bands of interest lie in the infrared and correspond to vibrational–rotational transitions. This will be discussed in detail in Chapter 3.

High sensitivity with good spatial resolution can be achieved by the combination of differential absorption and scattering (DAS). This technique was first suggested by Schotland (1966) for the purpose of remotely evaluating the water vapor content of the atmosphere. In this approach a comparison is made between the atmospheric backscattered laser radiation monitored when the frequency of the laser is tuned to closely match that of an absorption line (within the molecule of interest) and when it is detuned to lie in the wing of the line. In this way, the large Mie scattering cross section is employed to provide spatial resolution and to ensure a strong return signal at both frequencies, while the ratio of the signals yields the required degree of specificity due to differential absorption. These advantages appear to bestow on the DAS technique the greatest sensitivity for long-range monitoring of specific molecular constituents (Measures and Pilon, 1972). The acronym DIAL, standing for differential absorption lidar, has gained considerable popularity for all of the laser remote sensing techniques that rely on differential absorption. This powerful technique will be discussed further in Chapters 4 and 5.

The realization that short-wavelength lasers could broaden the spectrum of applications as a result of laser-induced fluorescence led to the development of a new form of remote sensor, termed *laser fluorosensor* (Measures and Bristow, 1971; Fantasia et al., 1971). This instrument was originally conceived to permit airborne detection and characterization of oil spills at sea. More recenly, it has also been used to measure the thickness of such oil spills from the air (O'Neil et al., 1981); see Chapter 6. The discovery of fluorescent decay spectra (Houston et al., 1973; Measures et al., 1974) offers the prospects of improving the identification capability of laser fluorosensors while allowing them to reduce their spectral resolution.

Originally, laser-induced fluorescence from natural bodies of water was considered to constitute a source of background emission that could interfere with the oil fluorescence signal (Measures and Bristow, 1971; Measures et al., 1973; Bristow et al., 1973). Further studies (Hoge and Swift, 1980; O'Neil et al., 1980, 1981) have not only diminished this concern but have also shown that this (apparent) water fluorescence signal might serve to indicate the presence of high organic contamination and thereby allow the dispersion of such effluent plumes to be mapped remotely (Measures et al., 1975).

The fluorescence of chlorophyll has long been known, and the possibility of employing a laser fluorosensor to remotely map the chlorophyll concentration a natural bodies of water has also been studied (Browell, 1977; Bristow et al., 1981; Hoge and Swift, 1981). This topic will be discussed at length in Chapter 6.

It is clear that the scope of lasers in environmental sensing is extensive. They can be used to undertake (1) concentration measurements of both major and minor constituents in the atmosphere; (2) detection and mapping of specific constituents as required in pollution monitoring; and (3) airborne mapping of fluorescent substances such as chlorophyll (or oil slicks) in (or on) lakes and oceans. Furthermore, these observations can be made remotely with both spatial and temporal resolution from the ground or mobile platforms such as boats, helicopters, aircraft, or in certain instances satellites.

1.1.2 Laser Remote Sensor Systems

The functional elements and manner of operation of a typical laser remote sensor is schematically illustrated in Figure 1.1. An intense pulse of optical energy emitted by a laser is directed through some appropriate output optics toward the target of interest. The output optics improve the beam collimation, provide spatial filtering, and block the transmission of any unwanted radiation. Often, a small fraction of this pulse is sampled to provide a zero-time marker (a temporal reference signal that can also be used to normalize the return signal in the event that the laser's output reproducibility is inadequate) and a check on the laser wavelength where this is important.

The radiation gathered by the receiver optics is passed through some form of spectrum analyzer on its way to the photodetection system. The spectrum analyzer serves to select the observation wavelength interval and thereby discriminates against background radiation at other wavelengths. It can take the form of a monochromator, a polychromator, or a set of narrow-band spectral filters together with a laser wavelength blocking filter (except in the case of elastic scattering). The choice of photodetector is often dictated by the spectral region of interest, which in turn is determined by the kind of application and the type of laser employed.

In principle, there are two basic configurations for laser remote sensors. The *bistatic* arrangement involves a considerable separation of transmitter and receiver to achieve spatial resolution. Today, this arrangement is rarely used, as nanosecond lasers are capable of providing spatial resolution of a few feet, and

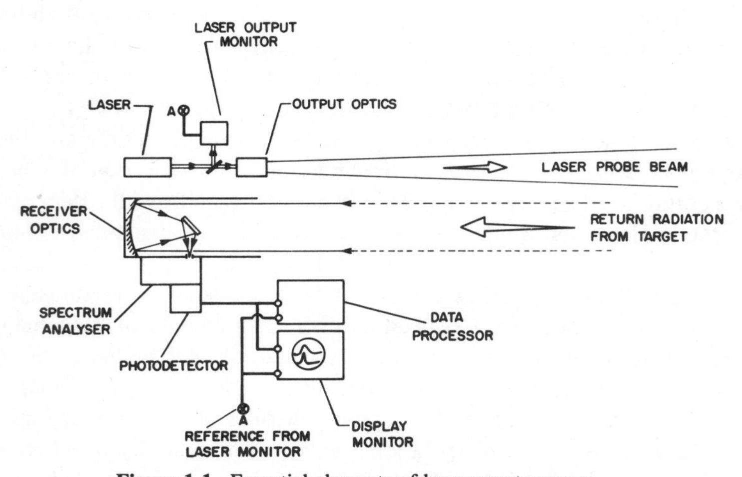

Figure 1.1. Essential elements of laser remote sensor.

so in most instances a *monostatic* configuration is employed. Under these circumstances, the transmitter and receiver are at the same location, so that in effect one has a single-ended system. A monostatic lidar can either be *coaxial* or *biaxial*. In a coaxial system the axis of the laser beam is coincident with the axis of the receiver optics, while in the biaxial arrangement the laser beam only enters the field of view of the receiver optics beyond some predetermined range. This configuration avoids the problem of near-field backscattered radiation saturating the photodetectors but is optically not quite as efficient as the coaxial approach. The near-field backscattering problem in a coaxial system can be overcome by either gating of the photodetector or use of a fast shutter (Poultney, 1972a,b). A relatively simple method of aligning the transmitter and receiver optics is described by Oppenheim and Menzies (1982).

Newtonian and Cassegrainian reflecting telescopes form the mainstay of the receiver optics to date. A biaxial Newtonian system is portrayed in Figure 1.1. The combined virtures of compact design and long focal length account for the growing popularity of the Cassegrainian system. Telescopes based on large plastic Fresnel lenses may offer some advantages with regard to cost, weight, and size (Grams and Wyman, 1972) and thereby be of particular interest in the development of airborne lidars. The size of the receiver's aperture depends to a large extent on the technique and range involved.

The signal from the photodetector may be processed via analog or digital techniques. The early work invariably involved the former (A-scope) approach where the backscattered signal intensity was displayed as a function of elapsed time (proportional to range) on a wide-bandwidth oscilloscope and photographed. The development of very fast dual-waveform digitizers, such as the Biomation 4500 and Tektronix 7612D, make real-time data processing possible. Uthe and Allen (1975) provided a brief review of the data-handling techniques employed in atmospheric probing, while Fredriksson et al. (1981) and Houston et al. (1986) detailed a representative lidar real-time data-recording system. A schematized view of the Mobile Raman lidar system of Houston et al. (1986) is presented as Figure 1.2.

The return signal can have an initial rise in the case of an offset configuration (i.e., the field of view of the receiver optics increasingly overlaps the path of laser excitation with increasing range). The subsequent fall of the signal is principally a manifestation of the inverse-square diminution of the signal with range. This $1/R^2$ dependence leads to many applications to a signal amplitude dynamic range that extends over several decades. Wide-band logarithmic amplifiers and gain-switching techniques can be used to compress this range so that the signals are compatible with recording and display electronics (Uthe and Allen, 1975; Frush, 1975). We shall see later that optical techniques may also be used to compress the dynamic range of the observed signal.

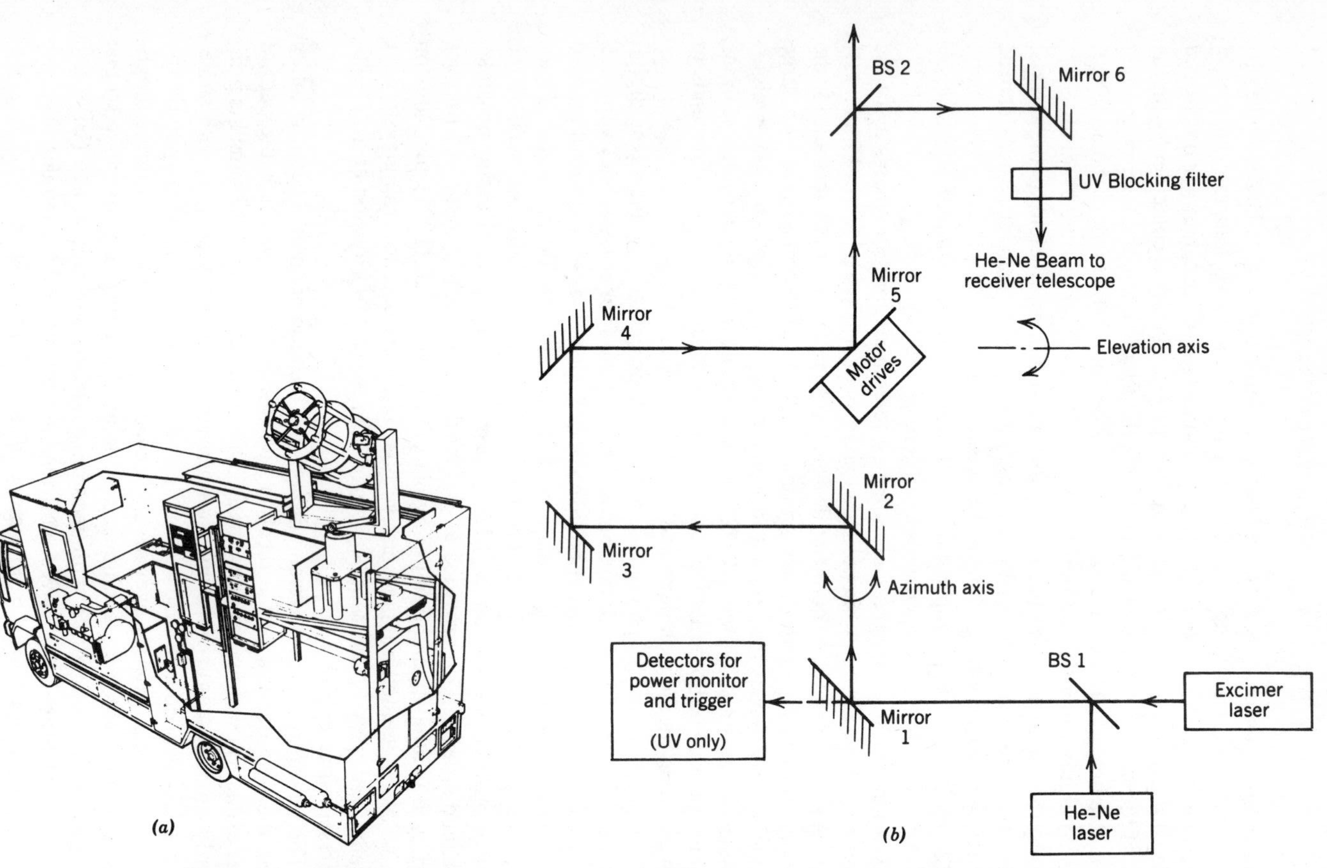

Figure 1.2. (*a*) Mobile Raman lidar system. (*b*) Schematic diagram of transmitter optical system. (From Houston et al., 1986.)

1.1.3 Receiver Systems

The characteristics involved in determining the choice of a photodetector include the spectral response, quantum efficiency, frequency response, current gain, and dark current. Sometimes, other considerations such as physical size, ruggedness, and cost may also be important. In most instances, the wavelength of the signal to be detected constitutes the primary factor in selecting the class of photodetector to be employed in any application. For wavelengths that lie between 200 nm and 1 μm [ultraviolet (UV) to near infrared (IR)], photomultipliers are generally preferred because of their high gain and low noise. Indeed, the single-photon detection capability of these devices has led to low-light-level detection schemes based on counting of individual photons (Poultney, 1972b).

In general, the performance of a photomultiplier is determined by (1) the spectral response of its photocathode, (2) the dark-current characteristics of its photocathode, (3) the gain of the dynode chain, (4) the time dispersal effects of the electrons moving through the dynode chain, and (5) the transit time of the electrons between the last dynode and the anode. Since a photomultiplier represents a current source, the observed signal is limited at high frequencies to the voltage that can be generated across the cable impedance $Z_0 < 100\ \Omega$ (usually). Although cables with impedance $Z_0 > 100\ \Omega$ are available, in general, their high-frequency loss characteristics limit their usefulness. Consequently, current gain constitutes an important consideration when high-frequency response is required.

Photomultipliers with high gain reach almost ideal quantum-noise-limited sensitivity for the detection of weak light signals. The large range of photocathode materials currently available offer a wide choice of spectral response characteristics, as illustrated by the selection of curves presented in Figure 1.3. Although fatigue in photomultipliers has long been recognized, the study of Lopez and Rebolledo (1981) has shown that in some photomultipliers the fatigue experienced in photon counting is comparable to that found with continuous light sources of the same mean irradiance.

Infrared detectors can be divided, broadly speaking, into two classes: photodetectors and thermal detectors. The most sensitive IR detectors are semiconductors in which the incident radiation creates charge carriers via a quantum interaction. These photodetectors may be further divided into photovoltaic and photoconductive devices. Of these, photovoltaic devices (photodiodes) are the more popular for laser remote sensing. Although some photodiodes can be used in the visible, they come into their own at longer wavelengths, where their high quantum efficiency (30–80%) becomes important. Unfortunately, the output of a photodiode must be externally amplified so that its sensitivity is often limited by thermal noise. Altmann et al. (1980) developed a fast current amplifier that enables a photovoltaic InSb detector operated at zero bias voltage for (optimum detectivity) to achieve background-limited performance.

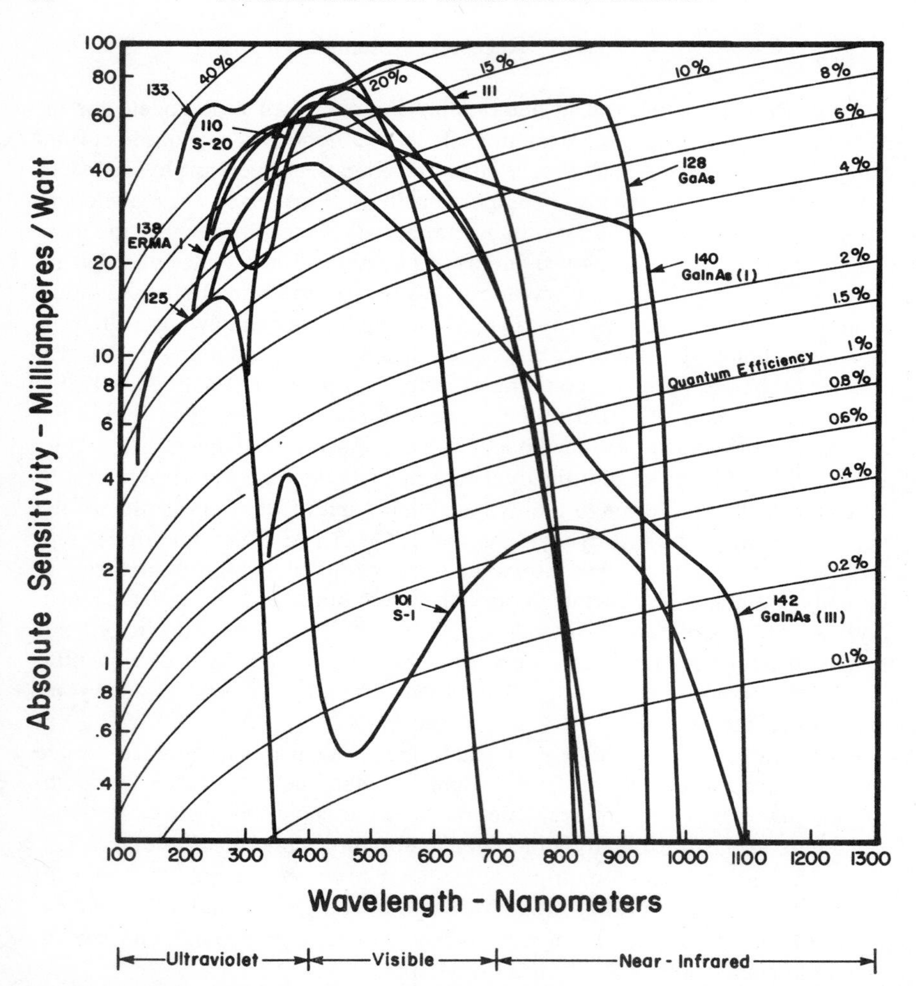

Figure 1.3. Typical photocathode spectral response characteristics (RCA photomultiplier brochure).

Some photodiodes, when operated at a high reverse bias, develop internal gain through a process of carrier multiplication. These avalanche photodiodes are similar to photomultipliers in the sense that their sensitivity is no longer determined by thermal noise of the detector and output circuit (Keyes and Kingston, 1972). One of the most widely used and most sensitive IR detectors for the 1–5.5-μm range appears to be the liquid-nitrogen-cooled InSb photodiode (Wang, 1974); Shewchun et al., 1976.) A useful overview of IR detectors has been prepared by Emmons et al. (1975), and Lussier (1976a) has compiled several tables that summarize various IR detector characteristics.

The spectrum analyzer is used to select the wavelength interval of observation and to provide adequate rejection of all off-frequency radiation, whether this be laser-scattered radiation, solar background radiation, or any other form of radiation having a wavelength different from that of the signal. In general, this is accomplished with the aid of one or more spectral components.

These components fall into one of three basic categories: absorption filters, interferometric elements, and dispersive systems. Absorption filters can take the form of colored glass, gelatin, or a liquid solution (Leonard, 1974) and are employed to attenuate the incident intensity, separate interfering spectral orders, or block all wavelengths that are shorter or longer than those of interest. Long-wavelength pass filters (sometimes called short-wavelength blocking filters) are available from 250 nm to 1 μm (Leonard, 1974; Parker, 1968).

The problem of laser-induced fluorescence of short-wavelength cutoff filters has been addressed by Bristow (1979). Short-wavelength pass filters, however, are less plentiful. The internal spectral transmittance characteristics of some 800 colored glasses presented in the comprehensive review of Dobrowolski et al. (1977) have provided the transmittance curves for a number of new bandpass filters in the visible. A guide to IR transmissive materials has been provided by Lussier (1976b).

An important component of many laser remote sensors is a vapor-deposited dielectric interference filter. These filters, made of alternating layers of high and low refractive index, are useful over most of the UV to intermediate IR region of the spectrum. Their transmission profiles are similar to those of a low-order Fabry–Perot interferometer, being dispersive in nature and having a bandwidth that can be close to 1 nm. In almost all instances interference filters are designed for use with collimated radiation incident normal to the surface of the filter. If the filter is tilted by 45°, the center passband can shift to shorter wavelengths by as much as 2–3%. These filters can also be temperature sensitive. The sideband transparency of such a filter is close to 10^{-3}. In many instances a stack of two or more such filters is used to improve this factor and narrow the spectral width of the passband. However, this is always achieved at the expense of the passband transmission coefficient and an enhanced sensitivity to tipping.

The best, but most expensive, narrow-band filters are made of birefringent materials and are called Lyott filters (Walther and Hall, 1970). The wide field of view of this kind of narrow-band filter is particularly useful in lidar work. These filters can also be tuned. Reviews of this capability of birefringent filters are available (Title and Rosenberg, 1981; Gunning, 1981).

For those applications that require high spectral resolution, the choice often lies between a Fabry–Perot interferometer and a grating monochromator (Born and Wolf, 1964). Of these, the Fabry–Perot etalon is usually the cheaper, can provide the higher resolving power, and has the greater light throughput. Indeed, it would be ideal for many applications if it were not for its major drawback—many overlapping orders. This difficulty can be overcome by prefiltering with an interference filter, with a second, wider passband Fabry–Perot interferometer, or with a dispersive element. Each approach has its own limitations. A detailed comparison of prism, grating, and Fabry–Perot etalon spectrometers has been given by Jacquinot (1954). The passband of a Fabry–Perot etalon can be scanned by varying the pressure of the gas between the interferometer plates (Girard and Jacquinot, 1967) or by displacing one plate relative to the other (Ramsay, 1962).

In those situations where measurement of a spectral profile is important or where many wavelengths are of interest, a grating monochromator offers some advantages. Of the wide array of monochromator systems available, the Czerny–Turner arrangement appears to be one of the most popular, being typically capable of providing a stray-light rejection ratio of 10^{-6}.

1.1.4 Lasers Relevant to Remote Sensing

Certain types of lasers are capable of emitting pulses of optical energy that possess very high peak power, narrow bandwidth, and short duration and propagate with a low degree of divergence. Lasers of this nature are close to ideal for probing the environment but must also be capable of operating at a high repetition rate for most airborne missions and for those atmospheric applications in which the return signal is very weak.

The range of lasers available for remote sensing is extensive, and an even wider spectrum of possibilities exists if use is made of second-, third-, or even fourth-harmonic generation and parametric conversion (Byer, 1986; Fan et al., 1984). The development of tunable organic dye lasers (Schäfer, 1973) provided the means to excite specific atomic and molecular electronic transitions and thereby exploit both resonance scattering and differential absorption for the purpose of remote sensing. A population inversion is created within the dye by optical pumping with either a flashlamp or another laser. Pulsed operation usually involves Nd–YAG or rare-gas halide excimer lasers, while continuous-wave (CW) operation is achieved by pumping with a tightly focused argon laser. Flashlamp-pumped dye lasers, in general, provide the greatest energy per pulse, but their duration is rather long (hundreds of nanoseconds) for measurements with reasonable spatial resolution. Spectral narrowing and tuning across the broad emission band of a dye is achieved by means of a dispersive element such as a prism or a grating. Hänsch (1972) demonstrated that good spectral condensation, better than 10 pm (10^{-2} nm), could be achieved with a diffraction grating and an intracavity beam-expanding telescope. Futher spectral narrowing can be achieved with the inclusion of a Fabry–Perot etalon in the dye laser cavity.

A small degree of tunability can also be achieved with many high-pressure gas lasers and with some semiconductor lasers. A comprehensive review of high-pressure pulsed molecular lasers is provided by Wood (1974), while Nill (1974) and Hinkley (1972) present useful reviews of tunable infrared lasers. For environmental sensing, tunable IR lasers have the advantage that most materials possess vibrational-rotational transitions that can be selectively excited by IR radiation (Murray et al., 1976). The new class of tunable solid-state lasers such as alexandrite (Byer, 1986) should have a significant effect on laser remote sensing, since these lasers are from a practical standpoint much more desirable than their liquid (organic dye) counterparts.

At short wavelengths (below 300 nm) there is again a rich selection of transitions for a variety of materials; however, the lack of convenient tunable lasers limits the scope of remote sensing applications. Tunable laser radiation down to 230 nm is available with a frequency-doubled dye laser, but such a system is somewhat elaborate and limited in its output power.

Rare-gas halide lasers (Bhaumik et al., 1976; Ewing, 1978; Loree et al., 1979; Sze, 1979) are capable of supplying high power, with high efficiency, at wavelengths below 337 nm. Such excimer lasers are inherently tunable, albeit over a small spectral interval. For example, 100-MW pulses at 248.4 nm from a krypton fluoride laser are commercially available. Although such developments offer new possibilities with regard to remote sensing, due to the absence of solar background at these wavelengths and the possibility of achieving resonance Raman scattering (Rosen et al., 1975), the extreme sensitivity of living material to this radiation (Koller, 1969) could prevent its realization except in rather limited situations.

1.1.5 Sources of Noise

Of crucial importance to any discussion of remote sensing is the question of signal-to-noise ratio. Noise, in this context, may be thought of as false signals that can reduce the accuracy of a given measurement or even obscure the true signal completely. Noise, in general can have either an optical or a thermal origin. In the context of laser remote sensing, there are four important kinds of noise. These are listed in Table 1.2. The first three represent different forms of

Table 1.2. Kinds of Noise Relevant to Laser Environmental Sensing

Kind of Noise	Physical Mechanism
Noise in signal (quantum noise)	Statistical fluctuations of signal radiation
Background-radiation noise	Statistical fluctuations of background radiation
Dark-current noise	Thermal generation of current carriers in absence of optical signal
Thermal (Johnson, Nyquist) noise	Thermal agitation of current carriers

shot noise. Under daytime operation scattered solar radiation from either the sky or the ground can often dominate all other forms of noise. The solar spectral irradiance as observed from space and from the ground is shown in Figure 1.4 and is available elsewhere in tabulated form (Mecherikunnel and Duncan, 1982). The spectral radiance of a clear sky is shown in Figure 1.5. It is also important to realize that for both Raman and fluorescence measurements the background radiation may include a laser-scattered component if adequate spectral rejection is not provided.

The increment of radiative energy (arising from natural sources) accepted by the receiving optics in the detection time τ_d can be expressed in the form (Pratt, 1969)

$$E_b^N(\lambda) = \int_{\Delta\lambda_0} S_b(\lambda')\xi(\lambda')\Omega_0 A_0 \tau_d \, d\lambda' \tag{1.1}$$

where $S_b(\lambda')$ represents the spectral radiance of the sky background (W cm^{-2} nm^{-1} sr^{-1}), and $\xi(\lambda')$ represents the receiver system transmission efficiency at the wavelength λ' and includes the influence of any spectrally selecting components. Here, $\Delta\lambda_0$ can be taken as the spectral window of the receiver system, A_0 its effective aperture, and Ω_0 its acceptance solid angle. For a good optical

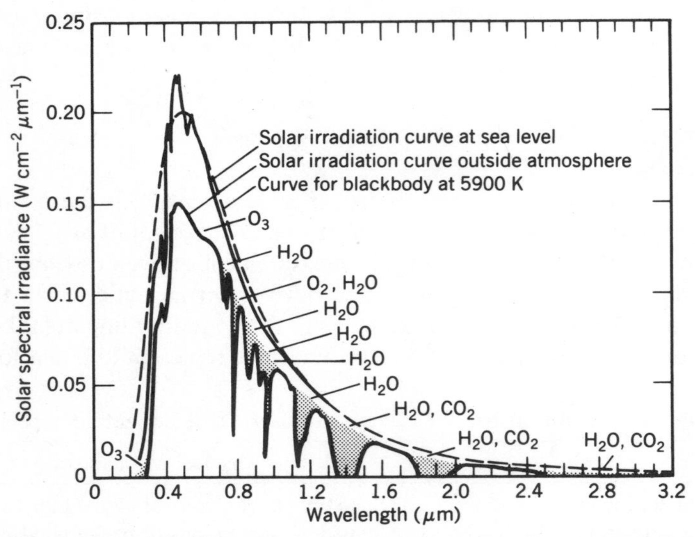

Figure 1.4. Spectral irradiance of direct sunlight before and after it passes through Earth's atmosphere. The stippled portion gives the atmospheric absorption. The sun is at the zenith (Shaw, 1953; Valley, 1965).

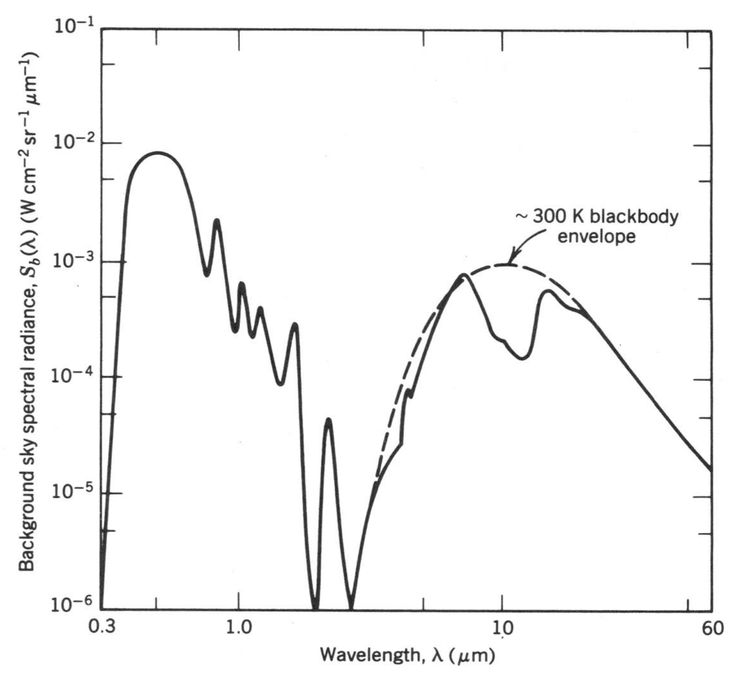

Figure 1.5. Diffuse component of typical background spectral radiance from sea level, zenith angle 45°, excellent visibility (Pratt, 1969).

system the *étendue matching condition* implies $\Omega_0 A_0 = \Omega_s A_s$, where Ω_s and A_s represent, respectively, the acceptance solid angle and the entrance aperture of the spectral element (such as a monochromator). For example, if a $f/7$ monochromator is used with an entrance slit of 0.1 cm^2 and $A_0 = 10^3$ cm^2, then for a well-matched system $\Omega_0 = 2 \times 10^{-6}$ sr. The spectral radiance of the clear daytime sky peaks in the visible (due to scattered solar radiation) and can attain a value of close to 10^{-5} W cm^{-2} nm^{-1} sr^{-1} (Ross, 1966; Pratt, 1969). Toward 300 nm, this background decreases rapidly because of attenuation within the ozone shield above Earth. The more gradual decline in the solar spectral radiance at the IR end of the spectrum is punctuated by many absorption bands (see Figure 1.5). The second hump, centered at about 10 μm, represents the thermal background radiation.

In the case of downward-pointing airborne lidars, reflected and scattered solar radiation from Earth provides the background radiation required in calculating $E_b^N(\lambda)$. Plass et al. (1976) have calculated the upward radiance from the ocean surface taking account of sun glitter, reflected sky radiance, and the upwelling scattered radiation from various depths. They attempted to make their model fairly realistic by including the effect of surface waves and have also calculated the upward radiation as seen from the top of the atmosphere. This would be of relevance to lidars operating from space platforms.

In order to reduce the solar background, the receiver system bandwidth $\Delta\lambda_0$ is always adjusted to be as small as compatible with the spectral width of the signal of interest. In almost all instances (scattering or fluorescence) the spectral window of the receiver system is sufficiently narrow that we can neglect the variation of the solar spectral radiance over the range of wavelength integration. In that case we can write

$$E_b^N(\lambda) = S_b(\lambda)\Omega_0 A_0 \tau_d K_0(\lambda) \tag{1.2}$$

where λ is the center wavelength of the radiation of interest and

$$K_0(\lambda) \equiv \int_{\Delta\lambda_0} \xi(\lambda')\, d\lambda' \tag{1.3}$$

is termed the *filter function* of the receiver system. It can be thought of as the effective receiver bandwidth at unit transmission efficiency.

The noise associated with a photomultiplier is determined by several factors: the type of photomultiplier, the characteristics of the photocathode, the operating gain, and the usage history of the particular photomultiplier selected. The predominant form of photomultiplier noise is associated with the release of single electrons from the photocathode in the absence of any incident light. This so-called *dark current* arises from thermal and field emission processes and can be anywhere from 10^{-13} to 10^{-17} A at room temperature. Cooling the device is a popular method of reducing this component of noise and has been reviewed by Foord et al. (1969). A good overall description of photomultiplier noise problems is provided by Poultney (1972b).

1.1.6 Signal-to-Noise Ratio

In a photomultiplier the arrival of the signal pulse of energy $E_s(\lambda)$ in the time interval t to $t + \tau_d$ gives rise to a momentary photocathode current

$$i_s(t) = \frac{\lambda\eta(\lambda)eE_s(\lambda)}{hc\tau_d} \tag{1.4}$$

where $\eta(\lambda)$ represents the quantum efficiency of the photocathode at wavelength λ, and e the electronic charge. It is worth noting that $\lambda E_s(\lambda)/hc\tau_d$ is the mean rate of arrival of signal photons (energy hc/λ) at the photocathode.

The equivalent noise photocathode current during the same time interval is given by

$$i_n = \frac{e\sqrt{\delta n_e^2}}{\tau_d} \tag{1.5}$$

where $\sqrt{\delta n_e^2}$ is the root-mean-square (rms) fluctuation of the total number of photocathode electrons created in the detection interval τ_d.

If, as is most often the case, this burst of photocathode electrons can be described by Poisson statistics, then we can write (Oliver, 1965)

$$[\delta n_e^2]^{1/2} = [\langle n_e \rangle]^{1/2} \tag{1.6}$$

where $\langle n_e \rangle$ is the mean number of photocathode-generated electrons produced in the interval τ_d, namely,

$$\langle n_e \rangle = \frac{1}{hc} \left\{ \lambda\eta(\lambda)[E_s(\lambda) + E_b^N(\lambda)] + \lambda_L \eta(\lambda_L) E_b^L(\lambda_L) \right\} + \langle n_e^d \rangle \tag{1.7}$$

Here $\langle n_e^d \rangle$ is the mean number of photocathode electrons created in the interval τ_d in the absence of any light; $E_b^N(\lambda)$ and $E_b^L(\lambda_L)$, respectively, represent the natural background and laser-scattered background pulses of radiant energy incident upon the photocathode in the period τ_d. If the wavelength of the photodetection system is set equal to λ_L, then of course the first and third terms within the braces of Equation (1.7) coalesce.

The photocathode signal-to-noise ratio can thus be expressed in the form

$$(\mathrm{SNR})_c = \frac{e\lambda\eta(\lambda)E_s(\lambda)}{hc\tau_d \left\{ \frac{e^2}{hc\tau_d^2} \left[\lambda\eta(\lambda)\left\{E_s(\lambda) + E_b^N(\lambda) + \nu_\eta E_b^L(\lambda_L)\right\}\right] + \frac{ei_d}{\tau_d} \right\}^{1/2}} \tag{1.8}$$

where i_d represents the mean photocathode dark current $(e\langle n_e^d \rangle / \tau_d)$ and $\nu_\eta \equiv \lambda_L \eta(\lambda_L) / \lambda\eta(\lambda)$. An alternative form of this expression is used by some authors (Inaba and Kobayasi, 1972; Poultney, 1972; Nakahara et al., 1972) who are concerned with electronically gating the photomultiplier in order to account for the natural-background and dark-current contributions to the noise of the system. This is of particular relevance in photon counting (Poultney, 1972a,b). Under these circumstances one assumes that the photomultiplier is gated on for a duration τ_g^s, during the period of signal return, and again for an additional period τ_g^b in the absence of any signal. In this case the gated signal-to-noise ratio is

$$(\mathrm{SNR})_c^g = e\lambda\eta(\lambda)E_s(\lambda)\frac{\tau_g^s}{\tau_d}\frac{1}{hc\tau_d}$$

$$\times\left(\frac{e^2}{hc\tau_d^3}\left\{\lambda\eta(\lambda)\left[E_s(\lambda)\tau_g^s + E_b^N(\lambda)(\tau_g^s + \tau_g^b)\right.\right.\right.$$

$$\left.\left.\left. + \nu_\eta E_b^L(\lambda_L)\tau_g^s\right]\right\} + \frac{ei_d}{\tau_d}\frac{\tau_g^s + \tau_g^b}{\tau_d}\right)^{-1/2} \tag{1.9}$$

As indicated in Equation (1.4), the equivalent signal current can be related to the signal energy received by the photodetector, and by the same reasoning the equivalent background current can be expressed in the form

$$i_b \equiv \frac{e\left[\lambda\eta(\lambda)E_b^N(\lambda) + \lambda_L\eta(\lambda_L)E_b^L(\lambda_L)\right]}{hc\tau_d} \tag{1.10}$$

If we introduce the photodetection bandwidth

$$B \equiv \frac{1}{2\tau_d} \tag{1.11}$$

then we can rewrite Equation (1.8) in the more conventional form

$$(\mathrm{SNR})_c = \frac{i_s}{\left[2eB(i_s + i_b + i_d)\right]^{1/2}} \tag{1.12}$$

where we see that the denominator represents the shot noise in the form first shown by Schottky (1918):

$$i_n = \left[2eB(i_s + i_b + i_d)\right]^{1/2} \tag{1.13}$$

If we introduce G as the *dynode chain gain*, then the anode signal current is

$$I_s = G\xi_e i_s \tag{1.14}$$

where ξ_e (<1) is the collection efficiency of the electrostatic focusing, that is to say, ξ_e defines what fraction of the photoelectrons created at the cathode arrive at the first dynode. The corresponding noise current at the anode is given by

$$I_n = eG\delta \frac{[\xi_e \langle n_e \rangle]^{1/2}}{\tau_d} \tag{1.15}$$

where δ (≈ 1) accounts for the statistical fluctuations in the emission of secondary electrons from the dynodes (Topp et al., 1969) and can be thought of as a noise factor associated with the gain. Consequently, we may write the signal-to-noise ratio for the anode current as

$$(\mathrm{SNR})_a = \frac{I_s}{I_n} = \frac{G\xi_e i_s}{[2eBG^2 F_G \xi_e (i_s + i_b + i_d)]^{1/2}} \tag{1.16}$$

where we have introduced F_G to represent δ^2.

It is clear that in the case of a photomultiplier the current multiplication factor G does not itself enter into the output SNR. However, a somewhat similar analysis in the case of an avalanche photodiode would produce an output SNR of the form (Melchior, 1972)

$$\mathrm{SNR} = \frac{G\xi_e i_s}{[(4kTB/R_{\mathrm{eq}}) + 2eBG^2 F_G \xi_e (i_s + i_b + i_d)]^{1/2}} \tag{1.17}$$

where the first term within the square brackets represents the Johnson or thermal noise current associated with the equivalent load resistance of the output circuit (R_{eq}), and i_d in this case refers to the bulk leakage current of the avalanche photodiode; ξ_e can be taken to be very close to unity. The magnitude of G is so large (10^5–10^7) in the case of a photomultiplier that thermal noise never imposes a real limit. However, the value of F_G can range from 1 to about 2.5 for both high-gain photomultipliers and photodiodes (Melchior, 1972). In general, the signal-to-noise ratio improves by averaging over many pulses. This improvement is proportional to the square root of the number of pulses averaged.

The expression for the output current signal-to-noise ratio [Eq. (1.17)] has found general application, although several effects are omitted. These include the contribution to the dark current from electrons that originate on the dynodes and the possibility of leakage current at the output socket of the device (particularly relevant to cooled photomultipliers). The influences of cosmic radiation and natural radioactivity have also been neglected—with justification, one might add, in most situations. These additional noise contributions can be taken into account by appropriately modifying the dark-current term (Topp et al., 1969).

In order to see where improvements in the SNR can be made, we shall express the output SNR (on the basis of a single pulse) in the general form

$$\mathrm{SNR} = \frac{E_s(\lambda)\left[\xi_e \lambda \eta(\lambda)\right]^{1/2}}{\left[F_G hc\left\{E_s(\lambda) + E_b^N(\lambda) + \nu_\eta E_b^L(\lambda_L) + (i_d + i_J)/2BS_d\right\}\right]^{1/2}} \tag{1.18}$$

where

$$S_d \equiv \frac{e\lambda\eta(\lambda)}{hc} \tag{1.19}$$

and

$$i_J \equiv \frac{2kT}{eG^2\xi_e R_{\mathrm{eq}} F_G} \tag{1.20}$$

represents the effective Johnson noise current. As we have seen, i_J is only important where there is negligible internal gain.

It is immediately apparent that a careful choice of the detector and its mode of operation can optimize the SNR, all other things being equal. In the case of a photomultiplier, the photocathode should be selected on the basis of maximum quantum efficiency at the wavelength of interest. The noise factor associated with the gain and the collection efficiency of the first dynode can both be optimized by application of a suitable voltage between the photocathode and the first dynode (Topp et al., 1969). Thermionic and field emission release of electrons from the dynodes can represent an important component of the dark-current noise in photomultipliers. Consequently, at very low light levels where dark-current noise is important, photon counting can significantly improve the SNR because it discriminates against single electron pulses that have not acquired the full gain (Poultney, 1972; Topp et al., 1969). A further increase in the SNR under these conditions can be obtained by focusing the incident light so that only a small area of the photocathode is illuminated while ensuring that all electrons that originate from the unilluminated regions fo the photocathode are defocused magnetically. Cooling also helps, since thermionic emission is strongly temperature dependent. Topp et al. (1969) demonstrated that use of these techniques enabled signals with a power of less than 10^{-17} W (35 photons s^{-1}) to be detected with a time constant of 1 s at 650 nm.

The general output SNR (1.18) can be expressed in another form:

$$(\mathrm{SNR})^2 = \frac{E_s^2(\lambda)/E(\lambda)}{E_s(\lambda) + E_b^T(\lambda) + (i_d + i_J)E(\lambda)/2eB^*} \tag{1.21}$$

where

$$E_b^T(\lambda) \equiv E_b^N(\lambda) + \nu_v E_b^L(\lambda_L) \tag{1.22}$$

$$B^* = \frac{BF_G}{\xi_e} \tag{1.23}$$

and

$$E(\lambda) \equiv \frac{hcF_G}{\lambda\eta(\lambda)\xi_e} \tag{1.24}$$

which corresponds to the signal energy that would just give a unit value to the SNR in the signal shot-noise limit to be defined shortly. Equation (1.21) represents a simple quadratic expression in terms of $E_s(\lambda)$ and has the solution

$$E_s(\lambda) = \tfrac{1}{2}(\mathrm{SNR})^2 E(\lambda)\left[1 + \left\{1 + \frac{4}{(\mathrm{SNR})^2}\left[\frac{E_b^T(\lambda)}{E(\lambda)} + \frac{i_d + i_J}{2eB^*}\right]\right\}^{1/2}\right] \tag{1.25}$$

In essence, there are four limiting situations that cover most laser-sensing situations. Three of them presuppose that the system's spectral discrimination is adequate to enable the laser-scattered components of the background signal to be neglected, that is, $E_b^T(\lambda) \approx E_b^N(\lambda)$. We shall consider these first. It is also reasonable to assume that for virtually all lidar work detectors with internal gain will be used, and consequently i_J is usually neglected. In this case the highest sensitivity is achieved if the detectable signal is limited only by the quantum fluctuations of the signal itself, namely,

$$E_b^N(\lambda) + \frac{i_d E(\lambda)}{2eB^*} \ll E_s(\lambda)$$

Under these circumstances (*signal shot-noise limit*) the minimum detectable energy (MDE) is

$$E_s^{\min}(\lambda) \approx \left[(\mathrm{SNR})^{\min}\right]^2 E(\lambda) \tag{1.26}$$

If we assume that $F_G/\xi_e \approx 1$ and that an acceptable value for $(\mathrm{SNR})^{\min}$ is about 1.5, we find that for $\eta(\lambda) \approx 0.2$ the minimum detectable number of photons is close to 10 per pulse.

In daytime operation, the level of background radiation can be so high that

$$E_b^N(\lambda) \gg \frac{i_d E(\lambda)}{2eB^*} + E_s(\lambda) \tag{1.27}$$

and we speak of the *background-noise limit*. Under these circumstances the MDE is

$$E_s^{\min}(\lambda) = \left\{\frac{F_G hc K_0(\lambda)\,\Omega_0 A_0 \tau_d S_b(\lambda)}{\xi_e \lambda \eta(\lambda)}\right\}^{1/2} (\mathrm{SNR})^{\min} \tag{1.28}$$

and it is evident that the filter function $K_0(\lambda)$ plays an important role in permitting small values of $E_s(\lambda)$ to be detected.

In the *dark-current-limited situation* we assume that

$$i_d \gg \frac{2eB^* E_b^N(\lambda)}{E(\lambda)} + \frac{2eB^* E_s(\lambda)}{E(\lambda)}$$

This leads to an MDE given by

$$E_s^{\min}(\lambda) \approx \frac{hc}{\lambda\eta(\lambda)}\left\{\frac{F_G i_d}{2eB\xi_e}\right\}^{1/2} (\mathrm{SNR})^{\min} \tag{1.29}$$

The typical value for the photocathode dark current of a photomultiplier is around 10^{-15} A, which means that the inequality expressed above is not likely to be satisfied unless the photon arrival rate is less than about $10^3\ \mathrm{s}^{-1}$. On the other hand, for practical bandwidths, most infrared detectors tend to be dark current limited. Thus, we shall introduce the term *detector detectivity* D^* (Kruse et al., 1963) defined by

$$D^* \equiv \frac{S_d}{(2ei_d/A_d)^{1/2}} \tag{1.30}$$

where S_d, defined by Equation (1.19), represents the detector sensitivity as defined by the relation

$$i_s \equiv \frac{S_d E_s(\lambda)}{\tau_d}$$

where A_d represents the detector area. In this case, using this equation with (1.19) and (1.30), we may express Equation (1.29) in the form

$$E_s^{\min}(\lambda) = \frac{1}{D^*}\left\{\frac{F_G A_d}{4B\xi_e}\right\}^{1/2}(\mathrm{SNR})^{\min} \tag{1.31}$$

For most solid-state devices the typical detector areas are only a few square millimeters, and the factor F_G/ξ_e can be replaced with unity. Representative detectivity (D^*) values for a number of such detectors have been given by Melchior (1972) and are reproduced here as Figure 1.6.

Clearly, Equations (1.28), (1.29), and (1.31) represent limiting cases that primarily serve to indicate the relative importance of the various factors in determining the minimum detectable signal. In general, Equation (1.25) should be employed, and under these circumstances the MDE for the background-dominated case characterized by

$$E_b^T(\lambda) \gg \frac{i_d E(\lambda)}{2eB^*} \tag{1.32}$$

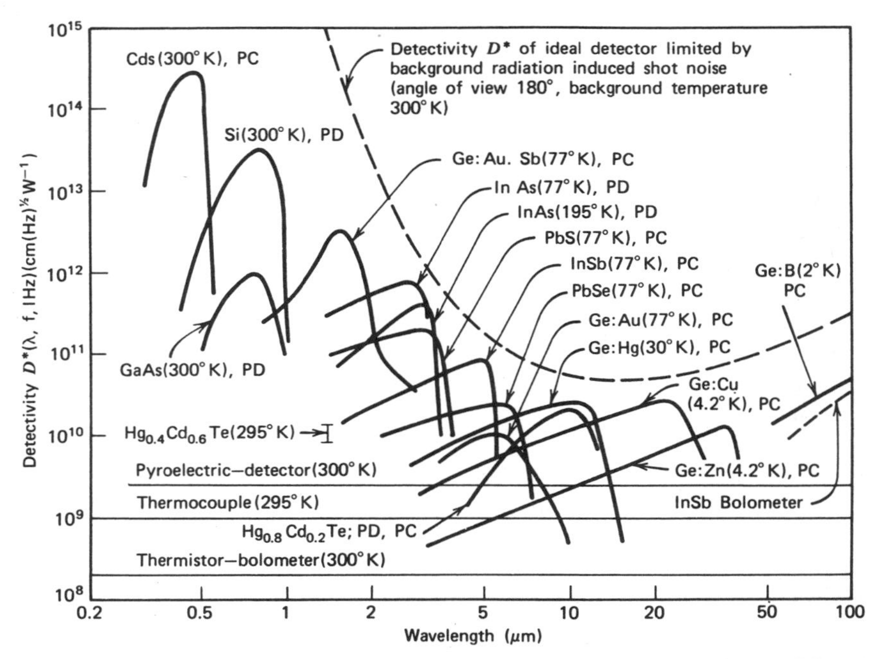

Figure 1.6. Spectral dependence of detectivity D^* for high-sensitivity photoconductors (PC) and photodiodes (PD). Representative values are given based on the literature and manufacturer's data (Melchior, 1972).

is given by the expression

$$E_s^{\min}(\lambda) \approx (\mathrm{SNR})^{\min}\left\{\tfrac{1}{2}(\mathrm{SNR})^{\min}E(\lambda) + \left[E(\lambda)E_b^T(\lambda)\right]^{1/2}\right\} \quad (1.33)$$

If laser-backscattered radiation dominates the natural-background component, then essentially $\nu_\eta E_b^L(\lambda_L)$ should replace $E_b^T(\lambda)$ in Equation (1.33). Alternatively, if spectral rejection of the laser-backscattered radiation is adequate, then

$$E_s^{\min}(\lambda) \approx (\mathrm{SNR})^{\min}\left\{\tfrac{1}{2}(\mathrm{SNR})^{\min}E(\lambda) + \left[E(\lambda)E_b^N(\lambda)\right]^{1/2}\right\} \quad (1.34)$$

A comparison of (1.28) with (1.34) reveals that in the more general situation the MDE includes an additional term that amounts to half of the signal shot-noise MDE. A similar result is found for the more general dark-current-dominated situation, where we can write

$$i_d \gg \frac{2eB^*E_b^T(\lambda)}{E(\lambda)} \quad (1.35)$$

and the MDE is

$$E_s^{\min}(\lambda) = (\mathrm{SNR})^{\min}\left[\frac{1}{2}(\mathrm{SNR})^{\min}E(\lambda) + \frac{1}{D^*}\left(\frac{F_G A_d}{4B\xi_e}\right)^{1/2}\right] \quad (1.36)$$

As we shall see, $E_S(\lambda)$ can be related through the appropriate form of the lidar equation to the density of the target species and the output energy of the laser. Consequently, these various MDE expressions can be used to evaluate either the minimum laser pulse energy required for detection of a given density of species at a specific range or the threshold density of a species that can be detected at a given range for a given transmitted laser energy—in various noise-dominated situations.

1.1.7 Noise Reduction Techniques

In many instances it is possible to improve the SNR by suppression of the noise. One of the first problems encountered in the early lidar work on the upper atmosphere involved near-field scattering of long-lived laser fluorescence being misinterpreted as return signals from high altitudes. Although sufficient separation of transmitter and receiver overcame this source of noise, fluorescence shutters tended to become the more usual method of avoiding this difficulty. A large separation (10 m) between the laser and the collection optics also prevents near-field scattering from overloading the photomultipliers. Alternatively, fast mechnical or electro-optical shutters can be used to avoid this problem (Poultney, 1972a). More recently, electronically gated photomultipliers have been used.

Spectral rejection of laser-backscattered radiation where the signal wavelength differs from that of the laser represents an important example of such noise suppression. Where absorption is used to measure the density of a specific molecular constituent, operation at two wavelengths (λ_0 and λ_w), corresponding to adjacent regions of strong and weak absorption by the molecule of interest, enables an allowance to be made for attenuation by the background constituents. Measurements at λ_0 and λ_w should in principle be performed simultaneously to avoid errors associated with temporal changes in the absorption and scattering properties of the atmosphere.

As an alternative to employing two lasers (an expensive and cumbersome business), Brassington (1978) proposed to switch the wavelength of operation of a tunable dye laser between λ_0 and λ_w on consecutive output pulses. For the detection of SO_2 this involved rocking the birefringent filter (the main wavelength-selective element) and the frequency-doubling crystal in synchronism with the laser pulses in order to provide an output at 300.1 and 299.4 nm. Alternative techniques for obtaining a pair of laser pulses from a nitrogen laser-pumped dye laser, with a wavelength separation that can range from 0 to 10 nm, were described by Inomata and Carswell (1977). In the IR part of the spectrum Kanstad et al., (1977) have developed a CO_2 laser that can intermittently operate on any two lines from either the 9.4- or the 10.4-μm band, and Stewart and Bufton (1980) proposed to use two CO_2 lasers in their DIAL experiments.

The range and detection limit of a Raman lidar is usually limited by the presence of luminescence at the same wavelength as the Raman signal. This luminescence can have a natural origin such as solar-scattered radiation (or moonlight), or it can arise as a result of the laser-induced fluorescence. Much of this broadband radiation can be made negligible compared to the Raman signal through the use of narrow-band filtering. Temporal discrimination is particularly effective against continuous radiation, and for local measurements some degree of suppression of laser-induced fluorescence can also be achieved in this manner (Burgess and Shepherd, 1977). Unfortunately, in laser remote sensing time-resolved measurements cannot, in general, be used to overcome laser-induced fluorescence; however, Morhange and Hirlimann (1976) have shown that considerable rejection of luminescence can be achieved through alternately switching between two laser lines while monitoring only at the Raman wavelength produced by one of them. If the wavelength separation of the two laser lines is small enough that the luminescence signal is nearly unchanged, then it can be subtracted from the other signal to yield the Raman-scattered component.

Long-path absorption of tunable infrared laser radiation has been found to be one of the most sensitive techniques for detecting low concentrations of molecular pollutants within the atmosphere (Hinkley, 1976; Hanst, 1976). Unfortunately, thermal fluctuations in the atmosphere were found to constitute the dominant source of noise in the received laser signal. In order to monitor trace concentrations of molecular pollutants over long atmospheric paths with high sensitivity, Ku et al. (1975) have developed a technique they term *fast derivative spectroscopy*. This involves taking the derivative of the absorption spectra through modulating the output wavelength of the laser. In the case of a semiconductor diode laser this can be accomplished by modulating the diode current about its steady value. This will be discussed further in Chapter 3.

1.2 LASER REMOTE SENSOR EQUATIONS

In this section we shall derive and study the basic equations of importance in the field of laser remote sensing. The form of equation to be used in any given situation depends on the kind of interaction invoked by the laser radiation. This in turn is determined by the nature of the measurement to be undertaken. For those applications in which backscattering (elastic or inelastic) of the laser beam is utilized, the form of the lidar equation is fairly simple and is derived first. Most atmospheric probing, including those instances where differential absorption is employed, is covered by this equation.

For those situations involving laser-induced fluorescence, finite relaxation effects of the laser-excited species has to be taken into consideration. This leads to a more complex form of the lidar equation and to an optical depth dependence of the target media, detector integration period, and laser pulse shape and duration. In the limit of large optical depth this form of the lidar equation becomes identical to the laser fluorosensor equation that was specifically developed to cover airborne lidars that probe natural bodies of water.

In general, interpretation of the lidar signal is further complicated by geometrical considerations that include the degree of overlap between the laser beam and the field of view of the receiver optics as well as the details of the telescope. Nevertheless, it is possible in many instances to use one simplified form of lidar equation under a fairly wide range of conditions.

1.2.1 Scattering Form of the Lidar Equation

In the case of a pulsed, monostatic lidar, the increment of signal power $\Delta P(\lambda, R)$ received by the detector in the wavelength interval $(\lambda, \lambda + \Delta\lambda)$ from the element of range located in the interval $(R, R + \Delta R)$ is given by

$$\Delta P(\lambda, R) = \int J(\lambda, R, \mathbf{r})\, \Delta\lambda\, \Delta R\, p(\lambda, R, \mathbf{r})\, dA\,(R, \mathbf{r}) \qquad (1.37)$$

where $J(\lambda, R, \mathbf{r})$ = laser-induced spectral radiance at wavelength λ at position $\mathbf{r}$ in target plane located at range R per unit range interval

$dA\,(R, \mathbf{r})$ = element of target area at position $\mathbf{r}$ and range R

$p(\lambda, R, \mathbf{r})$ = probability that radiation of wavelength λ emanating from position $\mathbf{r}$ at range R will strike detector

Many factors will affect this probability. These include geometric considerations, atmospheric attenuation, receiver optics, and spectral transmission characteristics. Fortunately, we can separate most of these influences and write

$$p(\lambda, R, \mathbf{r}) = \frac{A_0}{R^2} \times T(\lambda, R) \times \xi(\lambda) \times \xi(R, \mathbf{r}) \qquad (1.38)$$

where A_0/R^2 = acceptance solid angle of receiver optics (A_0 being area of objective lens or mirror)

$T(\lambda, R)$ = atmospheric transmission factor at wavelength λ over range R

$\xi(\lambda)$ = receiver's spectral transmission factor and includes influence of any spectrally selecting elements such as monochromator

$\xi(R, \mathbf{r})$ = probability of radiation from position $\mathbf{r}$ in target plane at range R detector, based on geometric considerations

For the moment we shall assume that $\xi(R, \mathbf{r})$ depends only on the *overlap* of the area of laser irradiation with the field of view of the receiver optics. Consequently, we shall refer to $\xi(R, \mathbf{r})$ as the *overlap factor*. Later, we shall return to this function and see that it is somewhat more involved and can depend quite critically on the details of the receiver optics. For this reason it is often referred to as the *geometric form factor* in the literature.

The *target spectral radiance* $J(\lambda, R, \mathbf{r})$ depends on the nature of the interaction between the laser radiation and the target medium. In this section we shall consider a *scattering* (elastic or inelastic) medium. In this instance we may write

$$J(\lambda, R, \mathbf{r}) = \beta(\lambda_L, \lambda, R, \mathbf{r}) I(R, \mathbf{r}) \qquad (1.39)$$

where $I(R, \mathbf{r})$ is the laser irradiance at position $\mathbf{r}$ and range R, and

$$\beta(\lambda_L, \lambda, R, \mathbf{r}) = \sum_i N_i(R, \mathbf{r}) \left\{ \frac{d\sigma(\lambda_L)}{d\Omega} \right\}_i^s \mathcal{L}_i(\lambda) \qquad (1.40)$$

is the *volume backscattering coefficient*, in which

$N_i(R, \mathbf{r})$ = number density of scatterer species i

$\{d\sigma(\lambda_L)/d\Omega\}_i^s$ = differential scattering cross section under irradiation with laser radiation at wavelength λ_L

$\mathcal{L}_i(\lambda)\,\Delta\lambda$ = fraction of scattered radiation that falls into wavelength interval $(\lambda, \lambda + \Delta\lambda)$.

The total signal power received by the detector at the instant t $(=2R/c)$, corresponding to the time taken for the leading edge of the laser pulse to propagate (at the velocity of light, c) to range R and the returned radiation to reach the lidar, can be expressed in the form

$$P(\lambda, t) = \int_0^{R=ct/2} dR \int_{\Delta\lambda_0} d\lambda \int J(\lambda, R, \mathbf{r}) p(\lambda, R, \mathbf{r})\, dA\,(R, \mathbf{r}) \quad (1.41)$$

The range integral is required to account for the fact that radiation reaching the detector at time t not only originates from the distance $ct/2$, but also from any position along the path of the laser pulse from which scattering arises. The range of wavelength integration extends over the lidar receiver's spectral window $\Delta\lambda_0$ centered about λ. If Equations (1.38) and (1.39) are employed, we can write

$$P(\lambda, t) = A_0 \int_0^{R=ct/2} \frac{dR}{R^2} \int_{\Delta\lambda_0} \xi(\lambda)\, d\lambda$$

$$\times \int \beta(\lambda_L, \lambda, R, \mathbf{r}) T(\lambda, R) \xi(R, \mathbf{r}) I(R, \mathbf{r})\, dA\,(R, \mathbf{r}) \quad (1.42)$$

In the case of a scattering medium the observed radiation is as narrow band as that of the laser radiation, and if we assume both to be much smaller than the receiver's spectral window $\Delta\lambda_0$, we can treat $\mathcal{L}_i(\lambda)$, and therefore β, as a delta function. If we also assume that the medium is homogeneous over the zone of overlap between the field of view and the laser beam, then we can write

$$P(\lambda, t) = A_0 \xi(\lambda) \int_0^{R=ct/2} \beta(\lambda_L, \lambda, R) T(\lambda, R) \frac{dR}{R^2}$$

$$\cdot \int \xi(R, \mathbf{r}) I(R, \mathbf{r})\, dA\,(R, \mathbf{r}) \quad (1.43)$$

As mentioned above, we shall assume at this point that the probability $\xi(R, \mathbf{r})$ is essentially unity where the field of view of the receiver optics overlaps the laser beam and zero elsewhere. We shall also assume that the lateral distribution of the laser pulse is uniform over an area $A_L(R)$ at range R. In this case

$$\int \xi(R, \mathbf{r}) I(R, \mathbf{r})\, dA\,(R, \mathbf{r}) = \xi(R) I(R) A_L(R) \tag{1.44}$$

and

$$P(\lambda, t) = A_0 \xi(\lambda) \int_{R=0}^{R=ct/2} \beta(\lambda_L, \lambda, R) T(\lambda, R) \xi(R) I(R) A_L(R) \frac{dR}{R^2} \tag{1.45}$$

An additional simplification that is usually made in the literature is to approximate the temporal shape of the laser pulse by a rectangle of duration τ_L. Then the limits of the range integration in Equation (1.45) are obviously $c(t - \tau_L)/2$ to $ct/2$. Furthermore, since the range of interest is generally much greater than the laser pulse length $c\tau_L$ (otherwise the resolution would be poor), we may treat the range-dependent parameters as constants over the small interval of range integration. Then the total scattered laser power received at a time $t = 2R/c$ can be expressed in the form

$$P(\lambda, t) = A_0 \xi(\lambda) \beta(\lambda_L, \lambda, R) T(\lambda, R) \xi(R) I(R) A_L(R) \frac{c\tau_L/2}{R^2} \tag{1.46}$$

More precisely, the last factor should be $(c\tau_L/2)/[R(R - c\tau_L/2)]$, but as stated, $R >> c\tau_L/2$.

For a rectangular-shaped laser pulse of duration τ_L,

$$I(R) = \frac{E_L T(\lambda_L R)}{\tau_L A_L(R)} \tag{1.47}$$

where E_L represents the output energy of the laser pulse, and $T(\lambda_L R)$ represents the atmospheric transmission factor at the laser wavelength to range R. It follows from the Beer–Lambert law, that the transmission factors are

$$T(\lambda_L, R) \equiv \exp\left(- \int_0^R \kappa(\lambda_L, R)\, dR \right) \tag{1.48}$$

and

$$T(\lambda, R) \equiv \exp\left(- \int_0^R \kappa(\lambda, R)\, dR \right) \tag{1.49}$$

where $\kappa(\lambda_L, R)$ and $\kappa(\lambda, R)$ represent the atmospheric attenuation coefficients at the laser and detected wavelengths, respectively. It is evident that combining these leads to the total atmospheric transmission factor

$$T(R) \equiv T(\lambda_L, R) T(\lambda, R) = \exp\left\{ - \int_0^R [\kappa(\lambda_L, R) + \kappa(\lambda, R)]\, dR \right\} \tag{1.50}$$

Although the instantaneous power falling upon the detector is a useful quantity to evaluate, a more pertinent entity is the increment of radiative energy at wavelength λ received by the detector during the interval $(t, t + \tau_d)$, where τ_d is the integration period for the detector and $t = 2R/c$:

$$E(\lambda, R) = \int_{2R/c}^{2R/c + \tau_d} P(\lambda, t)\, dt \tag{1.51}$$

Combining Equations (1.46), (1.47), and (1.48) with (1.49) yields the scattered laser energy received within the detector's response time τ_d:

$$E(\lambda, R) = E_L \xi(\lambda) T(R) \xi(R) \frac{A_0}{R^2} \beta(\lambda_L, \lambda, R) \frac{c\tau_d}{2} \tag{1.52}$$

This is often cited as the *basic scattering lidar equation*.

Implicit in the derivation of this equation is the assumption that $\tau_d \ll 2R/c$. The effective range resolution for such a system is limited to $c(\tau_d + \tau_L)/2$, as is clearly seen by reference to Figure 1.7. If one species dominates the scattering and its scattering cross section is isotropic, the lidar equation can be expressed in the form

$$E(\lambda, R) = E_L \xi(\lambda) T(R) \xi(R) \frac{A_0}{R^2} N(R) \frac{\sigma^s(\lambda_L, \lambda)}{4\pi} \frac{c\tau_d}{2} \tag{1.53}$$

where $\sigma^s(\lambda_L, \lambda)$ represents the total cross section for scattering at λ for incident radiation of wavelength λ_L. Equation (1.53) represents the basic single-constituent scattering form of the lidar equation.

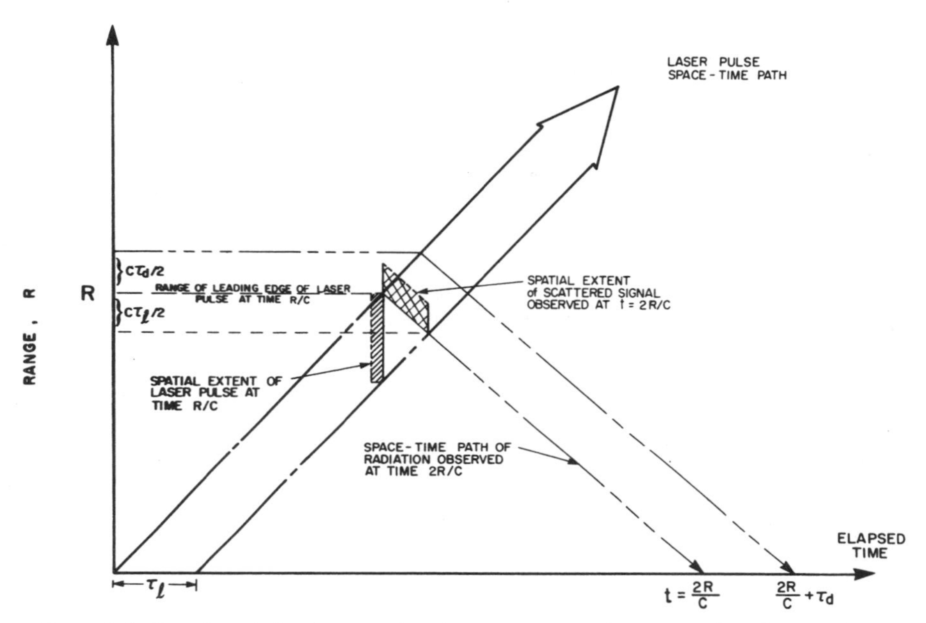

Figure 1.7. Spatial resolution for scattering phenomena as seen from space–time diagram of propagating rectangular-shaped laser pulse (Measures, 1977).

In the case of a more general laser pulse shape we can write

$$I(R^*) = \frac{E_L T(\lambda_L, R^*)}{A_L(R^*)} \mathcal{S}(t^*) \tag{1.54}$$

where $\mathcal{S}(t^*)$ describes the temporal behavior of the laser pulse in a frame of reference fixed to the leading edge of the laser pulse. Clearly,

$$t^* = \frac{2(R - R^*)}{c} \tag{1.55}$$

and represents the time taken for the leading edge of the laser pulse to propagate from R^* to R and the scattered radiation to return to R^*. In general,

$$\int_0^\infty \mathcal{S}(t^*)\, dt^* = 1 \tag{1.56}$$

Under these circumstances, the total signal power received after an elapsed time t corresponding to range R follows from Equation (1.45)

$$P(\lambda, R) = E_L A_0 \xi(\lambda) \int_{R=0}^{R=ct/2} \beta(\lambda_L, \lambda, R^*) T(R^*) \xi(R^*) \mathcal{G}(R^*) \frac{dR^*}{R^{*2}} \tag{1.57}$$

where $\mathcal{G}(R^*)$ is used to represent $\mathcal{G}(t^*)$.

If again we assume that the effective laser pulse length $c\tau_L$ is small compared to the range of interest, then we may treat the factor $\beta(\lambda_L, \lambda, R^*)T(R^*)\xi(R^*)/R^{*2}$ as constant over the small interval of range for which $\mathcal{G}(R^*)$ is finite:

$$P(\lambda, R) \cong E_L \frac{A_0}{R^2} \xi(\lambda) \beta(\lambda_L, \lambda, R) T(R) \xi(R) \int_{t=0}^{t=2R/c} \mathcal{G}(t^*) \frac{c\, dt^*}{2} \tag{1.58}$$

Since the upper limit of this integration corresponding to the time that scattered radiation from the leading edge of the laser pulse reaches the lidar system, it is evident from (1.56) that

$$\int_{t=0}^{t=2R/c} \mathcal{G}(t^*)\, dt^* \cong 1$$

and so we can write the total scattered laser power received at a time corresponding to the leading edge of the laser pulse propagating to a range R as

$$P(\lambda, R) = P_L \frac{A_0}{R^2} \xi(\lambda) \beta(\lambda_L, \lambda, R) \xi(R) \frac{c\tau_L}{2} \exp\left(- \int_0^R \kappa(R)\, dR \right) \tag{1.59}$$

where from Equation (1.50) we have introduced

$$\kappa(R) = \kappa(\lambda_L, R) + \kappa(\lambda, R) \tag{1.60}$$

as the two-way attenuation coefficient and $P_L \equiv E_L/\tau_L$ as the average power in the laser pulse.

In the event that we are interested in *elastic* (Mie or Rayleigh) scattering, then the wavelength of observation is invariably the same as that of the laser, and we may write

$$P(\lambda_L, R) = P_L \frac{A_0}{R^2} \xi(\lambda_L) \beta(\lambda_L, R) \xi(R) \frac{c\tau_L}{2} \exp\left(-2 \int_0^R \kappa(\lambda_L, R)\, dR \right) \tag{1.61}$$

In summary, Equation (1.59) is the scattering lidar equation most often quoted in the literature, and Equation (1.61) represents a special case that is of growing interest due to the increasing popularity of differential absorption techniques. With regard to the radiative energy received within the detector integration period, Equation (1.52) represents the more relevant form of the scattering lidar equation.

1.2.2 Differential Absorption Lidar (DIAL) Equation

Differential absorption of laser radiation by a particular molecular species represents both a selective and a sensitive method of measuring specific atmospheric constituents. There are two ways in which such measurements can be undertaken. Both involve using two laser pulses of slightly different wavelength (one chosen to coincide with a strong absorption feature of the specific constituent of interest, the other detuned into the wing of this feature) and comparing the attenuation of the two pulses. The difference in the techniques stems from the mechanism chosen to return the laser radiation to the lidar receiver system. In one case elastic scattering from atmospheric aerosols and particulates is employed, and consequently we shall refer to this as the DAS (differential absorption and scattering) technique. The other approach relies on scattering of the laser radiation from some conveniently located topographic target. An extreme example of this uses a strategically positioned retroreflector.

In the DAS approach, two laser wavelengths, λ_0 and $\lambda_0 + \delta\lambda$, are selected such that λ_0 corresponds to the center wavelength of some prominent absorption line of the molecule of interest, while $\lambda_0 + \delta\lambda$ lies in the wing of this line. If we write λ_W for $\lambda_0 + \delta\lambda$, and use Equation (1.63) (the elastic-scattering form of the lidar equation), the ratio of the return power signals at the two wavelengths is

$$\frac{P(\lambda_0, R)}{P(\lambda_W, R)} = \frac{\xi(\lambda_0)\beta(\lambda_0, R)}{\xi(\lambda_W)\beta(\lambda_W, R)} \exp\left\{-2\int_0^R [\kappa(\lambda_0, R) - \kappa(\lambda_W, R)]\, dR\right\} \tag{1.62}$$

where the output power of the laser is assumed to be the same at both wavelengths.

If we separate the absorption associated with the molecule of interest from the total attenuation coefficient, we can write

$$\int_0^R N(R)\sigma_A(\lambda_0\colon \lambda_W)\, dR = \frac{1}{2}\ln\frac{P(\lambda_W, R)\xi(\lambda_0)\beta(\lambda_0, R)}{P(\lambda_0, R)\xi(\lambda_W)\beta(\lambda_W, R)} - \int_0^R [\bar{\kappa}(\lambda_0, R) - \bar{\kappa}(\lambda_W, R)]\, dR \tag{1.63}$$

where we have introduced the *differential absorption cross section*,

$$\sigma_A(\lambda_0: \lambda_W) \equiv \sigma^A(\lambda_0) - \sigma^A(\lambda_W) \tag{1.64}$$

and have assumed that in general the total attenuation coefficient is given by

$$\kappa(\lambda, R) = \bar{\kappa}(\lambda, R) + N(R)\sigma^A(\lambda) \tag{1.65}$$

where $\bar{\kappa}(\lambda, R)$ is obviously the attenuation coefficient exclusive of the absorption contribution from the molecular species of interest, $N(R)$ represents the number density of these molecules at range R, and $\sigma^A(\lambda)$ their absorption cross section at wavelength λ.

In differential form, Equation (1.63) becomes

$$N(R) = \frac{1}{2\sigma_A(\lambda_0: \lambda_W)}\left[\frac{d}{dR}\left(\ln\frac{P(\lambda_W, R)}{P(\lambda_0, R)} - \ln\frac{\beta(\lambda_W, R)}{\beta(\lambda_0, R)}\right) + \bar{\kappa}(\lambda_W, R) - \bar{\kappa}(\lambda_0, R)\right] \tag{1.66}$$

where we have assumed that the receiver's spectral transmission factor is effectively independent of wavelength over the small interval $\delta\lambda$:

$$\xi(\lambda_0) \approx \xi(\lambda_W)$$

Additional simplification can be attained if we also assume that the volume backscattering coefficient β and the residual attenuation coefficient $\bar{\kappa}$ are independent of wavelength over this small interval $\delta\lambda$. We shall consider this in more detail later.

A considerable improvement in sensitivity can be achieved if this differential absorption technique is used in conjunction with a *topographic* scatterer. However, this gain in sensitivity is achieved at the expense of range resolution, so that this technique is only applicable in situations where the integrated concentration of the trace constituents along the path of the laser beam is worth evaluating. Under these circumstances, the signal power equation takes the form

$$P(\lambda_0, t) = \frac{E_L A_0}{\tau_L R_T^2}\xi(\lambda_0)\xi(R_T)\frac{\rho^s}{\pi}\exp\left(-2\int_0^{R_T}\kappa(\lambda_0, R)\,dR\right) \tag{1.67}$$

where ρ^s represents the scattering efficiency of the topographic target and R_T is the range to the topographic target. Values of ρ^s can range from 0.1 in the visible to 1 in the IR (Wolfe, 1966; Shumate et al., 1982; Grant, 1982).

The corresponding increment of radiative energy received within the detector's integration period τ_d is then

$$E(\lambda_0, R_T) = E_L \frac{A_0}{R_T^2} \xi(\lambda_0) \xi(R_T) \frac{\rho^s \tau_d}{\pi \tau_L} \exp\left(-2 \int_0^{R_T} \kappa(\lambda_0, R)\, dR \right) \quad (1.68)$$

provided $\tau_d \leq \tau_L$. In the event that $\tau_d > \tau_L$, the factor τ_d/τ_L is replaced with unity. In order to have optimum temporal discrimination against any solar background illumination of the target, τ_d should be chosen to be as close to τ_L as possible. At locations of known pollution emission, a retroflector might be positioned so as to maximize the system sensitivity. Under these conditions the factor $\rho^s A_0/\pi R_T^2$ in Equation (1.68) is replaced with ξ_0, the receiver collection efficiency. This can amount to an improvement of several orders of magnitude, depending primarily on the range.

As with the DAS technique, two closely spaced laser wavelengths must be employed if contributions other than from the species of interest are to be eliminated. With the same reasoning as above for the case of a distributed scatterer, we can express the integrated concentration of the constituent along the path of the laser beam as

$$\int_0^{R_T} N(R)\, dR = \frac{1}{2\sigma_A(\lambda_0 : \lambda_W)} \left(\ln \frac{E(\lambda_W, R_T)}{E(\lambda_0, R_T)} + \bar{\kappa}(\lambda_W, R_T) - \bar{\kappa}(\lambda_0, R_T) \right) \quad (1.69)$$

In many cases of pollution monitoring the constituent of concern is normally present in the atmosphere, so that it is the increased loading of the atmosphere that is the entity to be measured. Under these circumstances an additional measurement has to be undertaken either prior to the release of pollution or at a different orientation so that the path of the laser beam misses the effluent plume. This second measurement provides the reference background level that has to be subtracted from the measurement across the plume.

In such cases Byer and Garbuny (1973) have indicated that the criterion for minimum transmitted energy does not correspond to the use of a laser wavelength that coincides with the peak absorption cross section. Their results show that some degree of detuning may be necessary for optimization. On the other hand, Measures and Pilon (1972) have drawn attention to the severe attenuation of the laser beam that can occur if the laser wavelength is chosen to maximize absorption.

It is evident that the greatest sensitivity at a given range can be attained using a retroreflector. However, this technique normally limits the mode of measurements to evaluating the total burden of the constituent along the line joining the laser remote sensor with the retroreflector. Mounting a retroreflector on a mobile platform such as a balloon or a remote-piloted airplane could greatly extend the capabilities of this approach, permitting it to map the spatial distribution of a constituent.

1.2.3 Fluorescent Forms of Lidar Equation

If the laser is capable of exciting fluorescence within the target, then finite relaxation effects have to taken into consideration. Kildal and Byer (1971) were the first to recognize the significance of this effect and to illustrate its influence on the return signal. The following discussion stems from a more detailed analysis of this problem by Measures (1977, 1984).

In the case of fluorescence, the signal power received by the lidar photodetector is of a similar form to that indicated in Equation (1.41), except that the radiance of the target element arises from the emission of excited molecules:

$$J(\lambda, R, \mathbf{r}) = \sum_i \frac{N_i^*(R, \mathbf{r})\, hc\mathcal{L}_i^F(\lambda)}{4\pi\lambda\tau_{\text{rad}}^i} \tag{1.70}$$

where $N_i^*(R, \mathbf{r})$ = number density of laser-excited molecules (or atoms) of species i at position $\mathbf{r}$ of target plane at range R capable of undergoing fluorescence

$\mathcal{L}_i^F(\lambda)\,\Delta\lambda$ = fraction of fluorescence emitted by species i into wavelength interval $(\lambda, \lambda + \Delta\lambda)$

τ_{rad}^i = radiative lifetime for excited molecules (or atoms) of species i

h, c = Planck's constant and velocity of light in vacuum, respectively

We shall assume that only one molecular species is excited by the laser and that these molecules can return to the ground level (or some other low-lying level) with the subsequent emission of radiation at some wavelength $\lambda(>\lambda_L)$, or they may be deexcited by some nonradiative process. If the laser power density is very high, the excited molecule may be forced to return to its original level by the process of stimulated emission, and saturation effects can arise Measures (1984).

In the weak-beam limit, which is generally of interest for remote sensing, stimulated emission can be neglected, and the temporal variation of the excited-state number density $N^*(R, t)$ can be expressed (Measures, 1984) in the form

$$\frac{dN^*(R, t)}{dt} = \frac{\lambda_L \sigma^A(\lambda_L)}{hc} N(R, t) I(R, t) - \frac{N^*(R, t)}{\tau} \tag{1.71}$$

where $\sigma^A(\lambda_L)$ = absorption cross section per molecule for incident radiation at wavelength λ_L

$N(R, t)$ = ground-state number density in target plane at range R and time t (we have dropped the $\mathbf{r}$ dependence, as we shall assume that the target medium is homogeneous over the area of laser excitation)

τ = observed lifetime of excited population and is given by $1/\tau = 1/\tau_{\text{rad}} + C_Q$, where τ_{rad} represents radiative lifetime of excited state, and C_Q represents collision quenching rate for the excited state

We shall restrict our attention to situations where the level of irradiation is sufficiently low that nonlinear effects can be neglected and that no appreciable depletion of the ground-state number density is produced. In addition, we shall assume that the ground-state number density prior to irradiation is $N_0(R)$ and that $N^*(R, 0) \equiv 0$. Then the solution of Equation (1.71) yields the temporal variation of the excited-state population at range R and time t:

$$N^*(R, t) = \frac{\lambda_L N_0(R)\sigma^A(\lambda_L)}{hc} e^{-1/\tau} \int_0^t I(R, x) e^{x/\tau}\, dx \tag{1.72}$$

where $I(R, x)$ represents the laser irradiance at range R and at a *time* x after the leading edge of the laser pulse reaches this location (x is a dummy time variable).

At the instant that the leading edge of the laser pulse reaches range R, the target medium at range R' ($<R$) will have been exposed to laser radiation for a period $(R - R')/c$. Fluorescence induced by the leading edge of the laser pulse and propagating toward the detector will be reinforced by fluorescence emanating from the target medium at range R', having been exposed to laser radiation for a period $t' = 2(R - R')/c$. Consequently, the appropriate value of radiance to be used in the range integral of Equation (1.41) is obtained by combining Equations (1.70) and (1.72), namely,

$$J(\lambda, R') = \frac{N_0(R')\sigma^A(\lambda_L)\lambda_L \mathcal{L}^F(\lambda)}{4\pi\tau_{\text{rad}}\lambda} e^{-t'/\tau} \int_0^{t'} I(R', x) e^{x/\tau}\, dx \tag{1.73}$$

If we introduce the spectrally integrated fluorescence cross section,

$$\sigma^F(\lambda_L) \equiv \sigma^A(\lambda_L)\frac{\tau}{\tau_{\text{rad}}} \tag{1.74}$$

and Equations (1.73) and (1.74) are substituted into (1.41), and we make the same assumptions in regard to the probability factor $p(\lambda, R, \mathbf{r})$ as we did in the scattering section, we arrive at the fluorescence signal power received at the lidar detector after an elapsed time t ($=2R/c$):

$$P(\lambda, R) = \frac{A_0 \sigma^F(\lambda_L)\lambda_L}{4\pi\tau\lambda} \int_0^R dR'\, \xi(R') A_L(R') \frac{N_0(R')}{R'^2} e^{-t'/\tau} \times \int_0^{t'} I(R', x) e^{x/\tau}\, dx \int_{\Delta\lambda_0} \mathcal{L}^F(\lambda)\xi(\lambda)T(\lambda, R')\, d\lambda \quad (1.75)$$

The range integration in Equation (1.75) is along the space–time path of the observed ray, path AB of Figure 1.8.

In most cases of interest the variation of the atmospheric transmission factor $T(\lambda, R')$ over the spectral bandwidth $\Delta\lambda_0$ of the receiver system is small enough that it can be taken out of the wavelength integral in Equation (1.75). Most practical situations can be approximated by one of two limiting cases. If the bandwidth of the receiver system is made adequate to accept the entire fluorescence profile, we have a situation akin to that of scattering, where the wavelength integral can be replaced by the factor $T(\lambda, R')\xi(\lambda)$. Under these circumstances the wavelength λ corresponds to the center value of the fluorescence profile.

If, as is more often the situation, the spectral window of the receiver system is small compared to the spectral width of the observed fluorescence, the wavelength integral can be approximated in the following manner:

$$\int_{\Delta\lambda_0} \mathcal{L}^F(\lambda)\xi(\lambda)T(\lambda, R')\, d\lambda = T(\lambda, R')K_0(\lambda)\mathcal{L}^F(\lambda) \quad (1.76)$$

where

$$K_0(\lambda) \equiv \int_{\Delta\lambda_0} \xi(\lambda')\, d\lambda' \quad (1.77)$$

and is termed the *filter function*. The wavelength λ indicated on the right-hand side (RHS) of Equation (1.76) corresponds to the value at the center of the receiver bandwidth; $K_0(\lambda)$ represents the effective bandwidth that would transmit, with unit transmission efficiency, the same fraction of the fluorescence as achieved by the real system.

We shall again assume that the laser irradiance at range R' and time x is given by

$$I(R', x) = \frac{E_L T(\lambda_L, R')}{A_L(R')} \mathcal{G}(x) \quad (\mathrm{W\ cm^{-2}}) \quad (1.78)$$

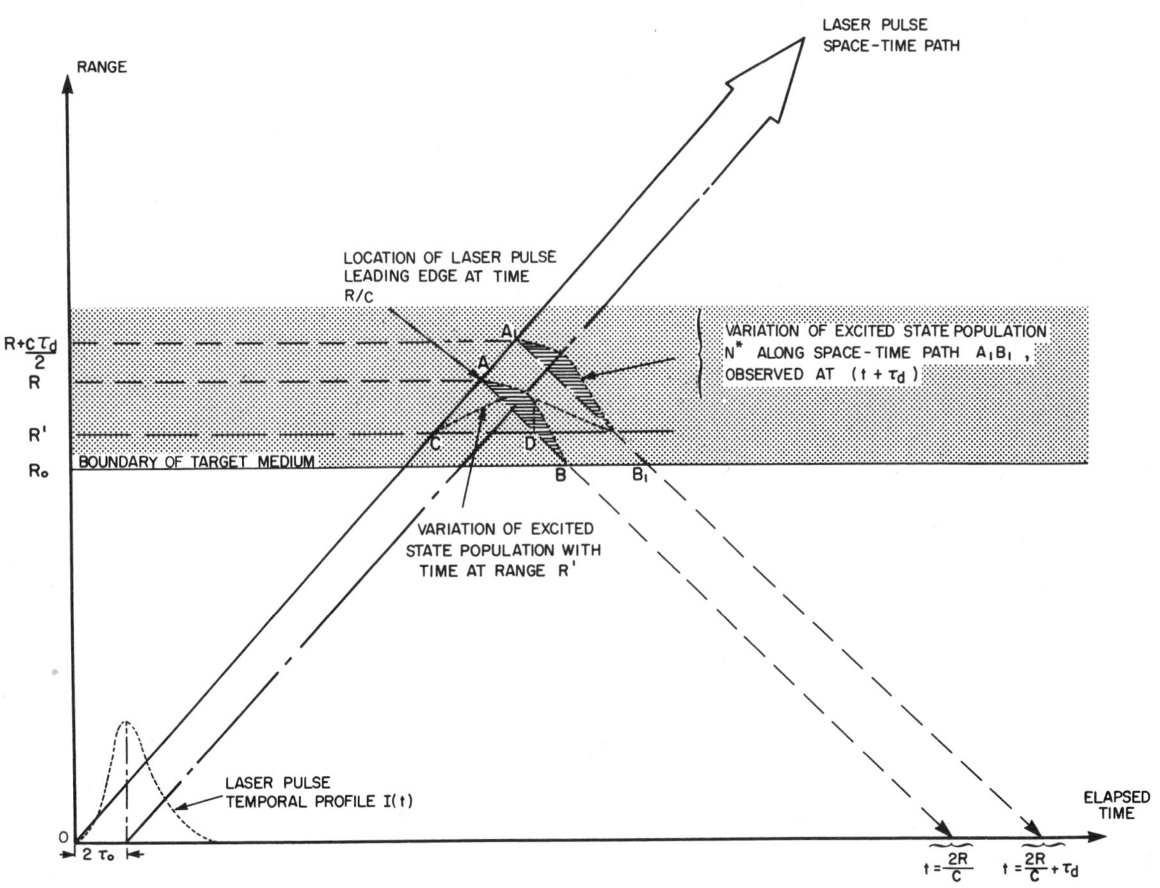

Figure 1.8. Space–time view of laser pulse propagation and excitation of fluorescent target medium (Measures, 1977).

where, as before,

$$\int_0^{\infty} \mathcal{G}(x)\, dx = 1$$

Furthermore, if the total attenuation coefficient $\kappa_T \equiv \kappa(\lambda_L) + \kappa(\lambda)$ is regarded as a constant within the target medium, the boundary of which is located at range R_0, the radiative energy received by the detector in the time interval $(t, t + \tau_d)$ is

$$E(\lambda, R) = E_L T(R_0) K_0(\lambda) \xi(R_0) \frac{A_0 N_0 \sigma^F(\lambda_L) \mathcal{L}^F(\lambda)}{4\pi R^2 \tau} \int_t^{t+\tau_d} H(R)\, dt' \tag{1.79}$$

where

$$H(R) = \int_{R_0}^{R} dR'\, e^{-\kappa_T(R'-R_0)} e^{-t'/\tau} \int_0^{t'} \mathcal{G}(x) e^{x/\tau}\, dx \tag{1.80}$$

To obtain this relation, we have expressed

$$T(\lambda, R') T(\lambda_L, R') = T(R_0) e^{-\kappa_T(R'-R_0)} \tag{1.81}$$

where

$$T(R_0) = \exp\left(-\int_0^{R_0} \kappa(R)\, dR \right) \tag{1.82}$$

and have assumed that the overlap factor $\xi(R')$ is only weakly dependent on the range, so that we can approximate its value by $\xi(R_0)$. We have also assumed that the physical extent of the fluorescent target is sufficiently small that the factor $1/R^2$ can be taken out of the range integration and that the factor λ_L/λ can be approximated by unity.

Clearly, Equation (1.79) can be written in a form that bears a close resemblance to the scattering lidar equation as expressed by Equation (1.53), namely,

$$E(\lambda, R) = E_L K_0(\lambda) T(R_0) \xi(R_0) \frac{A_0}{R^2} N_0 \frac{\sigma^F(\lambda_L, \lambda)}{4\pi} \frac{c\tau_d}{2} \gamma(R) e^{-\kappa_T(R-R_0)} \tag{1.83}$$

where we have introduced the *fluorescence cross section*

$$\sigma^F(\lambda_L, \lambda) = \sigma^F(\lambda_L)\mathcal{L}^F(\lambda) \tag{1.84}$$

and the *fluorescence* (or *lifetime*) *correction factor*

$$\gamma(R) = \frac{2}{c\tau_d \tau} e^{\kappa_T(R - R_0)} \int_t^{t+\tau_d} H(R)\, dt' \tag{1.85}$$

to account for the effects of the finite lifetime of the laser-excited molecules, the optical depth of the target medium, the detector integration period, and the laser pulse shape and duration.

It should also be evident that in the fluorescence case the filter function $K_0(\lambda)$ is used in place of $\xi(\lambda)$ and that since we have assumed the fluorescent medium only exists beyond range R_0, the atmospheric transmission factor $T(R)$ is replaced with its value at the boundary [i.e., $T(R_0)$] multiplied by the transmission factor within the target medium [i.e., $e^{-\kappa_T(R - R_0)}$].

The fluorescence correction factor $\gamma(R)$ can be evaluated once the shape of the laser pulse has been defined. Measures (1977, 1984) has calculated the variation of this correction factor for various-shaped laser pulses.

In the case of an optically thin target the lifetime correction factor $\gamma(z)$ tends to approach unity for large penetration depths. This is illustrated in Figures 1.9*a* and *b*, where the variation in the lifetime correction factor with the normalized effective penetration depth $z^* \equiv (R - R_0)/L$ is shown. Here $L \equiv c\tau_L^*/2$ and represents the laser effective pulse length, and $T^* \equiv \tau_L^*/\tau$ represents the ratio of the laser effective pulse duration to the fluorescence lifetime and can be seen to be an important parameter in determining the magnitude of $\gamma(z^*)$. It should be noted (Measures, 1977) that very similar results are obtained if we consider the simpler rectangular-shaped laser pulse of duration τ_L, and consequently we shall drop the asterisk from τ_L in the remaining discussion. For small values of T^* (i.e., $\tau_L \ll \tau$) it is apparent that $\gamma(z) \ll 1$ even for considerable penetration depths.

If $\tau_L \gg \tau$ (i.e., T^* is large), the value of $\gamma(z)$ approaches unity for target penetration depths of about one laser pulse length. Nevertheless, the fact that the correction factor $\gamma(z)$ can be much less than unity even for large values of T^*—corresponding to the *scattered limit*—indicates that in effect the simple form of the scattering lidar equation (1.53) overestimates the expected signal in situations where the return arises from a real laser pulse close to a sharp boundary. The results presented in Figure 1.9*a* assume $\tau_d/\tau_L \approx 0.2$. If, however, we set $\tau_d = \tau_L$, somewhat similar behavior is observed Figure 1.9*b*, except that $\gamma(z^*)$ has a finite value at $z = 0$. This can be understood in terms of the spatial resolution of the system.

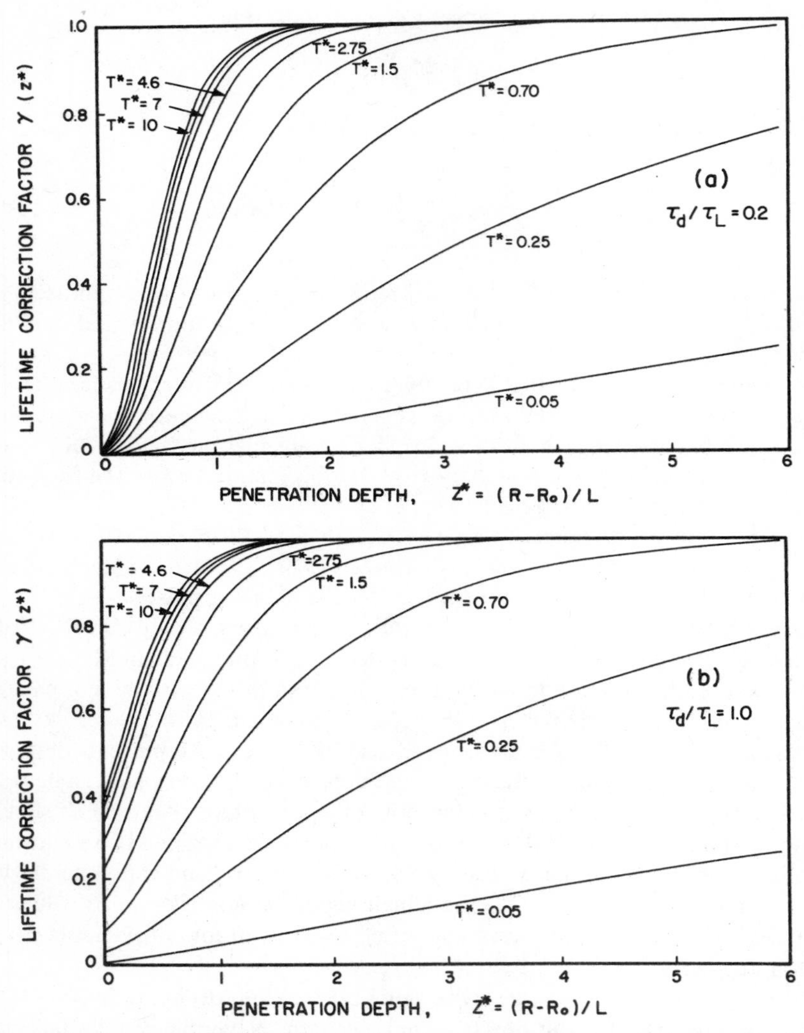

Figure 1.9. Variation of optically thin correction factor $\gamma(z^*)$ with normalized penetration z^* into a fluorescent target for several values of T^* (ratio of laser pulse duration to fluorescence lifetime, τ_L/τ); R is the range of the laser pulse leading edge, R_0 is the range of the target boundary, and L is the laser pulse length ($c\tau_L/2$). Ratio of laser pulse length to attenuation length is $\kappa_T L = 0.005$ for each case (Measures, 1977).

In the scattering limit (i.e., $\tau \approx 0$), we saw that the spatial resolution is determined by $c(\tau_L + \tau_d)/2$. For those situations where the excited-state lifetime has to be taken into account, the range resolution is approximated by $c(\tau_d + \tau_L + \tau)/2$. This is illustrated for the case of a rectangular-shaped laser pulse in Figure 1.10.

In the case of hydrographic work the optical depth of the target is usually large ($\kappa_T L \geq 1$). In this case, combining the exponential factor of Equation (1.83) with the lifetime correction factor $\gamma(z^*)$, we derive a new correction factor termed the *optically thick correction factor*:

$$\bar{\gamma}(z^*) = \gamma(z^*) e^{-\kappa_T L z^*} \tag{1.86}$$

The variation of $\bar{\gamma}(z^*)$ with penetration depth z^* has been evaluated for a similar range of T^*-values, but with $\tau_d = 0.2\tau_L$ and optical depth values $\kappa_T L$ of 2.5 and 25, respectively. The results indicate that whereas $\gamma(z^*)$ tends to unity for large penetration depths in an optically thin medium, $\bar{\gamma}(z^*)$ reaches a maximum and then decays to zero for large penetrations. It is also evident from a study of Figures 1.9*a* and *b* that the influence of long relaxation times is merely to prevent $\gamma(z^*)$ from reaching unity until considerable penetration of the target has been attained. By contrast, the maximum value of $\bar{\gamma}(z^*)$ is reduced considerably by long fluorescent lifetimes in an optically thick medium. For most atmospheric applications involving electronic transitions, $\gamma(z^*)$ will be close to unity, as typical values for τ are a few nanoseconds. This may not be true for infrared fluorescent studies, where the lifetime of vibrational–rotational transitions can be much longer.

In hydrographic work, the targets are nearly always optically thick, and Measures (1977) has shown that if $\kappa_T L > 5$, $\tau_d > 0.2\tau_L$, and $\tau_L > \tau$ (short fluorescence lifetimes compared to laser pulse duration), the limiting value of the correction factor is given by

$$\bar{\gamma}(z^*) \cong \frac{2}{\kappa_T c \tau_d} \tag{1.87}$$

Under these circumstances the lidar equation, represented by Equation (1.83), takes the form

$$E(\lambda, R) = E_L T(R_0) K_0(\lambda) \xi(R_0) \frac{A_0}{R^2} N_0 \frac{\sigma^F(\lambda_L)}{4\pi} \frac{\mathcal{L}^F(\lambda)}{\kappa(\lambda_L) + \kappa(\lambda)} \tag{1.88}$$

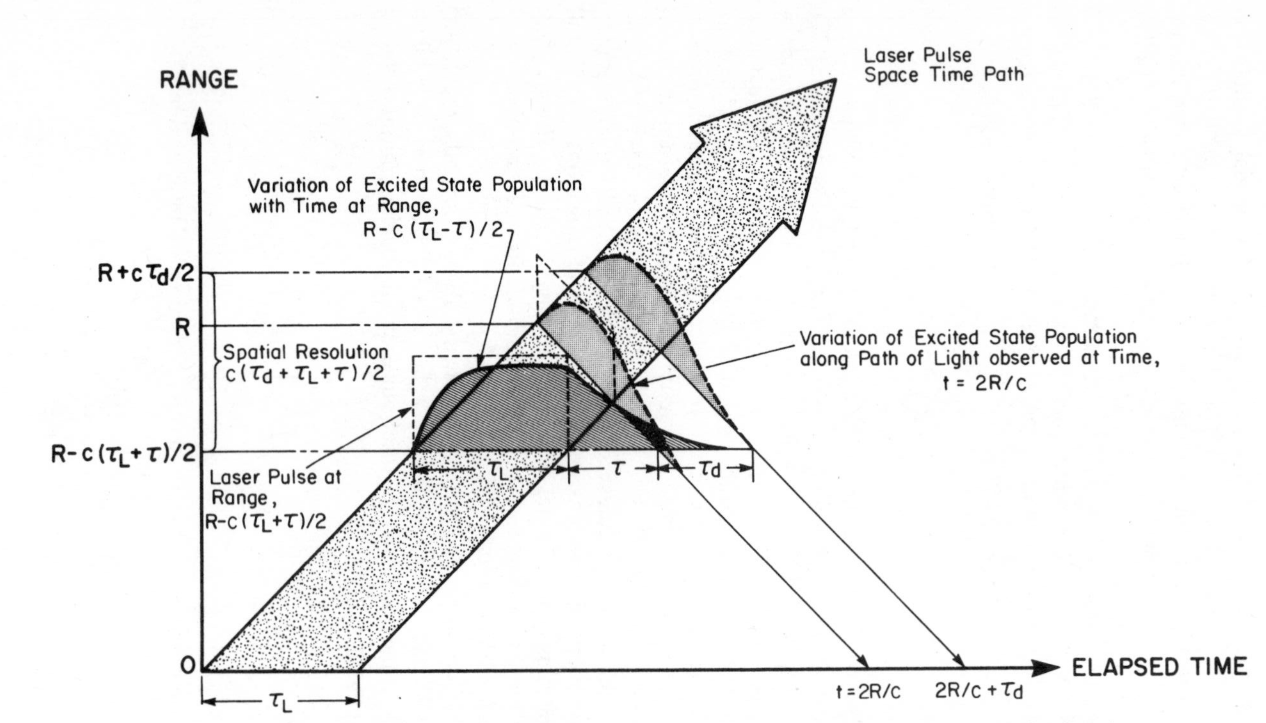

Figure 1.10. Space–time view of propagation and excitation of fluorescent medium by rectangular-shaped laser pulse. Spatial resolution attainable is clearly seen to be $c(\tau_d + \tau_L + \tau)/2$. Flattening of excitation curve is intended to imply saturation.

If we assume that $\kappa(\lambda_L) >> \kappa(\lambda)$, as is often the case, and further that $\kappa(\lambda_L) \approx N_0\sigma^A(\lambda_L)$, then we arrive at an equation of the form

$$E(\lambda, R) = E_L T(R_0) K_0(\lambda) \xi(R_0) \frac{A_0}{4\pi R^2} F(\lambda, \lambda_L) \tag{1.89}$$

where we have introduced the *target fluorescence efficiency* (Measures, 1984):

$$F(\lambda, \lambda_L) \equiv Q^F \mathcal{L}^F(\lambda) \tag{1.90}$$

Under such circumstances the lidar equation, as expressed by Equation (1.89), becomes identical to the *laser fluorosensor equation* as formulated by Measures et al. (1975), and the return signal is no longer capable of providing any information regarding the concentration of the fluorescent species within the target zone. However, since the target fluorescence efficiency $F(\lambda, \lambda_L)$ is proportional to the emission profile $\mathcal{L}^F(\lambda)$, identification of the target is still possible from a spectral scan of the fluorescence return.

In the event that fluorescence decay time measurements are required in order to better characterize the target medium (Measures et al., 1974), care must be used in selecting the values of τ_L and τ_d in relation to the expected values of τ, as the simple form of the lidar equation, expressed by Equation (1.89), may not be adequate, and the more complete equation may have to be employed. In the case of oil slicks, where the need for characterization is greatest, the extremely small penetration depth of the laser radiation [tens of microns (Measures et al., 1973)] ensures that Equation (1.89) is reasonable even when the detector integration time is made short enough for good decay time resolution.

In summary, the fluorescence form of the lidar equation (1.83) is more complex than for scattering through the introduction of the lifetime correction factor $\gamma(R)$. As a result, considerable care is needed in the interpretation of the return radiation, particularly where the fluorescence lifetime greatly exceeds the laser pulse duration and the fluorescent medium possesses a fairly sharp boundary.

In most situations, the value of this correction factor tends to unity for target penetration greater than a few laser pulse lengths, provided the medium is optically thin and the laser pulse duration is greater than, or of the same order of magnitude as, the lifetime of the laser-excited molecule. For atmospheric work these conditions will often be fulfilled even if good spatial resolution is required.

Under these circumstances, the fluorescence (energy) lidar equation, as given by Equation (1.83) can be rewritten in a form that makes it remarkably similar to its scattering counterpart, Equation (1.53), namely,

$$E(\lambda, R) = E_L K_0(\lambda) T(R) \xi(R) \frac{A_0}{R^2} N(R) \frac{\sigma^F(\lambda_L, \lambda)}{4\pi} \frac{c\tau_d}{2} \tag{1.91}$$

The essential differences are the use of the filter function $K_0(\lambda)$ in place of $\xi(\lambda)$ and of course the fluorescence cross section $\sigma^F(\lambda_L, \lambda)$ instead of the scattering cross section $\sigma^S(\lambda_L, \lambda)$. However, in the limit of great optical depth we have shown that all information on the conentration of the fluorescent species is lost, and the lidar equation becomes identical to the laser fluorosensor Equation (1.89), developed for hydrographic work.

1.2.4 Geometry of Receiver Optics

In the development of the lidar equations for both scattering and fluorescent targets we assumed that the geometric probability factor $\xi(R, \mathbf{r})$ was unity where the field of view of the receiver optics overlapped the laser beam and zero elsewhere. The distribution of laser irradiance across the target plane was also assumed to be uniform over the area of illumination $A_L(R)$. Although these assumptions are reasonable for long-range measurements, their soundness is highly questionable for short-range work, and in this section we shall not make these assumptions and study the consequences for the lidar equation. Although we shall employ the *scattering* lidar equation for this purpose, the results obtained and the conclusions drawn would be equally applicable in the case of fluorescence work because, as shown earlier, the form of the lidar equation is very similar.

We shall assume that the laser irradiance at position $\mathbf{r}$ in the target plane located at range R from the lidar is given by

$$I(R, r, \psi) = \frac{P_L T(\lambda_L, R)}{\pi W^2(R)} F(R, r, \psi) \tag{1.92}$$

where we assume azimuthal symmetry so that position $\mathbf{r}$ can be represented by (r, ψ). The origin of this coordinate system is taken to be the intersection of the target plane with the axis of the receiver optics (telescope). Here, r is the radial displacement of the point of interest in the target plane from the telescope axis, and ψ is the corresponding azimuthal angle from a vertical plane passing through this axis. This is illustrated in Figure 1.11, where $W(R)$ represents the radius of the laser pulse in the target plane at the instant of interest, P_L is the total output power of the laser at a time R/c earlier, and $F(R, r, \psi)$ describes the distribution of this laser power over the target plane at the instant of interest. The two most common distributions used in this context are

$$F(R, r, \psi) = \mathcal{G}(R, r, \psi) = e^{-[r^*/W(R)]^2} \tag{1.93}$$

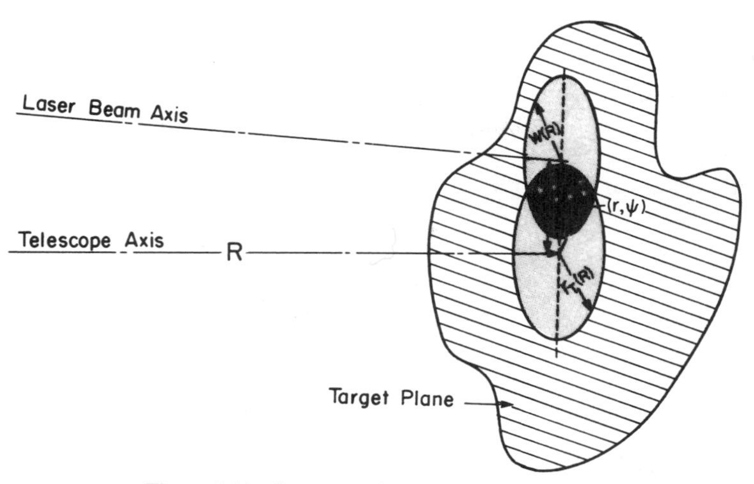

Figure 1.11. Geometry of target plane at range R.

and

$$F(R, r, \psi) = \mathcal{H}(R, r, \psi)$$
$$= \begin{cases} 1 & \text{where the receiver optics field of view and laser beam overlap} \\ 0 & \text{elsewhere} \end{cases} \tag{1.94}$$

The first corresponds to a Gaussian distribution about the laser beam axis, so that

$$r^* = (r^2 + d^2 - 2rd \cos \psi)^{1/2} \tag{1.95}$$

where d represents the separation of the laser and telescope axes in the target plane (see Figure 1.11). This kind of laser pulse distribution is expected where the output of the laser is predominantly in the TEM_{00} mode. In this situation the laser beam radius

$$W(R) = (W_0^2 + \theta^2 R^2)^{1/2} \tag{1.96}$$

where W_0 = laser output aperture radius
θ = laser's half divergence angle

Clearly, for the Gaussian beam, $W(R)$ corresponds to the exponential radius.

For a laser pulse that is dominated by high-order transverse modes, the *flat* distribution given by Equation (1.94) is a reasonable approximation. In this case $W(R)$ represents the actual radius of the beam; nevertheless, we shall use Equation (1.96) to define $W(R)$.

If we use Equation (1.92) with Equation (1.43) and assumes a rectangular temporal profile of duration τ_L for the laser pulse, the total scattered laser power received by the detector at the instant t $(=2R/c)$ is

$$P(\lambda, t) = P_L \xi(\lambda) r_0^2 \int_{R=c(t-\tau_L)/2}^{ct/2} dR \frac{\beta(\lambda_L, \lambda, R) T(R)}{W^2(R) R^2} \times \int_{r=0}^{r_T} \int_{\psi=0}^{2\pi} \xi(R, r, \psi) F(R, r, \psi) r\, dr\, d\psi \qquad (1.97)$$

In formulating this equation, we have assumed that the field of view of the receiver optics in the target plane is a circle of radius

$$r_T(R) = r_0 + \phi R \qquad (1.98)$$

where r_0 = effective radius of telescope lens (or mirror)
ϕ = receiver optics half opening angle

If a circular detector of radius r_D is positioned on the axis of the telescope of effective focal length f,

$$\phi = \frac{r_D}{f} \qquad (1.99)$$

This will be discussed further in the next section.

We can simplify Equation (1.97) in the same way that we had previously by taking account of the fact that the range of interest is always very much greater than the laser pulse length, that is, $ct/2 >> c\tau_L/2$. Under these circumstances, we can neglect dependence of all of the variables in comparison with the range integration, and the total scattered laser power received by the detector after an elapsed time t (corresponding to the leading edge of the laser pulse propagating to a range R) is given by

$$P(\lambda, t) = P_L \frac{c\tau_L}{2} \xi(\lambda) \frac{r_0^2}{R^2} \beta(\lambda_L, \lambda, R) \frac{T(R)}{W^2(R)} \times \int_{r=0}^{r_T} \int_{\psi=0}^{2\pi} \xi(R, r, \psi) F(R, r, \psi) r\, dr\, d\psi \qquad (1.100)$$

Evaluation of the double integral in this equation requires detailed knowledge of the geometric probability factor $\xi(R, r, \psi)$ as well as the laser irradiance distribution function $F(R, r, \psi)$. One of the approaches often adopted in the literature is to introduce the *effective telescope area*,

$$A(R) \equiv \frac{A_0}{\pi W^2(R)} \int_{r=0}^{r_T} \int_{\psi=0}^{2\pi} \xi(R, r, \psi) F(R, r, \psi) r \, dr \, d\psi \quad (1.101)$$

where A_0 $(=\pi r_0^2)$ represents the area of the telescope objective lens (or mirror). This enables the scattering (power) lidar equation to be written in the form

$$P(\lambda, R) = P_L \xi(\lambda) \beta(\lambda_L, \lambda, R) \frac{A(R)}{R^2} \frac{c\tau_L}{2} \cdot \exp\left\{ -\int_0^R [\kappa(\lambda_L, R) + \kappa(\lambda, R)] \, dR \right\} \quad (1.102)$$

which can be seen to be identical to Equation (1.59) except that $A(R)$ has replaced the product $A_0 \xi(R)$. Evidently, the geometric form factor $\xi(R)$ introduced earlier can be defined according to the relation

$$\xi(R) \equiv \frac{1}{\pi W^2(R)} \int_{r=0}^{r_T} \int_{\psi=0}^{2\pi} \xi(R, r, \psi) F(R, r, \psi) r \, dr \, d\psi \quad (1.103)$$

In the case of a rectangular-shaped laser, temporal profile P_L represents the average laser output power.

1.2.4.1 Simple Overlap Factor

The geometric form factor for a coaxial lidar having no apertures (other than the objective lens or mirror of the telescope) or obstructions is unity, provided the divergence angle of the laser beam is less than the opening angle of the telescope. In reality, the majority of lidar systems employ reflecting (Newtonian or Cassegrainian) telescopes. These invariably require some kind of mirror support structure that represents an obstruction to the return radiation. Furthermore, in the case of coaxial lidar systems a mirror is also required to merge the telescope and laser axes.

The geometric form factor can also be evaluated fairly easily for a biaxial lidar if the objective lens (or mirror) of the telescope represents the limiting aperture of the receiver optics, we neglect any obstruction, and we assume a flat laser distribution over the area of illumination. Under these circumstances, the geometric probability factor $\xi(R, r, \psi)$ is unity in the region of overlap between the laser beam and the field of view of the receiver optics and zero elsewhere. The geometrical form factor, as defined by Equation (1.103), becomes

$$\xi(R) = \frac{1}{\pi W^2(R)} \int_{r=0}^{r_T(R)} \int_{\psi=0}^{2\pi} \mathcal{H}(R, r, \psi) r \, dr \, d\psi \qquad (1.104)$$

Under these conditions, we can think of $\xi(R)$ as a simple overlap factor, and we can write

$$\xi(R) = \frac{\mathcal{A}\left\{r_T(R), W(R); d(R)\right\}}{\pi W^2(R)} \qquad (1.105)$$

where $\mathcal{A}$ represents the *area overlap function*. The radius of the receiver optics field of view in the target plane, $r_T(R)$, is given by Equation (1.98), and we shall assume that the radius of the laser pulse in the target plane, $W(R)$, is described by Equation (1.96). The separation of the telescope and laser axes in the target plane is

$$d = d_0 - R\delta \qquad (1.106)$$

where d_0 = separation of axes at lidar
δ = inclination angle between laser and telescope axes (see Figure 1.12)

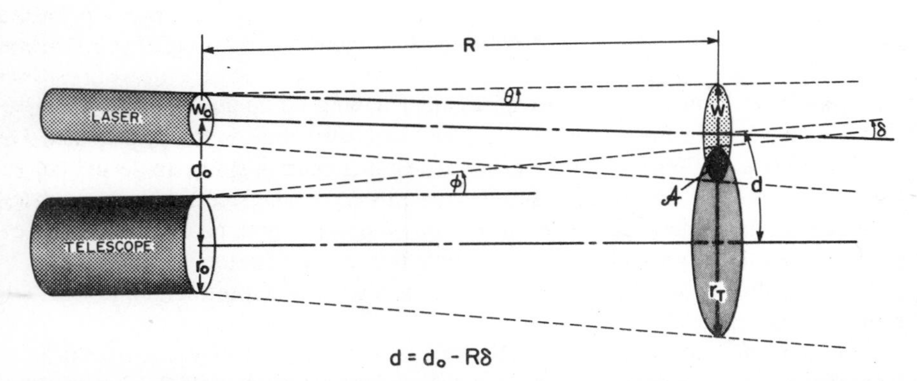

Figure 1.12. Geometry of biaxial lidar, where separation of laser and telescope axes is $d = d_0 - R\delta$ in target plane. r_T is radius of circular field of view and W is radius of circular region of laser illumination.

Three situations are possible:

1. The separation of the axes is too large for there to be any overlap between the receiver optics field of view and the area of laser illumination (see Figure 1.13*a*), that is, $\mathcal{A} = 0$ if $d > r_T + W$.
2. The separation of the axes is small enough that either the area of laser illumination lies totally within the receiver optics field of view or vice versa. The former case is illustrated in Figure 1.13*c*. This amounts to saying that if $d < |r_T - W|$, $\mathcal{A} = \pi \times$ (the smaller of r_T^2 or W^2).
3. The separation of the axes lies between these extremes:

$$|r_T - W| < d < r_T + W$$

This situation is illustrated in Figure 1.13*b*, and under these circumstances, the area overlap function is

$$\mathcal{A}\{r_T, W; d\} = W^2\psi_W + r_T^2\psi_r - r_T d \sin \psi_r \qquad (1.107)$$

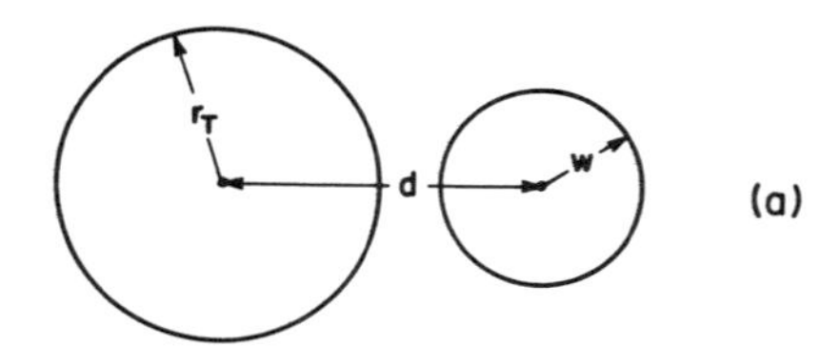

(a)

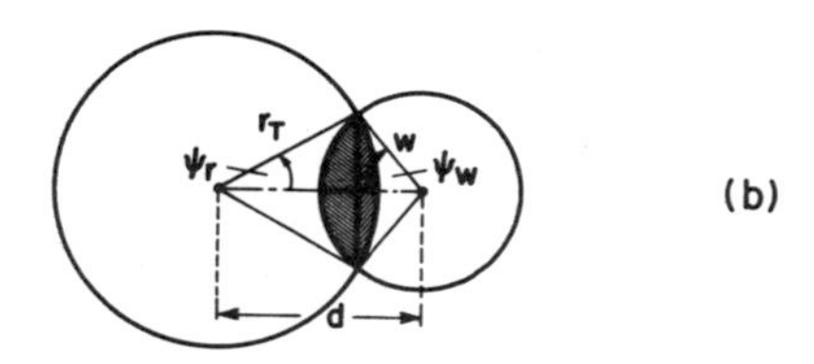

(b)

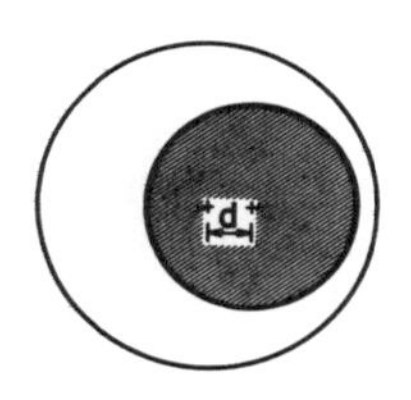

(c)

Figure 1.13. Three overlap situations possible for biaxial lidar, r_T is radius of circular field of view and W is radius of circular region of laser illumination.

where

$$\psi_W = \cos^{-1}\left[\frac{d^2 + W^2 - r_T^2}{2Wd}\right] \tag{1.108}$$

and

$$\psi_r = \cos^{-1}\left[\frac{d^2 + r_T^2 - W^2}{2r_T d}\right] \tag{1.109}$$

If we introduce the nondimensional parameters

$$z \equiv \frac{R}{r_0} \qquad A = \frac{r_0}{W_0} \qquad D = \frac{d_0}{r_0} \qquad \rho(r, \phi) \equiv \frac{r_T}{r_0} = 1 + z\phi$$

$$s(z, \delta) \equiv \frac{d}{r_0} = D - z\delta \qquad \omega(z, \theta) \equiv \frac{W}{W_0} = (1 + z^2\theta^2A^2)^{1/2}$$

and

$$y(z, \theta, \phi) \equiv \frac{\omega^2(z, \theta)}{\rho^2(z, \phi)A^2} \tag{1.10}$$

then the overlap factor can be expressed in the form

$$\xi(z) = \frac{\psi_W(z)}{\pi} + \frac{1}{\pi y(z)}\left[\psi_r(z) - \frac{s(z)}{\rho(z)}\sin\psi_r(z)\right] \tag{1.111}$$

where

$$\psi_W(z) = \cos^{-1}\left[\frac{s^2(z) + y(z)\rho^2(z) - \rho^2(z)}{2s(z)\rho(z)\sqrt{y(z)}}\right] \tag{1.112}$$

and

$$\psi_r(z) = \cos^{-1}\left[\frac{s^2(z) + \rho^2(z) - y(z)\rho^2(z)}{2s(z)\rho(z)}\right] \tag{1.113}$$

Clearly, the overlap factor $\xi(z)$ will depend on the normalized range z and will also depend on the values of the angular parameters θ, ϕ, and δ and the scale parameters A and D. First we shall consider the case of a biaxial lidar system in which the laser and telescope axes are parallel (i.e., $\delta = 0$) and both the telescope opening angle and the laser divergence angle are parallel (i.e., $\delta = 0$) and both the telescope opening angle and the laser divergence angle are 2 mrad (i.e., $\theta = \phi = 10^{-3}$). The variation of $\xi(z)$ with z for three values of D(1.0, 1.1, and 1.25) and two values of A(20 and 5) are presented in Figure 1.14. Although $\xi(z)$ tends to approach unity for large values of z, it is quite apparent that a separation of 10% more than the telescope lens (or mirror) radius (i.e., $D = 1.1$) can prevent the overlap factor from attaining this value until large values of z. Indeed, for $A = 20$ and $D = 1.1$ the value of $\xi(z)$ can be very small ($<1\%$) for $z < 50$. This feature of a biaxial arrangement can be useful in eliminating near-field scattering—an effect that can saturate or even damage the photodetector when the lidar system is designed for long-range measurements.

The sensitivity of $\xi(z)$ to a change in the opening angle of the telescope, ϕ [which can be thought of as a measure of the sensitivity of $\xi(z)$ to the detector radius] can be gauged by reference to Figure 1.15, where ϕ is varied between 10^{-2} and 10^{-3}. The laser divergence half angle θ was assumed to be 10^{-3}, A was set equal to either 20 or 5, and D was assigned the value of 1.25.

The behavior of $\xi(z)$ in the case of a biaxial lidar with an inclination between the telescope and laser axes is radically different from that of the parallel-axis arrangement discussed above. In this instance, $\xi(z)$ increases with z at first, attains the value of unity over some range interval, and then rapidly drops to zero for larger values of z. This is illustrated in Figure 1.16, where $\xi(z)$ is

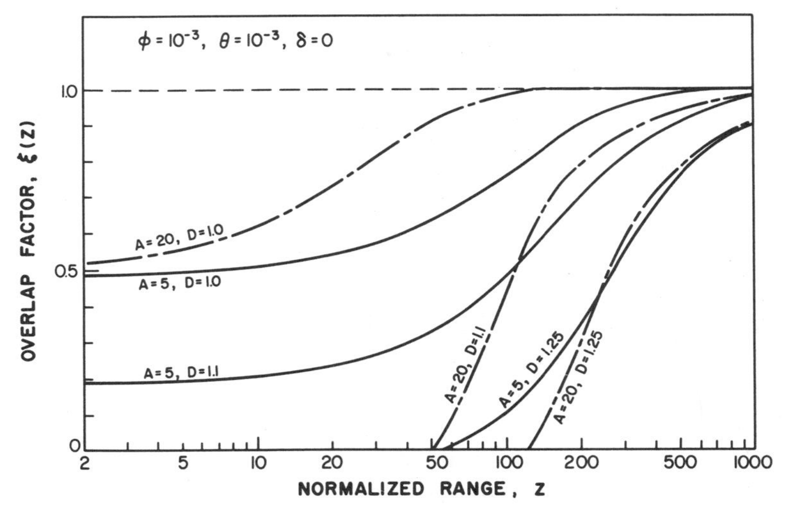

Figure 1.14. Variation of overlap factor $\xi(z)$ with normalized range z for A of 5 and 20 (with D of 1.0, 1.1, and 1.25 for each value of A).

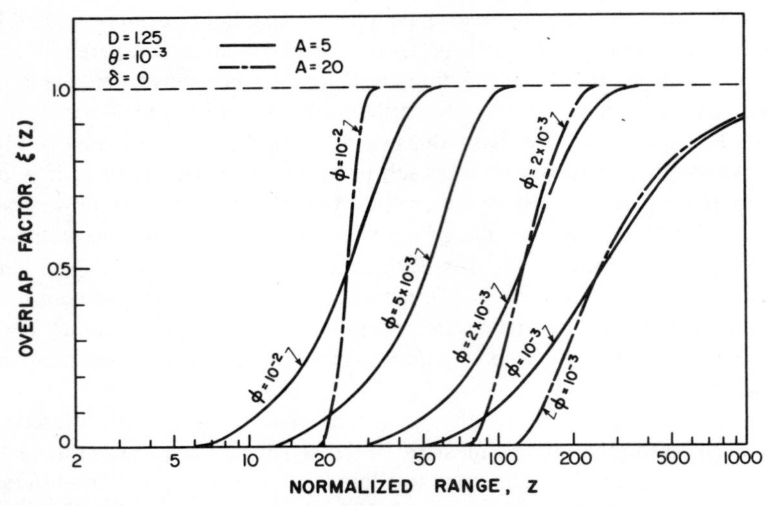

Figure 1.15. Variation of overlap factor $\xi(z)$ with z for A of 5 and 20 (with ϕ of 10^{-3}, 2×10^{-3}, 5×10^{-3}, and 10^{-2} for each value of A).

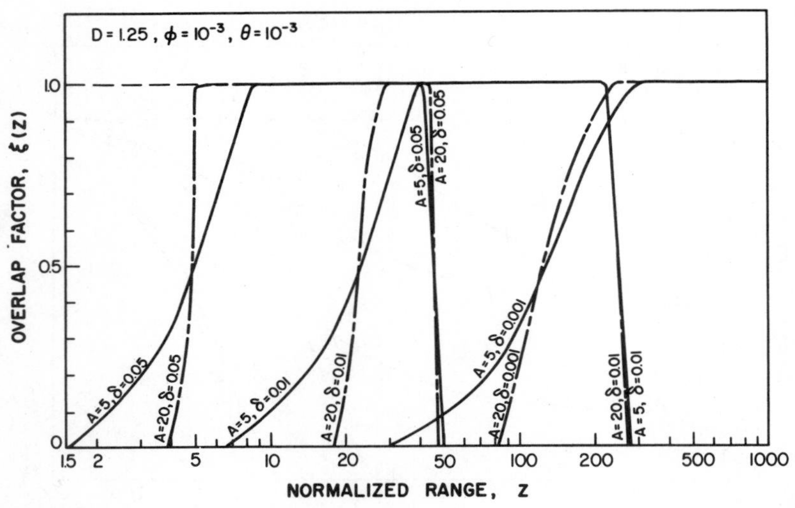

Figure 1.16. Variation of overlap factor $\xi(z)$ with z for A of 5 and 20 (with δ of 0.001, 0.01, and 0.05 for each value of A).

plotted against z for three values of the inclination angle ($\delta = 0.05, 0.01$, and 0.001) and two values of A (20 and 5). In this set of examples, $D = 1.25$ and $\theta = \phi = 10^{-3}$.

These results clearly indicate that interpretation of short-range lidar measurements must take proper account of the geometric factors involved. It is also quite apparent from Figure 1.16 that misalignment of the optical system can lead to misinterpretation of even long-range measurements. For example, a misalignment of the telescope and laser axes of only 10 mrad would lead to a lack of signal return for $z > 300$ (i.e., $R > 300r_0$) irrespective of the size of the telescope or the power of the laser, supposing the separation of the two axes at the lidar was about 25% more than the telescope radius (i.e., $D = 1.25$) and both the half opening angle ϕ of the telescope and the half divergence angle θ of the laser were 1 mrad.

1.2.4.2 Geometric Form Factor

As discussed earlier, the kinds of telescopes used for the receiver optics of most laser remote sensors have some form of central obstruction, and many also have limiting apertures that are smaller than the telescope objective lens (or mirror).

In order to be able to evaluate the geometric form factor $\xi(R)$ for these more general situations, it is necessary to consider the geometric optics involved. If we represent the telescope by a single lens of radius r_0 and focal length f, it is well known from geometric optics that each radiating point (r, ψ) in the target plane at range R will give rise to a circle of illumination (called the *circle of least confusion*) of radius

$$r_c = \frac{r_0 f}{R} \tag{1.114}$$

in the focal plane. The center of this circle of confusion is radially displaced a distance

$$r_f = \frac{rf}{R} \tag{1.115}$$

from the telescope axis. This is illustrated in Figure 1.17. From this it is fairly obvious that if a detector of radius r_D is centered on the telescope axis with its sensitive surface lying in the focal plane of the objective lens (or mirror) of the telescope, it will receive radiation from a target plane circle of radius $r_T(R) = r_0 + r_D R/f$, as given by Equation (1.98). This enables us to see that the receiver's field of view is a circle of radius $r_0 + \phi R$, as illustrated in Figure 1.18,

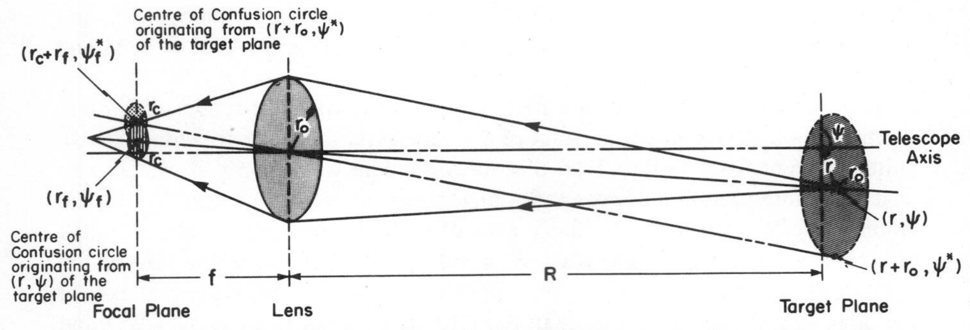

Figure 1.17. Ray diagram illustrating that each point within a circle of radius r_0 (that of telescope lens or mirror) centered about point (r, ψ) in target plane at range R will contribute radiation to point (r_f, ψ_f) in focal plane, where $r_c = r_0 f/R$ and $r_f = rf/R$.

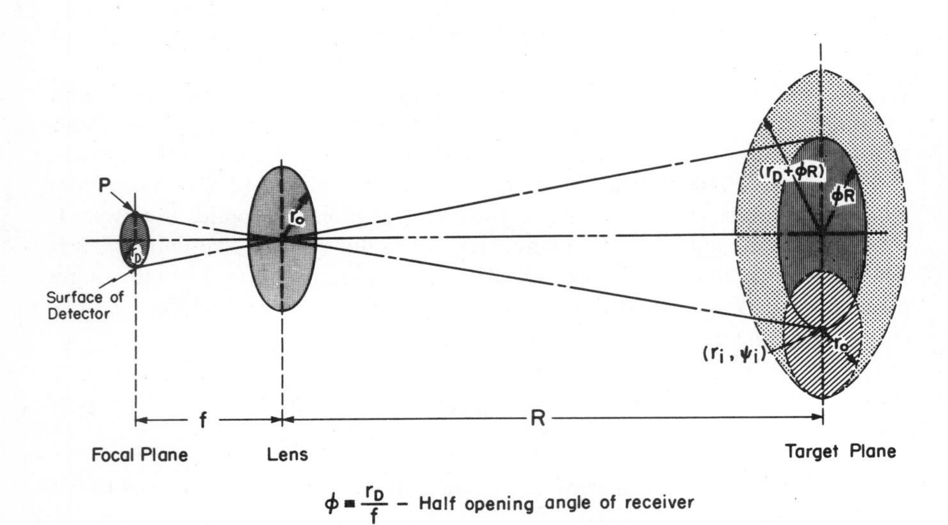

Figure 1.18. Ray diagram illustrating that each point on detector receives radiation from a circle of radius r_0 (that of telescope lens or mirror) in target plane. A point P on the edge of detector receives radiation from a circle of radius r_0 centered at (r_i, ψ_i) in target plane, where $r_i = r_D R/f = \phi R$. Thus, field of view of receiver is a circle of radius $r_T = r_0 + \phi R$.

where the half opening angle of the receiver optics is $\phi \equiv r_D/f$, as stated in Equation (1.99).

The calculation of the geometric form factor can be quite complicated for an actual lidar system. Nevertheless, a reasonable approximation can be obtained if a couple of simplifying assumptions are made: First, the limiting aperture is assumed to lie in the focal plane of the telescope, and second, the obstruction is imagined to reside in the plane of the objective lens (or mirror) of the telescope. Reference to the ray diagram presented as Figure 1.19 suggests that under these circumstances the geometric probability factor will be azimuthally symmetric and given by

$$\xi(R, r, r_a, r_b) = \frac{\mathcal{A}(r_a, r_c; r_f) - \mathcal{A}(r_a, r_b'; r_f)}{\pi r_c^2} \tag{1.116}$$

where $\mathcal{A}(r_1, r_2; r_f)$ = area overlap function described earlier

r_a = radius of aperture—which in certain instances corresponds to radius of detector's sensitive surface

r_c = *radius of confusion*

r_f = radial displacement from telescope axis of center of circle of confusion arising from position (r, ψ) in target plane, $= rf/R$

r_b' = radius of the obstruction's shadow in the focal plane, $= r_b f/R$

r_b = radius of central-mirror support structure of telescope

With this in mind, the telescope effective area $A(R)$, as defined by Equation (1.101), becomes

$$A(R) = \frac{A_0}{\pi W^2(R)} \int_{r=0}^{r_T} \int_{\psi=0}^{2\pi} \xi(R, r, r_a, r_b) F(R, r, \psi) r \, dr \, d\psi \tag{1.117}$$

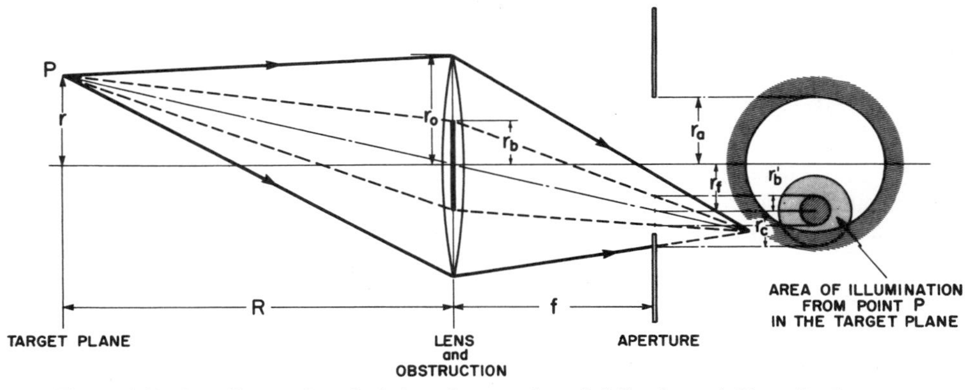

Figure 1.19. Ray diagram for calculation of geometric probability factor $\xi(R, r, \psi)$ where $r_f = rf/R$, $r_c = r_0 f/R$, and $r_b' = r_b f/R$.

It should be noted that if the aperture radius is the same as that of the lens and there is no obstruction (i.e., $r_b = 0$), $\xi(R, r, r_0) = \mathcal{A}(r_0, r_c; 0)/(\pi r_c^2) = 1$. If this is not the case, we can write

$$A(R) = \frac{A_0}{\pi W^2(R)} \int_{r=0}^{r_T} \int_{\psi=0}^{2\pi} \xi(R, r, r_a, r_b)\, \mathcal{G}(R, r, \psi) r\, dr\, d\psi \qquad (1.118)$$

for a Gaussian laser pulse distribution, where $\mathcal{G}(R, r, \psi)$ was described by Equation (1.93); and if this is incorporated into Equation (1.118), we have

$$A(R) = \frac{A_0}{\pi W^2(R)} \int_{r=0}^{r_T} \xi(R, r, r_a, r_b) \int_{\psi=0}^{2\pi} e^{-(r^2 + d^2 - 2rd\cos\psi)/W^2(R)}\, d\psi\, r\, dr \qquad (1.119)$$

A relation that is helpful to the evaluation of Equation (1.119) can be found in a number of good mathematical texts and takes the form

$$e^{x\cos\psi} = I_0(x) + 2I_1(x)\cos\psi + 2I_2(x)\cos\psi + \cdots \qquad (1.120)$$

where $I_n(x)$ is a modified Bessel function of the first kind of order n. Integrating both sides of Equation (1.120) with respect to ψ yields

$$\int_{\psi=0}^{2\pi} e^{x\cos\psi}\, d\psi = 2\pi I_0(x) \qquad (1.121)$$

If this is used in Equation (1.119), the telescope effective area is seen to be

$$A(R) = \frac{r_0^2}{W^2} \int_{r=0}^{r_T} \xi(R, r, r_a, r_b)\, e^{-(r^2 + d^2)/W^2} 2\pi I_0 \frac{2rd}{W^2}\, r\, dr \qquad (1.122)$$

Halldórsson and Langerholc (1978) have evaluated Equation (1.122) for a few representative situations and have also considered the case of uniform laser illumination. Their results indicate that for a coaxial lidar with a flat laser distribution, $A(R)$ becomes equal to the uncorrected receiver aperture,

$$A_{\mathrm{tel}} \equiv \pi(r_0^2 - r_b^2) \qquad (1.123)$$

for a range in excess of 300 m, provided the telescope opening angle is greater than the laser divergence angle. Smaller limiting values of $A(R)$ were found for a Gaussian laser distribution. If the opening angle is less than the divergence angle (e.g., $\phi = 10^{-4}$, while $\theta = 5 \times 10^{-4}$), $A(R)$ appears to level off at a

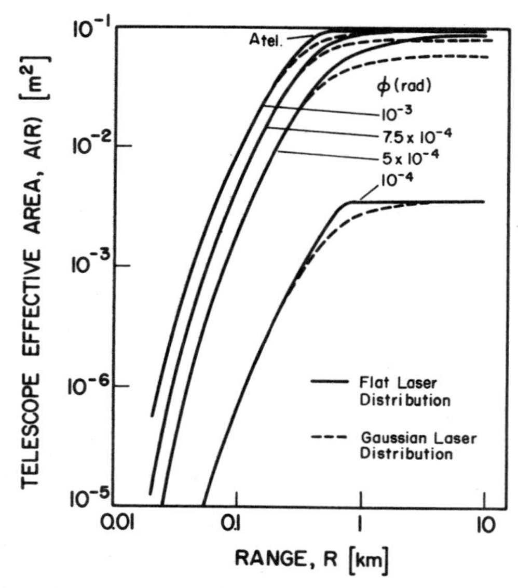

Figure 1.20. Variation of telescope effective area $A(R)$ with range R for several values of opening angle ϕ, in the case of a biaxial configuration with $r_0 = 0.175$ m, $r_b = 0.04$ m, $\delta = 0$, $\theta = 5 \times 10^{-4}$ rad, $d_0 = 0.2$ m, and $W_0 = 0.01$ m (Halldórsson and Langerholc, 1978).

value of about one-thirtieth that of A_{tel} for both kinds of laser distribution. On the other hand, $A(R)$ is found always to decrease rapidly with decreasing R. This is attributed to the shadowing effect of the central obstruction that these authors included in their calculations. Similar behavior was observed in the case of a biaxial lidar configuration with $D \approx 1$, except that the range for which $A(R)$ levels off is increased somewhat. This latter case is presented as Figure 1.20, where $r_0 = 0.175$ m, $r_b = 0.04$ m, $d_0 = 0.2$ m, $\theta = 5 \times 10^{-4}$ rad, $W_0 = 0.01$ m, and $\delta = 0$.

Halldórsson and Langerholc (1978) also investigated the effects of coplanar misalignment for the biaxial lidar arrangement and found that the telescope effective area remained considerably below A_{tel}, even for long-range observations, once the inclination of the telescope and laser axes exceed the laser divergence half angle.

1.2.4.3 Small Detectors and Geometric Compression of Lidar Return Signal

There are in fact two kinds of radiation loss mechanisms that affect the geometric probability factor $\xi(R, r, \psi)$. The first includes the overlap of the laser pulse with the receiver optics field of view and the shadowing of the detector by a secondary mirror support structure of the telescope and have been treated above. The second is concerned with the fact that radiation from a target plane located at short or intermediate ranges is not focused onto the focal plane. Instead it forms a diffuse region of illumination in the focal plane, so that a *small* detector positioned at the location will not receive all of the radiation expected on the basis of our above deliberations.

Both of these loss mechanisms can prevent the amount of scattered laser power incident on the detector from decreasing in accordance with the $1/R^2$ prediction of the lidar equation (1.102). This turns out to be of considerable value where lidar measurements are required over a large range interval. For example, a laser remote sensor that is intended to be operated between 100 m and 10 km must be capable of handling a dynamic range of five orders of magnitude. Although there are electronic methods of handling dynamic ranges of this magnitude, clearly there are some advantages in designing the receiver optics to perform the signal compression.

The large dynamic range predicted by the lidar equation presupposes that all of the backscattered radiation collected by the receiver optics is focused onto the detector. As we have indicated above, this may not always be the case, and we shall now determine the backscattered irradiance incident on the detector for the case where the detector size is important. If the lidar is to be efficient for long-range observations, the detector's sensitive surface should usually be located in the focal plane of the telescope.

Harms et al. (1978) have evaluated both the focal plane irradiance and the total signal power received by several detectors of different size. They assumed a coaxial lidar configuration and a Gaussian laser beam but neglected the shadowing effect of the secondary mirror structure needed for a coaxial arrangement. Their calculations of the focal plane irradiance are illustrated in Figure 1.21 and reveal that the irradiance decreases only slowly with increasing range for short and intermediate values and then drops dramatically at greater values of the range. Clearly, a small detector positioned on the axis will only intercept a tiny fraction of the incident radiation from short and intermediate ranges, while it receives most of the radiation backscattered from large distances. This geometric signal compression is seen in Figure 1.22, where it is evident that the optimum detector radius for the configuration considered by Harms et al. (1978) is probably around 0.4 mm. This size of detector would provide an almost flat response out to about 1 km and then a decreasing signal that would be in keeping with the predictions of the lidar equation beyond 4 km. Such a detector is providing both a *small dynamic range* and *maximum sensitivity* at large distances. The parameters used in these calculations are presented in Table 1.3. In addition, $d_0 = \delta = 0$ for the coaxial configuration assumed.

Harms (1979) has extended this work to biaxial lidar configurations and has also included the effect of a central obstruction in both coaxial and biaxial ar-

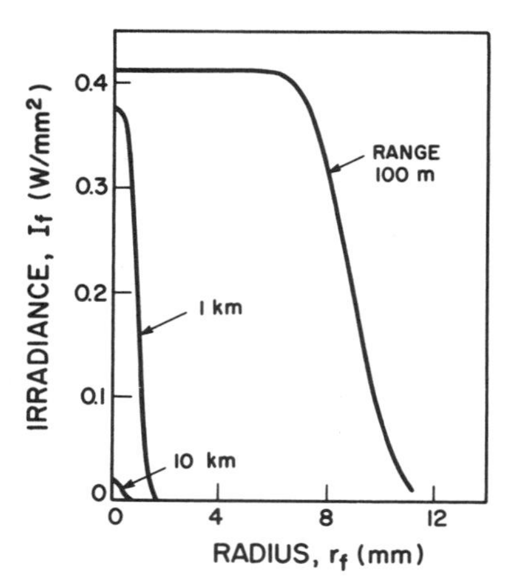

Figure 1.21. Irradiance in focal plane of lidar telescope with $r_0 = 0.3$ m, $r_b = 2$ mm (corresponding to $\phi = 0.67$ mrad), and $W_0 = 0.075$ m (corresponding to $\theta = 0.125$ rad) (Harms et al., 1978).

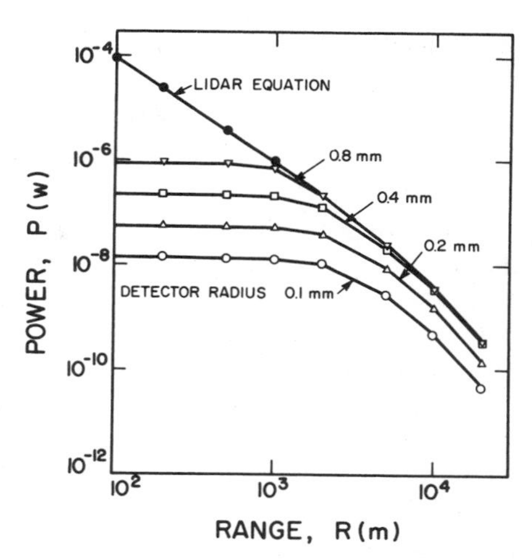

Figure 1.22. Power incident on detectors of radii 0.1–0.8 mm for $r_0 = 0.30$ m, $W_0 = 0.075$ m (corresponding to $\theta = 0.125$ mrad) of a coaxial lidar arrangement (Harms et al., 1978). Optimum detector radius 0.4 mm.

rangements. This work shows that a judicious choice of detector size can lead to a compression of the lidar signal dynamic range with almost no loss of sensitivity for long-range measurements. This appears to be true for both coaxial or biaxial lidar systems. In a coaxial configuration the shadow cast by the secondary-mirror support structure results in a reduction of the near signal, while in a biaxial configuration the effects of an obstruction are not very important. In a biaxial arrangement the separation and inclination of the telescope and laser axes are important parameters.

It should also be noted that a small detector has the added advantage of a small inherent noise level. It also provides a small field of view that allows good spatial resolution and matching of the receiver optics to the laser beam, thereby reducing the contribution from background radiation arising beyond the lateral extent of the laser beam. Lastly, small detectors are available as linear

Table 1.3. Parameters Assumed by Harms et al. (1978) and Harms (1979)

$r_0 = 0.30$ m	$\theta = 1.25 \times 10^{-4}$ rad	$\kappa(R) = 5 \times 10^{-2}$ km^{-1}
$f = 3$ m	$\tau_L = 500$ ns	$\beta(R) = 10^{-3}$ km^{-1} sr^{-1}
$W_0 = 0.075$ m	$P_L = 50$ kW	

photodiode arrays that can be used in simultaneous multiwavelength measurements.

1.2.5 Solutions of Lidar Equation

As we have seen, the appropriate lidar equation provides a means of relating the radiation returned from a probing laser beam to the relevant optical properties (such as the scattering or attenuation coefficients) of the target medium. It is implicitly assumed that these optical properties can be related to some physical property (such as the density of some specific constituent) of interest in the target. In order to evaluate these optical properties from the return signal, the lidar equation has to be solved. In this section we shall consider some of the most useful methods of solving the lidar equation.

In this regard we can introduce a form of the lidar equation that will reasonably apply to both long-range scattering and fluorescent measurements in the limit of good range resolution and optically thin targets, namely,

$$E(R) = E_D \frac{A(R)}{R^2} \beta(R) \exp\left(- \int_0^R \kappa(R)\, dR \right) \qquad (1.124)$$

where

$$E_D = E_L \frac{c\tau_d}{2} \{\xi(\lambda) \text{ or } K_0(\lambda)\} \qquad (1.125)$$

depending on whether scattering or fluorescence is involved [see Eqs. (1.53) and (1.83), respectively]. Here $\beta(R)$ represents either the volume scattering coefficient or the volume fluorescence coefficient.

It is evident that Equation (1.124) will have to be solved for one or both of the optical parameters $\beta(R)$ and $\kappa(R)$.† The simplest approach is to assume that the two-way atmospheric transmission factor can be approximated by unity. This effectively reduces the lidar equation to only one unknown and has been useful for some aerosol work. Another approach involves making use of some kind of atmospheric model to specify $\kappa(R)$.

In general, however, both $\kappa(R)$ and $\beta(R)$ are unknown, and it is necessary to assume some kind of relation between them. To eliminate system constants (such as the area or the spectral transmission factor of the receiver) and uncer-

†Recall that $\kappa(R)$ represents the two-way attenuation coefficient, see Equation (1.60).

tainties associated with the laser pulse irreproducibility, we introduce a new, *range-normalized signal variable* defined by the relation[‡]

$$S(R) \equiv \ln\{E(R)R^2\} \tag{1.126}$$

The virtue of this new variable is seen by subtracting from it the S value appropriate to some reference range $R_\#$ (this might, e.g., correspond to the location of some monitor of β):

$$\begin{aligned} S(R) - S(R_\#) &= \ln \frac{E(R)R^2}{E(R_\#)R_\#^2} \\ &= \ln \frac{\beta(R)A(R)}{\beta(R_\#)A(R_\#)} - \int_{R_\#}^{R} \kappa(R)\, dR \end{aligned} \tag{1.127}$$

In the case of long-range work with a lidar system that is well aligned ($\delta \approx 0$), we saw earlier that $A(R) \approx A_0$, so that

$$S(R) - S(R_\#) = \ln \frac{\beta(R)}{\beta(R_\#)} - \int_{R_\#}^{R} \kappa(R)\, dR \tag{1.128}$$

and this difference in the S values only depends on the atmospheric-dependent factors. The differential form of Equation (1.128) is

$$\frac{dS(R)}{dR} = \frac{1}{\beta(R)} \frac{d\beta(R)}{dR} - \kappa(R) \tag{1.129}$$

and we can see that for a homogeneous atmosphere, where $d\beta(R)/dR \approx 0$,

$$\kappa_{\text{hom}} \approx -\frac{dS(R)}{dR} \tag{1.130}$$

[‡]In much of the literature (Davis, 1969; Johnson and Uthe, 1971), the function S is defined in terms of the range-corrected signal ratio in decibels:

$$S(R) = 10 \log \frac{E(R)\,R^2}{E(R_\#)\,R_\#^2}$$

This definition is closely related to the form of measurement using logarithmic amplifiers (Johnson and Uthe, 1971).

This is the basis of the *slope method* of inversion, which is appropriate in good visibility conditions, provided multiple scattering is unimportant (Collis and Russell, 1976). Viezee et al. (1973) attempted to employ this technique to assess slant visibility for aircraft landing operations through lidar observations at 694.3 nm, and Murray et al. (1978) made use of this approach in the measurement of atmospheric IR extinction using a CO_2 laser. Unfortunately, the slope method becomes inappropriate for certain conditions that are sometimes important for lidar work, that is, remotely measuring atmospheric inhomogeneities like fogs, clouds, and smoke plumes (Klett, 1981).

This is a reasonable body of evidence that suggests that, in the case of elastic scattering (i.e., $\lambda = \lambda_L$), $\beta(R)$ can often be related to $\kappa(R)$ through a relation of the kind

$$\beta(\lambda_L, R) = \text{const } \kappa^{\mathcal{g}}(\lambda_L, R) \tag{1.131}$$

where $\mathcal{g}$ depends on the lidar wavelength and the specific properties of the constituent of interest (Collis and Russell, 1976). Klett (1981) has indicated that the value of this exponent is generally in the range $0.67 \leq \mathcal{g} \leq 1.0$. A relationship similar to (1.131), with $\mathcal{g} = 1.0$, would also be expected in the case of fluorescence work.

Eliminating β in terms of κ, through Equation (1.131), enables us to rewrite Equation (1.129) in terms of only one unknown:

$$\frac{dS(R)}{dR} = \frac{\mathcal{g}}{\kappa_L(R)} \frac{d\kappa_L(R)}{dR} - 2\kappa_L(R) \tag{1.132}$$

where we have introduced

$$\kappa_L(R) = \kappa(\lambda_L, R) \tag{1.133}$$

The factor of 2 in Equation (1.132) is required by virtue of Equation (1.60). Although this is a nonlinear ordinary differential equation, it has a well-known form, namely, that of the Bernoulli or homogeneous Ricatti equation (Klett, 1981). The solution of Equation (1.132) is made easier by introducing the reciprocal of the attenuation coefficient,

$$\mathcal{u} \equiv \frac{1}{\kappa_L(R)} \tag{1.134}$$

On substitution of Equation (1.134) into (1.132), we obtain

$$\frac{d\mathcal{u}}{dR} = -\frac{1}{\mathcal{g}}\left(2 + \mathcal{u}\frac{dS}{dR}\right) \tag{1.135}$$

The solution of this equation is quite straightforward; however, Klett (1981) has shown that the limits of integration make a difference to the stability and accuracy of the solution. In particular, the *reference range* $R_{\#}$ should be taken as the upper limit of range integration rather than as the lower limit as assumed by earlier researchers. With this in mind, the solution of (1.135) can be written in the form

$$\mathcal{u}(R_{=}) - \mathcal{u}(R)\exp\left(-\frac{1}{\mathcal{g}}\int_{R}^{R_{=}} \frac{dS}{dR^*}\,dR^*\right)$$

$$= -\frac{2}{\mathcal{g}}\int_{R}^{R_{=}} dR^*\exp\left(-\frac{1}{\mathcal{g}}\int_{R^*}^{R_{=}} \frac{dS}{dR^{**}}\,dR^{**}\right) \tag{1.136}$$

where $\mathcal{g}$ is assumed to be independent of range.

If we then draw on Equation (1.134), we can arrive at an equation for the attenuation coefficient,

$$\kappa_L(R) = \frac{e^{-[S(R_{=})-S(R)]/\mathcal{g}}}{1/\kappa_L(R_{\#}) + \dfrac{2}{\mathcal{g}}\displaystyle\int_{R}^{R_{\#}} dR^*\, e^{-[S(R_{=})-S(R)]/\mathcal{g}}} \tag{1.137}$$

Although this solution appears to work reasonably well for conditions ranging from clear air to haze, in a turbid atmosphere (such as a fog or cloud) multiple scattering occurs, and the above solution is not valid. In these instances a more sophisticated formulation of the lidar equation is necessary (Collis and Russell, 1976). Improved algorithms have recently been discussed by Klett (1986) and Bissonnette (1986).

Lastly, a word of caution should be added in connection with the assumption that the range dependence of the effective telescope area $A(R)$ can be neglected. As we have seen earlier, this is highly questionable for short-range measurements or if the telescope and laser axes are inclined to each other.

1.3 ANALYSIS AND INTERPRETATION OF LIDAR SIGNALS

In order to aid in the interpretation of lidar return signals and to draw attention to some of their less obvious features, it has sometimes been found useful to calculate the expected signal for a given situation. Simulations of this kind have also served as the basis of comparisons between the different lidar techniques used in the detection of trace atmospheric molecular constituents. One of the first such comparative studies was undertaken by Measures and Pilon (1972). Their analysis clearly indicated the potential superiority of differential absorption and scattering (DAS) over laser-induced fluorescence (F) or Raman backscattering (R) for the purpose of remotely detecting small concentrations of molecules in the lower regions of the atmosphere.

1.3.1 Spectral Rejection of Laser-Backscattered Radiation

In certain instances, namely, when fluorescence or Raman scattering is being employed, laser-backscattered radiation can limit the sensitivity of the lidar system in two ways. If insufficient spectral rejection is used, some fraction of this laser return will arrive coincident with the signal. On the other hand, if inadequate care is given to the design of the lidar, it is possible at short wavelengths for near-field laser-backscattered radiation to induce fluorescence within some optical component (such as a lens) that is positioned ahead of the spectral analyzer or to saturate the photodetector. These situations can be avoided by the inclusion of a narrow-band laser-blocking filter ahead of all vulnerable components. A biaxial lidar configuration is generally less susceptible to the latter kind of problem. Nevertheless, it is clear that considerable care must be given to it in the design of a laser remote sensor.

An estimate of the degree of spectral rejection necessary to avoid the problem of laser-backscattered radiation arriving with the signal, and thereby reducing the achievable signal-to-noise ratio, can be made on the basis of equation (1.18). The criterion relevant to any given situation will of course depend on the predominant source of noise competing with the backscattered laser radiation. Under daylight operating conditions, where background radiation tends to constitute the dominant source of noise, the necessary criterion is simply

$$E_b^L(\lambda_L) \ll \frac{E_b^N(\lambda)}{\nu_\eta} \tag{1.138}$$

Using Equations (1.2) and (1.52), we can establish a suitable criterion in terms of the *spectral rejection ratio* $\xi(\lambda_L)/\xi(\lambda)$, namely,

$$\frac{\xi(\lambda_L)}{\xi(\lambda)} \ll \frac{2S_b(\lambda)\Omega_0 R^2\, \Delta\lambda_0}{E_L \beta(\lambda_L, R)\nu_n \xi(R)c} \exp\left(2\int_0^R \kappa(\lambda_L, R)\, dR\right) \tag{1.139}$$

where we have assumed that we can express the filter function $K_0(\lambda)$ in terms

Table 1.4. Lidar Minimum Detectable Energy for Three Limiting SNR Cases[a]

Limit	Criterion	MDE ($E_s^{\min} \approx$)
Signal shot noise	$E_s \gg E_b^N + E_d$	$[(\mathrm{SNR})^{\min}]^2 E(\lambda)$
Background noise	$E_b^N \gg E_s + E_d$	$[E(\lambda) E_b^N(\lambda)]^{1/2} (\mathrm{SNR})^{\min}$
Dark-current noise	$E_d \gg E_s + E_b^N$	$[\frac{1}{2}\tau_d A_d]^{1/2} (\mathrm{SNR})^{\min}/D^*$

[a] Neglecting laser-backscattered radiation.

of the spectral transmission factor at the wavelength of the signal and an appropriate linewidth $\Delta\lambda_0$, namely,

$$K_0(\lambda) = \xi(\lambda)\Delta\lambda_0 \tag{1.140}$$

The range dependence of the criterion given by the inequality (1.134) arises as a result of using the backscattered laser radiation return that originates from the same location as the signal. In principle, any earlier component can be discriminated against by temporal means. Alternatively, the spectral rejection ratio could be calculated on the basis that the laser-backscattered radiation has to be much less than the signal arriving at the same time.

1.3.2 Signal-to-Noise Limits of Detection for Molecular Constituents

As we saw earlier [Eq. (1.21)], it is possible to express the lidar detector (output) signal-to-noise ratio in terms of the increment of radiative signal energy $E_s(\lambda, R)$ and the total radiative background energy $E_b^T(\lambda)$ received by the detector within its response time:

$$(\mathrm{SNR})^2 = \frac{E_s^2(\lambda, R)/E(\lambda)}{E_s(\lambda, R) + E_b^T(\lambda) + (i_d + i_J)E(\lambda)/2eB^*} \tag{1.141}$$

The last term in the denominator represents the sum of the dark-current and Johnson current noise contributions. For most laser remote sensors the detector will have some degree of internal gain, and consequently, the Johnson noise term can be neglected. As indicated earlier, there are three limiting SNR cases for lidar systems monitoring a return signal wavelength sufficiently displaced from the laser wavelength for successful spectral discrimination against laser-backscattered radiation. These are summarized in Table 1.4.[†]

[†] It should be noted that although the expressions provided in Table 1.4 are approximate, for most situations they are nevertheless adequate.

1.3.2.1 Signal Shot-Noise Limit

The greatest sensitivity is achieved when the minimum detectable energy (MDE) is limited only by the quantum fluctuations of the signal itself. This *signal shot-noise limit* arises when

$$E_s(\lambda, R) >> E_b^T(\lambda) + E_d \tag{1.142}$$

where

$$E_d = \frac{i_d E(\lambda)}{2eB^*} \tag{1.143}$$

and $E(\lambda)$ was defined by Equation (1.24). Under these conditions, the MDE is

$$E_s^{\min}(\lambda, R) \approx \left[(\mathrm{SNR})^{\min}\right]^2 E(\lambda) \tag{1.144}$$

where $(\mathrm{SNR})^{\min}$ represents the smallest credible value for the SNR. By using the appropriate form of the lidar equation, Equation (1.144) can be translated into a relation for the threshold number density of a molecular species that can be detected with a given lidar configuration.

By way of illustration, let us consider the case of Raman scattering from a rectangular-shaped (both spatially and temporally) laser pulse of energy E_L and duration τ_L. In this instance Equation (1.53) represents the relevant form of lidar equation, and we can write

$$E_s(\lambda, R) = E_L \xi(\lambda) \frac{A_0 \xi(R)}{R^2} N(R) \frac{\sigma^R(\lambda_L, \lambda)}{4\pi} \frac{c\tau_d}{2} \cdot \exp\left(-\int_0^R \kappa(R)\, dR\right) \tag{1.145}$$

where we have also used Equations (1.50) and (1.60) and $\sigma^R(\lambda_L, \lambda)$ is taken as the appropriate Raman scattering cross section for the molecule of interest. Substitution of Equation (1.145) into (1.144) yields the threshold number density $N^{\min}$ that can be detected at range R averaged over the range interval $c(\tau_L + \tau_d)/2$ under conditions of signal shot-noise limit, namely,

$$[N(R)]^{\min} \cong \frac{R^2 \left[(\mathrm{SNR})^{\min}\right]^2 \exp\left(\int_0^R \kappa(R)\, dR\right)}{E_L \xi(R) \left[\sigma^R(\lambda_L, \lambda)/4\pi\right] U(\lambda)} \tag{1.146}$$

where we have introduced the (wavelength-sensitive) *system parameter*

$$U(\lambda) \equiv \frac{A_0 \xi(\lambda) \tau_d \lambda \eta(\lambda) \xi_e}{2hF_G} \tag{1.147}$$

Clearly, this parameter should be made as large as possible if a high sensitivity is to be achieved.

The differential Raman cross section $\sigma^R(\lambda_L, \lambda)/4\pi$ for a wide range of constituents has a value that lies between 10^{-30} and 10^{-29} cm^2 sr^{-1} for excitation at 337 nm (Measures, 1984). In order to indicate the kind of performance that might be expected from a representative lidar system that is operating in the signal shot-noise limit, we shall evaluate the threshold number density of carbon monoxide molecules that may be detected with a lidar system having the following characteristics: $A_0 = 5000$ cm^2, $\tau_d = 20$ ns, $\xi(\lambda) = 0.5$, $\eta(\lambda) = 0.2$, $\xi_e = 1.0$, and $F_G = 1.0$. For $\lambda_L = 308$ nm (XeCl laser), we have $\lambda = 330$ nm, $\sigma^R(\lambda_L, \lambda) = 5.8 \times 10^{-30}$ cm^2 sr^{-1} for CO (Measures, 1984). If we also assume the receiver's optical efficiency corresponding to range R [i.e., $\xi(R) = 1$] and $(\text{SNR})^{\min} = 1.5$, then $U(\lambda) = 2.49 \times 10^{23}$ cm^3 J^{-1}, and

$$[N(R)]_{\text{CO}}^{\min} \approx \frac{1.70 \times 10^{10} R^2}{E_L} \exp\left(\int_0^R \kappa(R)\, dR \right) \tag{1.148}$$

where R is in meters, E_L in Joules, and $\kappa(R)$ in reciprocal minutes.

If we assume reasonably clear weather conditions (i.e., visibility of about 10 km), then $\kappa(R) \cong 10^{-3}$ m^{-1} at 308 nm, (Measures, 1984). In Figure 1.23 we have plotted the threshold concentration (in ppm) of CO that could be detected as a function of range R for a XeCl laser having an output energy E_L of 0.30, 0.75, and 1.90 J (full curves). We have taken the sea-level density of air to be 2.55×10^{19} cm^{-3} in this set of calculations. It is apparent from Figure 1.23 that Raman scattering can be viewed as potentially useful for remote monitoring of pollution sources where the concentrations are likely to be fairly high and ranges of 100–300 m would be acceptable. Alternatively, a Raman lidar could be employed to measure the density of major atmospheric constituents, such as N_2, O_2, and H_2O, for altitudes of a few kilometers.

1.3.2.2 Background-Noise Limit

A laser remote sensor operating under daylight conditions is likely to be subject to levels of background illumination so high that

$$E_b^N(\lambda) >> E_s(\lambda, R) + E_d \tag{1.149}$$

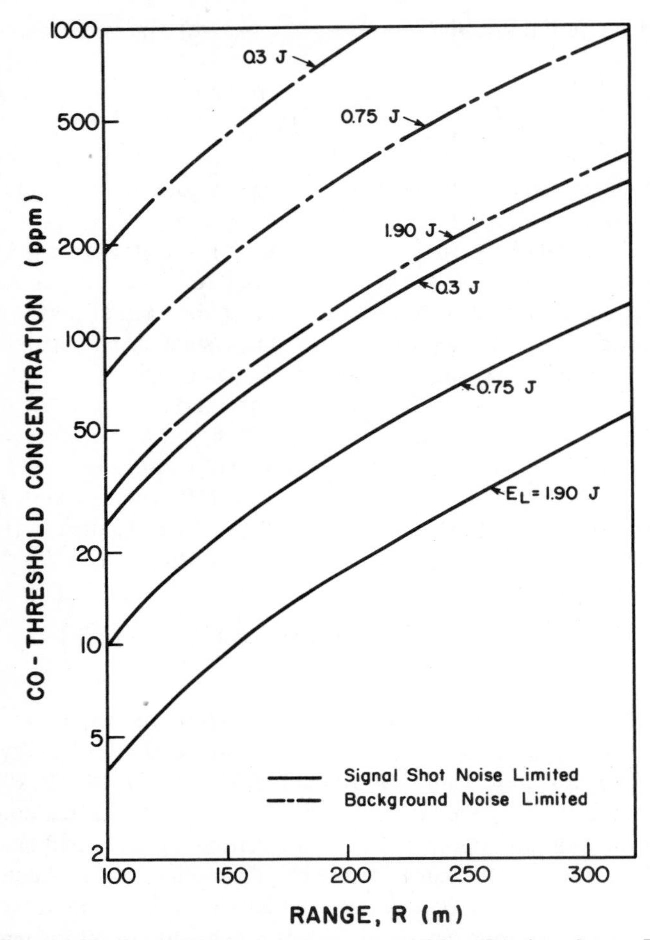

Figure 1.23. Threshold concentration (ppm) of CO as function of range *R*.

Nevertheless, signal extraction is still possible provided the fluctuation in this background radiation is sufficiently small and the time-averaged level of the radiation is not so great as to saturate the detector. The MDE under these circumstances was given by Equation (1.28) and can be expressed in the form

$$E_s^{\min}(\lambda, R) \approx \left[E(\lambda)E_b^N(\lambda)\right]^{1/2}(\mathrm{SNR})^{\min} \tag{1.150}$$

This can be translated into a threshold density for a particular atmospheric molecular constituent through the use of the appropriate lidar equation. For example, if we again imagine using the Raman lidar discussed earlier, we use Equation (1.145) together with (1.150) to yield the threshold number density,

$$[N(R)]^{\min} \approx \frac{2R^2 \exp\left(\int_0^R \kappa(R)\, dR\right)}{E_L \xi(R)[\sigma^R(\lambda_L, \lambda)/4\pi]} \left\{\frac{S_b(\lambda)\Omega_0 \Delta\lambda_0}{2U(\lambda)c}\right\}^{1/2} (\mathrm{SNR})^{\min} \tag{1.151}$$

where we have again used the system parameter $U(\lambda)$, defined by Equation (1.147), and Equation (1.140) for the filter function $K_0(\lambda)$.

It is evident that improvement in sensitivity for the background-noise-limited situation necessitates using as small as possible values of the receiver's acceptance solid angle Ω_0 and the effective spectral window $\Delta\lambda_0$. The system parameter should also be made as large as practical.

If the same Raman lidar is considered, the system parameter $U(\lambda)$ is again 2.49×10^{23} cm^3 J^{-1}. However, in the background-noise-limited situation we also have to know the receiver optics acceptance solid angle and system's effective spectral window. We shall assume $\Omega_0 = 2 \times 10^{-6}$ sr, $\Delta\lambda_0 = 1$ nm, and again $(\mathrm{SNR})^{\min} \cong 1.5$. If we also take a fairly conservative value of 4×10^{-6} W cm^{-2} nm^{-1} sr^{-1} for the sky spectral radiance $S_b(\lambda)$ at 330 nm, we obtain

$$[N(R)]_{\mathrm{CO}}^{\min} \approx \frac{1.3 \times 10^{11} R^2}{E_L} \exp\left(\int_0^R \kappa(R)\, dR\right) \tag{1.152}$$

where, as before, R is in meter, E_L is in Joules, and $\kappa(R)$ is in reciprocal minutes. If we assume the same weather conditions so that $\kappa(R) \approx 10^{-3}$ cm^{-1} at 308 nm, we arrive at the broken curves in Figure 1.23 for the same set of laser energies. A comparison of the carbon monoxide threshold detection concentration curves for the same lidar system operating under signal shot-noise- and background-noise-limited situations clearly illustrates the superiority of the former. Unfortunately, the radiance of sunlit clouds can, for longer wavelengths, be close to one order of magnitude larger than the above figure (Pratt, 1969). On the other hand, the radiance of a clear, moonless night sky is typically 10^{-7} smaller.

1.3.2.3 *Dark-Current Limit*

In the dark-current limited situation,

$$i_d >> \frac{2eB^*}{E(\lambda)} [E_s(\lambda, R) + E_b^T(\lambda)] \tag{1.153}$$

The dark current associated with photomultipliers is rarely large enough to satisfy (1.153), and even in those situations where the dark current is excessive, cooling the photomultiplier will more often than not alleviate the problem. The MDE of solid-state detectors, on the other hand, is often determined by the criterion

$$E_s^{\min}(\lambda, R) \approx \frac{1}{D^*}\left(\frac{F_G A_d}{4B\xi_e}\right)^{1/2} (\mathrm{SNR})^{\min} \qquad (1.154)$$

Representative values of the detectivity D^*, defined by Equation (1.30), of a number of solid-state detectors range from 10^{13} cm $\mathrm{Hz}^{1/2}\ \mathrm{W}^{-1}$ in the visible to 10^{10} cm $\mathrm{Hz}^{1/2}\ \mathrm{W}^{-1}$ in the IR part of the spectrum Melchior (1972). In most instances these devices are employed only for lidar systems that operate in the IR, and consequently, it is unlikely that they would be used in laser remote sensors based on Raman scattering or fluorescence.

1.3.3 Differential Absorption Detection Limit

Probably the most important role played by solid-state detectors will be in the expanding area of IR differential absorption. This technique has considerable potential for laser remote sensors due to the wide range of molecular species that are detectable in this part of the spectrum. In the case of the DIAL technique it is essentially the difference between two backscattered laser signals at slightly different wavelengths that is related to the density of a specific molecular constituent. In fact, the incremental decrease in the observed signal at λ_0, associated with the range increment ΔR and arising from the attenuation of the specific molecular constituent for which the laser is tuned, is given by

$$\Delta E_0^* = \Delta E_0 - \Delta E_w \qquad (1.155)$$

where

$$\Delta E_0 = E(\lambda_0, R) - E(\lambda_0, R + \Delta R) \qquad (1.156)$$

represents the total decrease in the signal at λ_0, and

$$\Delta E_w = E(\lambda_w, R) - E(\lambda_w, R + \Delta R) \qquad (1.157)$$

effectively represents the incremental decrease in the signal due to nonspecific attenuation. Here, λ_0 represents the laser wavelength chosen to be close to the peak of the absorption line of the molecule of interest, and λ_w the wavelength of the laser pulse detuned to lie in the wing of the absorption line; see Figure 1.24.

As a first-order approximation we shall assume that the change in the "off" energy signal, ΔE_w, is small compared to the change in the "on" energy signal over the small range interval ΔR. Under these circumstances, detection of the incremental decrease of the backscattered energy signal at λ_0 requires that

$$E(\lambda_0, R) - E(\lambda_0, R + \Delta R) > \frac{E(\lambda_0, R + \Delta R)}{\text{SNR}} \tag{1.158}$$

which states, in effect, that this incremental energy signal should be greater than the noise in the λ_0 signal from range $R + \Delta R$. If we use the backscattering lidar equation with (1.65),

$$E(\lambda_0, R) = E_L \xi(\lambda_0) \frac{A_0 \xi(R)}{R^2} \beta(\lambda_0, R) \frac{c\tau_d}{2} \cdot \exp\left(-2 \int_0^R [\bar{\kappa}(R) + N(R)\sigma^A(\lambda_0)]\, dR\right) \tag{1.159}$$

in the equality (1.158), we can arrive at a criterion for the threshold number density of the molecular constituent that can be detected by means of the differential absorption and scattering:

$$[N(R)]^{\min} = \frac{1}{2\sigma^A(\lambda_0)\Delta R} \ln\left[\left(1 + \frac{1}{\text{SNR}}\right)\left(\frac{R}{R + \Delta R}\right)^2\right] \tag{1.160}$$

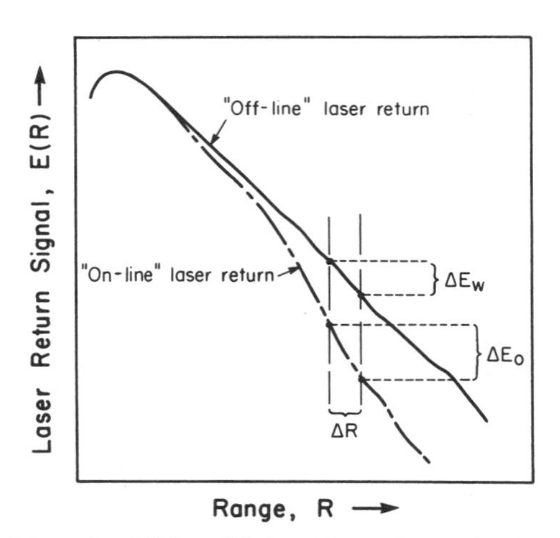

Figure 1.24. Schematic of differential absorption and scattering laser return signals.

In arriving at this result, we have assumed that ΔR is sufficiently small that the factor $\xi(R)\beta(\lambda_0, R)$ can be regarded as a constant over this range increment and that the term $\bar{\kappa}(\lambda_0, R)/\sigma^A(\lambda_0)$ is very much less than the right-hand side of Equation (1.160), where $\bar{\kappa}(\lambda_0, R)$ represents the residue attentuation coefficient after allowing for absorption by the molecular constituent of interest. This assumption is tantamount to stating that $N(R)\sigma^A(\lambda_0) >> \bar{\kappa}(\lambda_0, R)$ at the point of interest.

Clearly, the smaller the signal-to-noise ratio, the larger the threshold density given by Equation (1.160). A conservative form of Equation (1.160), which also allows for good spatial resolution, can be written

$$[N(R)]^{\min} = \frac{1}{2\sigma^A(\lambda_0)\Delta R} \ln\left[1 + \frac{1}{(\mathrm{SNR})^{\min}}\right] \tag{1.161}$$

If $(\mathrm{SNR})^{\min} \cong 1.5$, we have

$$[N(R)]^{\min} \approx \frac{1}{4\sigma^A(\lambda_0)\Delta R} \tag{1.162}$$

It may at first seem rather surprising that the DIAL threshold density appears to be independent of virtually everything other than the absorption cross section and the range increment. If one reflects for a moment on the means by which the presence of the molecular constituent of interest is detected, one will understand this state of affairs. In essence, detection of the species is assured if the incremental change in the signal is discernible against the noise in the signal. An implicit assumption in all of this is that the laser-backscattered energy is always larger than the MDE for the detector, so that, from Equation (1.154),

$$E(\lambda_0, R + \Delta R) \gtrsim \frac{1}{D^*}\left(\frac{F_G A_d}{4B\xi_e}\right)^{1/2} (\mathrm{SNR})^{\min} \tag{1.163}$$

This, in effect, provides us with a criterion for the minimum laser output energy required for a given lidar system to comply with this assumption. The magnitude of this threshold laser energy is determined by using Equation (1.159):

$$E_L^{\min} \approx \frac{2R^2(\mathrm{SNR})^{\min}}{\beta(\lambda_0, R)\xi(R)U^*(\lambda_0)} \exp\left(2\int_0^R \kappa(\lambda_0, R)\, dR\right) \tag{1.164}$$

where we have introduced the modified system parameter,

$$U^*(\lambda_0) = A_0\xi(\lambda_0)c\tau_d D^*\left(\frac{4B\xi_e}{F_G A_d}\right)^{1/2} \tag{1.165}$$

This can be related to the system parameter previously introduced, in Equation (1.147), through the substitution of Equations (1.19) and (1.30). That is,

$$U^*(\lambda_0) = 2U(\lambda_0)\left(\frac{2eBF_G}{i_d\xi_e}\right)^{1/2} \tag{1.166}$$

In the case of a solid-state detector we can generally set $F_G/\xi_e \approx 1$, and if we draw on Equation (1.11), we can write

$$U^*(\lambda_0) = \frac{A_0\xi(\lambda_0)cD^*}{(A_d B)^{1/2}} \tag{1.167}$$

In order to gain some experience with the magnitude of these quantities, we shall consider the representative DIAL system of Murray (1977). The characteristics of this system are listed in Table 1.5. In general, the threshold concentration (in ppm) of a molecular constituent of interest can be written

$$C_i^{\min} = \frac{N_i^{\min} \times 10^6}{N_{\text{atm}}} \tag{1.168}$$

Table 1.5. Characteristics of a Representative Infrared Dial System[a]

DF Lasers	
Energy, E_L	150 mJ
Pulse width, τ_L	1.0 μs
Beam divergence, 2θ	1.0 mrad
Wavelength, λ_L	3.6 to 3.9 μm
Receiver	
Telescope area, A_0	792 cm^2
Field of view, 2ϕ	3.0 mrad
HgCdTe detector	
Size	1 × 1 mm
D^* (3.7 μm)	10^{10} cm Hz$^{-1/2}$ W^{-1}
τ_d	75 ns

[a]From Murray (1977).

where N_i^{min} is the threshold number density of species i, and N_{atm} is the total number density of molecules in the atmosphere under conditions of interest. Nominally, at sea level, $N_{atm} = 2.55 \times 10^{19}$ cm^{-3} (where the Loschmidt number is 2.69×10^{19} cm^{-3}). The attenuation coefficient for the molecule of interest is

$$\kappa_A^i(\lambda_0) = \sigma_i^k(\lambda_0) N_{atm} \quad (\text{cm}^{-1}\ \text{atm}^{-1}) \tag{1.169}$$

1.3.4 Lidar Error Analysis

Any quantitative measurement has errors and uncertainties associated with it. The purpose of an error analysis is therefore twofold. First, we require to estimate the possible error incurred in a given measurement in order that the accuracy of that measurement can be ascertained, and second, the dependence or errors and uncertainties on the measurement system parameters needs to be identified so that the technique can be optimized.

The terms *accuracy* and *precision* as related to experimental observations are sometimes confused (Bevington, 1969) has distinguished between them as follows: accuracy is the measure of how close an experimental result comes to the true value, while precision is a measure of how exactly a result is determined, without reference to any true value. At this point it may also be well to recall the difference between systematic errors and random errors. The former arise when faulty equipment or techniques are used, while the latter can be thought of as a measure of the irreproducibility of a given observation.

Russell et al. (1979) discussed various sources of lidar scattering equation error and evaluated the range dependence of both the backscattering ratio and the aeroticulate backscattering coefficient expected for two lidar systems based on several atmospheric models. They also calculated the range dependence of the error associated with each of these parameters. The two lidars were assumed to be very similar, except that one employed a ruby laser operating at 694.3 nm and the other a neodymium glass laser operating at 1.06 μm.

In the case of fluorescence the wide spectral interval required to capture an appreciable fraction of the emitted radiation tends to make background radiation the dominant source of noise. In order to separate the fluorescence signal from the background radiation, two measurements are undertaken. First, the detector is permitted to sample the background radiation for some period prior to the moment the laser fires, and then the detector is permitted to sample the radiation during the time interval chosen to coincide with the return of fluorescence emanating from a target range $R(ct/2)$.

If this background radiation arises from laser-induced fluorescence in an interfering species, a different approach is needed that involves firing the laser at two wavelengths: one selected to provide the maximum signal from the species of interest, and the other selected to minimize this signal. Heaps (1981) has made an initial attempt at considering the error associated with this kind of measurement. Unfortunately, his analysis is somewhat limited in that he assumes no change of the fluorescence efficiency from the interfering species when the laser wavelength is considered to be shifted sufficiently for there to be negligible fluorescence from the species of interest.

As we have seen earlier, laser remote sensing based on differential absorption and scattering appears to have very little competition when it comes to remote three-dimensional mapping of atmospheric trace constituents. Schotland (1974), in the first detailed error analysis of a differential absorption lidar, revealed that accurate spatial distribution measurements of a gas density required knowledge regarding the uncertainties of many parameters, including the absorption coefficient, the power measurement, the exact laser wavelength, and the atmospheric parameters involved. In the case of water vapor measurements, Schotland indicated that for observations below 2 km the greatest source of error was instability in the laser wavelength, while above that altitude it was the uncertainty in the power measurements.

In a DIAL system it is very important that the time between consecutive laser pulses at λ_1 and λ_2 be kept at a minimum in order to avoid the influence of atmospheric turbulence on both the backscattering coefficient and the extinction coefficient. In the early development of such differential absorption lidar systems a single laser was used for measurements at one wavelength; it was then returned to the second wavelength and other series of measurements undertaken. The averages of the two series of measurements were then compared. Unfortunately, this procedure led to substantial errors due to variations in the atmospheric properties. Furthermore, uncertainties in the knowledge of the absorption coefficients, the atmospheric backscattering coefficient and the laser system combined with interference from other molecular species, speckle effects, and atmospheric turbulence tend to make DIAL measurements particularly susceptible to error. Indeed, there has been some speculation that averaging over N measurements might not lead to the expected $N^{-1/2}$ improvement in the signal-to-noise ratio (Menyuk et al., 1985). However, recent experiments (Milton and Woods, 1986; Fukuda et al., 1984; Grant, 1987) have suggested that this $N^{-1/2}$ rule applies in the case of well-designed systems.

Killinger and Menyuk (1981) discussed this problem and undertook a comparison of single- and dual-laser DIAL systems. Their results indicated that the observed increase in the measurement accuracy was in agreement with that predicted from a theory that considered the statistical and temporal character of DIAL returns. A further discussion of lidar error analysis and optimization is presented in Chapter 6, and a useful overview is provided by Measures (1984).

REFERENCES

Altmann, J., S. Kohler, and W. Lahmann (1980). Fast Current Amplifier for Background-Limited Operation of Photovoltaic InSb Detectors. *J. Physics. 6: Sci. Instrum.* **13,** 1275–1279.

Bevington, P. R. (1969). *Data Reduction and Error Analysis for the Physical Sciences*, McGraw-Hill, New York.

Bhaumik, M. L., R. S. Bradford, and A. R. Roit (1976). High Efficiency KrF Excimer Laser. *Appl. Phys. Lett.* **28,** 23–24.

Bissonnette, L. R. (1986). Sensitivity Analysis of Lidar Inversion Algorithms. *Appl. Optics* **25,** 2122–2126.

Born, M., and E. Wolf (1964). *Principles of Optics*, 2nd ed., Pergamon Press, New York.

Brassington, D. J. (1978). Alternate-Pulse Wavelength Switching of a Dye Laser. *J. Phys. E: Sci. Instrum.* **2,** 119–120.

Bristow, M. P. (1979). Fluorescence of Short Wavelength Cutoff Filters. *Appl. Optics* **18,** 952–955.

Bristow, M. P. F., W. R. Houston, and R. M. Measures (1973). Development of a Laser Fluorosensor for Airborne Surveying of the Aquatic Environment. NASA Conference on the Use of Lasers for Hydrographic Studies, Wallops Island, Sept. 1973, SP-375, 197–202.

Bristow, M., D. Nielsen, D. Bundy, and R. Furtek (1981). Use of Water Raman Emission to Correct Airborne Laser Fluorosensor Data for Effects of Water Optical Attenuation. *Appl. Optics* **20,** 2889–2906.

Browell, E. V. (1977). Analysis of Laser Fluorosensor Systems for Remote Algae Detection and Quantification, NASA TN D-8447.

Burgess, S., and I. W. Shepherd (1977). Fluorescence Suppression in Time-Resolved Raman Spectra. *J. Physics E: Sci. Instrum.* **10,** 617–620.

Byer, R. L. (1986). *Tunable Solid State Lasers for Remote Sensing*, Springer-Verlag, New York.

Byer, R. L., and M. Garbuny (1973). Pollutant Detection by Absorption Using Mie Scattering and Topographic Targets as Retroreflectors. *Appl. Optics* **12,** 1496–1505.

Collis, R. T. H., and P. B. Russell (1976). Lidar Measurement of Particles and Gases by Elastic Baskscattering and Differential Absorption, in *Laser Monitoring of the Atmosphere*, E. D. Hinkey, ed., Springer-Verlag, New York.

Davis, P. A. (1969). The Analysis of Lidar Signatures of Cirrus Clouds. *Appl. Optics* **8,** 2099–2102.

Dobrowolski, J. A., G. E. Marsh, D. G. Charbonneau, J. Eng. and P. D. Josephy (1977). Colored Filter Glasses: An Intercomparison of Glasses Made by Different Manufacturers. *Appl. Optics* **16,** 1491–1521.

Emmons, R. B., S. R. Hawkins, and C. F. Cuff (1975). Infrared Detectors: An Overview. *Opt. Eng.* **14,** 21–30.

Ewing, J. J. (1978). Rare-Gas Halide Lasers. *Physics Today*, May, 32–39.

Fan, Y. X., R. C. Eckardt, and R. L. Byer (1984). $AgGaS_2$ Infrared Parametric Oscillator. *Appl. Phy. Let.* **45,** 313–315.

Fantasia, J. F., T. M. Hard, and H. C. Ingrao (1971). An Investigation of Oil Fluorescence as a Technique for Remote Sensing of Oil Spills. Report No. DOT-TSC-USCG-71-7, Transportation Systems Center, Dept. of Transportation, Cambridge, MA.

Fiocco, G., and L. D. Smullin (1963). Detection of Scattering Layers in the Upper Atmosphere (60–140 km) by Optical Radar. *Nature* **199,** 1275–1276.

Foord, R., R. Jones, C. Oliver, and E. Pike (1969). *Appl. Optics* **8,** 1975.

Fredriksson, K., B. Galle, K. Nystrom, and S. Svanberg (1981). Mobile Lidar System for Environmental Probing. *Appl. Optics* **20,** 4181–4189.

Frush, C. L. (1975). A New Lidar Signal Processing and Display System. *Optical Quantum Electronics* **7,** 179–185.

Fukuda, T., Y. Matsuura, and T. Mori (1984). Sensitivity of Coherent Range-Resolved Differential Absorption Lidar. *Appl. Optics* **23,** 2026–2030.

Girard, A., and P. Jacquinot (1967). Principles of Instrumental Methods in Spectroscopy, in *Advanced Optimal Techniques*, A. van Heel, ed., North-Holland, Amsterdam, Chapter 3.

Grams, G. W., and C. M. Wyman (1972). Compact Laser Radar for Remote Atmospheric Probing. *J. Appl. Meteor.* **11,** 1108–1113.

Grant, W. B. (1982). Effect of Differential Spectral Reflectance on DIAL Measurements Using Topographical Targets. *Appl. Optics* **21,** 2390–2394.

Grant, W. B., J. R. Bogan, and A. M. Brothers (1987). $N^{-1/2}$ Dependence of DIAL Lidar Average Signal Standard Deviation. *Appl. Optics*.

Gunning, W. J. (1981). Electro-Optically Tuned Spectral Filters: Review. *Opt. Eng.* **20,** 837–845.

Halldórsson, T., and J. Langerholc (1978). Geometrical Form Factors for the Lidar Function. *Appl. Optics* **17,** 240–244.

Hänsch, T. W. (1972). Repetitively Pulsed Tunable Dye Laser for High Resolution Spectroscopy. *Appl. Optics* **11,** 895.

Hanst, P. L. (1976). Optical Measurement of Atmosphere Pollutants: Accomplishments and Problems. *Optical Quantum Electronics* **8,** 87–93.

Harms, J., W. Lahmann, and C. Weirkamp (1978). Geometrical Compression of Lidar Return Signals. *Appl. Optics* **17,** 1131–1135.

Harms, J. (1979). Lidar Return Signals for Coaxial and Noncoaxial Systems with Central Obstruction. *Appl. Optics* **18,** 1559–1566.

Heaps, W. S. (1981). Selection of Fluorescence Lidar Operating Parameters for SNR Maximization. *Appl. Optics* **20,** 583–587.

Hinkley, E. D. (1972). Tunable Infrared Lasers and Their Applications to Air Pollution Measurements. *Opto-electronics* **4,** 69–86.

Hinkley, E. D. (1976). Laser Spectroscopic Instrumentation and Techniques: Long Path Monitoring by Resonance Absorption. *Optical Quantum Electronics* **8,** 155–167.

Hoge, F. E., and R. N. Swift (1980). Oil Film Thickness Measurement Using Airborne Laser-Induced Water Raman Backscatter. *Appl. Optics* **19,** 3269–3281.

Hoge F. E., and R. N. Swift (1981). Absolute Tracer Dye Concentration Using Airborne Laser-Induced Water Raman Backscatter. *Appl. Optics* **20,** 1191–1202.

Houston, J. D., S. Sizgoric, A. Ulitsky, and J. Banic (1986). Raman Lidar System for Methane Gas Concentration Measurements. *Appl. Optics* **25,** 2115–2121.

Houston, W. R., D. G. Stephenson, and R. M. Measures (1973). LIFES: Laser Induced Fluorescence and Environmental Sensing. NASA Conference on the Use of Lasers for *Hydrographic Studies*, NASA SP-375, 153–169.

Inaba, H., and T. Kobayasi (1972). Laser-Raman Radar. *Opto-electronics* **4,** 101–123.

Inomata, H., and A. I. Carswell (1977). Simultaneous Tunable Two-Wavelength Ultraviolet Dye Lasers. *Optics Comm.* **19,** 5–6.

Jacquinot, P. J. (1954). The Luminosity of Spectrometers with Prisms, Gratings or Fabry–Perot Etalons. *J. Opt. Soc. Am.* **44,** 761.

Johnson, W. B., and E. E. Uthe (1971). Lidar Study of the Keystone Stack Plume. *Arm. Environ.* **5,** 730–724.

Kane, T. J., and R. L. Byer (1985). Monolithic, Unidirectional Single-Mode Nd : YAG Ring Laser. *Optics Lett.* **10,** 65–67.

Kanstad, S. O., A. Bjerkestrand, and T. Lund (1977). Tunable Dial-Line CO_2 Laser for Atmospheric Spectroscopy and Pollution Monitoring. *J. Phys. Sci. Instrum.* **10,** 998–1000.

Keyes, R. J., and R. H. Kingston (1972). A Look at Photon Detectors. *Phys. Today*, Mar., 48–54.

Kildal, H., and R. L. Byer (1971). Comparison of Laser Methods for the Remote Detection of Atmospheric Pollutants. *Proc. IEEE* **59,** 1644–1663.

Killinger, D. K., and N. Menyuk (1981). Remote Probing of the Atmosphere Using a CO_2 DIAL System. *IEEE J. Quant. Electr.* **QE-17,** 1917–1929; Effects of Turbulence-Induced Correlation on Laser Remote Sensing Errors. *Appl. Phys. Lett.* **38,** 968–970.

Klett, J. D. (1981). Stable Analytical Inversion for Processing Lidar Returns. *Appl. Optics* **20,** 211–220.

Klett, J. D. (1986). Extinction Boundary Value Algorithms for Lidar Inversion. *Appl. Optics* **25,** 2462–2464.

Koller, L. R. (1969). *Ultraviolet Radiation*, 2nd ed., Wiley, New York.

Kruse, P. W., L. D. McGlauchlin, and R. B. McQuistan (1963). *Elements of Infrared Technology*, Wiley, New York.

Ku, R. T., E. D. Hinkley, and J. O. Sample (1975). Long-Path Monitoring of Atmospheric Carbon Monoxide with a Tunable Diode Laser System. *Appl. Optics* **14,** 854–861.

Leonard, D. A. (1974). Measurement of Aircraft Turbine Engine Exhaust Emissions. *Laser Raman Gas Diagnostics*, Plenum Press, New York, pp. 45–61.

Ligda, M. G. H. (1963). Proc. Conf. Laser Technol., 1st, San Diego, Calif., 63–72.

Lopez, R. J., and M. A. Rebolledo (1981). Fatigue in Photomultipliers due to Excitation by Pulsed Light Sources. *Rev. Sci. Instrum.* **52,** 1852–1854.

Loree, T. R., R. C. Sze, D. L. Barker, and P. B. Scott (1979). New Lines in the UV: SRS of Excimer Laser Wavelengths. *IEEE J. Quant. Elec.* **QE-15,** 337–342.

Lussier, F. M. (1976a). Choosing an Infrared Detector. *Laser Focus*, Oct. 66–71.

Lussier, F. M. (1976b). Guide to IR-Transmissive Materials. *Laser Focus*, Dec. 47–50.

McClung, F. J., and R. W. Hellwarth (1962). Giant Optical Pulsations from Ruby. *J. Appl. Phys.* **33,** 828–829.

Measures, R. M. (1977). Lidar Equation Analysis—Allowing for Target Lifetime, Laser Pulse Duration, and Detector Integration Period. *Appl. Optics* **16,** 1092–1103.

Measures, R. M. (1984). *Laser Remote Sensing*, Wiley, New York.

Measures, R. M., and M. Bristow (1971). The Development of a Laser Fluorosensor for Remote Environmental Probing. *Can. Aeron. Space J.* **17,** 421–422.

Measures, R. M., and G. Pilon (1972). A Study of Tunable Laser Techniques for Remote Mapping of Specific Gaseous Constituents of the Atmosphere. *Opto-Electronics* **4,** 141–153.

Measures, R. M., W. R. Houston, and M. Bristow (1973). Development and Field Tests of a Laser Fluorosensor for Environmental Monitoring. *Can. Aeron. Space J.* **19,** 501–506.

Measures, R. M., H. R. Houston, and D. G. Stephenson (1974). Laser Induced Fluorescence Decay Spectra-A New Form of Environmental Signature. *Opt. Eng.* **13,** 494–450.

Measures, R. M., J. Garlick, W. R. Houston, and D. G. Stephenson (1975). Laser Induced Spectral Signatures of Relevance to Environmental Sensing. *Can. J. Remote Sensing* **1,** 95–102.

Mecherikunnel, A., and C. Duncan (1982). Total and Spectral Solar Irradiance Measured at Ground Surface. *Appl. Optics* **21,** 554–556.

Melchior, H. (1972). Demodulation and Photodetection Techniques, in *Laser Handbook*, Vol. 1, F. T. Arecchi and E. O. Schultz-Dubois, eds., North-Holland, Amsterdam, Chapter 7.

Melfi, S. H., M. L. Brumfield, and R. M. Storey, Jr. (1973). Observation of Raman Scattering by SO_2, in a Generating Plant Stack Plume. *Appl. Phys. Lett.* **22,** 402–403.

Menyuk, N., and D. K. Killinger (1983). Assessment of Relative Error Sources in IR DIAL Measurement Accuracy. *Appl. Optics* **22,** 2690.

Menyuk, N., D. K. Killinger and C. R. Menyuk (1985). Error Reduction in Laser Remote Sensing: Combined Effects of Cross Correlation and Signal Averaging. *Appl. Optics*, **24,** 118.

Milton, M. J., and P. T. Woods (1986). Atmospheric Contributions to the Column Variance in Direct-Detection DIAL, in Abstracts, Thirteenth International Laser Radar Conference, Toronto, Aug. 11–15, 1986. NASA Conference Publication No. 2431, p. 73.

Morhange, J. F., and C. Hirlimann (1976) Luminescence Rejection in Raman Spectroscopy. *Appl. Optics* **15,** 2969–2970.

Murray, E. R. (1977). Remote Measurement of Gases Using Discretely Tunable Infrared Lasers. *Optical Engineering* **16,** 284–290.

Murray, E. R., R. D. Hake, Jr., J. E. Van der Laan, and J. G. Hawley (1976). Atmospheric Water Vapor Measurements with a 10 Micrometer DIAL System. *Appl. Phys. Lett.* **28,** 542–543.

Murray, E. R., M. F. Williams, and J. E. van der Laan (1978). Single-Ended Measurement of Infrared Extinction Using Lidar. *Appl. Optics* **17,** 296–299.

Nakahara, S., K. Ito, S. Ito, A. Fuke, S. Komatsu, H. Inaba, and T. Kobayasi (1972). Detection of Sulphur Dioxide in Stack Plume by Laser Raman Radar. *Opto-electronics* **4,** 169–177.

Nill, K. W. (1974). Tunable Infrared Lasers. *Opt. Eng.* **13,** 516–522.

Oliver, B. M. (1965). Thermal and Quantum Noise. *Proc. IEEE* **53,** 436–454.

O'Neil, R. A., L. Buje-Bijunas, and D. M. Rayner (1980). Field Performance of a Laser Fluorosensor for the Detection of Oil Spills. *Appl. Optics* **19,** 863–870.

O'Neil, R. A., F. E. Hoge, and M. P. F. Bristow (1981). The Current Status of Airborne Laser Fluorosensing. 15th International Symposium on Remote Sensing of Environment, Ann Arbor, Michigan, May, 379–398.

Oppenheim, U. P., and R. T. Menzies (1982). Aligning the Transmitter and Receiver Telescopes of an Infrared Lidar: A Novel Method. *Appl. Optics* **21,** 174–175.

Parker, C. A. (1968). *Photoluminescence of Solutions*, Elsevier, New York.

Plass, G. N., G. W. Kattawar, and J. A. Guinn, Jr. (1976). Radiance Distribution over a Ruffled Sea: Contributions from Glitter, Sky, and Ocean. *Appl. Optics* **15,** 3161–3165.

Poultney, S. K. (1972a). Laser Radar Studies of Upper Atmosphere Dust Layers and the Relation to Temporary Increases in Dust to Cometary Micrometeoroid Streams. *Space Research* **12,** 403–421.

Poultney, S. K. (1972b). Single Photon Detection and Timing: Experiments and Techniques. *Advances in Electronics and Electron Physics* **31,** 39–117.

Pratt, W. K. (1969). *Laser Communication Systems*, Wiley, New York.

Ramsay, J. V. (1962). A Rapid-Scanning Fabry-Perot Interferometer with Automatic Parallelism Control. *Appl. Optics* **1,** 411–413.

Rosen, H., P. Robish, and O. Chamberlain (1975). Remote Detection of Pollutants Using Resonance Raman Scattering. *Appl. Optics* **14,** 2703–2706.

Ross, M. (1966). *Laser Receivers (Devices, Techniques, Systems)*, Wiley, New York.

Russell, P. B., T. J. Swissler, and M. P. McCormick (1979). Methodology for Error Analysis and Simulation of Lidar Aerosol Measurements. *Appl. Optics* **18,** 3783–3797.

Schäfer, F. P. (1973). *Dye Lasers*, Topics in Applied Physics, Springer-Verlag, New York.

Schotland, R. M. (1966). Some Observation of the Vertical Profile of Water Vapor by a Laser Optical Radar. Proc. 4th Symposium on Remote Sensing of the Environment 12–14 April 1966, Univ. of Michigan, Ann Arbor, 273–283.

Schotland, R. M. (1974). Errors in Lidar Measurements of Atmospheric Gases by Differential Absorption. *J. Appl. Meteorology* **13,** 71–77.

Schottky (1918). *Ann. Phys Leipzig* **57,** 54.

Shaw, J. H. (1953). Solar Radiation. *Ohio J. Sci.* **53,** 258–271.

Shewchun, J., B. K. Garside, E. A. Ballik, C. C. Y. Kwan, M. M. Elsherbiny, G. Hogenkamp, and A. Kazandjian (1976). Pollution Monitoring Systems Based on Resonance Absorption Measurements of Ozone with a 'Tunable' CO_2 Laser; Some Criteria. *Appl. Optics* **15,** 340–346.

Shumate, M. S., S. Lundqvist, V. Persson, and S. T. Eng (1982). Differential Reflectance of Natural and Man Made Materials at CO_2 Laser Wavelengths. *Appl. Optics* **21,** 2386–2389.

Stewart, R. W., and J. L. Bufton (1980). Development of a Pulsed 9.5 Micrometer Lidar for Regional Scale O_3 Measurement. *Optical Engineering* **19,** 503–507.

Sze, R. C. (1979). Rare Gas Halide Avalanche Discharge Lasers. *IEEE J. Quant. Electr.* **QE-15,** 1338–1347.

Title, A. M., and W. J. Rosenberg (1981). Tunable Birefringent Filters. *Optical Eng.* **20,** 815–823.

Topp, J. A., H. W. Schrotter, H. Hacker, and J. Brandmuller (1969). Improvement of the Signal-to-Noise Ratio of Photomultipliers for Very Weak Signals. *Rev. Sci. Instrum.* **40,** 1164–1169.

Uthe, E. E., and R. J. Allen (1975). A Digital Real-Time Lidar Data Recording Processing and Display System. *Optical Quantum Electronics* **7,** 121–129.

Valley, S. L., Ed. (1965). *Handbook of Geophysics and Space Environments,* McGraw-Hill, New York.

Viezee, W., J. Obianas, and R. T. H. Collis (1973). *Evaluation of the Lidar Technique of Determining Slant Range Visibility for Aircraft Landing Operations*, SRI Report AFCRL-TR-0708.

Walther, H., and J. L. Hall (1970). Tunable Dye Laser with Narrow Spectral Output. *Appl. Phys. Lett.* **17,** 239.

Wang, C. P. (1974). Application of Lasers in Atmospheric Probing. *Acta Astronaut.* **1,** 105–123.

Weitkamp, C. (1981). The Distribution of Hydrogen Chloride in the Plume of Incineration Ships: Development of New Measurements Systems. *Wastes in the Ocean,* Vol. 3, Wiley; also GKSS 81/E/57.

Wolfe, W. L., ed. (1966). *Handbook of Military Infrared Technology,* ONR Cat. No. 65-62266, U.S. Government Printing Office, Washington, D.C.

Wood, O. R., II (1974). High Pressure Pulsed Molecular Lasers. *Proc. IEEE* **62,** 355–397.

Zhou, B., T. J. Kane, G. J. Dixon, and R. L. Byer (1985). Efficient, Frequency-Stable Laser-Diode-Pumped Nd:YAG Laser. *Opt. Lett.* **10,** 62–64.

CHAPTER

2

TRANSMISSION THROUGH THE ATMOSPHERE

REINHARD BEER

Division of Earth and Space Sciences
The Jet Propulsion Laboratory
Pasadena, California

2.1 INTRODUCTION

All measurements of, or through, Earth's atmosphere are affected by its properties. Thus, it is of great importance both to the planning and to the reduction of field measurements to have a clear understanding of the influence of the ambient atmosphere. This chapter addresses the topic of atmospheric transmission over the spectral interval from the ultraviolet ozone/oxygen/Rayleigh scattering cutoff to the mid-IR (700 cm^{-1}) where water vapor and carbon dioxide effectively blanket the residual transmission. The topics to be discussed are molecular absorption; molecular emission; aerosol and particulate extinction; aerosol and particulate scattering; molecular scattering; and clouds.

Molecular absorption and emission comprise both line and continuum effects; scattering sources are both natural and anthropogenic. The distinction between "aerosols" and "clouds" is, of course, arbitrary, but in fact the section on clouds deals not with their physical properties but with the likelihood of their interfering with observations.

Two problems not covered are atmospheric turbulence and phase coherence. Turbulence is not inherently a factor in atmospheric transmission, but the jitter of the image of a target point with respect to a focal plane detector can mimic a transmission loss. Scattering, besides causing a loss of transmission, can influence the coherence of the radiation. Since both of these areas are fields of active research, the reader is referred to the voluminous literature if these topics are of importance.

Although it is of little value for detailed analysis on the frequency scale required for laser remote sensing, Figure 2.1 provides a useful overview of the general characteristics of the vertical transmittance of the atmosphere from the UV cutoff to the mid-IR. It is derived from the well-known LOWTRAN code that has been widely used in studies of remote sensing and is displayed on a wavelength scale for convenience. However, throughout most of this chapter the spectroscopic unit to be employed is the reciprocal centimeter or wavenumber (cm^{-1})—the temporal frequency in hertz divided by the velocity of light.

The LOWTRAN code employs the technique of *band models* to treat molecular absorption. While making for very rapid computation, the method is generally unsatisfactory for investigations at high spectral resolution. In particular, laser analyses will normally be conducted in spectral regions that are, by and large, transparent but can be very narrow ("microwindows"). LOWTRAN

Figure 2.1. Overview of general characteristics of transmittance of Earth's atmosphere from UV cutoff to mid-IR. Labels indicate various climatological model atmospheres used in computations. Letters are standard astronomical nomenclature for IR windows. Data derived from LOWTRAN code. Figure courtesy of the Air Force Geophysics Laboratory.

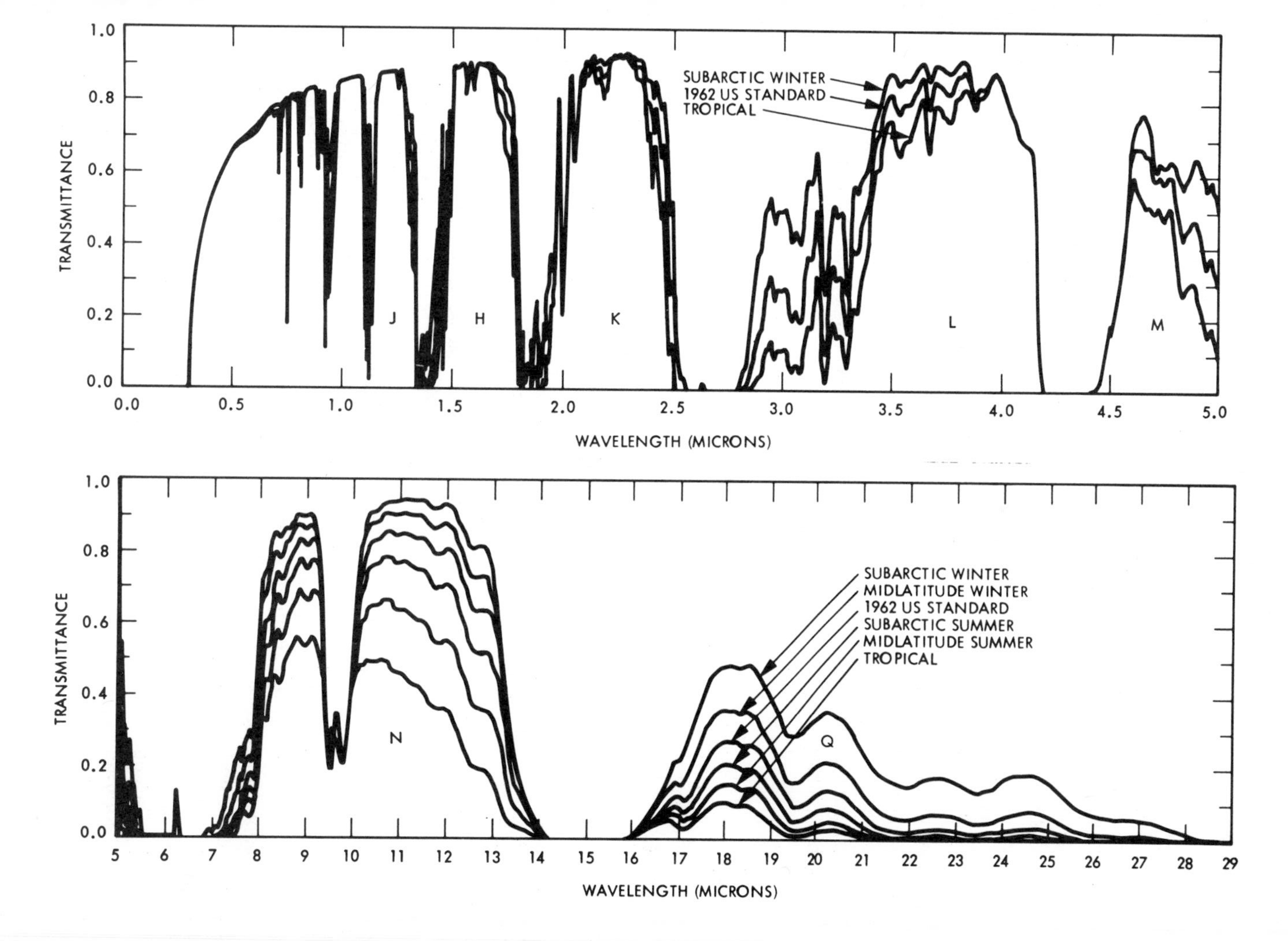

SUBARCTIC WINTER
1962 US STANDARD
TROPICAL
TRANSMITTANCE
J
H
K
L
M
WAVELENGTH (MICRONS)
SUBARCTIC WINTER
MIDLATITUDE WINTER
1962 US STANDARD
SUBARCTIC SUMMER
MIDLATITUDE SUMMER
TROPICAL
TRANSMITTANCE
N
Q
WAVELENGTH (MICRONS)

computes in 20-cm^{-1} wide averages (5 cm^{-1} in some versions), whereas laser studies are often confined to intervals 0.01 cm^{-1} wide or less. Thus, this chapter concentrates on methodologies that account for molecular absorption on a line-by-line basis wherever this is appropriate or possible. However, to avoid redundancy, the topic of line *shape* is not covered. An excellent introduction to this topic will be found in Chapter 3. All the computations displayed in this chapter were made using the Voigt function for the lineshape.

The primary emphasis of this chapter is utilitarian. Thus, rigorous analysis is replaced by graphs, tables, and "rules of thumb" whenever possible. In this way it is hoped that the experimenter will be able to make a reasonable assessment of the likely line-of-sight atmospheric transmittance without the need to delve into a small library of texts and handbooks. Nevertheless, to keep this chapter within bounds, whenever data tabulations become too extensive, they have either been abstracted or referenced.

2.2 RADIATIVE TRANSFER

The governing equation for the time-independent transport of radiation through the atmosphere can be expressed as

$$dJ(\tilde{\nu}, s, \Omega)/ds = -\kappa(\tilde{\nu}, s)J(\tilde{\nu}, s, \Omega) + \kappa(\tilde{\nu}, s)S(\tilde{\nu}, s, \Omega) \qquad (2.1)$$

where $J(\tilde{\nu}, s, \Omega)$ = spectral radiance $[\mathrm{W \cdot cm^{-1} \cdot sr^{-1} \cdot (cm^{-1})^{-1}}]$ at wavenumber $\tilde{\nu}$ cm^{-1} from a point s along a line element ds comprehending solid angle Ω sr

$\kappa(\tilde{\nu}, s)$ = extinction coefficient (cm^{-1}) at wavenumber $\tilde{\nu}$

$S(\tilde{\nu}, s, \Omega)$ = source function

This deceptively simple equation merely states that the change in spectral radiance along a path ds is the sum of an attenuation term proportional to the radiance itself and a source term adding additional photons into the directed solid angle Ω. The solution, analytical or numerical, can (and does) fill textbooks because each individual term is itself the solution to an endless variety of possibilities. Since the primary purpose of this section is to provide definitions for a number of terms to be used later, only an outline of the formal solution follows.

The source function can be generalized for the case of local thermodynamic equilibrium (LTE) by

$$\kappa(\tilde{\nu}, s)S(\tilde{\nu}, s, \Omega) = [\sigma(\tilde{\nu}, s)/4\pi] \oint_{4\pi} p(\tilde{\nu}, s, \Omega, \Omega') \times J(\tilde{\nu}, s, \Omega')\, d\Omega' + a(\tilde{\nu}, s)B(\tilde{\nu}, s, T) \qquad (2.2)$$

where $p(\tilde{\nu}, s, \Omega, \Omega')$ = phase function representing probability that radiation at s, $J(\tilde{\nu}, s, \Omega')$ incoming from direction Ω', is elastically scattered into direction Ω

$\sigma(\tilde{\nu}, s)$ = elastic scattering coefficient (cm^{-1})

$B(\tilde{\nu}, s, T)$ = Planck function for temperature T at point s (assumed to be within atmosphere)

$a(\tilde{\nu}, s)$ = molecular absorption coefficient (cm^{-1})

In addition,

$$\kappa(\tilde{\nu}, s) = \sigma(\tilde{\nu}, s) + a(\tilde{\nu}, s) \tag{2.3}$$

The formal solution to Equation (2.1) is

$$J(\tilde{\nu}, s, \Omega) = J(\tilde{\nu}, s_0, \Omega)\exp\left[-\int_{s_0}^{s} \kappa(\tilde{\nu}, s')\, ds'\right] + \int_{s_0}^{s} S(\tilde{\nu}, s', \Omega)\exp\left[-\int_{s'}^{s} \kappa(\tilde{\nu}, s'')\, ds''\right]\kappa(\tilde{\nu}, s')\, ds' \tag{2.4}$$

where $J(\tilde{\nu}, s_0, \Omega)$ is the boundary condition at the source or target point s_0 and both ds' and ds'' are in the directed solid angle Ω.

The argument of the exponential terms in (2.4) is often called the *optical depth*, defined by

$$\tau(\tilde{\nu}, s_1, s_2) = \int_{s_1}^{s_2} \kappa(\tilde{\nu}, s')\, ds' \tag{2.5}$$

To lead to tractable solutions, it is necessary to investigate the detailed nature of the terms in (2.4) to approximate or eliminate unnecessary elements.

2.2.1 Clear Atmosphere

If the line of sight is free of aerosols, particulates, or clouds and the frequency is low enough that molecular (Rayleigh) scattering can be ignored ($\tilde{\nu} < 15000$ cm^{-1}, say), the extinction coefficient $\kappa(\tilde{\nu}, s)$ depends only on gaseous absorption. The boundary condition $J(\tilde{\nu}, s_0, \Omega)$ can then be expanded as

$$J(\tilde{\nu}, s_0, \Omega) = \sum_{n=1}^{N} J^{(n)}(\tilde{\nu}, s_0, \Omega) \tag{2.6}$$

where

$$J^{(1)}(\tilde{\nu}, s_0, \Omega) = P(\tilde{\nu}, s_0, \omega) \tag{2.7a}$$

the radiance emitted by the source (laser) in the case of direct, point-to-point, reception, and

$$J^{(1)}(\tilde{\nu}, s_0, \Omega) = P(\tilde{\nu}, s_1, \Omega_1)\exp[-\tau(\tilde{\nu}, s_1, s_0)]A(\tilde{\nu}, s_0)\rho(\tilde{\nu}, s_0, \Omega, \Omega_1) \tag{2.7b}$$

for the case of backscattered radiation where the source is located at s_1 and propagates in the direction Ω_1. Here, $A(\tilde{\nu}, s_0)$ is the albedo of the backscattering surface (the ratio of total reflected radiation to that incident) and $\rho(\tilde{\nu}, s_0, \Omega, \Omega_1)$ is a function describing the distribution of the backscattered radiation from the direction Ω_1 to Ω. Then

$$J^{(2)}(\tilde{\nu}, s_0, \Omega) = \epsilon(\tilde{\nu}, s_0, \Omega)B[\tilde{\nu}, T(s_0)] \tag{2.7c}$$

the thermal emission from the target area, temperature $T(s_0)$, and emissivity $\epsilon(\tilde{\nu}, s_0, \Omega)$. Note that emissivity is direction dependent for most real surfaces. Then

$$J^{(3)}(\tilde{\nu}, s_0, \Omega) = A(\tilde{\nu}, s_0)\int_{s0}^{\infty} \oint_{4\pi} \rho(\tilde{\nu}, s_0, \Omega, \Omega')B[\tilde{\nu}, T(s')] \times \exp[-\tau(\tilde{\nu}, s', s_0)]\kappa(\tilde{\nu}, s')\, ds'\, d\Omega' \tag{2.7d}$$

the atmospheric emission propagating to the target at s_0 and backscattered to the receiver. Also,

$$J^{(4)}(\tilde{\nu}, s_0, \Omega) = A(\tilde{\nu}, s_0)\rho(\tilde{\nu}, s_0, \Omega, \Omega_0)I_s\exp[-\tau(\tilde{\nu}, s_0, \Omega_0)] \tag{2.7e}$$

the solar contribution to the backscatter. Here, I_s is the solar irradiance at the "top" of the atmosphere from direction Ω_0 and $\tau(\tilde{\nu}, s_0, \Omega_0)$ is the opacity applicable to the *inbound* radiation. Then

$$J^{(5...N)}(\tilde{\nu}, s_0, \Omega) = A(\tilde{\nu}, s_0)\rho(\tilde{\nu}, s_0, \Omega, \Omega_n)I_n\exp[-\tau(\tilde{\nu}, s_0, s_n, \Omega)] \tag{2.7f}$$

the contribution from all other extraneous near-collimated sources (moon, street lighting, local hot spots, etc.)

Presumably, an objective in any laser experiment is to make $J^{(1)}(\tilde{\nu}, s_0, \Omega)$ very much greater than all other terms. While this may be feasible for direct-reception measurements (or those using retroreflectors), experiments employing backscatter from the natural terrain may well need to account for some or all of the terms depending on the exact emission–detection technique employed.

Under these circumstances, Equation (2.4) simplifies to

$$J(\tilde{\nu}, s, \Omega) = \left[\sum_{n=1}^{N} J^{(n)}(\tilde{\nu}, s_0, \Omega)\right] \exp\left\{-\left[\tau(\tilde{\nu}, s_0, \Omega) - \tau(\tilde{\nu}, s, \Omega)\right]\right\} + \exp\left[\tau(\tilde{\nu}, s, \Omega)\right] \times \int_{\tau(\tilde{\nu}, s, \Omega)}^{\tau(\tilde{\nu}, s_0, \Omega)} B\left[\tilde{\nu}, T(s')\right] \times \exp\left[-\tau(\tilde{\nu}, s', \Omega)\right] d\tau(\tilde{\nu}, s', \Omega) \tag{2.8}$$

with

$$\tau(\tilde{\nu}, s, \Omega) \equiv \int_{s}^{\infty} \kappa(\tilde{\nu}, s')\, ds'$$

Defining

$$t(\tilde{\nu}, s, \Omega) = \exp\left[-\tau(\tilde{\nu}, s, \Omega)\right] \tag{2.9}$$

as the transmittance, the general case of transatmospheric sounding further simplifies to

$$J(\tilde{\nu}, s, \Omega) = \left[\sum_{n=1}^{N} J^{(n)}(\tilde{\nu}, s_0, \Omega)\right] t(\tilde{\nu}, s_0, \Omega)/t(\tilde{\nu}, s, \Omega) + t(\tilde{\nu}, s, \Omega) \int_{z(s_0)}^{z(s)} B\left[\tilde{\nu}, T(s')\right]\left[\partial t(\tilde{\nu}, s', \Omega)/\partial z\right] dz(s') \tag{2.10}$$

where a change of variable to altitude z has been made. The term $\partial t(\tilde{\nu}, s, \Omega)/\partial z$ is the weighting function familiar in passive remote temperature sounding of the atmosphere, and its product with the Planck function $B[\tilde{\nu}, T(s)]$ indicates where in the atmosphere the thermal contribution chiefly originates.

2.2.2 Scattering

When scattering is important, Equation (2.4) takes the form

$$\begin{aligned} J(\tilde{\nu}, s, \Omega) &= J(\tilde{\nu}, s_0, \Omega)t(\tilde{\nu}, s_0, \Omega)/t(\tilde{\nu}, s, \Omega) + t(\tilde{\nu}, s, \Omega) \\ &\times \int_{\tau(\tilde{\nu}, s, \Omega)}^{\tau(\tilde{\nu}, s_0, \Omega)} [\omega(\tilde{\nu}, s')/4\pi] \\ &\times \left[\oint_{4\pi} p(\tilde{\nu}, s', \Omega, \Omega')J(\tilde{\nu}, s', \Omega')\, d\Omega' \right] \\ &\times t(\tilde{\nu}, s', \Omega)\, d\tau(\tilde{\nu}, s', \Omega) + t^{-1}(\tilde{\nu}, s, \Omega) \\ &\times \int_{\tau(\tilde{\nu}, s, \Omega)}^{\tau(\tilde{\nu}, s_0, \Omega)} [1 - \omega(\tilde{\nu}, s')] \\ &\times B[\tilde{\nu}, T(s')]t(\tilde{\nu}, s', \Omega)\, d\tau(\tilde{\nu}, s', \Omega) \end{aligned} \tag{2.11}$$

where $\omega(\tilde{\nu}, s') = \sigma(\tilde{\nu}, s')/\kappa(\tilde{\nu}, s')$, the so-called single-scattering albedo.

Even numerical calculations now become fraught with difficulty because the summations for the intensities must now incorporate terms such as ones for radiation scattered *into* the line of sight from nearby (or not so nearby) sources (Diner and Martonchik, 1984); radiation backscattered before it even reaches the desired target (Menzies et al., 1984; Menzies and Kavaya, 1985); radiation originating within the atmosphere and being scattered into the line of sight either directly or via the target; and so on. Accounting for no more than the important terms can be a nightmare. Worse yet, the phase function $p(\tilde{\nu}, s', \Omega, \Omega')$ is generally unknown or ill-defined except for a few simple cases, and $\omega(\tilde{\nu}, s')$ may also be uncertain or indeterminate. The most widely practiced approaches employ sweeping assumptions such as simple Rayleigh scattering or Mie scattering by dielectric spheres. Again, this is a field that breeds textbooks such as the classic of Van de Hulst (1957) or the more recent (and utilitarian) work of Bohren and Huffman (1983). A review more specifically aimed at atmospheric scattering is provided by Hansen and Travis (1974). This topic is revisited in Section 2.5.

2.3. MOLECULAR LINE ABSORPTION

The most obvious characteristics of the atmosphere that affects laser remote sensing is the phenomenon of molecular line absorption. Although the number of "natural" and "anthropogenic" species known to produce identifiable absorptions in the free atmosphere is large and growing, seven are of particular importance: water vapor (H_2O), carbon dioxide (CO_2), ozone (O_3), nitrous oxide (N_2O), methane (CH_4), carbon monoxide (CO), and oxygen (O_2); these will be discussed in some detail. Interference by other species can generally be expected to be weak except in long horizontal or limb-sounding paths.

2.3.1 Water Vapor

Water vapor is perhaps the greatest enemy of the remote sensor. Not only are its line absorptions widespread and strong, but also it is responsible for substantial continuum absorption, a phenomenon to be discussed later. Furthermore, atmospheric water is a principal constituent of aerosols, another source of atmospheric extinction reserved for later consideration.

There is scarcely any region of the IR that is totally free of H_2O line absorptions. Even its rare isotopic form, HDO, produces significant absorption around 2720 cm^{-1} in a region that is otherwise one of the most transparent available.

Accounting for H_2O absorption (either in analysis or in planning for measurements) is always difficult because of its dramatic variability in time, location, and altitude. Indeed, this section would be incomplete without remarking that dealing with this one species alone usually requires a significant effort and expense on the part of the experimenter. A substantial fraction of the cost of modern astronomical observations[†] can be attributed to the efforts needed to get above as much of the atmospheric water vapor as possible. The typical scale height for H_2O in the troposphere is 1–2 km, markedly less than the pressure scale height ($\sim$8 km), so even modest elevations above sea level can be of great assistance in remote sensing. Unfortunately, this stratagem is not available for observation paths that must cover low altitudes.

The principal IR bands of H_2O are shown in Table 2.1. In this and later tables, the list contains only a subset of the known bands, but the contents were selected on the rough criterion that the band should have a noticeable effect on the overall transmittance. Thus, there are numerous unlisted bands stronger than some listed in Table 2.1; they have been left out because nearby bands are even stronger. However, as will be described later, every transition must be incorporated into detailed transmittance calculations.

The left column in the tables shows the band origin (noting that the "origin" of pure rotation bands is a zero wavenumber). A rough rule of thumb suggests that bands of light molecules produce significant effects over $\pm$50–100 cm^{-1} from the origin, although identifiable members of, for example, the 1594.75-cm^{-1} H_2O band can be detected more than 1000 cm^{-1} away.

[†]Surely, the classic field of remote sensing!

Table 2.1. **Principal Infrared Bands of Water Vapor (H_2O)**

Band Origin (cm^{-1})	Isotope Code[a]	Upper Vibration State, V'	Lower Vibration State, V''	Integrated Band Strength (cm^{-1}/molecule cm^{-2}) at 296 K
764.[b]	162	000	000	1.16E-20
905.[b]	171	000	000	1.94E-20
976.[b]	181	000	000	1.07E-19
1029.[b]	161	010	010	2.23E-20
1648.[b]	161	000	000	5.27E-17
1594.75	161	010	000	1.04E-17
2062.31	161	100	010	1.92E-22
2161.18	161	001	010	2.63E-22
2723.68	162	100	000	6.34E-22
3151.63	161	020	000	7.54E-20
3657.05	161	100	000	4.86E-19
3707.46	162	001	000	1.42E-21
3755.93	161	001	000	6.93E-18
4666.79	161	030	000	3.40E-22
5234.98	161	110	000	3.72E-20
5276.78	161	021	010	6.62E-22
5331.27	161	011	000	8.04E-19
6775.10	161	120	000	3.05E-21
6871.51	161	021	000	5.06E-20
7201.54	161	200	000	4.58E-20
7249.81	161	101	000	6.43E-19
7445.07	161	002	000	5.84E-21
8273.95	161	130	000	2.33E-22
8373.82	161	031	000	8.94E-22
8761.58	161	210	000	4.15E-22
8807.00	161	111	000	4.95E-20
9000.13	161	012	000	1.56E-21
10329.7	161	121	000	2.08E-21
10613.	161	201	000	2.12E-20
10869.	161	102	000	5.65E-22
11032.	161	003	000	2.38E-21
12151.	161	211	000	9.28E-22
13652.65	161	221	000	1.77E-22
13820.92	161	301	000	1.08E-21
14318.80	161	103	000	2.05E-22
15347.95	161	311	000	8.78E-23
16898.83	161	401	000	8.55E-23

[a] 161 = $H_2\,^{16}O$; 171 = $H_2\,^{17}O$; 181 = $H_2\,^{18}O$; 162 = HD ^{16}O. Vibrational state code = $v_1v_2v_3$.
[b] High-wavenumber limits of pure rotation bands.

The second column of Table 2.1 (and subsequent tables) contains a simple code identifying the molecular isotope responsible for the band; columns 3 and 4 contain the upper and lower vibrational states (essential elements in the determination of temperature effects), and column 5 contains the integrated band strength in units of reciprocal centimeter per molecule per square centimeter column computed by summing over all the individual rotational–vibrational linestrengths. Its principal use is to indicate, in general terms, which transitions are likely to be dominant.

Figure 2.2 shows a "standard" volume (number) mixing ratio of H_2O for the first 100 km of the atmosphere. In order to relate this to the spectroscopic unit of column density, it is necessary to multiply by the local number density of the free atmosphere and the pathlength. The number density is a variable that depends on the local temperature and pressure but has been tablulated for a number of average conditions into so-called climatological model atmospheres such as those shown in Figure 2.3. While these are very useful for computational purposes, there can be no substitute for direct measurements whenever possible. It should also be observed that for very long transatmospheric paths,

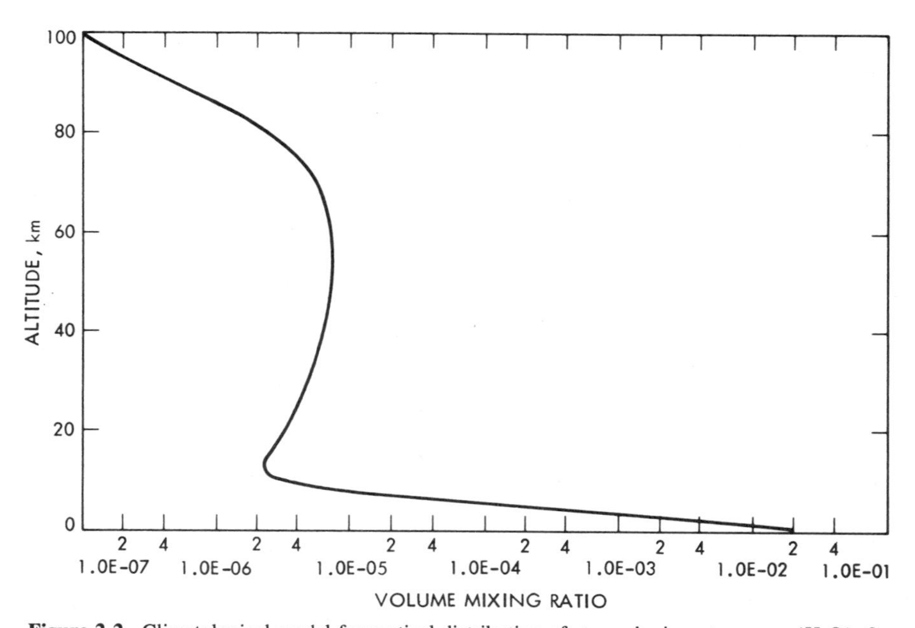

Figure 2.2. Climatological model for vertical distribution of atmospheric water vapor (H_2O), 0–100 km. Horizontal axis is volume mixing ratio. Absolute number density may be obtained by reference to Figure 2.3. Note that, in the troposphere, large variations from this model are to be expected.

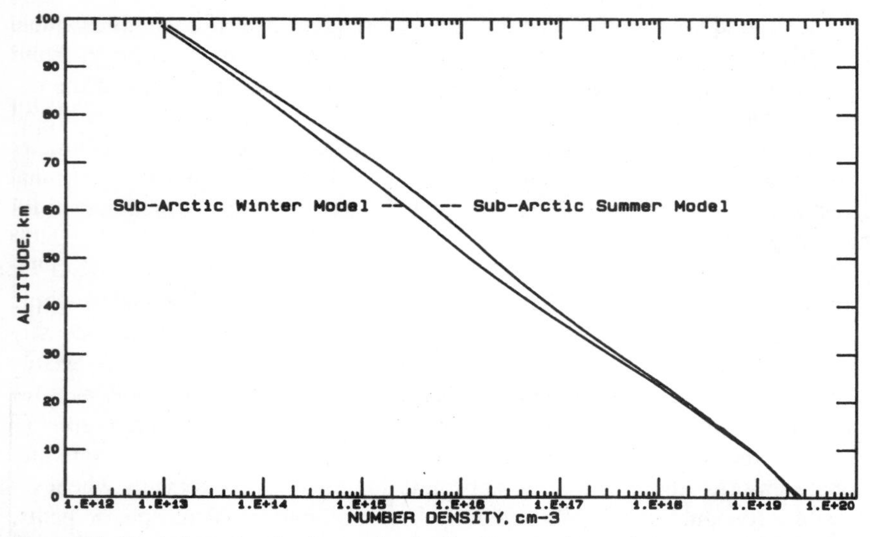

Figure 2.3. Vertical number density profiles for Earth's atmosphere. Only two extreme cases are shown: subarctic summer and winter. All other standard models fall between these extremes.

refraction can introduce nonnegligible effects that are also dependent on local conditions (see Section 2.6). For now, it should be observed that a "standard" H_2O vertical profile is useful only in the very crudest sense of advanced planning. It would be totally worthless for even the most cursory spectral analysis, and it is absolutely essential that the experimenter obtain direct information on the water vapor content in the line of sight as close to the time and place of observation as possible. In the reasonable expectation that such measurements will only be available near the surface, Allen (1973) offers the approximate expression for the total vertical water vapor column through the atmosphere above altitude z km:

$$N_{\text{tot}} \sim 730N(z)T(z)\exp[-0.105z] \quad \text{molecules cm}^{-2} \qquad (2.12)$$

where $N(z)$ is the locally estimated number density (cm^{-3}, see below) and $T(z)$ is the local absolute temperature in kelvins.

The previous paragraph leads directly into the problem of the Babel of units that has arisen as a result of the melding of two separate disciplines with long, independent, histories—spectroscopy and meteorology. Meteorological measurement techniques give temperature and pressure, not number density, and their methods for measuring water vapor content give results in terms of precipitable millimeters, relative humidity, or dew point. Thus, conversions are necessary.

Temperature, pressure, and density are related through the perfect-gas law:

$$P = NRT \tag{2.13}$$

where P is pressure, N is a number density, T is the absolute temperature, and, R is a gas constant ($=1.362565 \times 10^{-22}$) for P in atmospheres (1 atm $=$ 1013.25 mbar $= 1.01325 \times 10^5$ N m^{-2} $=$ 760 mm Hg), N is in molecules cm^{-3}, and T is in kelvins.

Precipitable millimeters of water are directly convertible to column density: 1 pr mm $= 3.3423 \times 10^{21}$ molecules cm^{-2}.

The relation between dewpoint and relative humidity is tabulated in handbooks such as the *Smithsonian Meteorological Tables* and the *CRC Handbook*. For rough calculations, the local number density of water vapor in molecules cm^{-3} is

$$\log N_{H_2O} = (\mathrm{RH}) \times [17.1834 + 3.25067 \times 10^{-2}\theta - 1.50311 \times 10^{-4}\theta^2] \tag{2.14}$$

where RH is the relative humidity and θ is temperature in celsius.

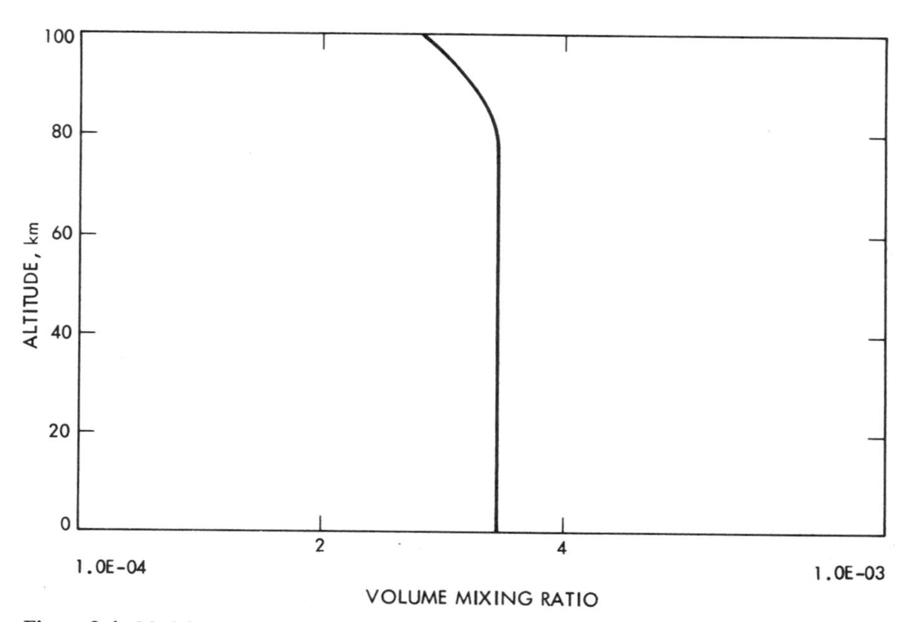

Figure 2.4. Model vertical distribution for carbon dioxide (CO_2). Note that strong evidence exists for a secular increase in CO_2 abundance.

Table 2.2. **Principal Infrared Bands of Carbon Dioxide (CO_2)**

Band Origin (cm^{-1})	Isotope Code[a]	Upper Vibration State, V'	Lower Vibration State, V''	Integrated Band Strength (cm^{-1}/molecule cm^{-2}) at 296 K
597.34	626	11102	02201	4.91E-21
618.03	626	10002	01101	1.36E-19
647.06	626	11102	10002	2.12E-20
648.48	636	01101	00001	8.24E-20
648.79	636	02201	01101	7.05E-21
667.38	626	01101	00001	7.95E-18
667.75	626	02201	01101	6.26E-19
688.67	626	11101	10001	1.43E-20
720.81	626	10001	01101	1.80E-19
741.73	626	11101	02201	7.67E-21
791.45	626	11101	10002	1.10E-21
960.96	626	00011	10001	6.80E-22
1063.74	626	00011	10002	8.97E-22
1932.47	626	11102	00001	1.06E-21
2076.86	626	11101	00001	4.87E-21
2093.35	626	12201	01101	4.86E-22
2129.76	626	20001	01101	1.84E-22
2224.66	626	10012	10001	1.27E-22
2260.05	636	02211	02201	3.46E-21
2261.91	636	10012	10002	2.03E-21
2262.85	636	10011	10001	1.23E-21
2271.76	636	01111	01101	8.17E-20
2283.49	636	00011	00001	9.60E-19
2324.14	626	02211	02201	3.08E-19
2326.60	626	10011	10001	1.18E-19
2327.43	626	10012	10002	1.93E-19
2332.11	628	00011	00001	3.33E-19
2336.63	626	01111	01101	7.66E-18
2349.14	626	00011	00001	9.59E-17
2429.38	626	10011	10002	1.06E-22
3339.34	626	21102	00001	1.10E-22
3498.76	636	11112	01101	7.27E-22
3527.74	626	10012	00001	9.42E-21
3552.85	626	12212	02201	3.12E-21
3568.22	626	20013	10002	3.38E-21
3571.84	628	10012	00001	5.22E-21
3580.33	626	11112	01101	8.03E-20
3589.65	626	20012	10002	1.78E-21
3612.84	626	10012	00001	1.03E-18
3632.91	636	10011	00001	1.60E-20

Table 2.2. (*Continued*)

Band Origin (cm^{-1})	Isotope Code[a]	Upper Vibration State, V'	Lower Vibration State, V''	Integrated Band Strength (cm^{-1}/molecule cm^{-2}) at 296 K
3639.22	636	11111	01101	1.50E-21
3675.13	628	10011	00001	4.78E-21
3714.78	626	10011	00001	1.50E-18
3723.25	626	11111	01101	1.13E-19
4807.70	626	21113	01101	6.64E-22
4853.63	626	20013	00001	8.07E-21
4887.39	636	20012	00001	2.98E-22
4965.39	626	21112	01101	2.66E-21
4977.84	626	20012	00001	3.50E-20
5099.66	626	20011	00001	1.12E-20
5123.20	626	21111	01101	1.06E-21
6227.92	626	30013	00001	4.27E-22
6347.86	626	30012	00001	4.27E-22
6935.14	626	01131	01101	1.13E-22
6972.58	626	00031	00001	1.49E-21

[a] 626 = $^{16}O\,^{12}C\,^{16}O$; 627 = $^{16}O\,^{12}C\,^{17}O$; 628 = $^{16}O\,^{12}C\,^{18}O$; 636 = $^{16}O\,^{13}C\,^{16}O$.
Vibrational state code = $v_1 v_2 l v_3 r$.

It must, however, be emphasized that all such conversions are valid only for homogeneous paths. Inhomogeneity must usually be handled on a piecemeal basis.

Two other conversions are also frequently encountered: For experimental convenience, laboratory spectroscopists often record their column densities in units of centimeter amagat, where 1 amagat is the atmospheric density at STP and the physical pathlength is measured in centimeters. The conversion is 1 cm amagat = 2.68675×10^{19} molecules cm^{-2}.

Field measurements, on the other hand, are frequently reported in terms of parts per million meters, a unit that can only be approximately converted unless the local atmospheric density is known (i.e., parts per million of what?). Roughly, 1 ppm meter = 2.37×10^{15} molecules cm^{-2} at the surface.

With the foregoing digression, the impact of a band or line on the transmittance through a given path can be roughly assessed through the product of the strength and the column density (for weak absorptions) or the square root of the product of strength, column density, and pressure-broadening coefficient (typically between 0.05 and 0.1 cm^{-1} at 1 atm) for "strong" features. The results, for historical reasons, is termed the *equivalent width*. The calculation is useful because if the equivalent width of a band is greater than about 10% of its half-width (i.e., the spectral range over which the apparent absorption lies within 50% of the peak), it may be expected that many of the individual lines in the band will be totally opaque.

2.3.2 Carbon Dioxide

Although, as has been widely reported, the abundance of CO_2 in the atmosphere is steadily increasing and, in addition, shows significant geographic and seasonal variation, for most purposes, it is adequate to assume a uniform volume mixing ratio of 3.30×10^{-4} up to at least 80 km (see Figure 2.4). The major bands of CO_2 are shown in Table 2.2. Because of the near constancy of the CO_2 profile, the strong bands around 667 and 2340 cm^{-1} are widely exploited for temperature sounding of the atmosphere. Furthermore, CO_2 is strongly evident in the atmospheres of Mars and Venus. Thus, its spectrum is one of the best known of all atmospheric species; more than 500 bands have been measured for this one molecule alone. Therefore, Table 2.2 shows but a small fraction of those known.

If the wings of the bands around 2340 cm^{-1} are important, it has been suggested by several authors that a sub-Lorentzian line shape should be employed. Discussion of this topic can be found in Le Doucen et al. (1985).

Table 2.3. Principal Infrared Bands of Ozone (O_3)

Band Origin (cm^{-1})	Isotope Code[a]	Upper Vibration State, V′	Lower Vibration State, V″	Integrated Band Strength (cm^{-1}/molecule cm^{-2}) at 296 K
1007.65	666	101	100	6.15E-20
1008.00	686	001	000	2.50E-20
1015.81	666	002	001	1.74E-19
1025.60	666	011	010	4.50E-19
1028.10	668	001	000	5.07E-20
1042.08	666	001	000	1.39E-17
1095.33	666	110	010	1.10E-20
1103.14	666	100	000	6.71E-19
1726.53	666	011	000	5.37E-20
1796.26	666	110	000	2.27E-20
2057.89	666	002	000	1.11E-19
2110.79	666	101	000	1.13E-18
2201.16	666	200	000	3.00E-20
2785.24	666	111	000	2.32E-20
3041.20	666	003	000	1.11E-19

[a] 666 = $^{16}O\,^{16}O\,^{16}O$; 686 = $^{16}O\,^{18}O\,^{16}O$; 668 = $^{16}O\,^{16}O\,^{18}O$. Vibrational state code = $v_1v_2v_3$.

2.3.3 Ozone

Ozone is another strongly absorbing species. Normally confined to the upper troposphere–lower stratosphere region (where it is, however, strongly variable with time and place), a feature of the modern urban environment can be a marked enhancement of O_3 concentration near ground level. Furthermore, O_3 is at least partially responsible for significant temperature fluctuations near the tropopause. Thus, observers in spectral regions that might be impacted by atmospheric thermal emission should use caution in the employment of climatological model atmospheres (which generally ignore the effect) for computation of transmission–emission.

Table 2.3 shows the major IR O_3 bands, and Figure 2.5 shows an average vertical profile.

Ozone is also the principal opacity source for the UV cutoff of solar radiation by the atmosphere, although Rayleigh scattering also plays a significant role. Allen (1973) gives an empirical relation for the important cutoff region that, in wavelength and column density terms, becomes

$$t = \exp[-6.186 \times 10^{19} \times N_{O_3}\gamma] \tag{2.15}$$

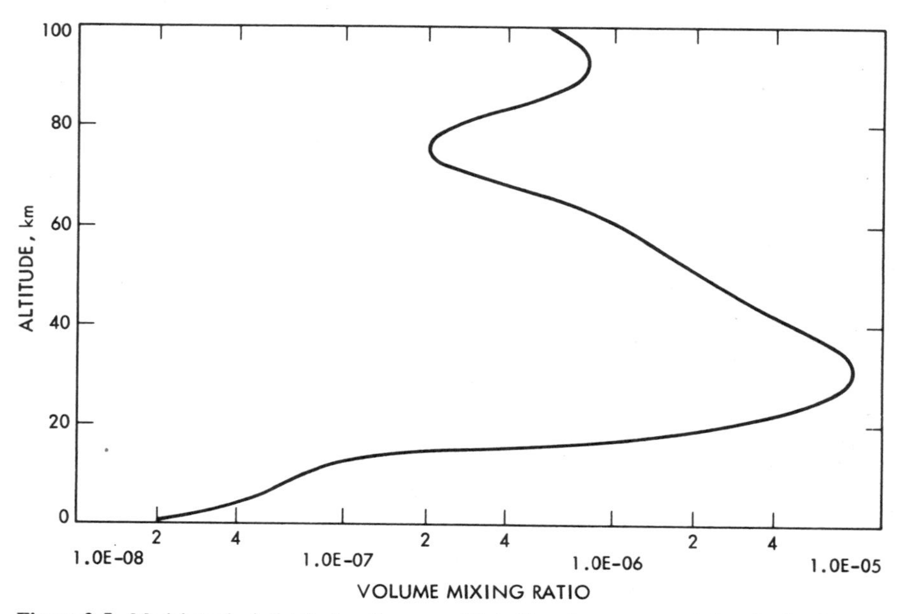

Figure 2.5. Model vertical distribution for ozone (O_3). Note that strong variations in distribution near peak are common, and significant enhancements in near-surface concentration occur in urban environments.

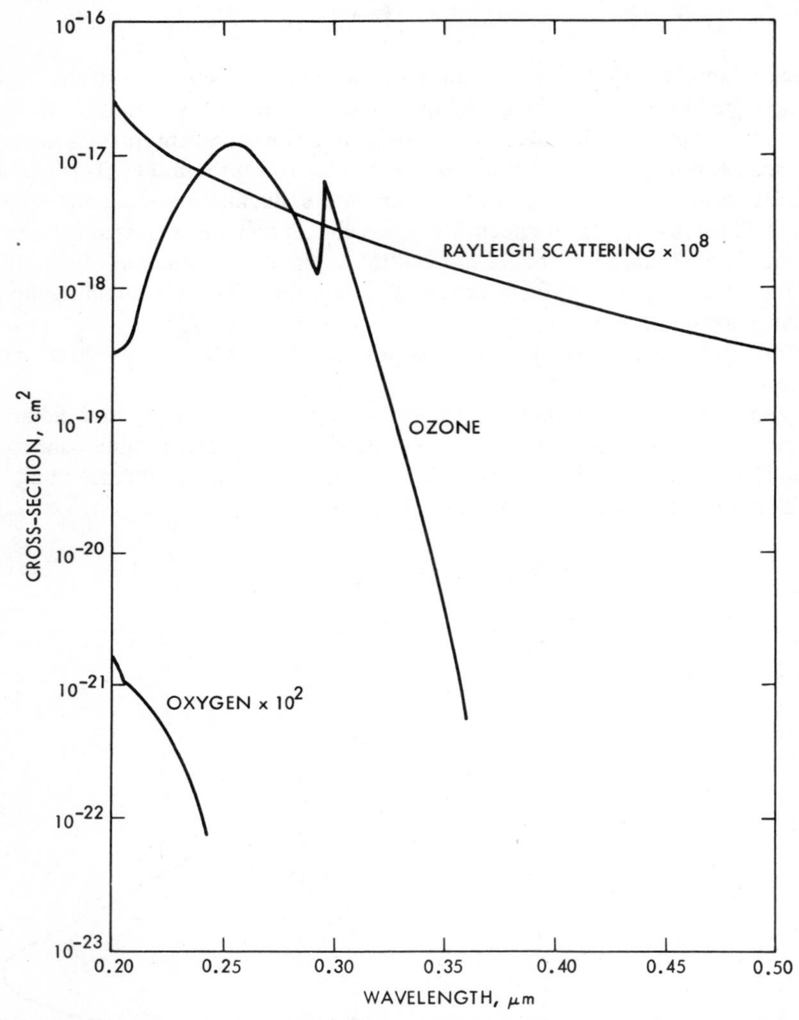

Figure 2.6. Visible–UV cross sections for Rayleigh (molecular) scattering and ozone and oxygen extinction. Note different scale factors for three curves.

where

$$\log(\gamma) = 17.58 - 56.4\lambda \quad (\lambda \text{ in } \mu\text{m}) \tag{2.16}$$

and N_{O_3} is the ozone column density ($\sim 8 \times 10^{18}$ molecules cm^{-2} in a standard vertical transatmospheric column). Figure 2.6 shows extinction cross sections for O_3 in the visible and near UV measured over relatively broad bands (~ 500 cm^{-1}). Higher resolution data can be found in Hudson (1971), for example.

An alternative approach is available through the LOWTRAN code, but the authors of this code warn that the results can be as much as 10% in error. Unfortunately, no line-by-line data for the ultraviolet ozone bands are known (to the author).

2.3.4 Nitrous Oxide

Nitrous oxide is reasonably well mixed in the troposphere (0.35 ppm), but the mixing ratio declines steadily above about 20 km (see Figure 2.7). Being a linear molecule, its spectrum is relatively sparse so that even in regions of strong

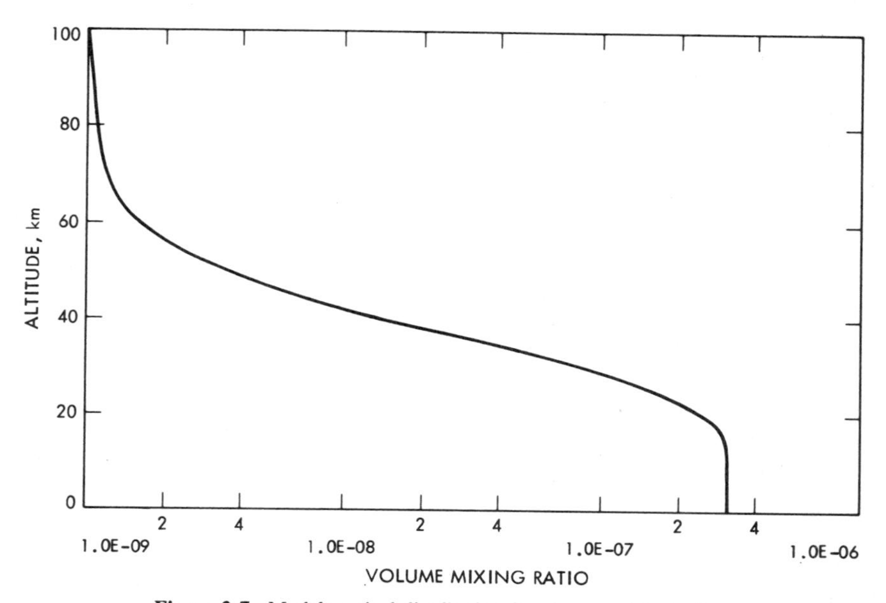

Figure 2.7. Model vertical distribution for nitrous oxide (N_2O).

Table 2.4. Principal Infrared Bands of Nitrous Oxide (N_2O)

Band Origin (cm^{-1})	Isotope Code[a]	Upper Vibration State, V'	Lower Vibration State, V''	Integrated Band Strength (cm^{-1}/molecule cm^{-2}) at 296 K
588.77	446	0110	0000	9.86E-19
588.98	446	0220	0110	1.11E-19
1160.30	446	0310	0110	1.19E-19
1168.13	446	0200	0000	3.52E-19
1284.90	446	1000	0000	8.80E-18
1291.50	446	1110	0110	9.93E-19
2177.66	456	0001	0000	1.81E-19
2195.39	446	0221	0220	3.15E-19
2195.84	446	0201	0200	1.68E-19
2195.91	446	1001	1000	1.02E-19
2209.52	446	0111	0110	5.67E-18
2223.76	446	0001	0000	5.03E-17
2462.00	446	1200	0000	2.76E-19
2563.34	446	2000	0000	1.20E-18
2577.09	446	2110	0110	1.21E-19
3473.21	446	1111	0110	1.92E-19
3480.82	446	1001	0000	1.73E-18

[a]446 = $^{14}N\ ^{14}N\ ^{16}O$; 456 = $^{14}N\ ^{15}N\ ^{16}O$. Vibrational state code = $v_1 v_2 l v_3$.

N_2O bands it is often possible to find microwindows suitable for remote sensing. Table 2.4 lists the major bands.

2.3.5 Methane

The mixing ratio of CH_4, currently about 1.3–1.6 ppm in the troposphere, is reported to be increasing at a rate of 1 or 2% per annum. Furthermore, significant local enhancements are observed over sanitary landfills (Grant, 1982) and other areas of decaying organic matter such as swamps and wetlands. Actually, no complete inventory of sources and sinks of atmospheric methane exists, so to some extent, its very presence in the free oxidizing atmosphere is mysterious. It is, however, a very well studied molecule both because of its unusual spectroscopically spherical symmetry and also because of its dominant influence on the spectra of outer-planet atmospheres where even the rare monodeuterated version is readily detected. Figure 2.8 shows a model vertical profile for CH_4, and Table 2.5 lists the band prominent in Earth's atmosphere.

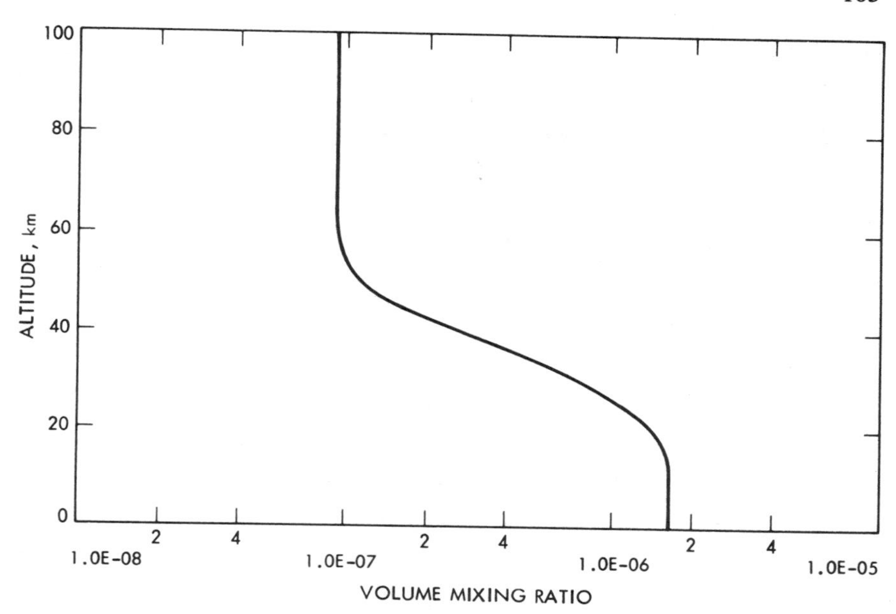

Figure 2.8. Model vertical distribution for methane (CH_4). Note that evidence exists for secular increase in CH_4 abundance, and significant enhancements are observed over areas of decaying organic matter.

2.3.6 Carbon Monoxide

Carbon monoxide, structurally the simplest of the major atmospheric absorbers, is well mixed in the troposphere (10 ppb), but as any city dweller knows, it is subject to dramatic enhancements in urban environments. However, the general background level is itself sufficient for a number of lines of the 1-0 band near 2120 cm^{-1} to be saturated. The overtone bands (see Table 2.6) are much weaker. Furthermore, the line spacings are enough that significant microwindows are available. A model vertical profile for CO is shown in Figure 2.9.

2.3.7 Oxygen

Oxygen, being a homonuclear diatomic molecule, has no permanent electric dipole moment; thus, there are no conventional IR vibrational–rotational bands. However, its electronic transitions are significant at the red end of the visible spectrum and, for lines of sight free of O_3, is a contributor to the ultimate UV cutoff (at about 0.2 μm for a 1-km horizontal, clear, path at sea level).

Table 2.5. Principal Infrared Bands of Methane (CH_4)

Band Origin (cm^{-1})	Isotope Code[a]	Upper Vibration State, V'	Lower Vibration State, V''	Integrated Band Strength (cm^{-1}/molecule cm^{-2}) at 296 K
1310.76	211	00000111	00000000	5.04E-18
1533.34	211	01100001	00000000	5.50E-20
2612	211	00000222	00000000	5.37E-20
2822	311	01100112	00000000	4.30E-20
2830	211	01100112	00000000	3.80E-19
3009.05	311	00011001	00000000	2.93E-19
3018.92	211	00011001	00000000	1.02E-17
3062	211	02200002	00000000	1.64E-19
4223.50	211	10000111	00000000	2.40E-19
4340	211	00011112	00000000	4.08E-19
4540	211	01111002	00000000	6.25E-20
6004.99	211	00022002	00000000	2.05E-20

[a] 211 = $^{12}CH_4$; 311 = $^{13}CH_4$. Vibrational state code = $v_1 v_2 l_2 v_3 l_3 v_4 l_4$. *Note:* There are also two unidentified bands of CH_4, one between 2430 and 3195 cm^{-1}, strength, 1.20E-20, and another at 4136–4666 cm^{-1}, strength 8.46E-20.

Table 2.6. Principal Infrared Bands of Carbon Monoxide (CO)

Band Origin (cm^{-1})	Isotope Code[a]	Upper Vibration State, V'	Lower Vibration State, V''	Integrated Band Strength (cm^{-1}/molecule cm^{-2}) at 296 K
2092.12	28	1	0	1.91E-20
2096.07	36	1	0	1.05E-19
2143.27	26	1	0	9.81E-18
4260.06	26	2	0	7.52E-20

[a] 26 = $^{12}C\ ^{16}O$; 36 = $^{13}C\ ^{16}O$; 28 = $^{12}C\ ^{18}O$.

Table 2.7 shows the origins of the near-IR electronic bands of O_2, generally known as the A, B, and γ atmospheric bands. The volume-mixing ratio is 0.20947. Broadband UV absorption cross sections are shown in Figure 2.6, and higher resolution data can be found in Hudson (1971).

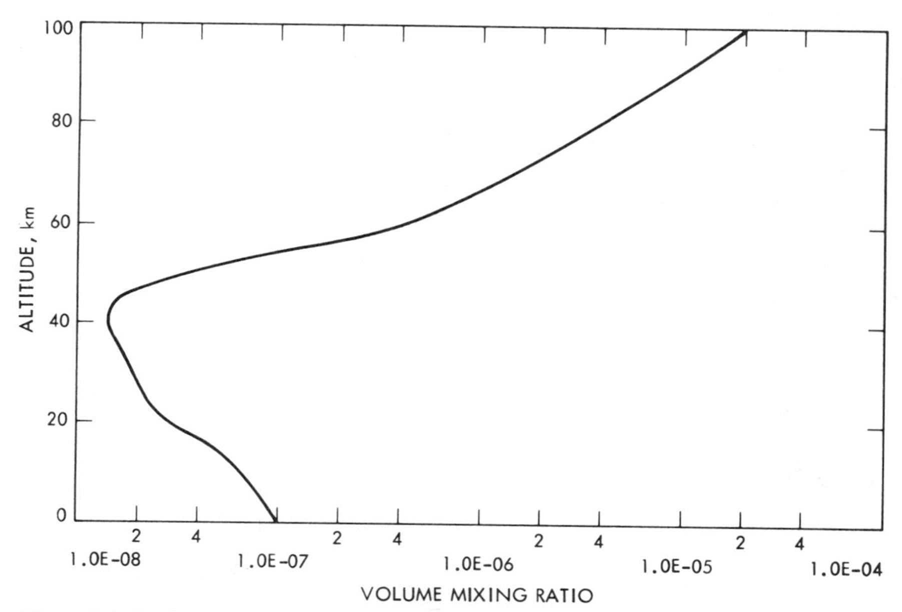

Figure 2.9. Model vertical distribution for carbon monoxide (CO). Significant enhancements are discernible in urban environments.

Table 2.7. Principal Bands of Oxygen (O_2)

Band Origin (cm^{-1})	Isotope Code[a]	Upper Vibration State, V'	Lower Vibration State, V''	Integrated Band Strength (cm^{-1}/molecule cm^{-2}) at 296 K
7882.43	66	AX 0	AX 0	1.82E-24
13120.91	66	BX 0	BX 0	1.95E-22
13122.97	68	BX 0	BX 0	7.92E-25
14525.66	66	BX 1	BX 0	1.22E-23
15902.42	66	BX 2	BX 0	3.78E-25

[a] 66 = $^{16}O\ ^{16}O$; 68 = $^{16}O\ ^{18}O$.

2.4 MOLECULAR CONTINUUM ABSORPTION

Molecular continuum absorption has traditionally incorporated the effects of the far wings of very strong lines and the cumulative effects of the myriad of lines individually too weak to be discernible. The advent of efficient line-by-line codes has rendered this definition obsolete (insofar as far wings are well rep-

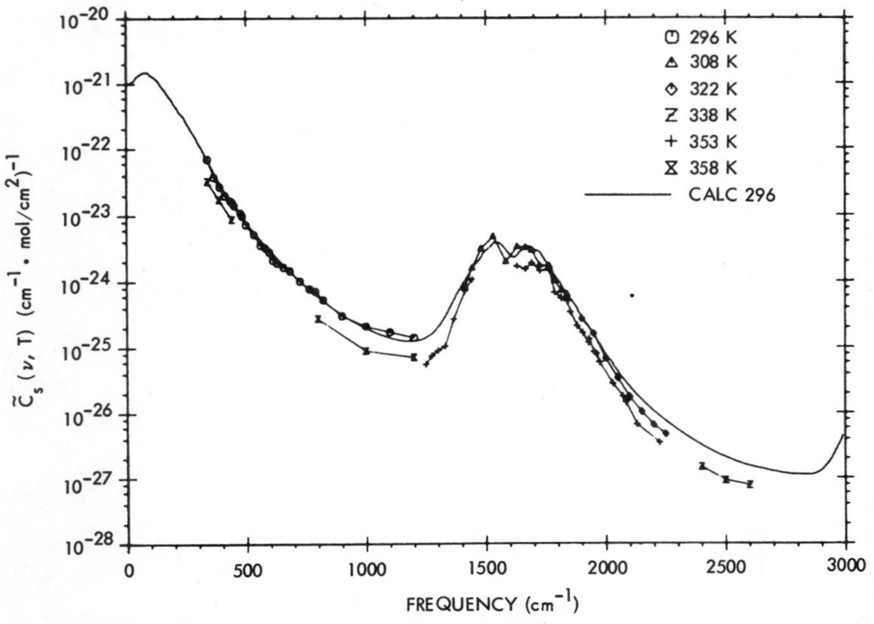

Figure 2.10. Tabulated self-broadened absorption coefficient for water vapor continuum compared to measurements. Data from LOWTRAN 6 manual. Figure courtesy of the Air Force Geophysics Laboratory.

resented by the chosen lineshape function). The term is now reserved for the "density-squared"-dependent phenomena of which the two most important are due to water vapor and molecular nitrogen.

2.4.1 Water Vapor Continuum

Infrared remote sensing through the lower troposphere is strongly impacted by the water vapor continuum. Under some conditions (e.g., humid tropical climates), the continuum absorption can reach 80% for a transatmospheric path.

The origin of the effect is highly controversial. Some authors (e.g., Dianov-Klokov et al., 1981) believe it to be due to dimers of H_2O, others (e.g., Burch and Gryvnak, 1980) that the origin lies in abnormally strong wings of known lines of normal H_2O vapor, and yet others that both phenomena are active (Loper et al., 1983). In any event, the consequence is that there is no sound theoretical basis to describe the effect. Thus, its temperature dependence, which is substantial, depends on the interpolation of empirical data. Kneizys et al. (1983) offer the expression

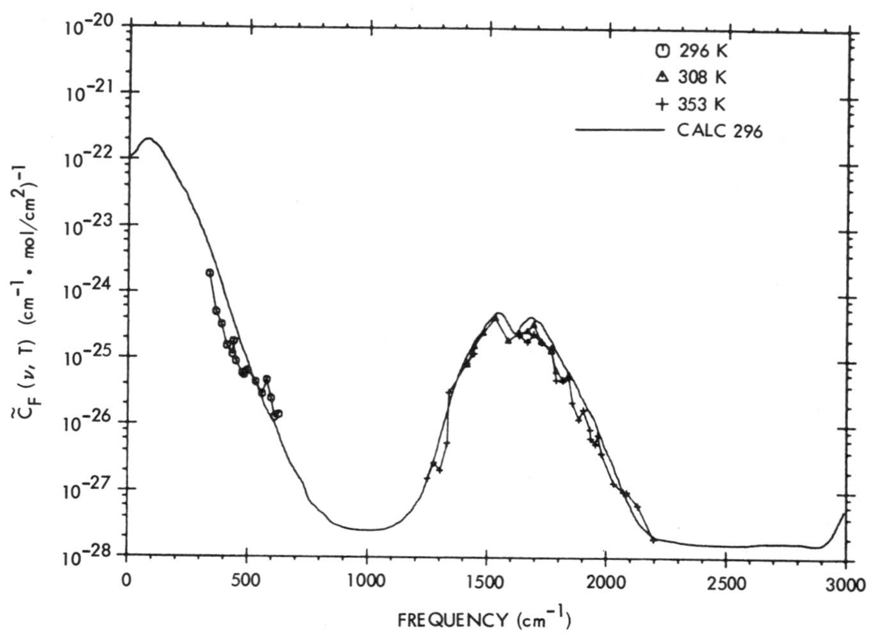

Figure 2.11. Tabulated air-broadened absorption coefficient for water vapor continuum compared to measurements. Data from LOWTRAN 6 manual. Figure courtesy of the Air Force Geophysics Laboratory.

$$\kappa(\tilde{\nu}, T) = N_s\tilde{\nu} \tanh(c_1\tilde{\nu}/2T)[(N_s/N_0)C_s(\tilde{\nu}, T) + (N_f/N_0)C_f(\tilde{\nu}, T)] \tag{2.17}$$

where $\kappa(\tilde{\nu})$ is the absorption coefficient at wavenumber $\tilde{\nu}$ and temperature T, $c_1 = 1.43879$, N_s is the water vapor number density, N_f the number density of all other species, and $N_0 = N_s + N_f$. The terms C_s and C_f are empirical parameters for the self and foreign coefficients that are derived from laboratory measurements. LOWTRAN 6 contains massive tables of values for C_s and C_f at two different temperatures (260 and 296 K). Figures 2.10 and 2.11, from Kneizys et al., shows how some of the tabulated data fit the experimental values. While a considerable improvement over earlier versions (such as appeared in LOWTRAN 5), other workers (Ben-Shalom et al., 1985) suggest that the results in some spectral regions may be in error by as much as 20%. For the important

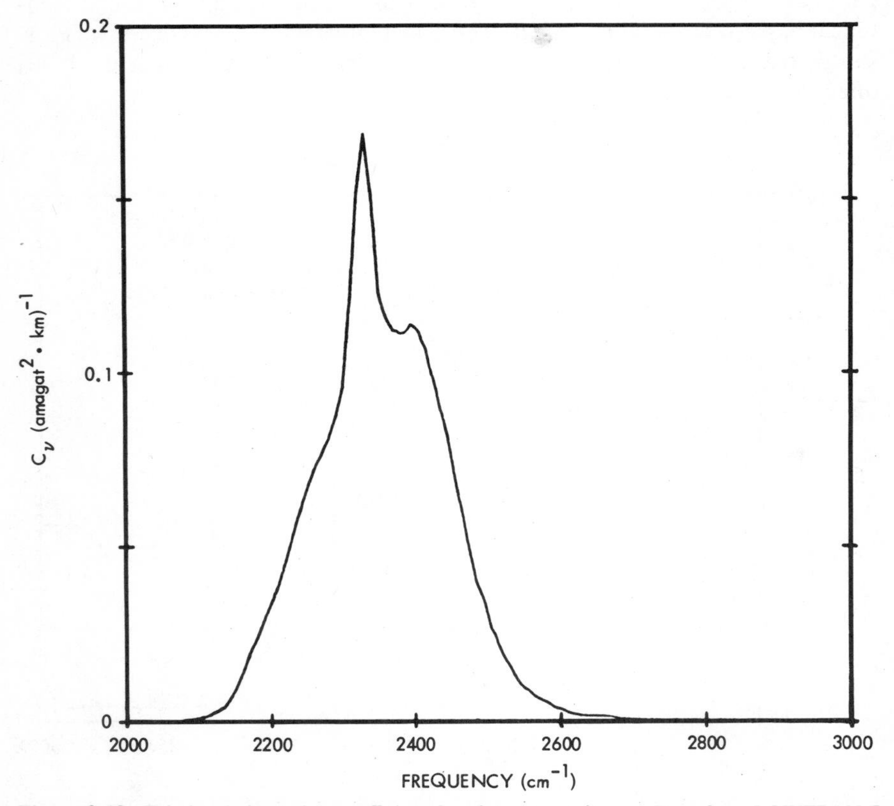

Figure 2.12. Tabulated absorption coefficient for nitrogen continuum. Data from LOWTRAN 5 manual. Figure courtesy of the Air Force Geophysics Laboratory.

region 900–1000 cm^{-1}, Loper et al. (1983) give coefficients for specific CO_2 laser lines that might be substituted in particular applications.

Clearly, much more experimental and theoretical work is needed on this most important topic. A recent review may be found in Clough et al. (1986).

2.4.2 Nitrogen Continuum

The absorption continuum due to molecular nitrogen has a different origin: Collisions induce a transient dipole moment into the molecule that can therefore interact with the radiation field. Peak absorption occurs near 2350 cm^{-1}, and the range of activity is roughly 2100–2700 cm^{-1} (see Figure 2.12). Tables from which absorption coefficients can be determined accompany the LOWTRAN code. These should be adequate even for line-by-line codes because the absorption is relatively unstructured. However, superimposed on the continuum are sharp features of the 1-0 N_2 quadrupole spectrum (notably near 2400 cm^{-1}) that, while weak, are clearly evident in high-resolution solar spectra. Parameters for these lines can be found in the Air Force Geophysical Laboratory (AFGL) compilation.

2.5 SCATTERING

In the utilitarian spirit of this chapter, scattering will not be treated in an exact fashion (insofar as "exact" has any meaning in this field, see Section 2.2). Furthermore, multiple scattering will be ignored on the grounds that the prevailing observation conditions wherein multiple scattering would be important are unlikely to be acceptable. It must be recognized, however, that multiple scattering always increases the attenuation so the results presented below will generally be upper limits for the atmospheric transmittance.

2.5.1 Rayleigh (Molecular) Scattering

For remote sensing in the visible and UV spectral regions, molecular scattering can be an important source of extinction. The classic expression for the Rayleigh scattering cross section is

$$\sigma_s(\tilde{\nu}) = (8\pi^3\tilde{\nu}^4/3)[(n^2 - 1)^2/N_0^2][(6 + 3\Delta)/(6 - 7\Delta)] \quad \text{cm}^2 \quad (2.18)$$

for $\tilde{\nu}$ in reciprocal centimeters. According to Young (1980), the depolarization ratio $\Delta = 0.0279$ for dry air and unpolarized incident light.

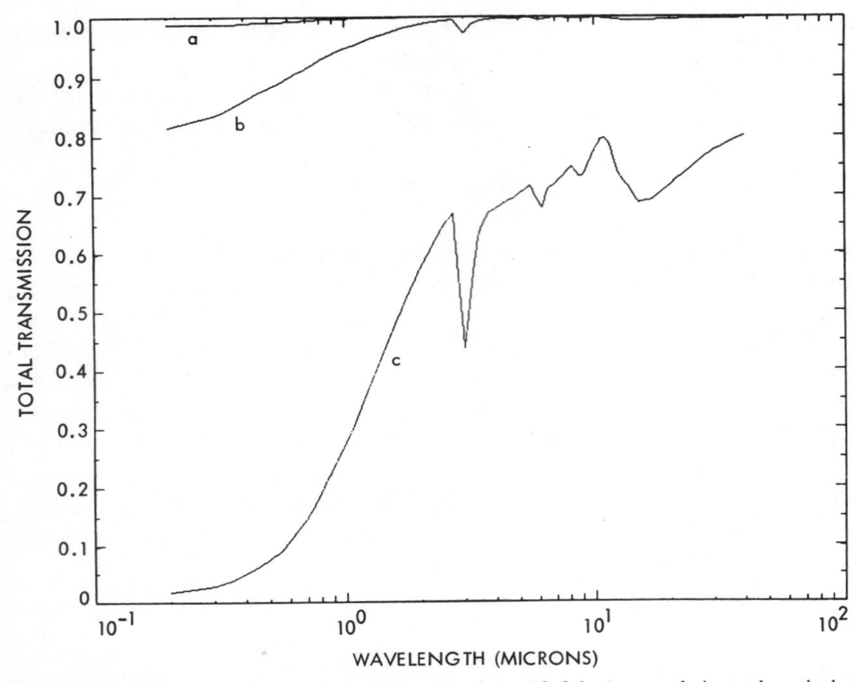

Figure 2.13. Illustration of importance of boundary layer (0–2 km) aerosols in total vertical atmospheric column. Shown are transmission spectra for LOWTRAN 5 "rural" aerosol model; no molecular extinction is included. Conditions: spring–summer, "normal" stratospheric and upper atmospheric aerosol loading, no fog, relative humidity (RH) 100%, and horizontal surface visibility $V(\lambda = 0.55\ \mu m) = 2$ km. Three vertical paths are 10–100 km (curve a), 2–100 km (curve b), and 0–100 km (curve c). Prominent feature at 3 μm in c arises from water condensation on surface of aerosols.

The standard formula for the refractive index of moist air is given by Edlen (1966):

$$(n-1) \times 10^6 = \left(88.43 + \frac{185.08}{1 - (\tilde{\nu}/1.14 \times 10^5)^2} + \frac{4.11}{1 - (\tilde{\nu}/6.24 \times 10^4)^2}\right) \frac{P - P_w}{P_0} \frac{296.15}{T} + \left[43.49 - \left(\frac{\tilde{\nu}}{1.7 \times 10^4}\right)^2\right] \frac{P_w}{P_0} \qquad (2.19)$$

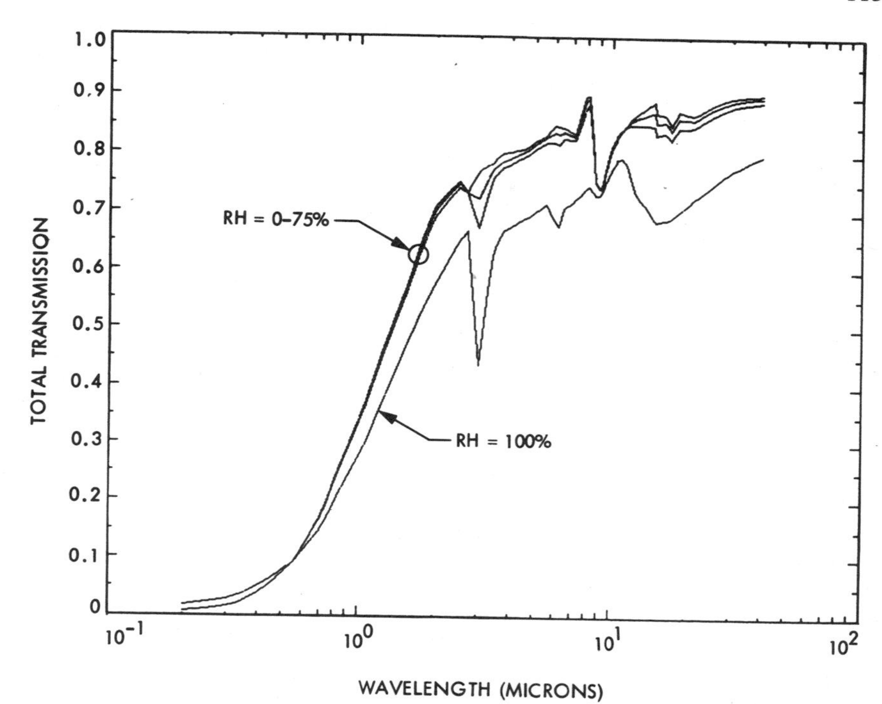

Figure 2.14. Effects of changes in relative humidity (RH) on atmospheric aerosols for same model as in Figure 2.13. Thus, curve *c* from Figure 2.13 (RH 100%) is reproduced here. Other models correspond to full-column vertical transmittances for relative humidities between 0 and 75%. Note development of 3- and 6-μm liquid water features and masking by water of strong 9-μm sulfate dispersion when aerosols are "wet" (RH 100%).

where P and T are the ambient pressure and temperature, $P_0 = 1$ atm $= 1013.25$ mbars $= 760$ mm Hg, and P_w is the partial pressure of water vapor. Note that N_0, the number density of air in Equation (2.18), is to be evaluated at the same P and T as the refractive index. Figure 2.6 shows a plot of σ_s through the visible and near UV.

The transmittance, then, is

$$t(\tilde{\nu}) = \exp[-N\sigma_s(\tilde{\nu})] \tag{2.20}$$

where $N \sim 2.2 \times 10^{25}$ for a vertical transatmospheric column
$\sim 2.7 \times 10^{24}$ for a 1-km horizontal path at 500 m elevation

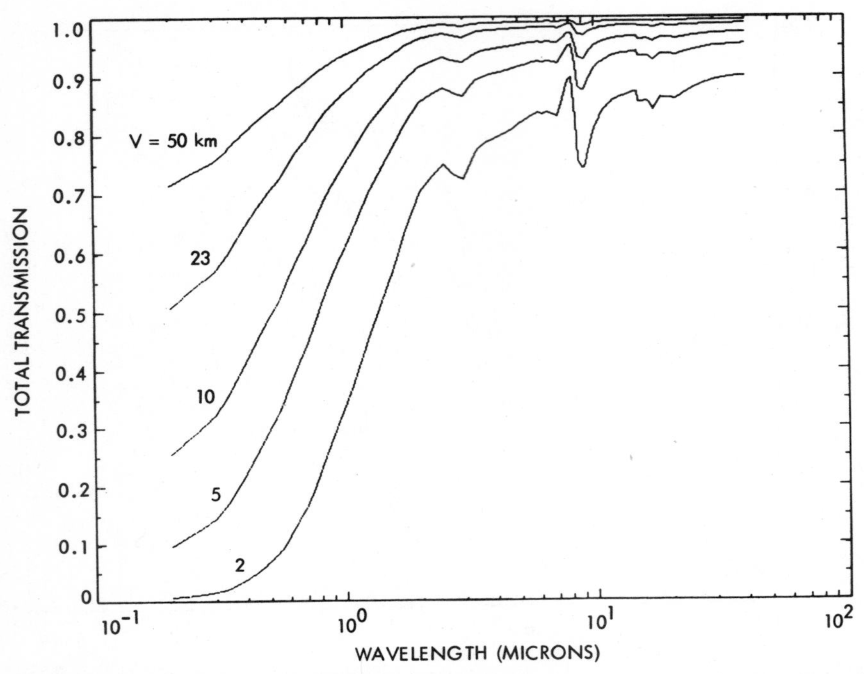

Figure 2.15. Effects of variations in sea-level horizontal visibility V ($\lambda = 0.55\ \mu$m) for rural model with RH 50% (other parameters as in Figure 2.13).

2.5.2 Aerosol and Particulate Scattering

Figures 2.13–2.21 show the wavenumber dependence of atmospheric transmittance based on LOWTRAN 5 models of various aerosols, hazes, and particulates. The parameterization is in terms of *meteorological range*, V, defined as

$$V = \frac{\ln(1/\epsilon)}{\beta} \tag{2.21}$$

where β is the extinction coefficient (km^{-1}) and ϵ is a threshold contrast. LOWTRAN sets $\epsilon = 0.02$, whence

$$V = 3.912/\beta \tag{2.22}$$

Generally, a value of V will not be available; however, according to the LOW-

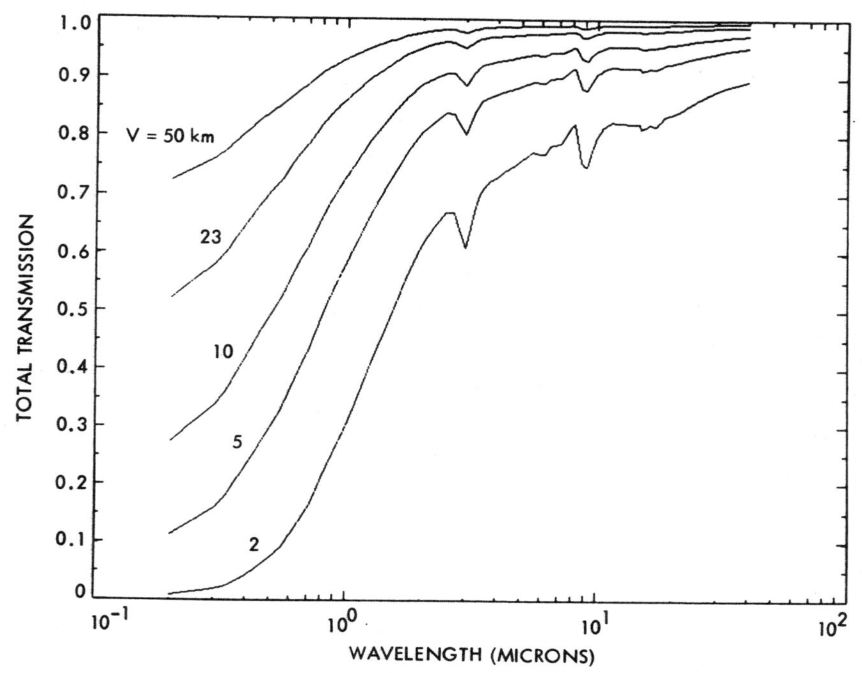

Figure 2.16. Parameters as in Figure 2.15 except LOWTRAN 5 "urban" aerosol model was used.

TRAN 5 manual, V can be estimated from a subjective measure of *horizontal visibility* V_n by

$$V \sim (1.3 \pm 0.3)V_n \tag{2.23}$$

where V_n is a naked-eye observation limit of a dark object against the horizon sky in daylight and a moderately intense light at night. Consequently, Figures 2.13–2.21 should be used for general reference only. For more specific, or purely horizontal, observing conditions, LOWTRAN itself can be invoked. By and large, aerosol extinction shows no very sharp features (on the scale of a molecular linewidth) so the low resolution of the LOWTRAN code should be a problem.

If necessary, recourse may be had to Mie scattering theory or its extensions for other, regularly shaped particles. Hansen and Travis (1974) and Bohren and Huffman (1983) offer complete FORTRAN listings of programs for calculating the scattering by homogeneous dielectric spheres. In addition, Bohren and Huff-

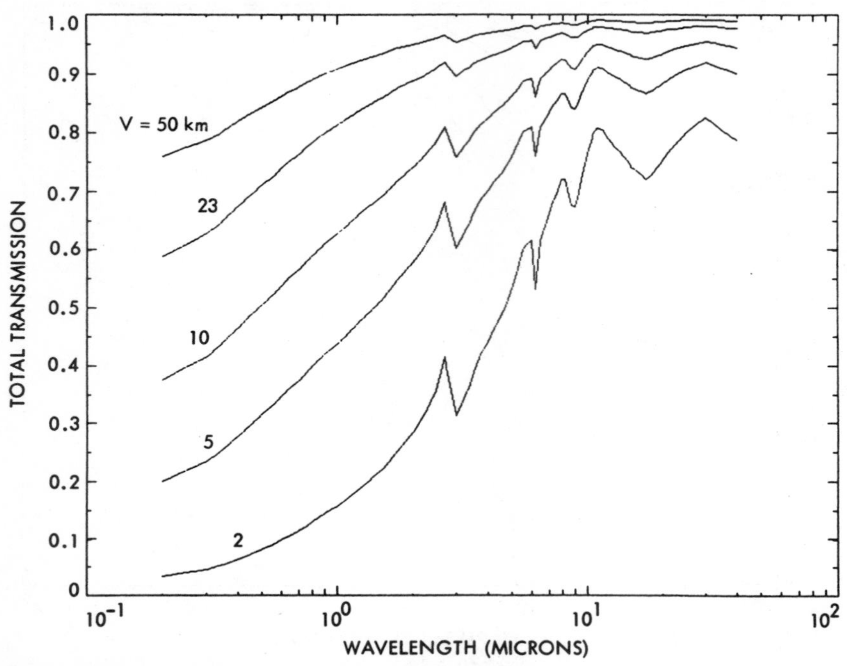

Figure 2.17. Parameters as in Figure 2.15 except LOWTRAN 5 "maritime" aerosol model was used.

man list programs to treat coated spheres and normally illuminated infinite cylinders. If a simple approximation will suffice (Mie scattering programs tend to be very time consuming and computationally intensive), Gordon (1985) offers the expression

$$I(\theta) = I(0)\left[\left|3J_1(X)\right|/X + X^{-3/2}\right]^2[1 + \cos^2\theta]/2 \qquad (2.24)$$

where θ is the scattering angle (forward = 0), $J_1(X) = (\sin X - X\cos X)/X$, $X = ka(1 + n - 2n\cos\theta)^{1/4}$, $k = 2\pi\tilde{\nu}$, a is the sphere radius, and n is the refractive index. The expression results in errors of up to a factor of 3 for $\theta < 30°$ but seems quite good for large scattering angles.

No general solution exists for particles of arbitrary size, shape, and complex refractive index. However, particles that are not *too* different from spheres and are comparable in size to the wavelength can be treated with a semiempirical theory developed by Pollack and Cuzzi (1980). Their theory fits reasonably well

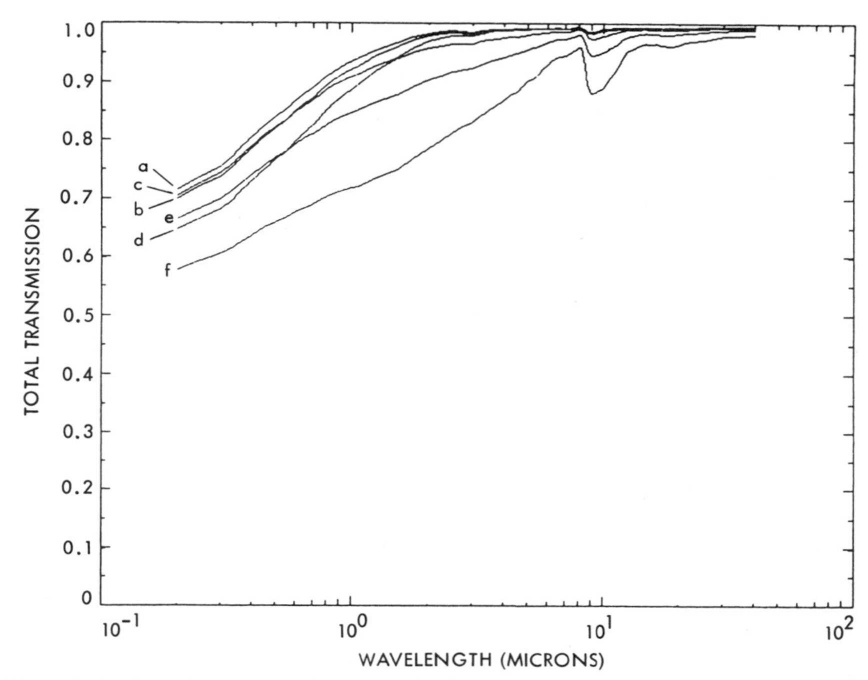

Figure 2.18. Possible influence of volcanic dust injected into stratosphere (10–30 km). Conditions: rural boundary layer model, RH 50%, $V = 50$ km, spring–summer, and normal upper atmosphere ($z > 30$ km) aerosol extinction. Curve *a* represents "background" (nonvolcanic) aerosol loading assuming, respectively, aged and fresh particle size distribution. Models *a* and *e* show "high" loading for two-size regimes, and *f* is "extreme" case with fresh size distribution.

to a suprising variety of shapes, but they caution that a combination of extreme shape (e.g., needle shaped, like some cirrus cloud particles) and a large imaginary part to the refractive index will give erroneous results.

2.6 CLOUDS

In the planning of transatmospheric experiments, an important consideration must be the likelihood that clouds will interfere with the measurements. Maps of the probability that a given line of sight will be free of clouds (commonly called the CFLOS probability) have been published by Lund et al. (1977) for much of the world for various time of day, season, and view angle. A more

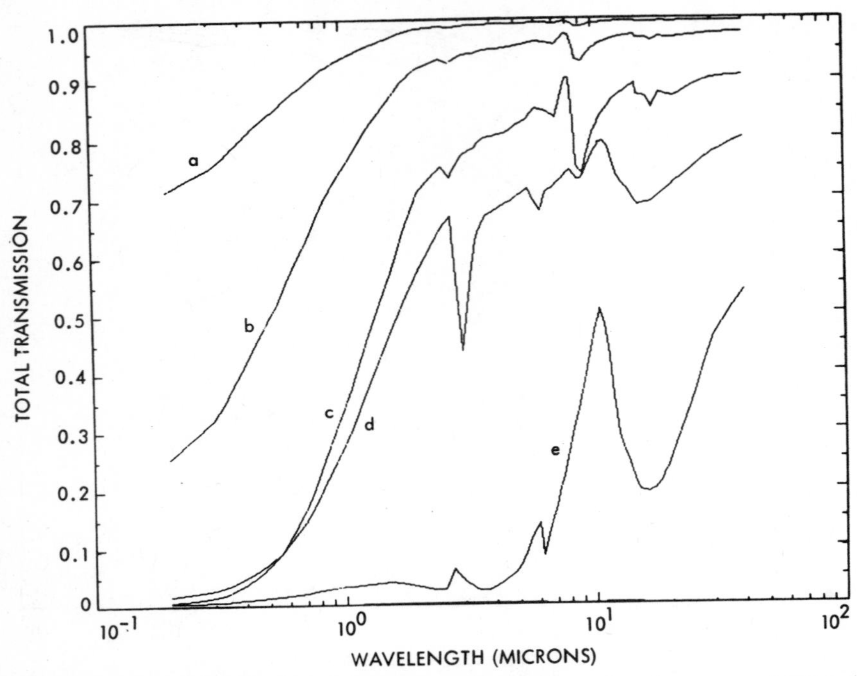

Figure 2.19. Rural model spectra for range of conditions: curves a, b, and c correspond to "dry" particles (RH 0%) and three horizontal visibilities V of 50, 10, and 2 km, respectively. Curve d is the same as c but with relative humidity increased to 100%. A "radiation" fog was then added to obtain model e, in which $V_{\text{fog}} = 0.5$ km, and the vertical extent was assumed to be 0.3 km. All models here assume spring–summer and normal (or background) stratospheric and upper atmospheric aerosol loading.

extensive (and current) three-dimensional cloud morphology 3DNEPH is maintained by the U.S. Air Force Environmental Technical Applications Center.

An additional important factor for remote sensing is the probability that a break in the cloud cover will be large enough (or last long enough) for an observation sequence to be completed. This topic has been addressed by Malick et al. (1979). They provide empirical expressions that relate observations of fractional cloud cover S to the angular dependence of CFLOS probability:

$$\log[C(\theta)]/\log[C(0)] = 1 + b \tan\theta \qquad (2.25)$$

where $C(\theta)$ is the CFLOS probability at the nadir/zenith angle θ, $C(0)$ is the vertical CFLOS probability given by

$$C(0) = 1 - S(1 + 35)/4 \qquad (2.26)$$

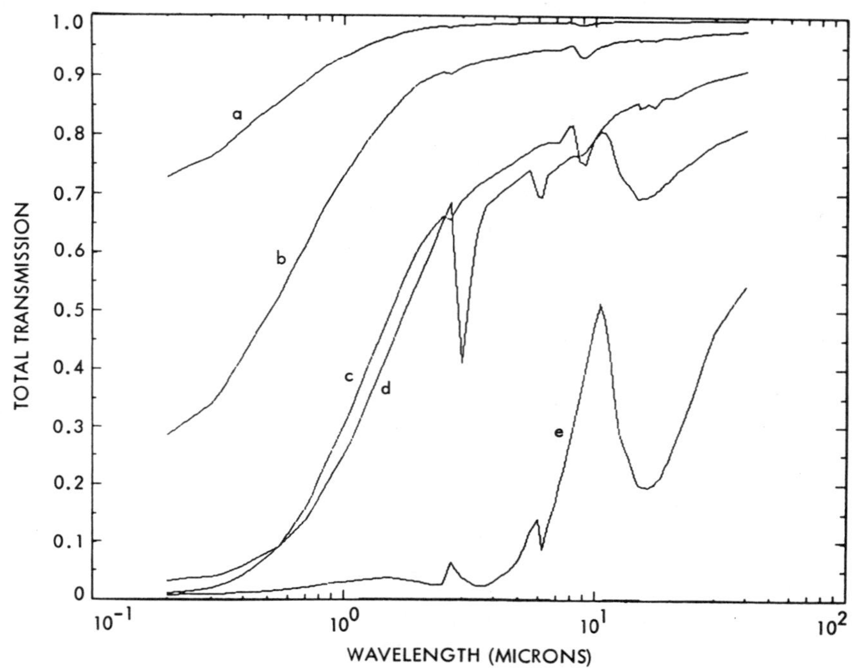

Figure 2.20. Parameters as in Figure 2.19 except LOWTRAN 5 urban aerosol model was used.

and b is a factor,

$$b = 0.275S \tag{2.27}$$

It is claimed that Equation (2.25) fits observed data to good accuracy.

In turn, the CFLOS probability C is related to the size L (in kilometers) of average clear intervals by two expressions:

$$L(0 \leq C \leq 0.5) = 8C + 1.2 \tag{2.28}$$

and

$$L(0.5 < C \leq 1.0) = C(9.2 - C)/(1 - C) \tag{2.29}$$

From these expressions we may deduce that with a cloud cover of 50% ($S = 0.5$), the CFLOS probability is 0.6875 in the vertical and 0.5142 at 45°, and the corresponding average (apparent) clear intervals are 18.7 and 9.2 km, respectively.

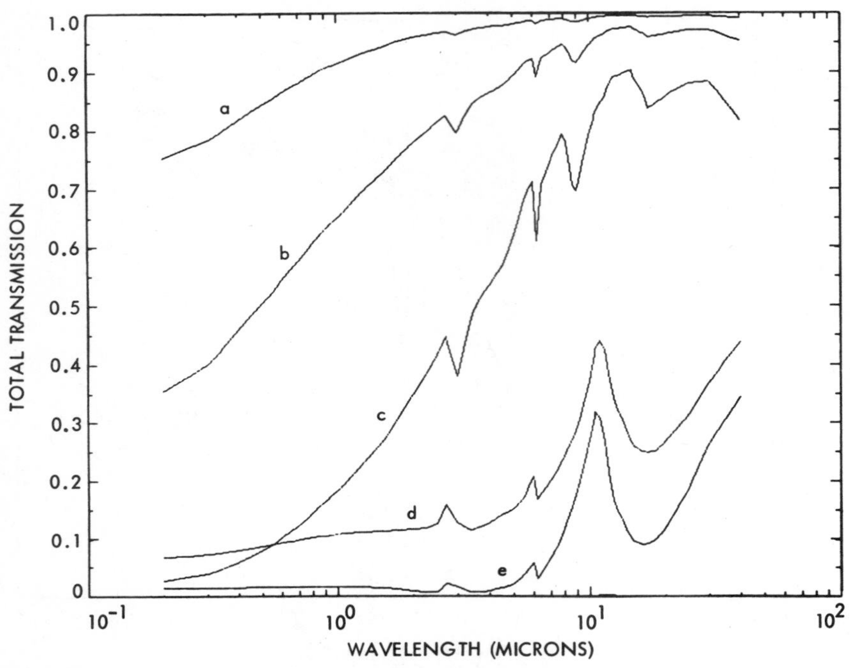

Figure 2.21. Parameters as in Figure 2.19 except LOWTRAN 5 maritime aerosol model was used.

2.7 CALCULATION OF ATMOSPHERIC TRANSMITTANCE

There are many line-by-line programs in current use; AFGL, for example, distributes one called FASCODE. Described in this section is a JPL code called EMISSION-SPECTRA (ES) that differs from most in that it is a reasonably user-friendly, interactive program and is specifically designed not only to predict atmospheric spectra with any viewing geometry but also to compare the model spectra with real data. It was developed from a simpler program called DISPLAY-SPECTRA (DS) that is primarily designed to model stratospheric spectra obtained in the "solar occultation" mode from balloons and spacecraft. ES and DS do, however, have many features in common and, above all, share the same data bases. ES incorporates atmospheric emission, target emission, and solar reflection from the target and therefore runs significantly slower than DS. Assuming, however, that these "background" effects may be significant even in laser remote sensing, the more extended code is outlined. Figure 2.22 shows a general flowchart for the program.

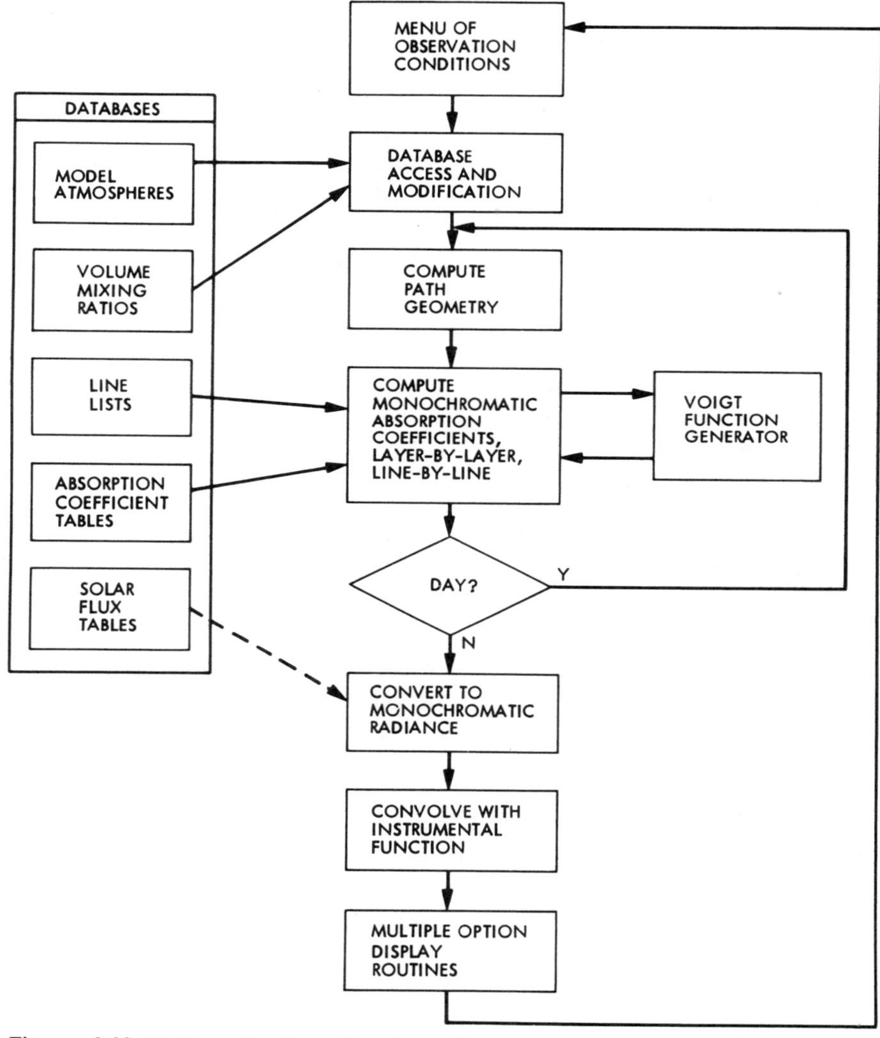

Figure 2.22. Outline flowchart for interactive spectrum modeling program EMISSION__SPECTRA. Various elements are described in text.

The major elements of the program are entry, data base access and modify, path geometry, absorption coefficients, monochromatic radiance calculation, convolution with instrumental function, and display options.

Entry. Entry is via a terminal menu that basically requests the observation conditions (observer's position and initial zenith angle of the line of sight, atmospheric model, solar zenith angle, target brightness temperature and geometric albedo, spectral range, and resolution and display dispersion). Some of these parameters may well have required some prior analysis, but defaults are provided for all values, and changes can be made at any time.

Data Base Access and Modify. The data base has five major elements:

1. *Model atmospheres.* Five models are standard with the option of replacement by a user model or real-time (but temporary) modification of any of the foregoing. The models tabulate temperature, pressure, and total number density as a function of altitude in 1-km increments for 0–100 km.
2. *Volume mixing ratios.* Some 40 known constituents are tabulated in 1-km altitude increments, and space exists for up to 20 more user constituents. All can be modified or replaced on-line.
3. *Line lists.* The basic line list of frequency, strength, pressure-broadening coefficient, lower state energy, isotopic identification, and upper and lower state quantum numbers is the well-known AFGL compilation (Rothman et al., 1983). The list has been expanded to incorporate extensions such as the AFGL trace gas compilation and updates on specific molecules. The current content is about 300,000 lines. In addition, the user may add his or her own line list. The list is searched, at each atmospheric level, for every line within the chosen region whose peak absorption coefficient exceeds a user-selected limit. In addition, account is taken of the wings of lines positioned up to 50 cm^{-1} above and below the region. Beyond this range, the expression describing the far wings of lines is insufficiently reliable and is, in any case, rarely of consequence.
4. *Absorption coefficient tables.* These are tables of H_2O and N_2 continuum absorption (see Section 2.4) plus tables of absorption coefficients for certain trace gases (mainly organic species) for which no line data are available. In addition, continuum extinction (e.g., hazes) can be incorporated in 1-km increments. Scattering, as such, is not yet implemented.
5. *Solar flux tables.* Derived from the tabulation of Labs and Neckel (1968, 1970), these provide the solar irradiance data when required.

Path Geometry. For the specific model atmosphere employed, the spherically refracted geometry is calculated in order to determine the pathlength through each atmospheric layer. The founding expression is the form of Snell's law for a spherically stratified medium at radius r,

$$n(r)r \sin\theta = \text{const} \tag{2.30}$$

A complete exposition may be found in the LOWTRAN 6 manual. However, the potential user is cautioned that lines of sight that graze a layer boundary can lead to singularities in the equations if a *brute-force* approach is employed. In these cases, continuity should be imposed by smoothing and/or interpolating as necessary. Incidentally, if zenith angles between about 75° and 105° are of no concern, a *plane-parallel* approximation can be used without significant error.

Absorption Coefficients. This is the heart of the program. First, a null matrix is established with one row for each layer in the atmosphere and a column for each wavenumber bin. The width in wavenumber of each bin is a matter of importance. Basically, the bins should be narrow (about one-tenth of the narrowest linewidth expected to be significant in the final result). Thus, paths confined to the lower troposphere may have their spectra successfully computed with a bin size $\sim 0.01\ \text{cm}^{-1}$. Bins for purely stratospheric paths and IR frequencies must be orders of magnitude narrower. Furthermore, if a new program proposes to use Scott's method (see below), the minimum bin size may need to be further reduced. The price paid for narrow bins is, of course, computation time, which will increase inversely as the bin size.

In ES, a brute-force approach to the determination of line-by-line absorption coefficients is used because the program requirements do not permit Scott's method to be used. For each line, at each layer, a Lorentz width and a Doppler width is calculated and a Voigt table look-up routine (see Chapter 3) is used to determine the central absorption coefficient in each bin. The computation begins at the line center and proceeds in both directions until the contribution from the line is below a user-set limit (typically 10^{-4}–10^{-7}in absorption coefficient).

Scott (1974) introduced a number of possible simplifications to (and hence acceleration of) this procedure. First, she uses a variable bin width that reduces with increasing altitude (but should bear an integral relationship with the largest bin). Second, she artificially shifts each line in frequency so that it is centered on a bin. This permits her to compute the absorption coefficient only half the number of times because the line is now symmetric in bin space. By making additional assumptions about constancy of linewidth within any one layer (the degree to which this can be done depends on the user's personal prejudices and the specific molecule mix assumed), she can set up simplified generating functions for the line profiles that greatly reduce the computation time. The problems with the method are (a) the assumptions about linewidth constancy are dubious, especially in the lower atmosphere; (b) the shifting of lines to enforce symmetry can result in obvious errors, especially if Doppler shifts are a required outcome or pressure shifts are important; and (c) a repeated reduction of bin width to avoid the error in (b) can result in *increased* rather than decreased computation time if the path must encompass a significant fraction of both the upper and lower atmosphere.

The critical element in the brute-force approach must be an informed judgment on the necessary bin size. As suggested earlier, paths through the lower troposphere can usually be successfully treated with a bin size as coarse as 0.01 cm^{-1} unless O_3 is a significant absorber in the chosen region. Then, of course, a variable bin size could be beneficial. An alternative, however, might be to test whether the O_3 line center fell within a given bin and, in that bin, to replace the bin-centered absorption coefficient by the peak value.

Into the same bins are summed the absorption coefficients derived from table look-ups and interpolation. If sunlight is to be incorporated, the whole procedure is repeated for the line of sight to the sun so that, in the next stage, the solar irradiance can be added to the total.

Monochromatic Radiance Calculation. It is at this stage that the numerical solution to the appropriate version of the equation of radiative transfer (see Section 2.2) occurs. The computation proceeds in the direction receiver to source, but for simplicity only a down-looking computation is described.

First, the absorption coefficients are converted to layer-by-layer transmittance and multiplied columnwise to the lower boundary. If the transmittance of the nth layer is $t_n(\nu)$, the contribution of the layer to the outgoing radiance is

$$J(\tilde{\nu}, n) = B(\tilde{\nu}, n)[1 - t_n(\tilde{\nu})] \prod_{m=n+1}^{N} t_m(\tilde{\nu}) \tag{2.31}$$

where $B(\tilde{\nu}, n)$ is the Planck function for the nth layer and the product proceeds layer to space. The emergent atmospheric radiance is thus the sum of this function over all layers. To this term is added the target radiance and the solar-reflected radiance multiplied by the inbound transmittance.

This part of the program can be greatly simplified if only the total one-way transmittance is of interest:

$$t(\tilde{\nu}) = \prod_{m=1}^{N} t_m(\tilde{\nu}) \tag{2.32}$$

Convolution with Instrumental Function. Any receiving system impresses its own spread function on the wavenumber resolution of the final outcome.

Since a prerequisite of the preceding calculations is a wavenumber bin width that is small compared to the final wavenumber resolution, the effect is properly computed by convolving the monochromatic radiances with the wavenumber spread function. ES was devised for use with Fourier transform spectrometers (FTS); therefore, a standard set of analytic spread functions is employed (Norton and Beer, 1976). The option exists, however, to replace these with user-supplied functions.

Convolution is done most readily using an array processor. Individuals not possessing such a device should give serious thought to operating in Fourier transform space, where convolution simplifies to multiplication. With modern fast Fourier transform routines, the double transformation required can still be faster than a direct convolution.

The basic computation is now complete. Stored in the computer is a tabulation of the spectrum *as it would be seen by the detector*. If more than one model spectrum was requested in the initial menu setup (ES accommodates up to four), the program repeats as necessary.

Multiple-Option Display Routines. The basic (default) display in ES is the spectrum (or spectra) on the terminal screen. At this time, a "soft-key" menu is enabled that permits manipulation of the displayed data. After each option is exercised, the program cycles back to the default display to await further options or until forced back to the input menu.

Currently available options are as follows:

1. Screen Hard Copy. This option exists for all the display options outlined below. Choice of this option also prints the input menu plus some derived data such as peak and integrated radiance and the total line-of-sight water vapor content in precipitable millimiters.
2. Off-Line Plot. Creates a file for subsequent plotting on a pen plotter.
3. Next/Previous Frame. Computes the next higher or lower wavenumber spectral region with a 12.5% overlap with the current frame.
4. Contribution Function. At a user-selected wavenumber within the current frame, the screen displays a plot of radiance versus altitude of the atmospheric emission (i.e., the contribution function). Also marked are the surface emission and solar reflection contributions, the vertical temperature profile (on a radiance scale), and a horizontal grid whose lines are spaced by one pressure scale height. A hard copy of this option also provides a tabulation of the layer-to-space transmittance and radiance.
5. Cross-Correlation. Computes and displays the cross-correlation of any two selected spectra.

6. Difference/Ratio. Plots the difference or ratio of any two spectra.
7. Modify Continuum. Permits a piecewise multiplication of the displayed spectrum by an arbitrary continuum. Useful for matching computed spectra to real data which incorporate the system broadband spectral response.
8. Screen Dump. Numerical tabulation of relative intensity versus wavenumber.
9. Mean and RMS. Self-explanatory.
10. Tick. Useful for annotating the display for features of particular interest.
11. Zoom. Horizontal scale expansion. Vertical offset and scale expansion are also separately available.
12. Noise. Temporarily adds filtered Gaussian random noise to the spectra. Useful for investigating the influence of noise on detectivity.
13. Image Transfer. Copies a spectrum into another display file. Useful for comparing the same spectrum under different degrees of manipulation of continuum or noise.

Future Expansion. The current display is monochrome. However, an extension to the use of color is presently being incorporated. Color is an excellent discriminant when comparing complex spectra, real or computed.

Eventually, ES will become, in turn, an element in a larger routine in which the computation parameters will be driven by a nonlinear least-squares algorithm (Ralston and Jennrich, 1978; Chang and Shaw, 1977; Niple, 1980) in order to provide a semiautomatic fit of model spectra to real data. Work on this task is in progress.

2.8 EXEMPLAR TRANSMITTANCE CALCULATIONS

Figures 2.23–2.52 show examples of transmittance calculations for two important spectral regions: 900–1100 and 9300–9500 cm^{-1}. Computations were made for two cases: a vertical transatmospheric case to sea level and a horizontal 1-km path at 500 elevation. The calculations were made using the EMISSION-SPECTRA program; however, for these illustrations the program was driven to provide only transmittance, not radiance. All calculations are for a clear nonscattering atmosphere; molecular continuum effects are, however, included. The 9300–9500-cm^{-1} region was computed using a midlatitude summer (MLS) model atmosphere with line-of-sight water vapor contents of 23.8 pr mm (vertical) and 13.4 pr mm (horizontal). The 900–1100-cm^{-1} region was also computed for a midlatitude winter (MLW) model because of the markedly different H_2O continuum at the two seasons. The vertical water vapor contents were 29.8

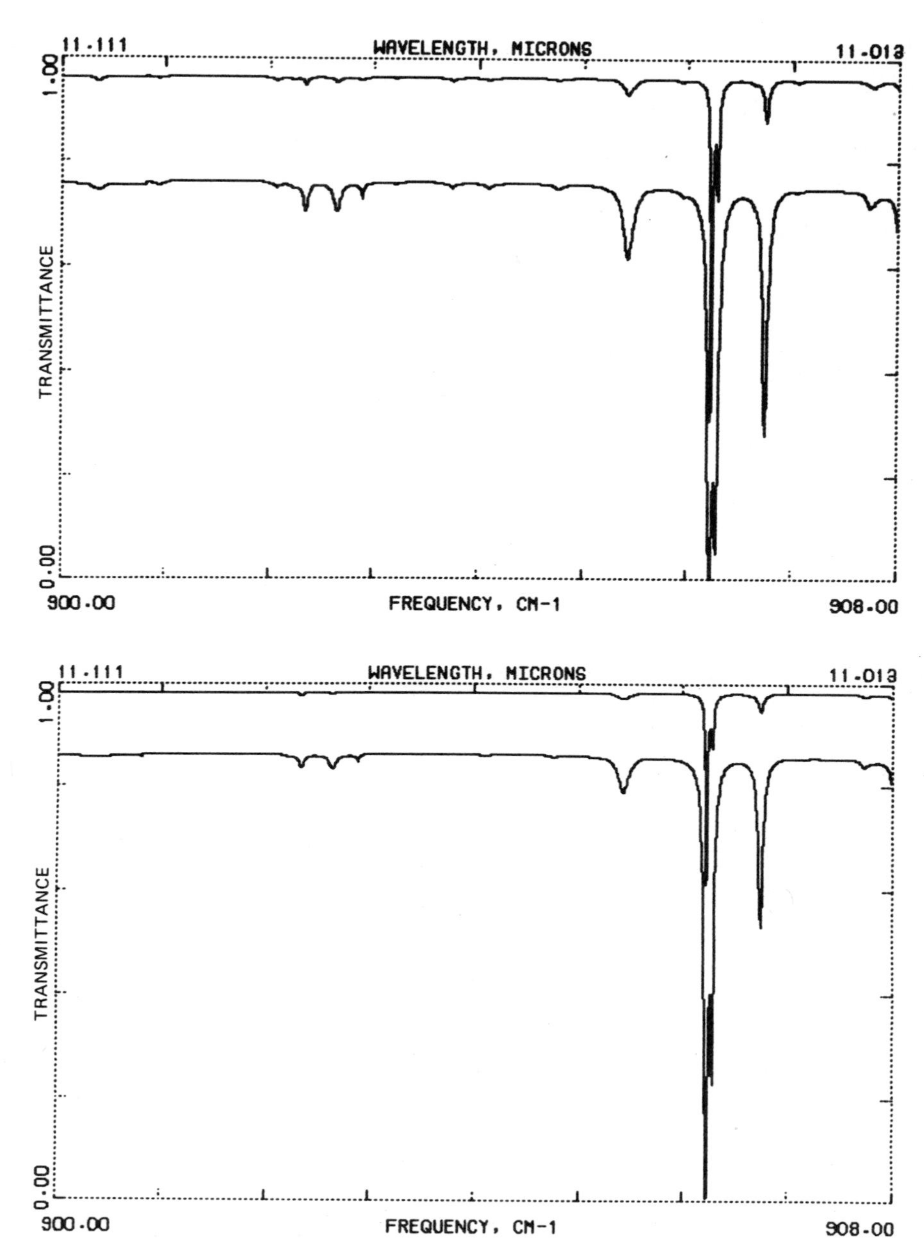

Figure 2.23. Computed transmission spectra for region 900–908 cm^{-1}. Upper frame is for vertical, clear, transatmospheric path to sea level; lower frame for 1-km horizontal path at 500 m elevation. In each frame, upper curve is for midlatitude winter model; lower for midlatitude summer. Major difference is attributable to influence of water vapor continuum.

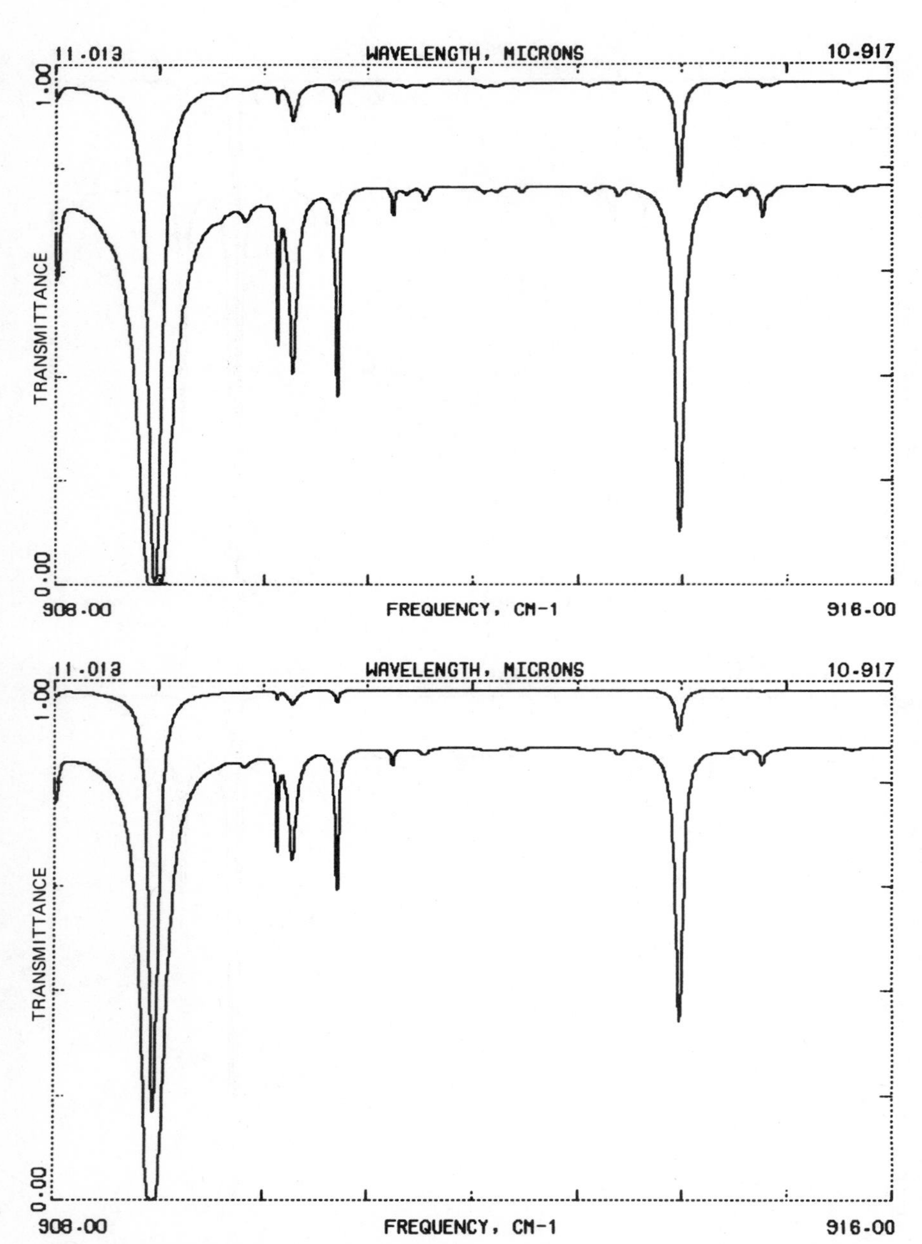

Figure 2.24. Computed transmission spectra for region 908–916 cm^{-1}. Other information as for Figure 2.23.

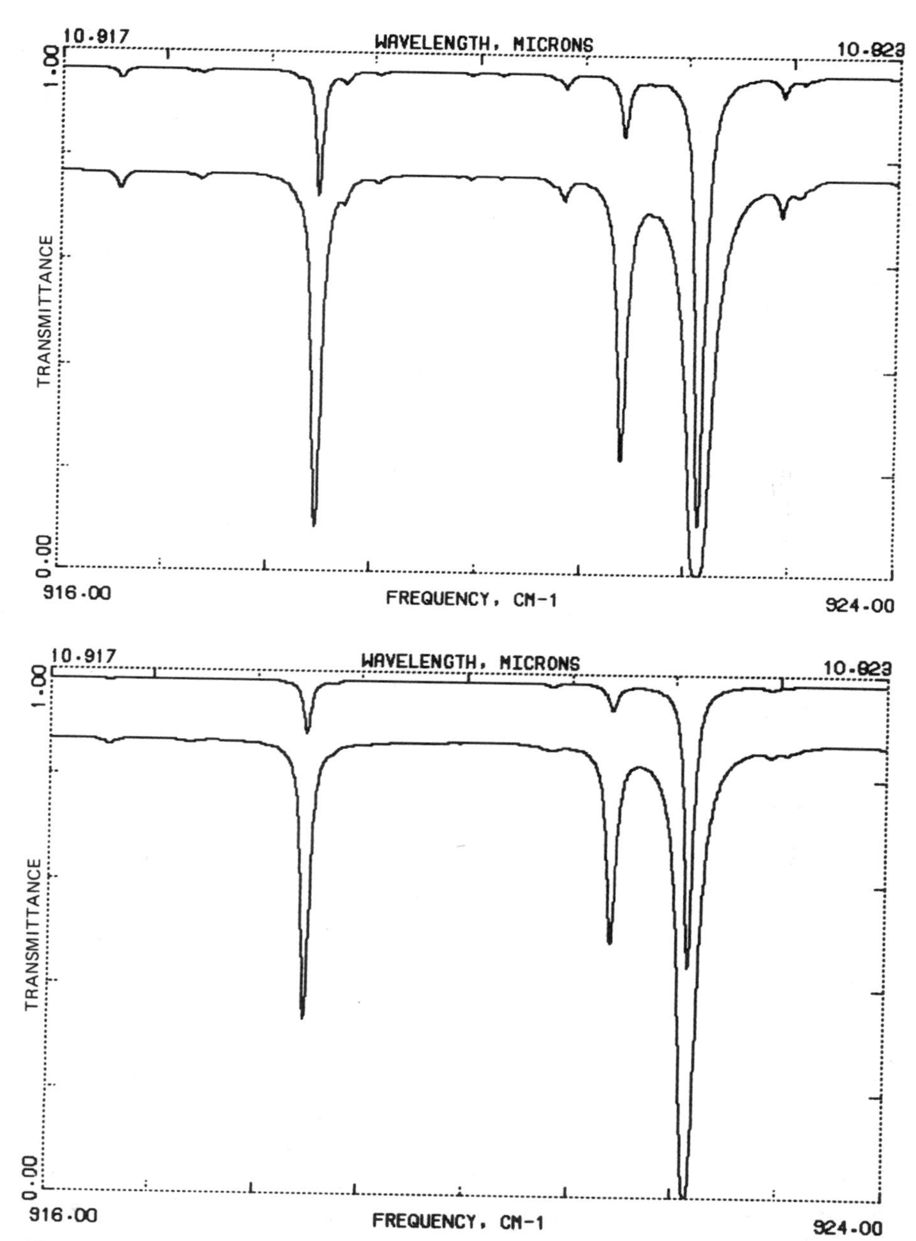

Figure 2.25. Computed transmission spectra for region 916–924 cm^{-1}. Other information as for Figure 2.23.

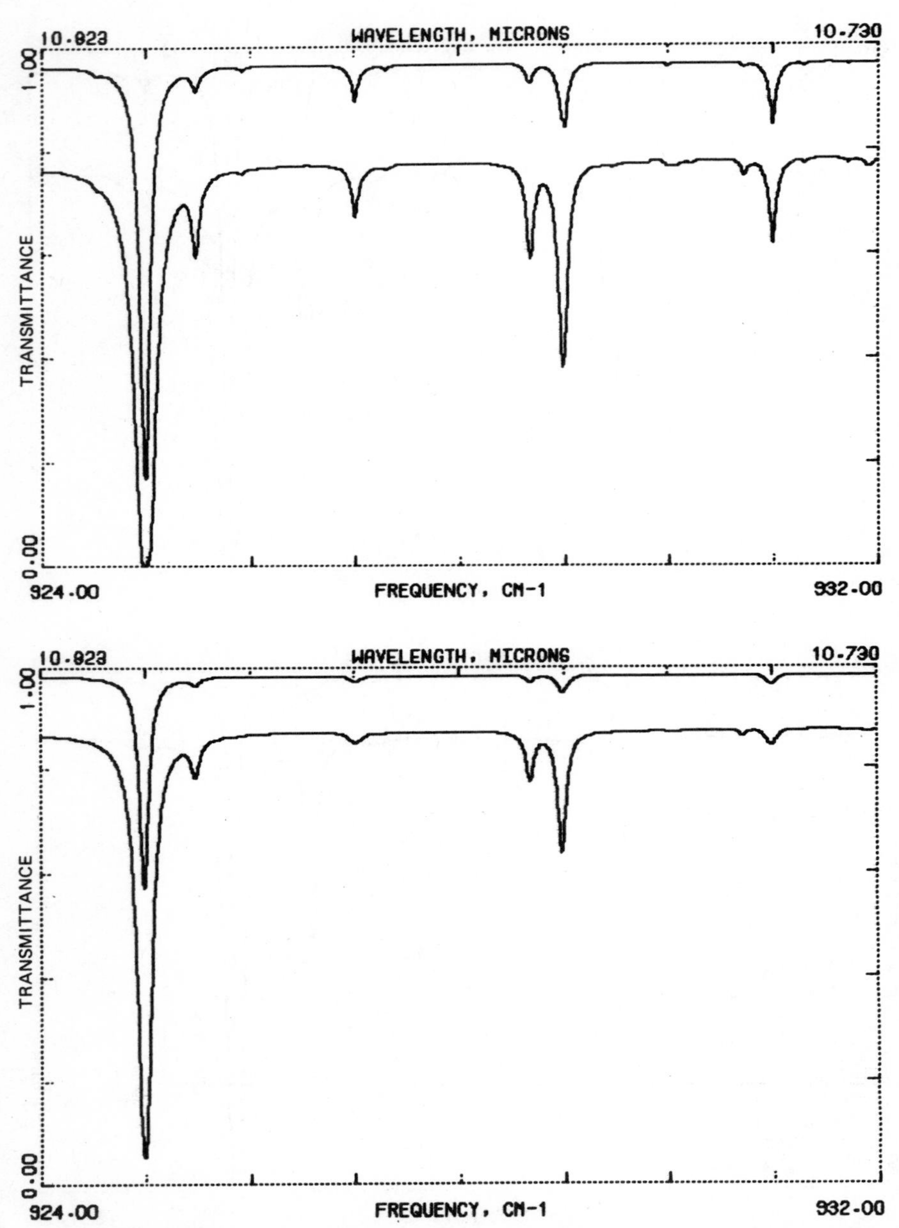

Figure 2.26. Computed transmission spectra for region 924–932 cm^{-1}. Other information as for Figure 2.23.

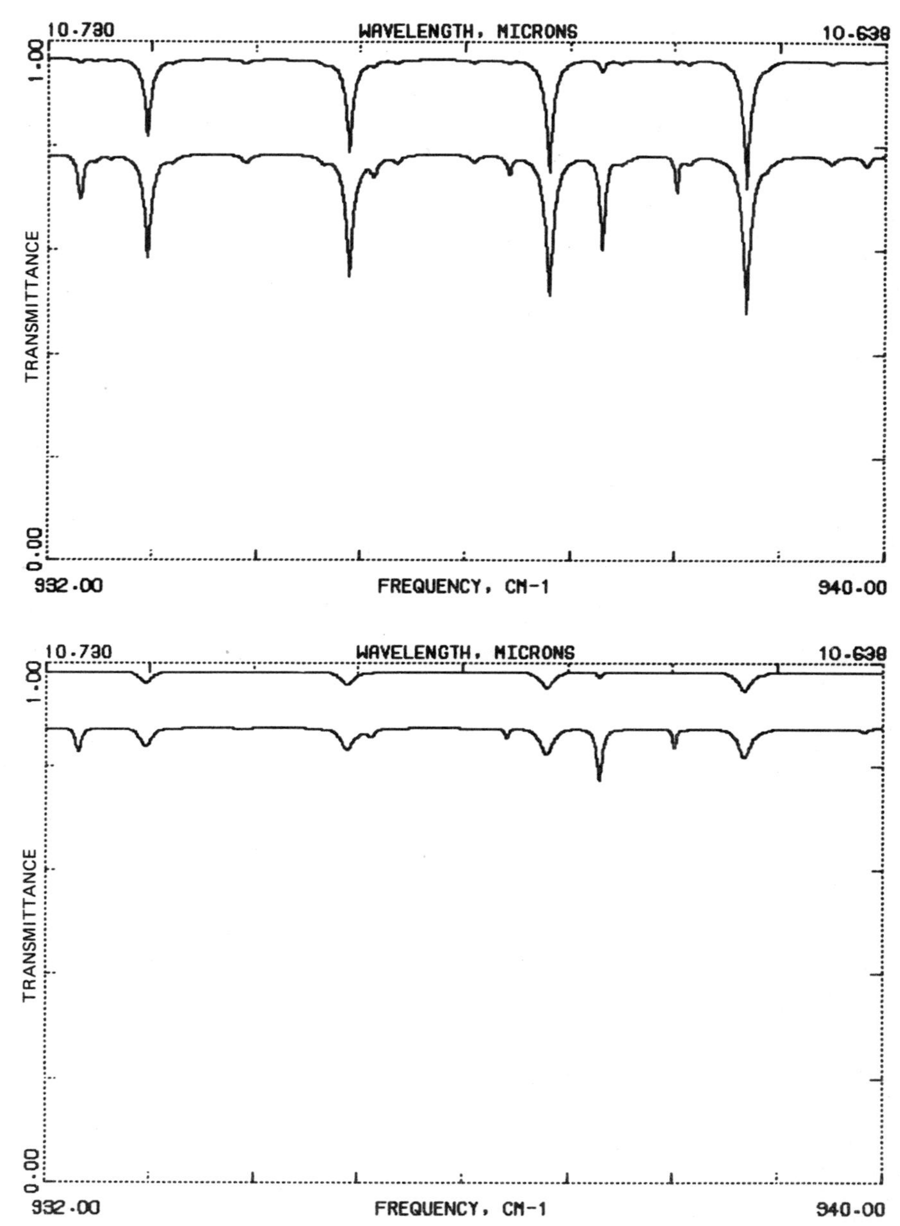

Figure 2.27. Computed transmission spectra for region 932–940 cm^{-1}. Other information as for Figure 2.23.

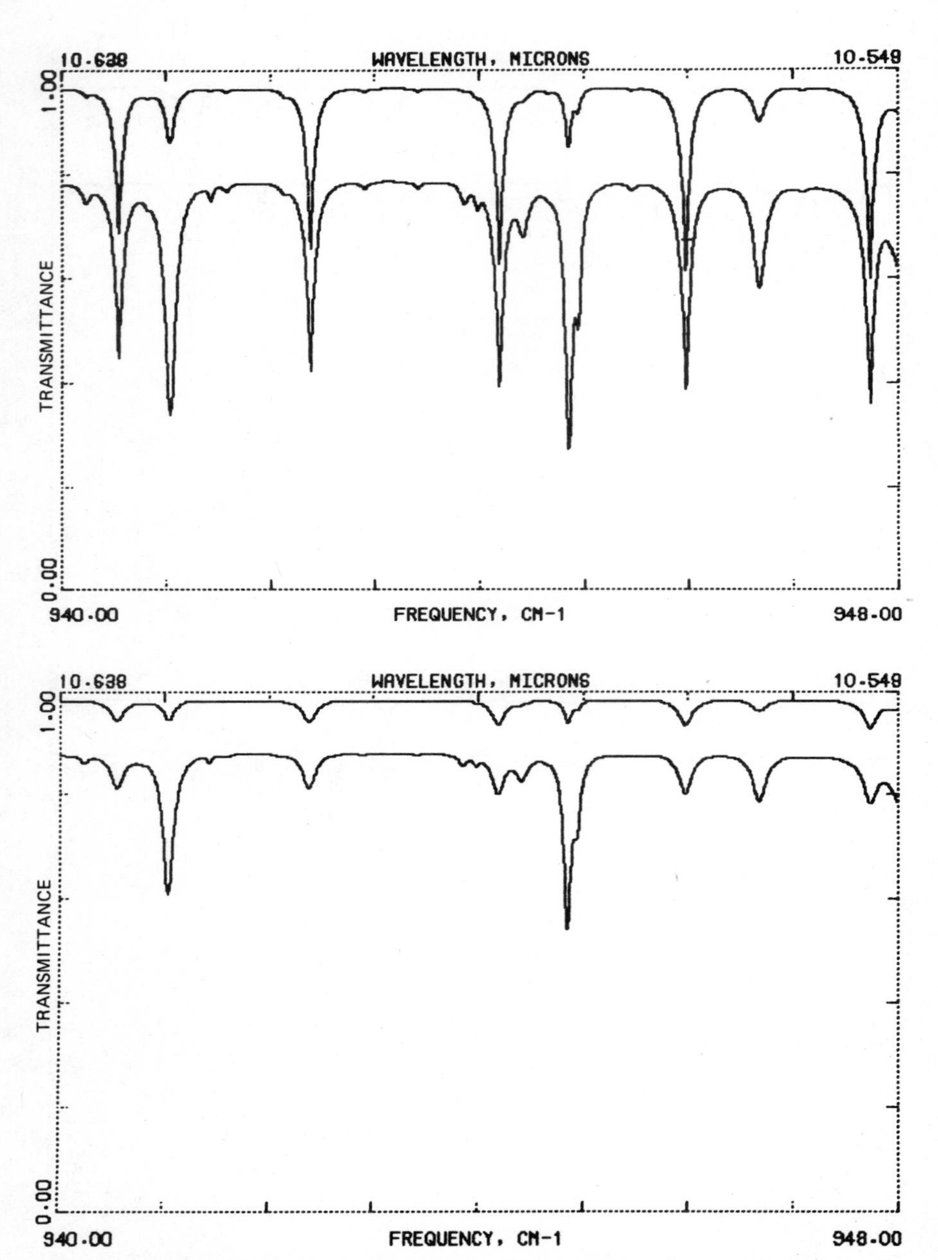

Figure 2.28. Computed transmission spectra for region 940–948 cm^{-1}. Other information as for Figure 2.23.

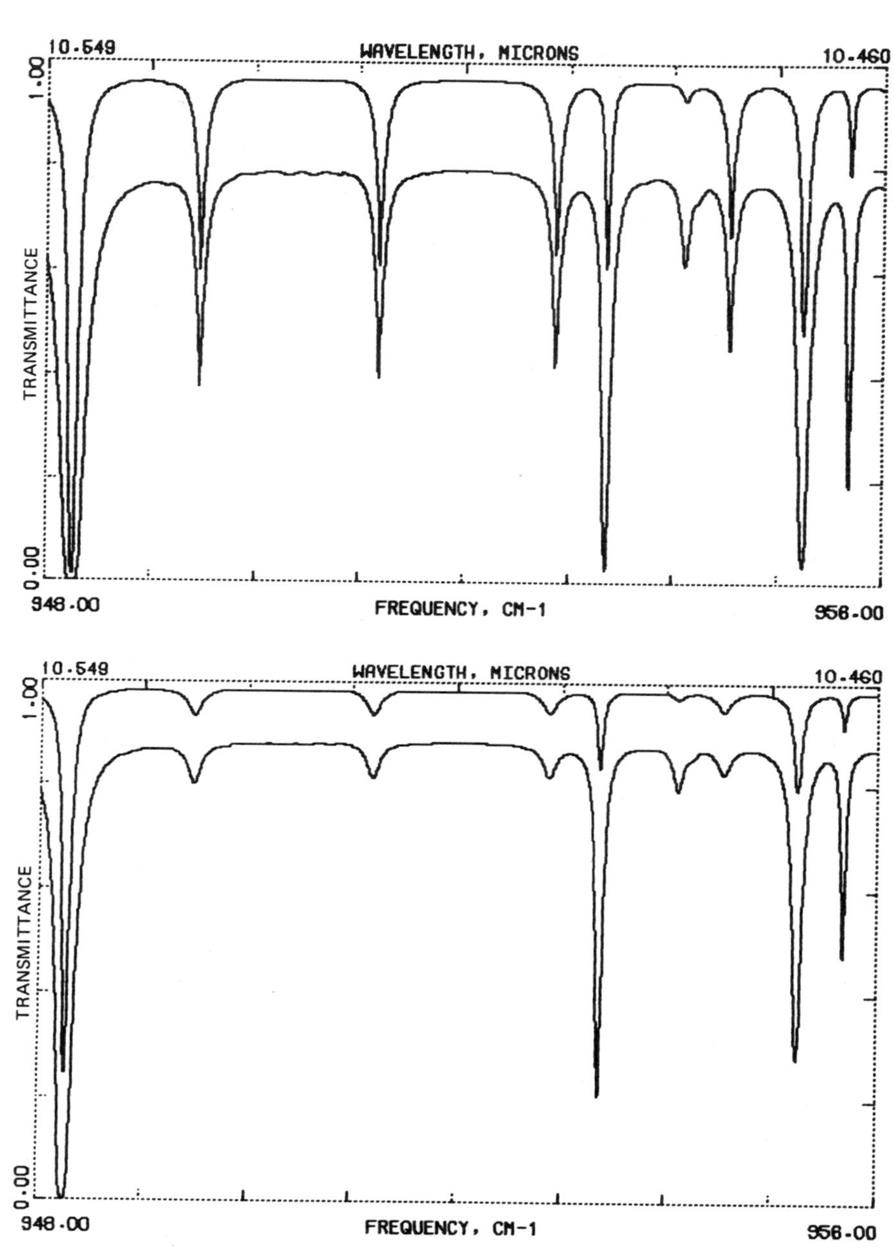

Figure 2.29. Computed transmission spectra for region 948–956 cm^{-1}. Other information as for Figure 2.23.

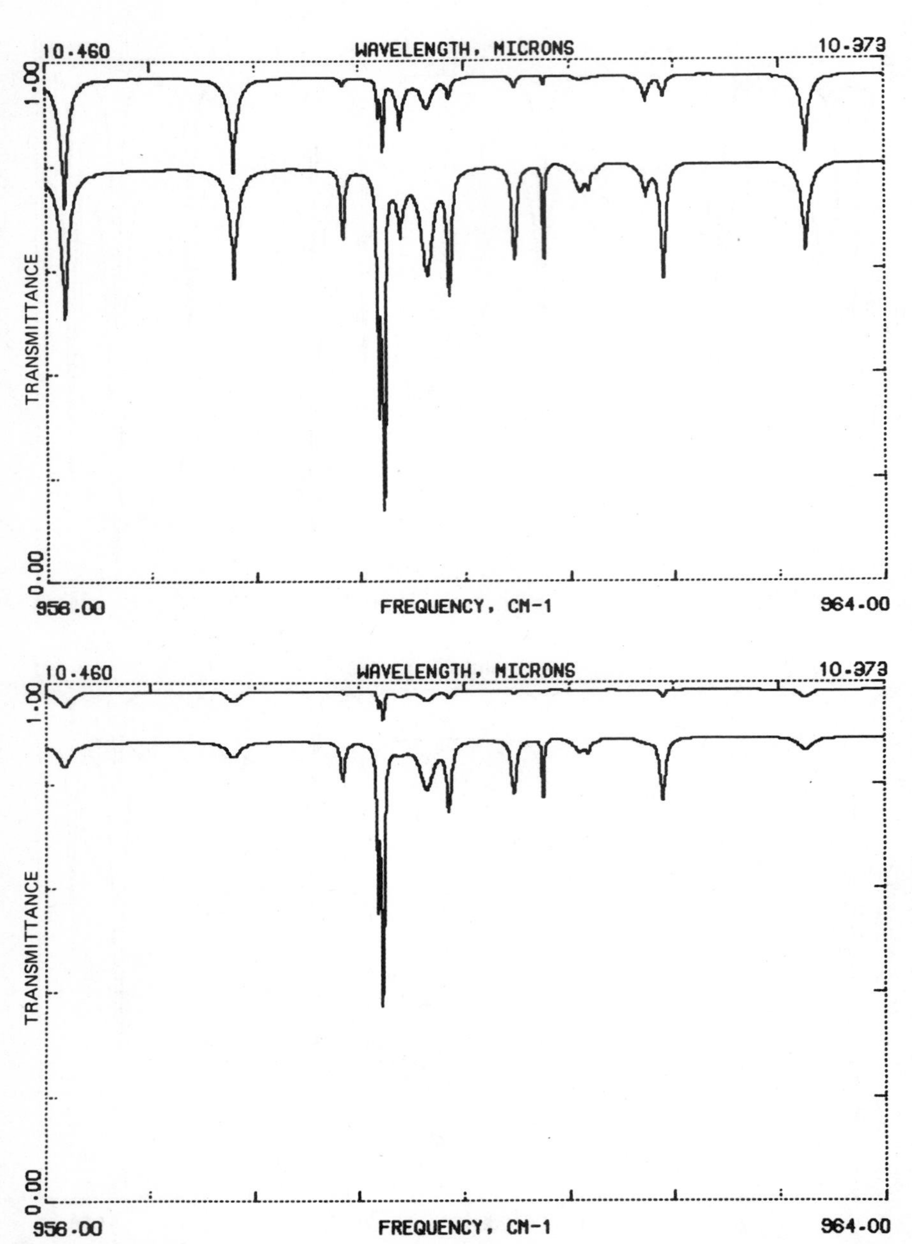

Figure 2.30. Computed transmission spectra for region 956–964 cm^{-1}. Other information as for Figure 2.23.

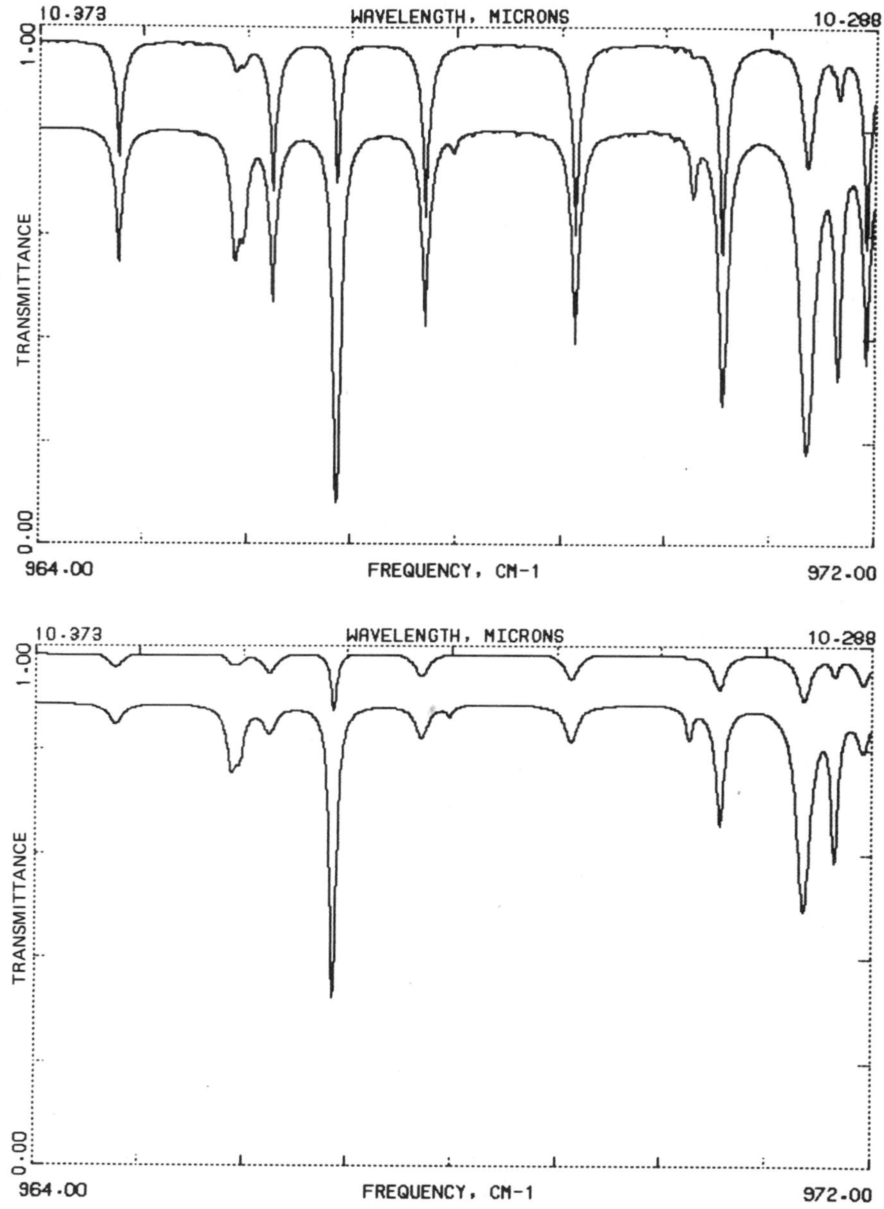

Figure 2.31. Computed transmission spectra for region 964–972 cm^{-1}. Other information as for Figure 2.23.

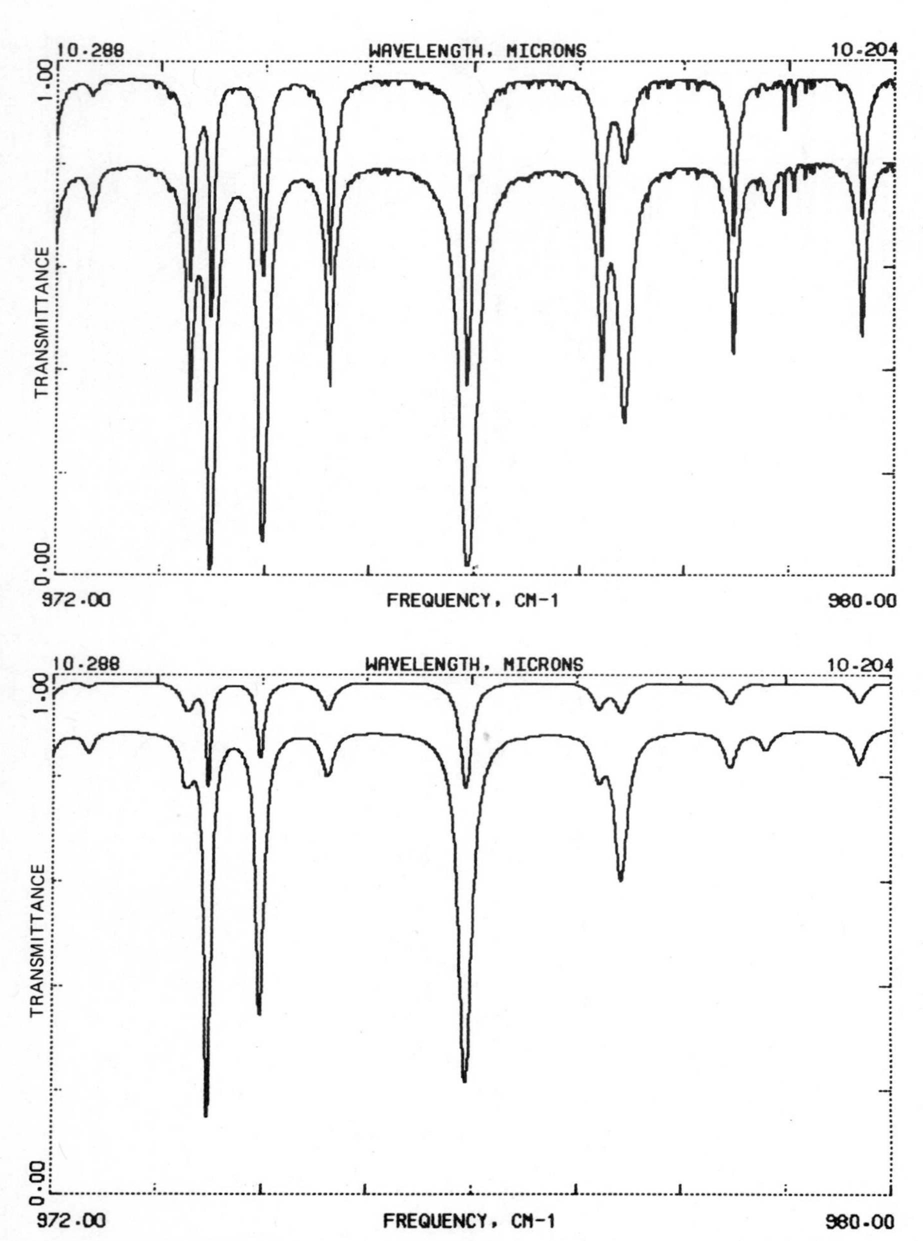

Figure 2.32. Computed transmission spectra for region 972–980 cm^{-1}. Other information as for Figure 2.23.

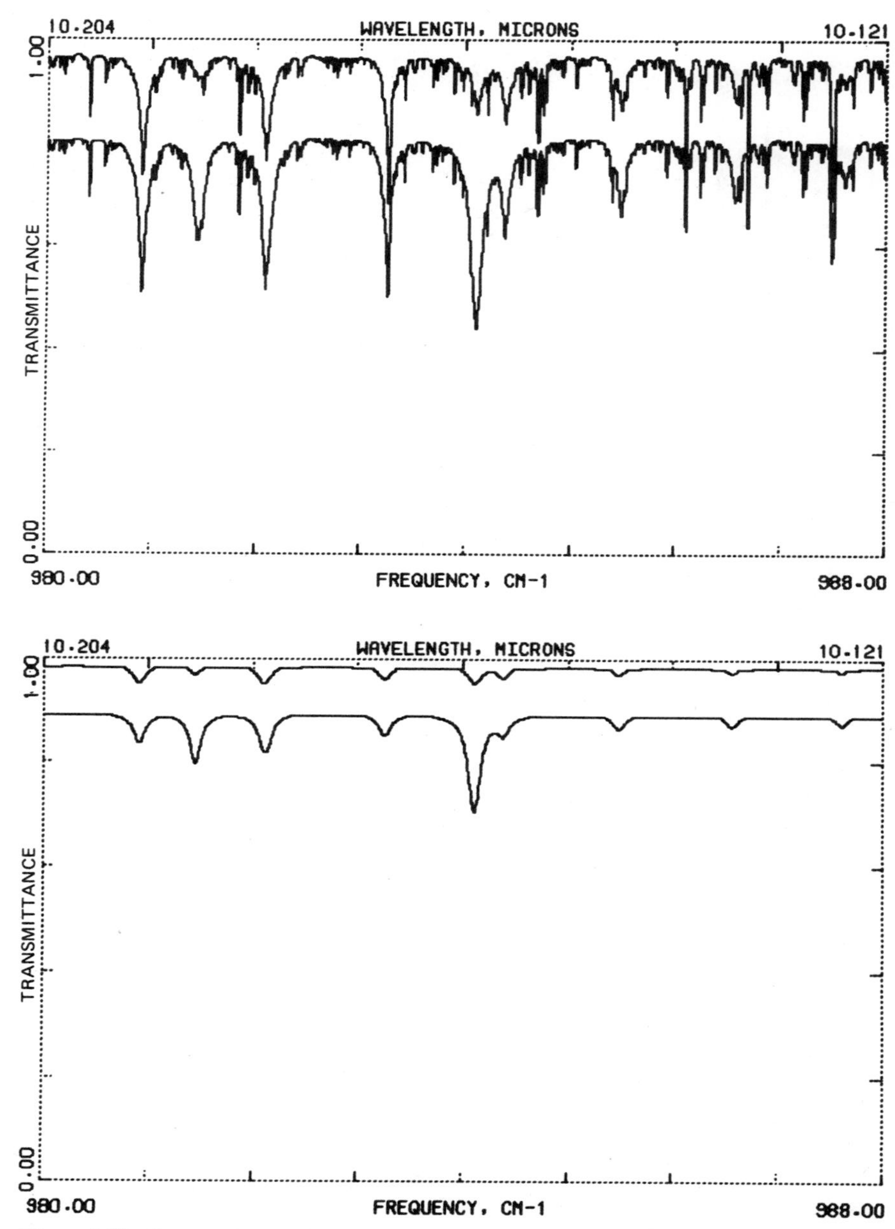

Figure 2.33. Computed transmission spectra for region 980–988 cm^{-1}. Other information as for Figure 2.23.

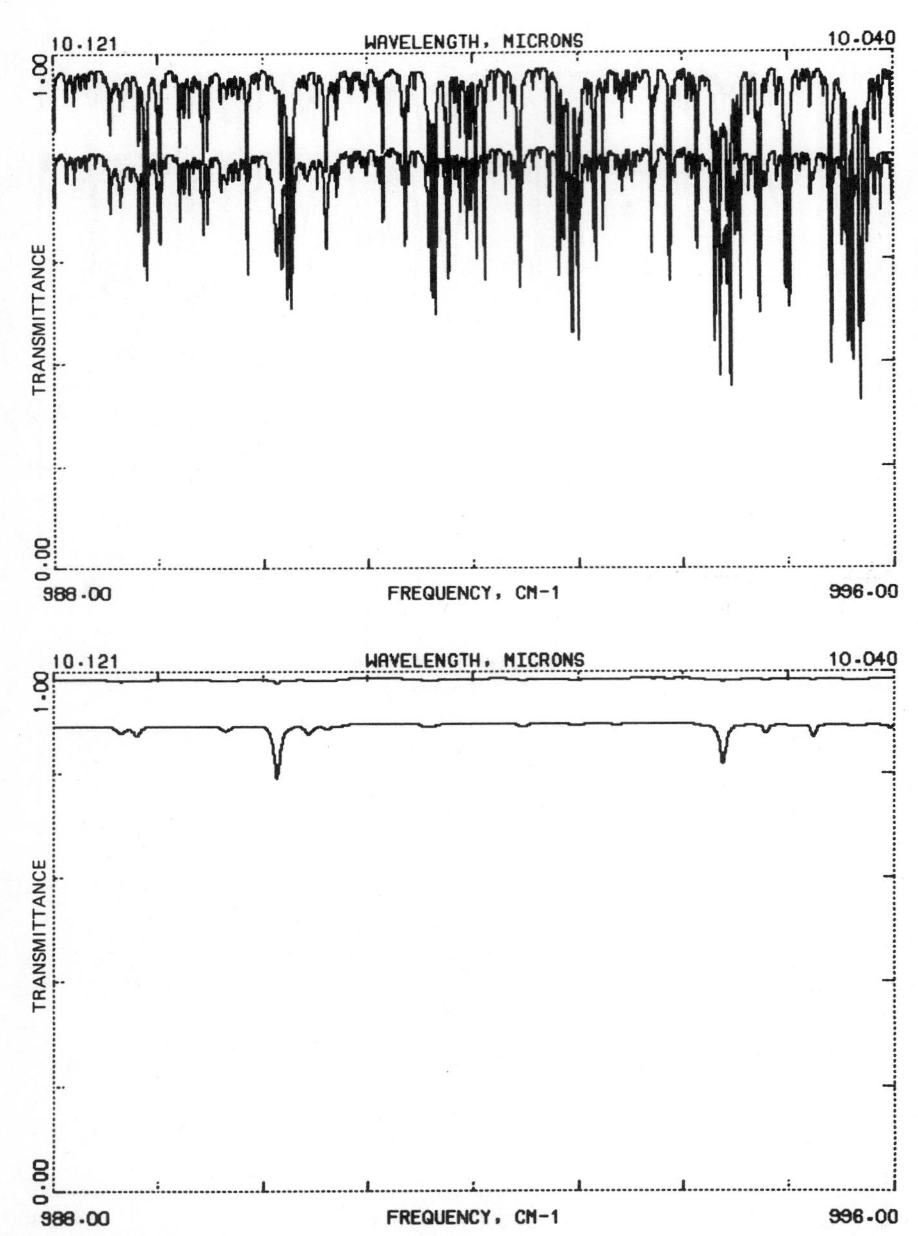

Figure 2.34. Computed transmission spectra for region 988–996 cm^{-1}. Other information as for Figure 2.23.

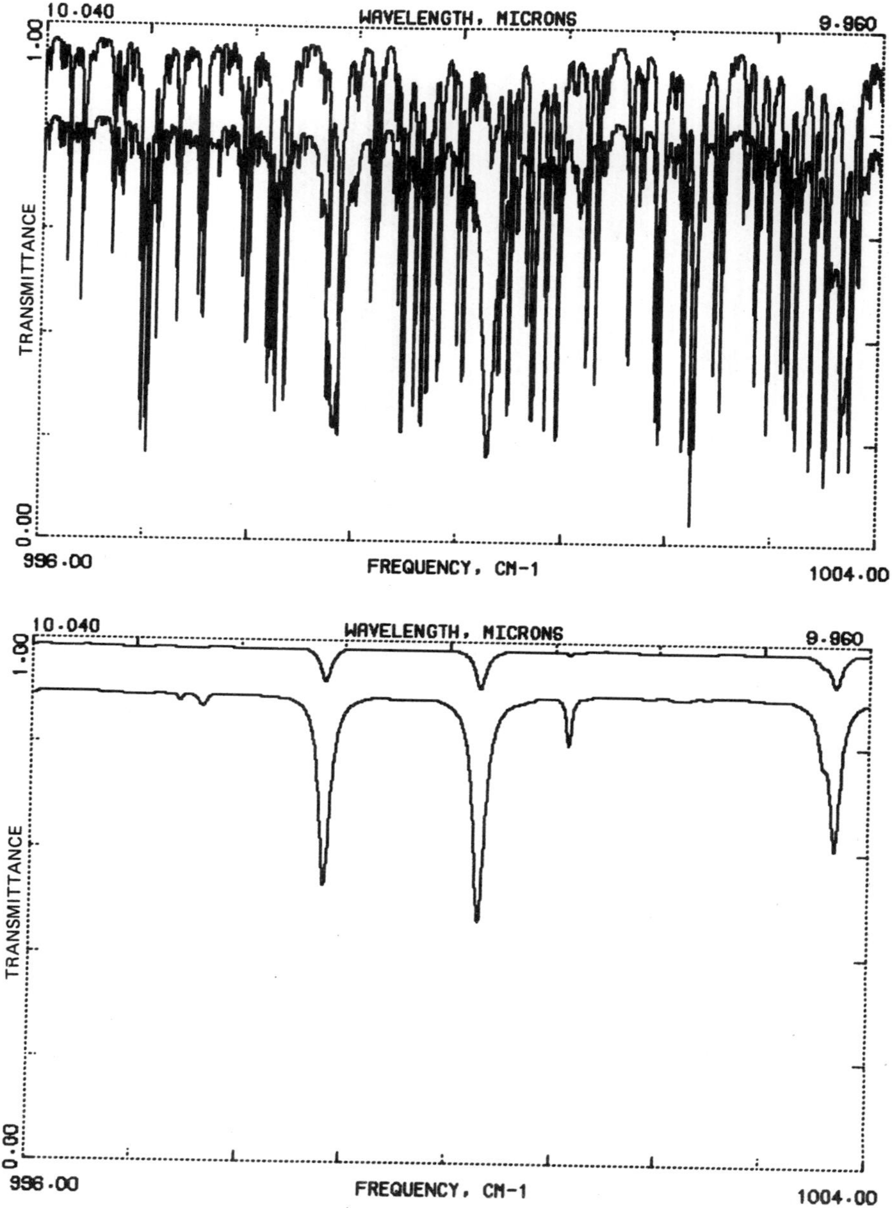

Figure 2.35. Computed transmission spectra for region 996–1004 cm^{-1}. Other information as for Figure 2.23.

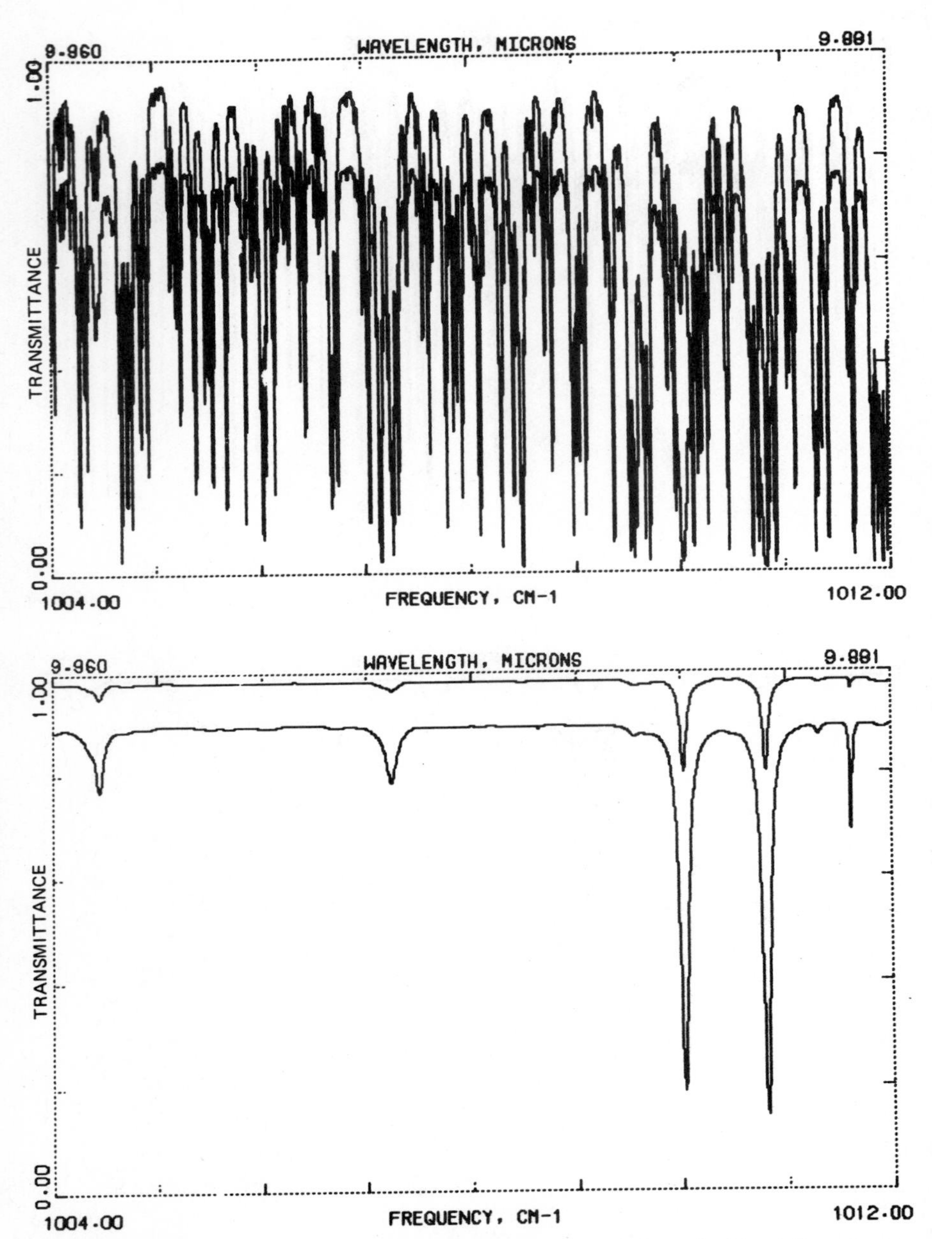

Figure 2.36. Computed transmission spectra for region 1004–1012 cm^{-1}. Other information as for Figure 2.23.

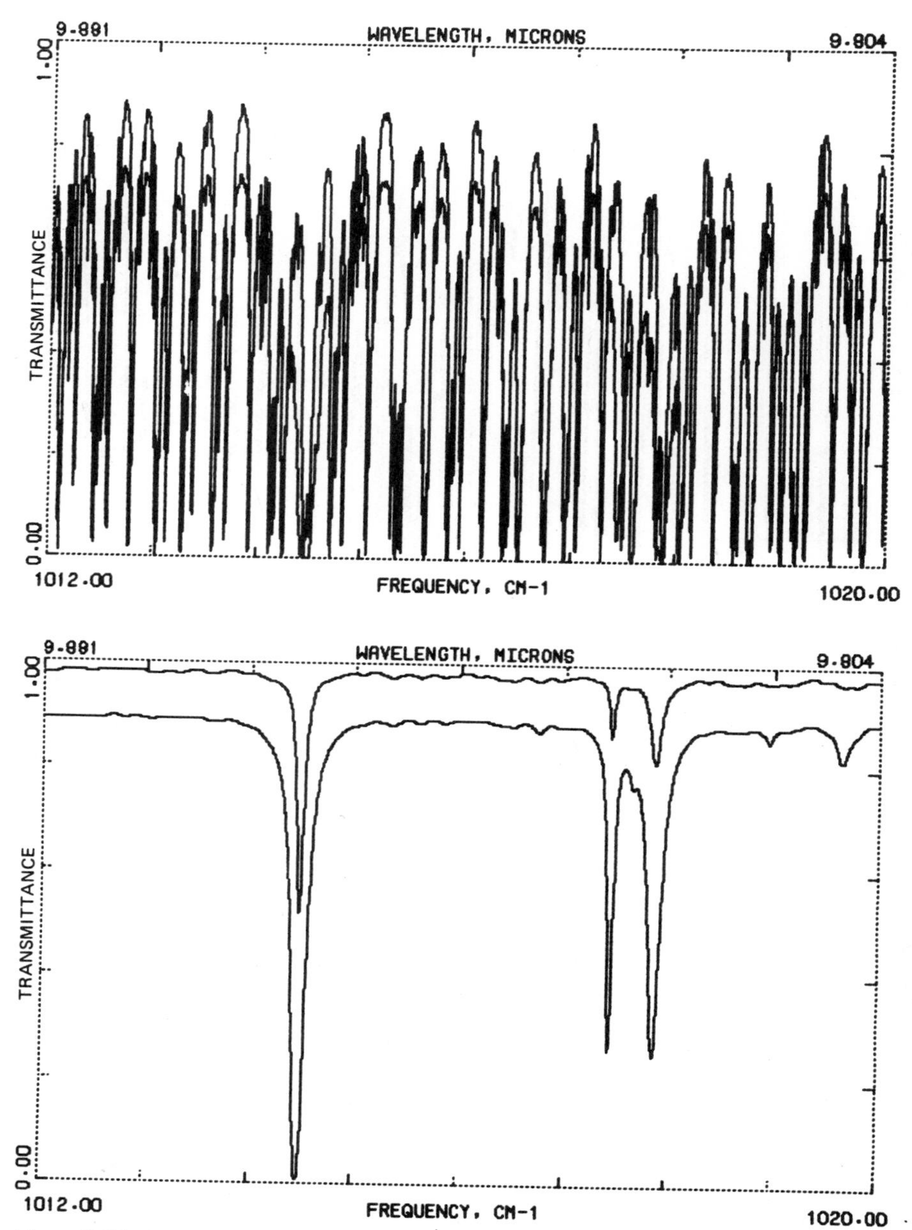

Figure 2.37. Computed transmission spectra for region 1012–1020 cm^{-1}. Other information as for Figure 2.23.

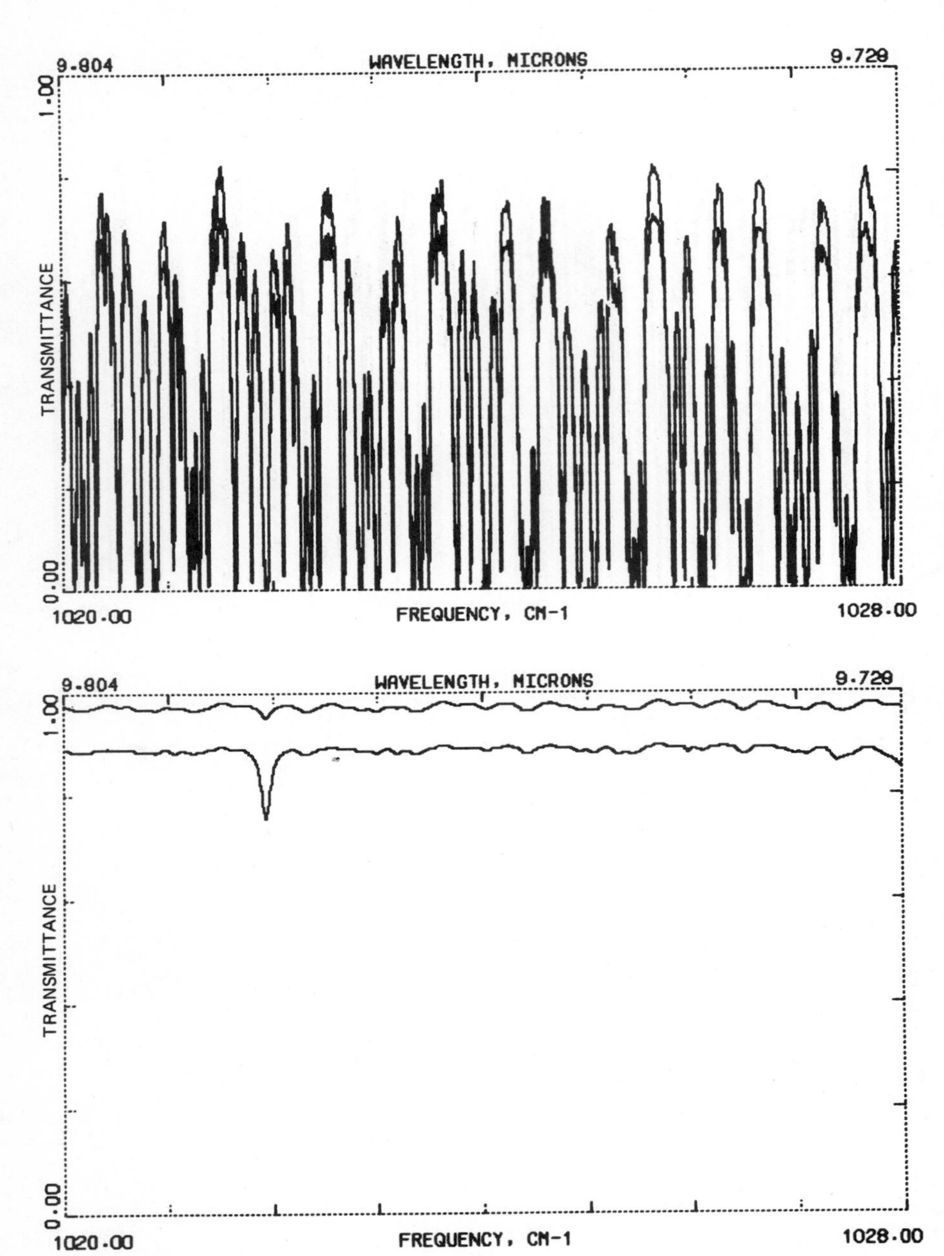

Figure 2.38. Computed transmission spectra for region 1020–1028 cm^{-1}. Other information as for Figure 2.23.

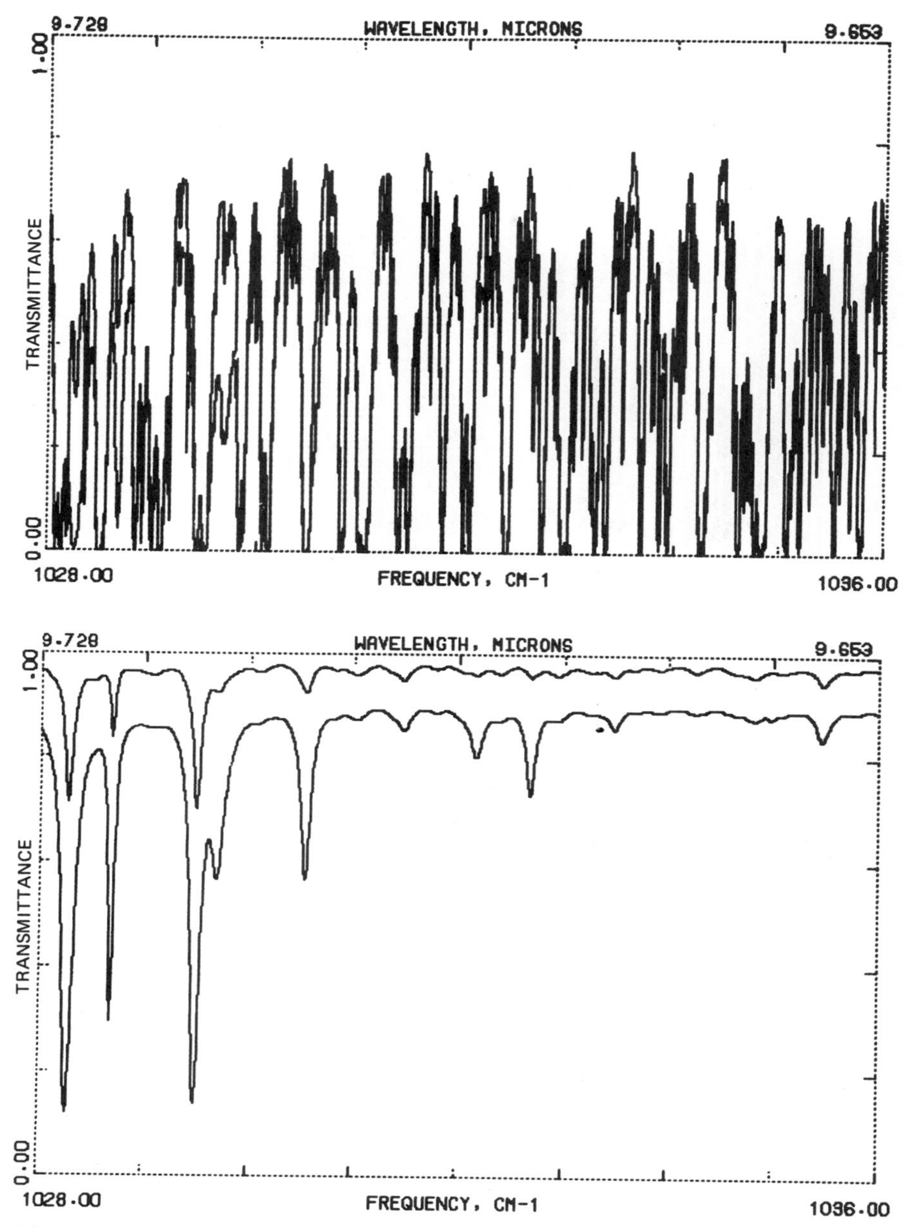

Figure 2.39. Computed transmission spectra for region 1028–1036 cm^{-1}. Other information as for Figure 2.23.

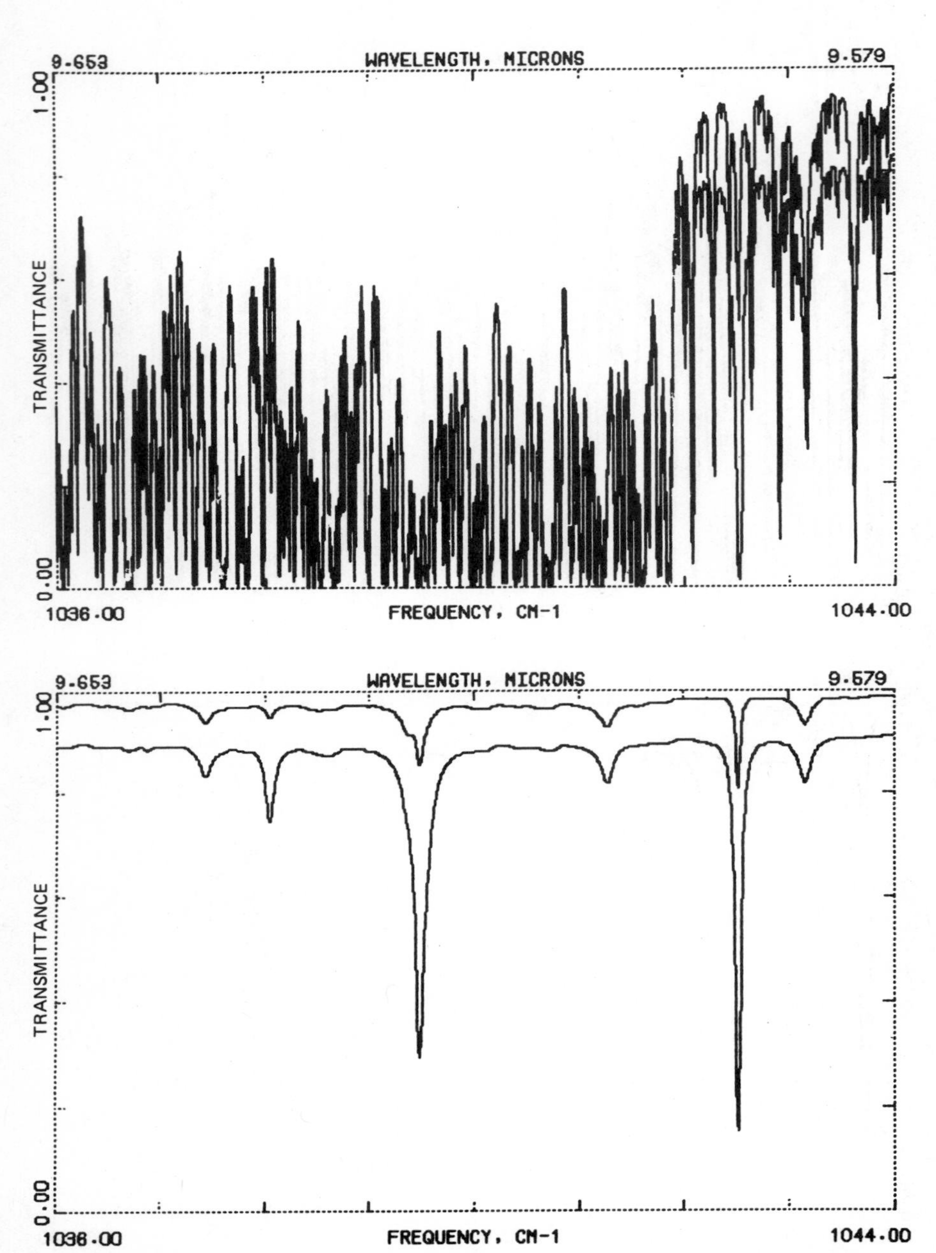

Figure 2.40. Computed transmission spectra for region 1036–1044 cm^{-1}. Other information as for Figure 2.23.

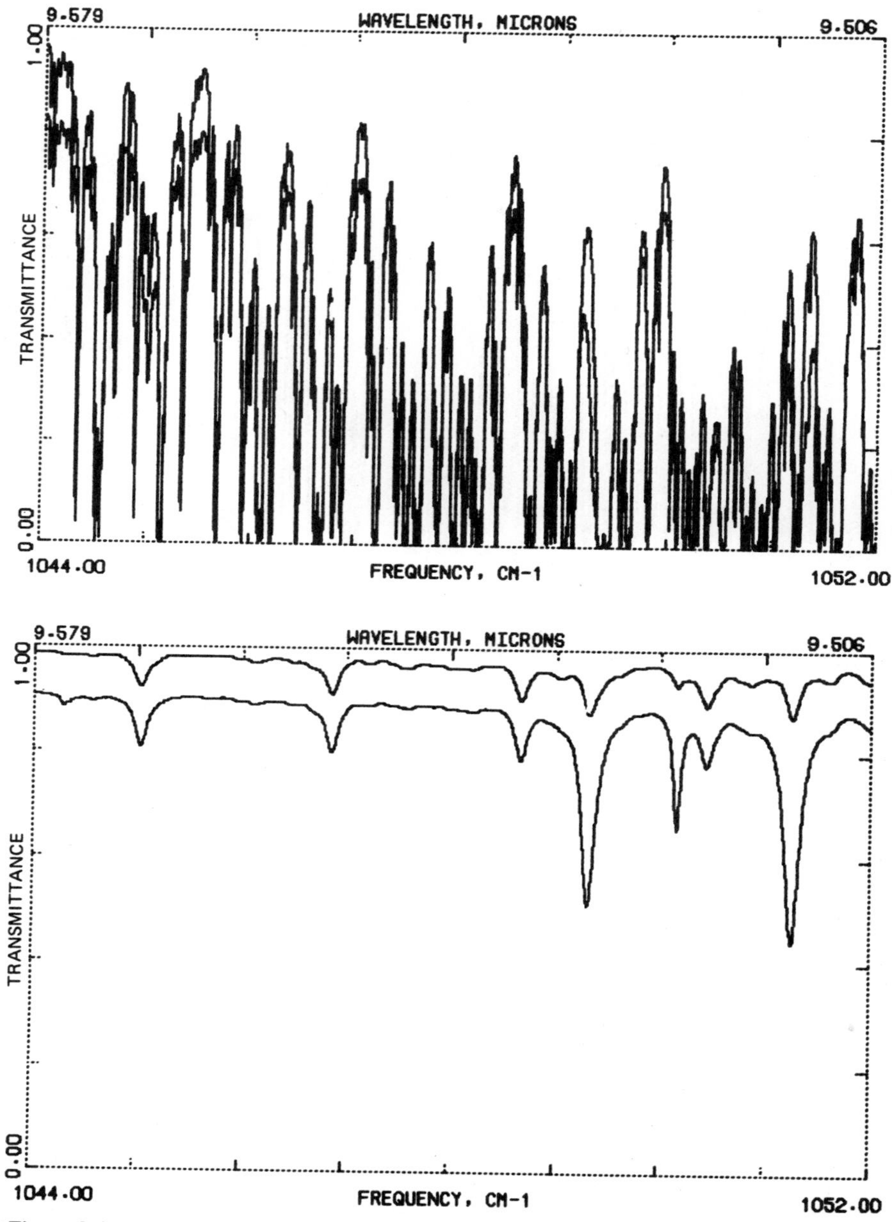

Figure 2.41. Computed transmission spectra for region 1044–1052 cm^{-1}. Other information as for Figure 2.23.

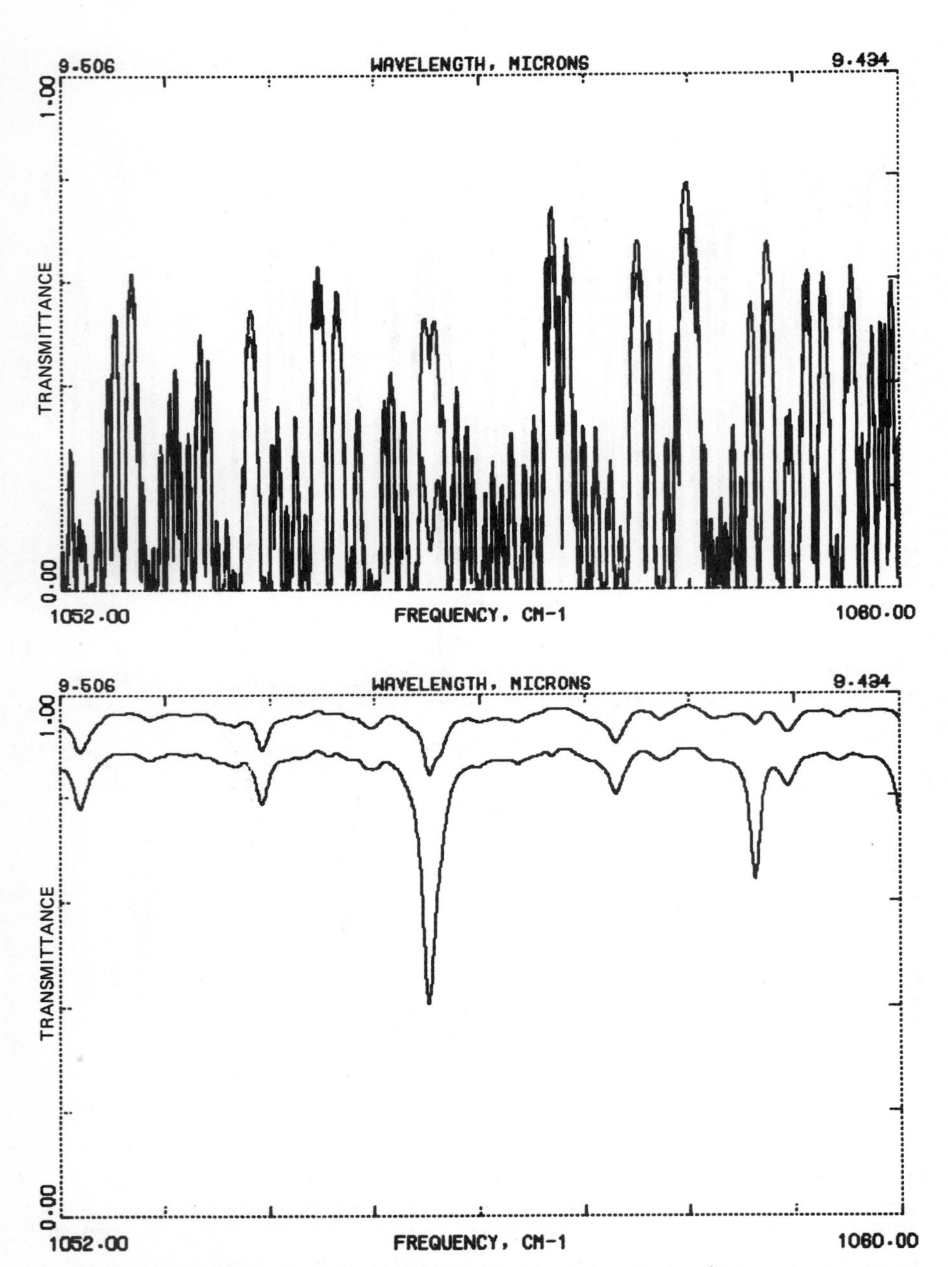

Figure 2.42. Computed transmission spectra for region 1052–1060 cm^{-1}. Other information as for Figure 2.23.

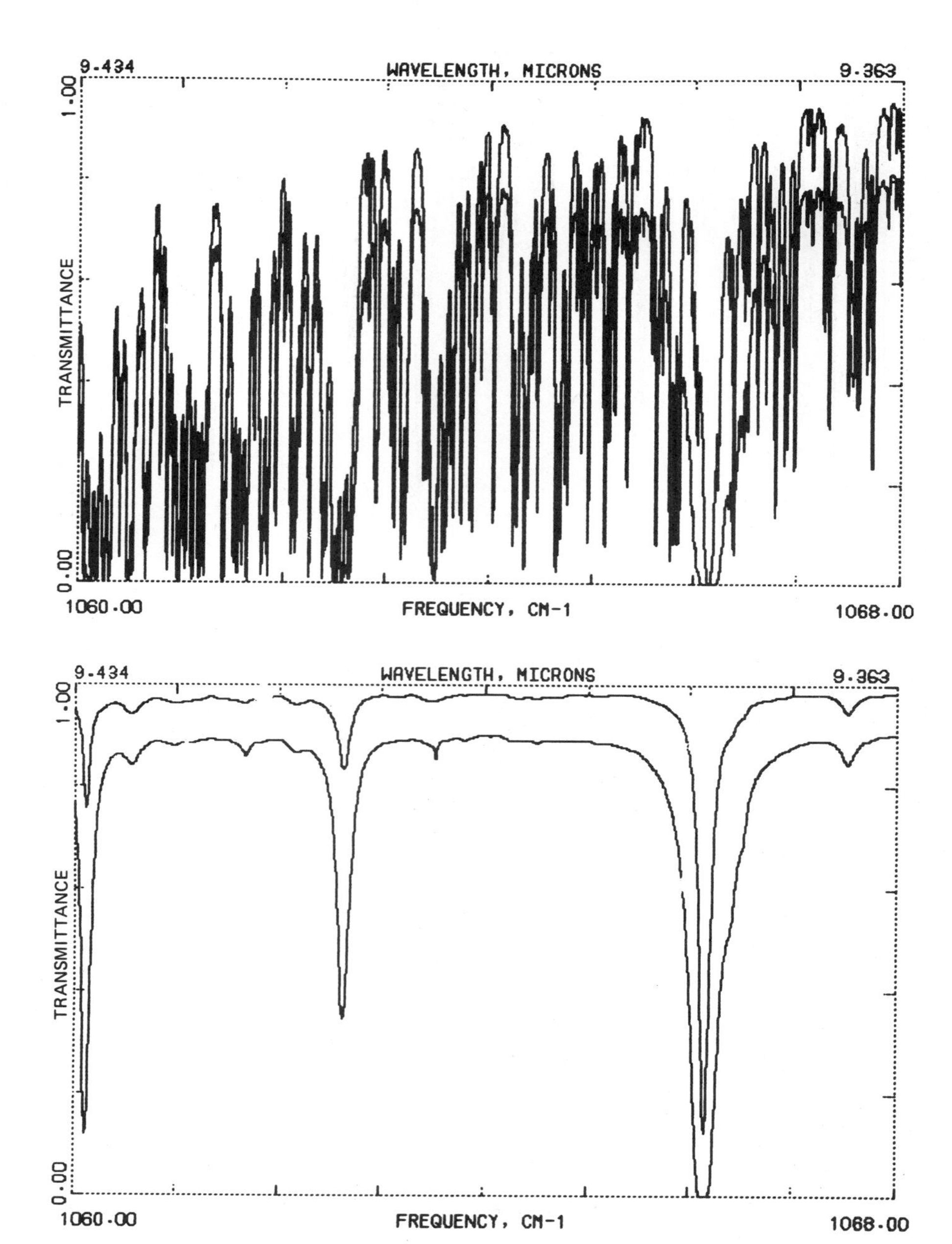

Figure 2.43. Computed transmission spectra for region 1060–1068 cm^{-1}. Other information as for Figure 2.23.

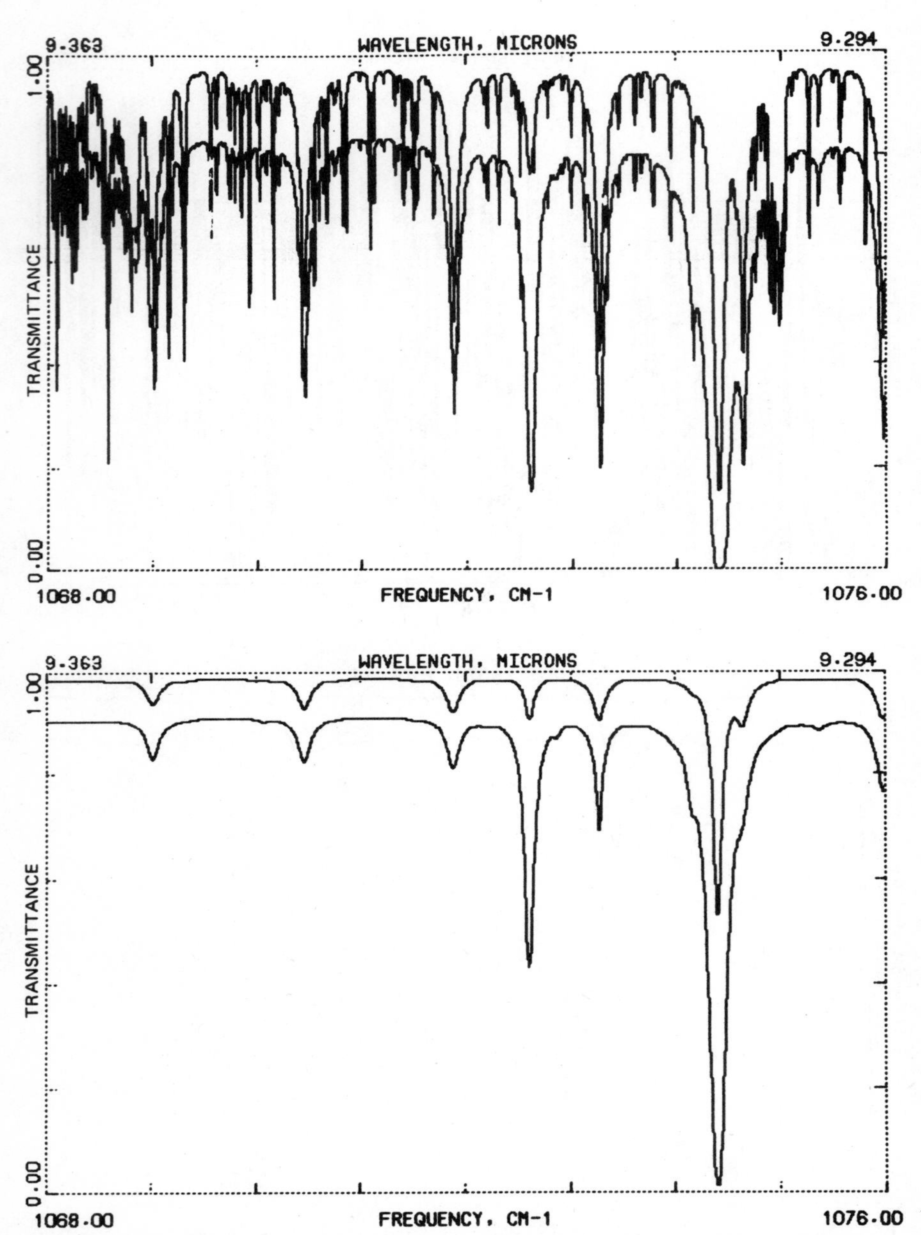

Figure 2.44. Computed transmission spectra for region 1068–1076 cm^{-1}. Other information as for Figure 2.23.

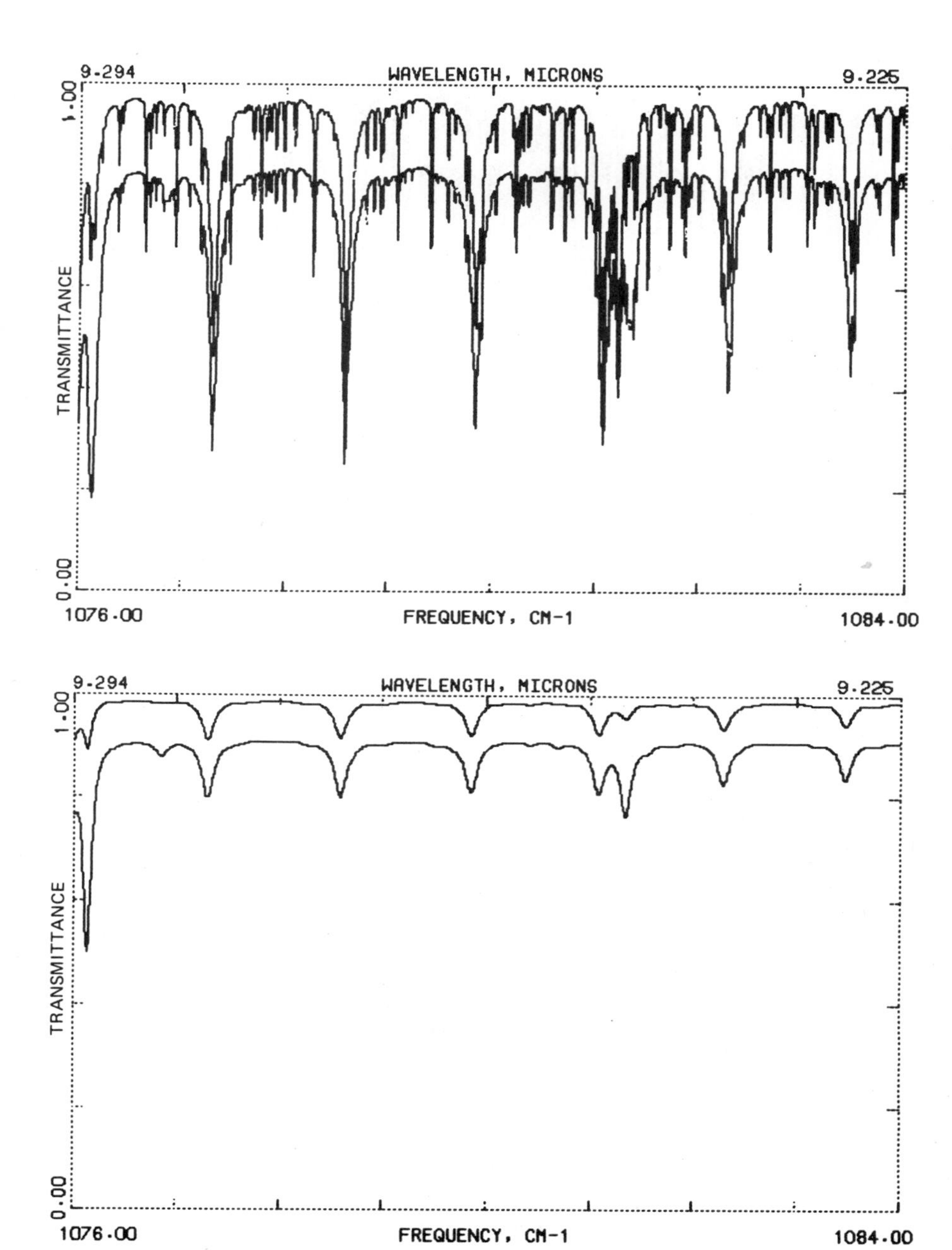

Figure 2.45. Computed transmission spectra for region 1076–1084 cm^{-1}. Other information as for Figure 2.23.

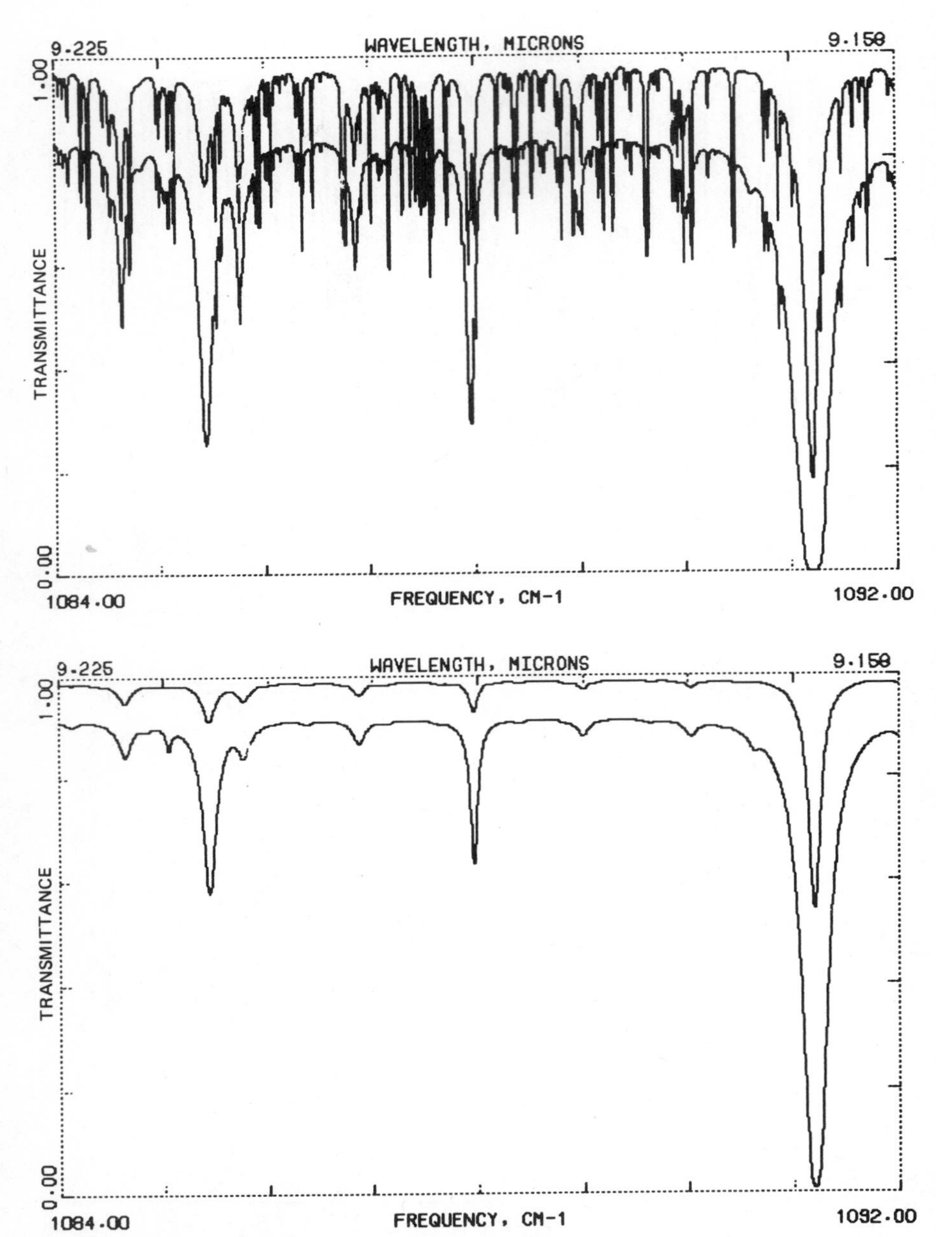

Figure 2.46. Computed transmission spectra for region 1084–1092 cm^{-1}. Other information as for Figure 2.23.

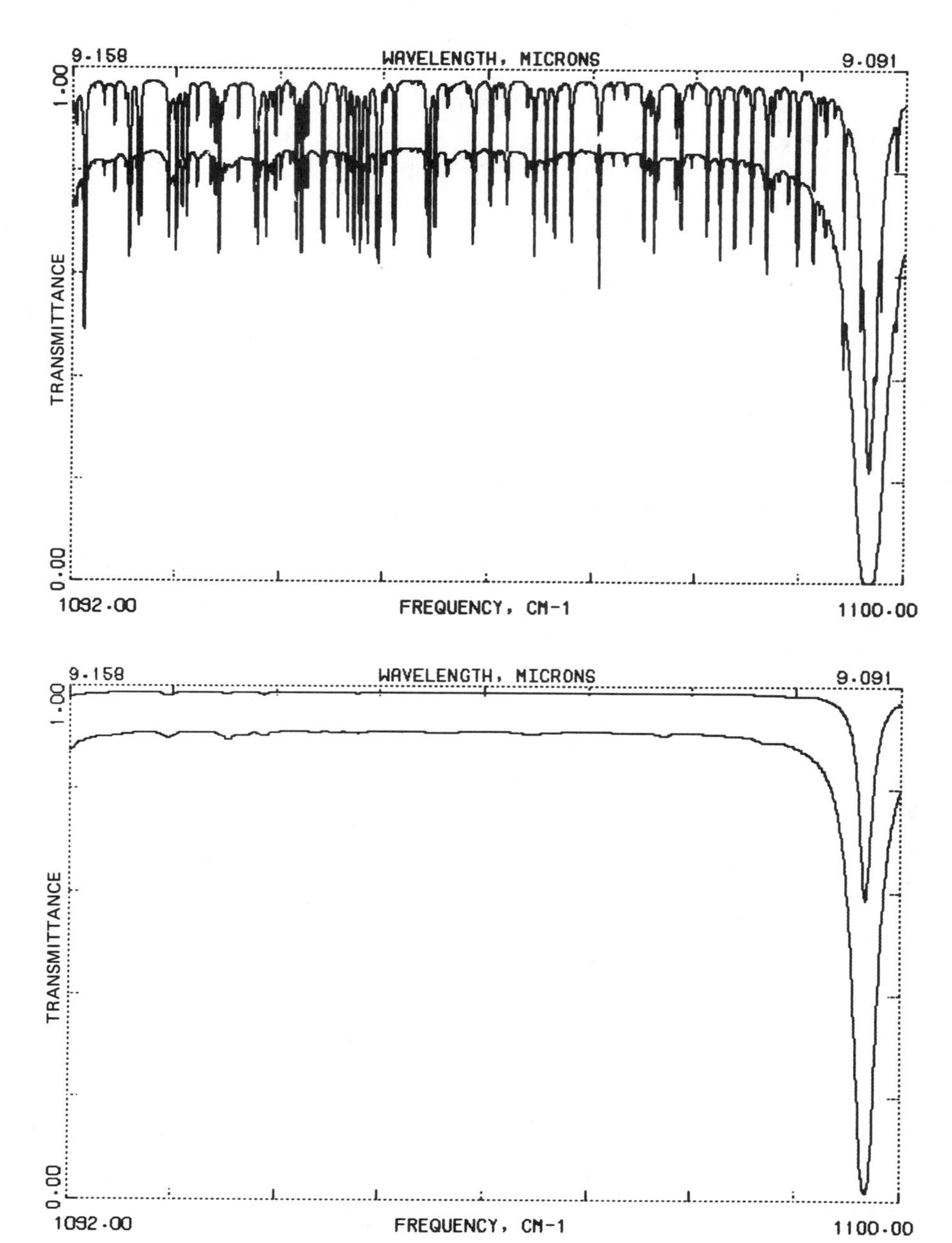

Figure 2.47. Computed transmission spectra for region 1092–1100 cm^{-1}. Other information as for Figure 2.23.

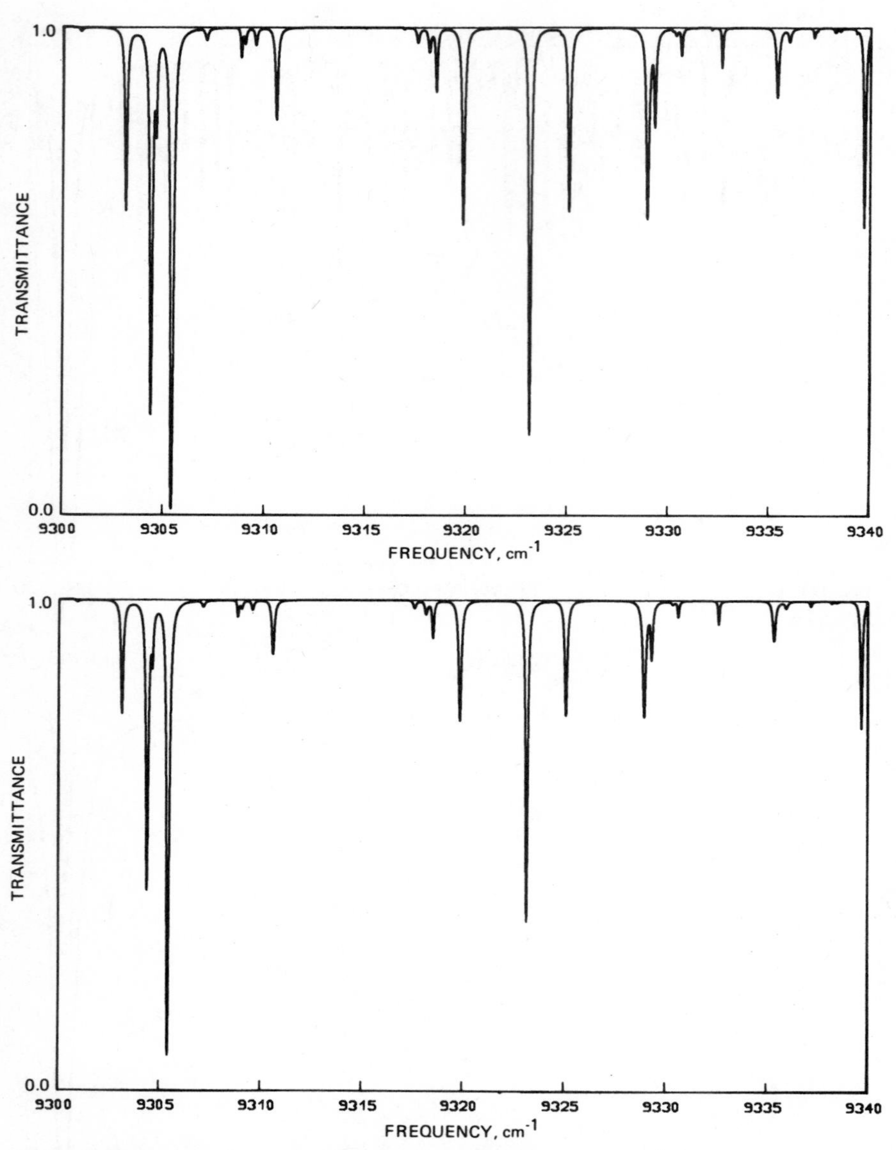

Figure 2.48. Computed transmission spectra for region 9300–9340 cm^{-1}. Upper frame is for vertical, clear, transatmospheric path to sea level; lower frame for 1-km horizontal path at 500 m elevation. Spectra computed for midlatitude summer model (midlatitude winter is not significantly different.)

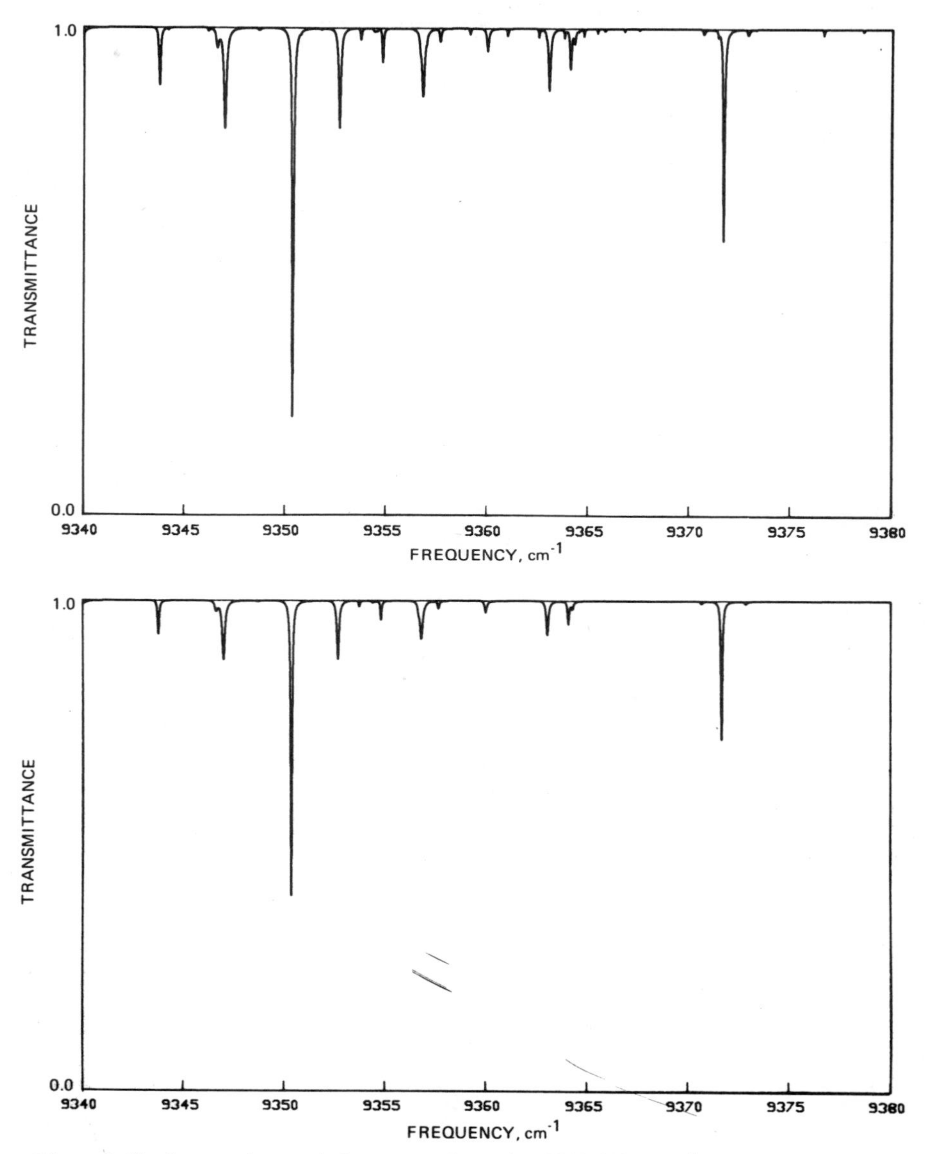

Figure 2.49. Computed transmission spectra for region 9340–9380 cm^{-1}. Other information as for Figure 2.48.

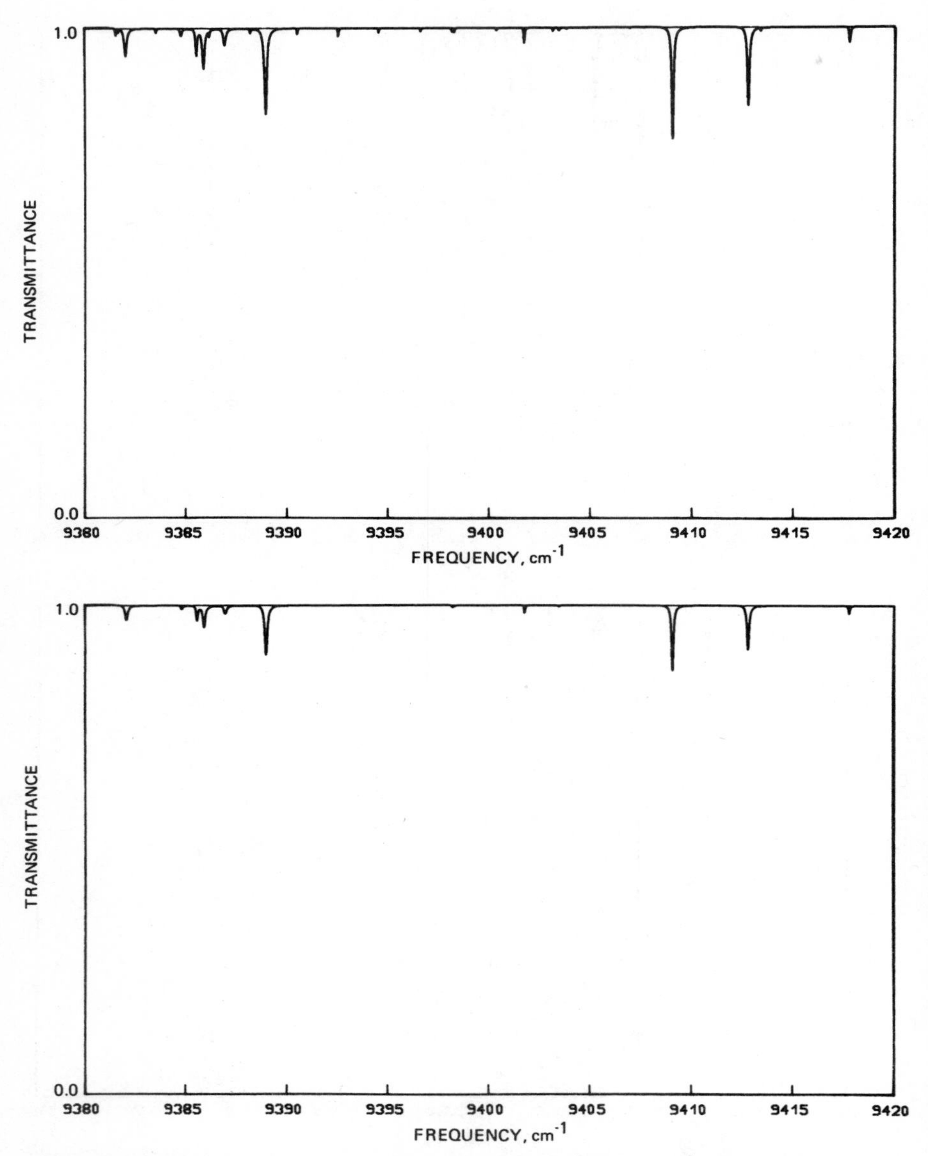

Figure 2.50. Computed transmission spectra for region 9380–9420 cm^{-1}. Other information as for Figure 2.48.

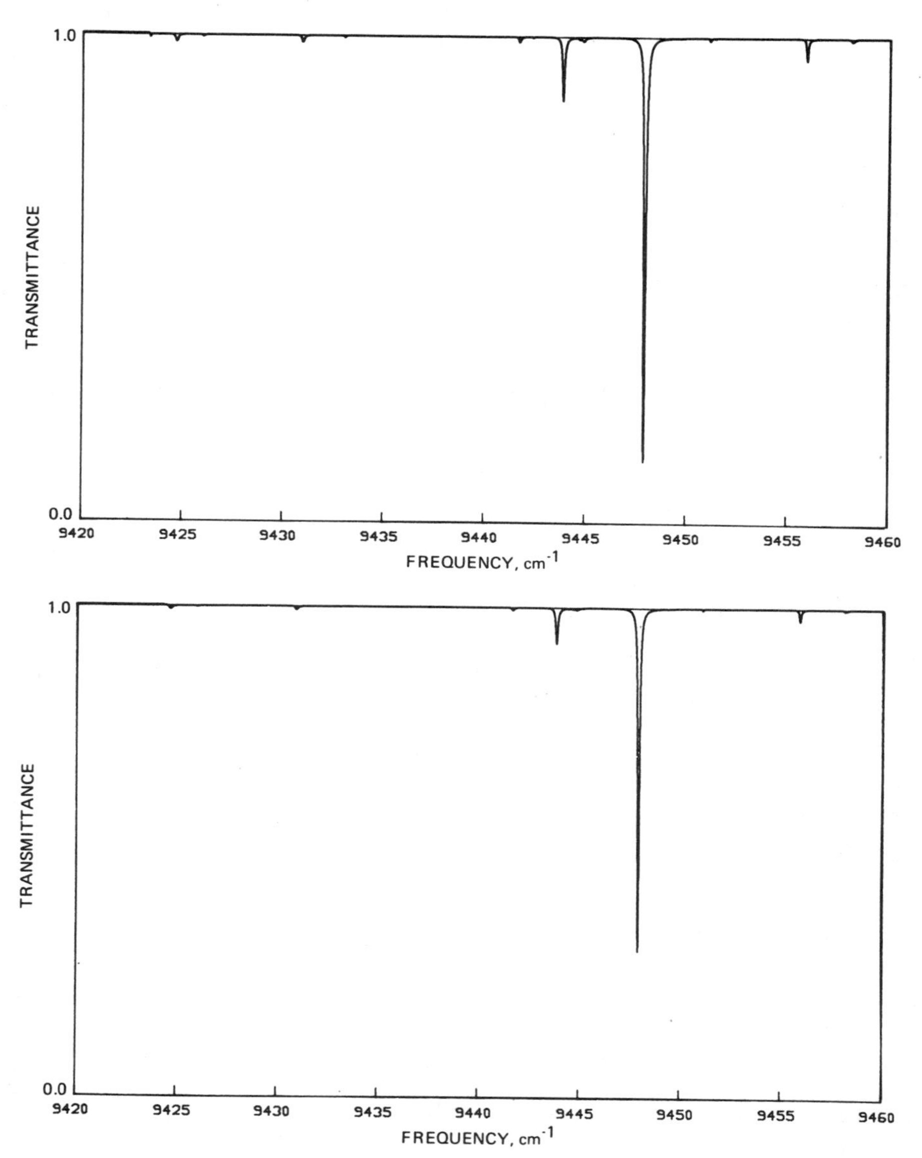

Figure 2.51. Computed transmission spectra for region 9420–9460 cm^{-1}. Other information as for Figure 2.48.

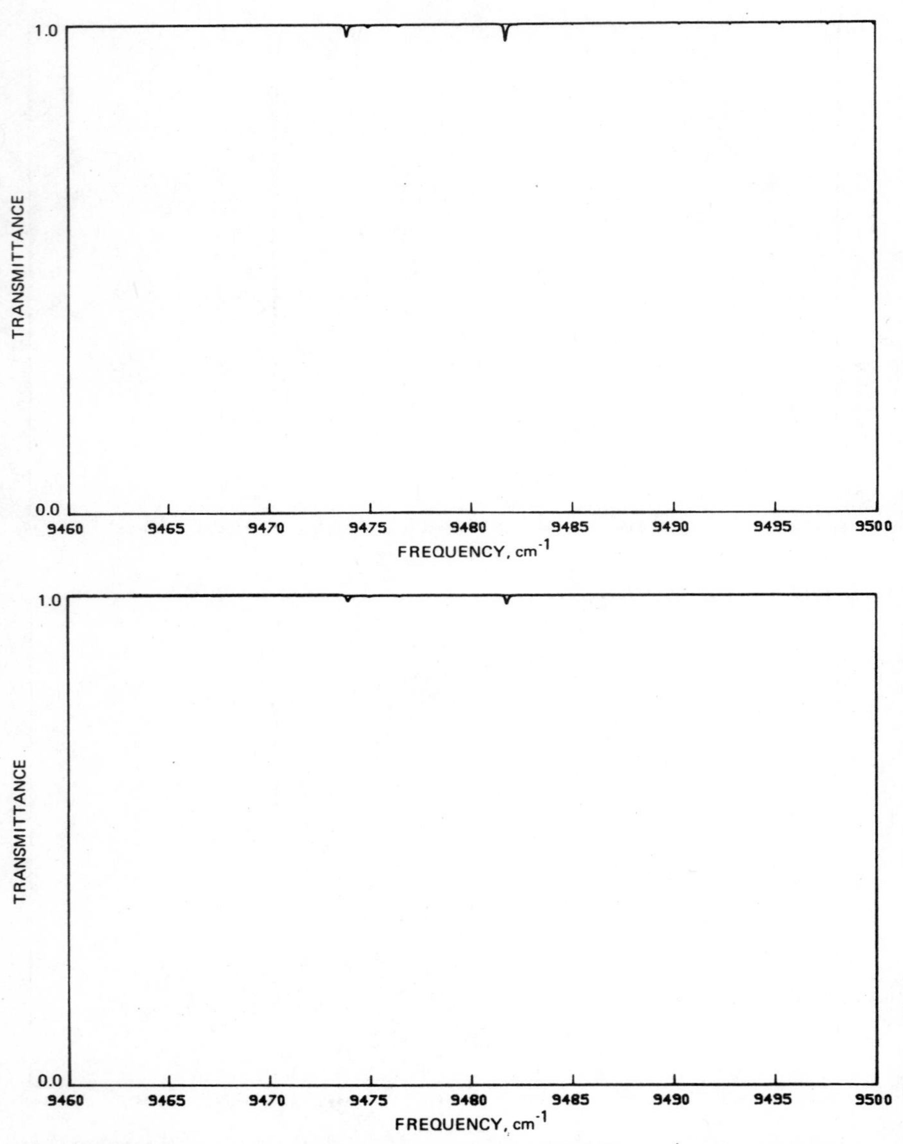

Figure 2.52. Computed transmission spectra for region 9460–9500 cm^{-1}. Other information as for Figure 2.48.

pr mm (MLS) and 8.7 pr mm (MLW). The horizontal values were 11.6 pr mm (MLS) and 3.0 pr mm (MLW).

Figures 2.23–2.52 must be used with caution; they are intended only for illustration of likely, clear, and obscured regions. Furthermore, while monumental, the line compilations used in these calculations are deficient in two respects: (a) there are inevitable errors in the data base, although every effort is made at minimization; and (b) the compilations are incomplete. Although most strong lines are reasonably well reproduced, comparison to experimental data acquired in the field and laboratory show that large numbers of weak features of almost every major absorber are as yet missing. Again, this situation is being remedied, but the task is lone and arduous.

The figures for the two regions are displayed at the same spectral resolution (0.01 cm^{-1}) but at different dispersions: the 9300–9500-cm^{-1} region at 40 cm^{-1} per frame and the 900–1100-cm^{-1} region at 8 cm^{-1} per frame because of the much greater line density in this latter interval. It is hoped that the data will be more useful to investigators than similar plots by other workers (e.g., Traub and Stier, 1976) because these are usually reproduced at too low a dispersion for laser investigations.

2.9 EXPERIMENTAL TRANSMITTANCE DATA

As stated at the outset, there is no substitute for transmission data acquired at the time and place of observation. However, if this is impossible, other experimental data may need to be used. Unfortunately, the vast majority of recent efforts at experimental measures of atmospheric transmission are based on trans-atmospheric solar spectra. While generally having excellent signal-to-noise ratio, they are rarely well calibrated for transmittance because the sun is not, itself, a well-calibrated source. Indeed, in the region of the solar $\Delta V = 1$ CO bands around 2100 cm^{-1}, the uncertainties in the solar irradiance exceed a factor of 2 (Vernazza et al., 1976). Furthermore, most of these data are taken from dry mountain locations (e.g., Kitt Peak, Arizona) and only rarely from sea level. Consequently, correction for other lines of sight at other locations is difficult.

For those investigators interested only in stratospheric and mesospheric transmittance, the situation is somewhat improved. Line densities are much lower than in the troposphere, and scattering is much less important. Most of the available data are solar spectra taken from balloons near sunrise or sunset since the investigators are usually interested in obtaining very long paths for trace constituent analysis. Furthermore, by following the sun to high angular elevations, a first-order correction for the solar continuum is possible.

The bibliography at the end of this chapter lists a selection of the available data; no claims are made for completeness. For the future (mid 1987) a much more extensive solar occultation atlas covering from the upper troposphere to well above 150 km altitude is expected as a product of the SPACELAB 3 ATMOS experiment is 1984. The ATMOS spectrometer (an FTS) covered the range 600–4800 cm^{-1} at 0.013 cm^{-1} resolution with spectra being acquired at 2-km-tangent-height intervals.

This chapter has attempted to provide a brief overview of the phenomena affecting atmospheric transmission in such a way as to be of immediate use to remote sensing investigators. Most topics have been treated cursorily, but space precludes greater depth.

ACKNOWLEDGMENT

All errors of omission and commission are entirely the author's, but acknowledgment for assistance is due to many colleagues, among whom are: Dr. C. B. Farmer, for permission to use the ATMOS Data Analysis Facility and data bases for the production of many of the illustrations; Cynthia A. Welty, who performed the computations used to make Figures 2.23–2.47, helped to edit the text, and is the guardian of EMISSION-SPECTRA; Dr. John V. Martonchik, for many helpful discussions on the vexed question of scattering; Dr. John F. Appleby and James E. Knighton, for the LOWTRAN 5 calculations shown in Figures 2.13–2.21; and Dr. Laurence S. Rothman and colleagues of the Air Force Geophysical Laboratories without whose unstinting efforts over many years there would be no AFGL Line Parameters Compilation and this chapter could not have been written.

The work described in this chapter was carried out at the Jet Propulsion Laboratory, California Institute of Technology, under contract with the National Aeronautics and Space Administration.

BIBLIOGRAPHY

LOWTRAN

Kneizys, F. X., E. P. Shettle, W. O. Gallery, J. H. Chetwynd, Jr., L. W. Abreu, J. E. A. Selby, R. W. Fenn, and R. A. McClatchey (1980). Atmospheric Transmittance/Radiance: Computer Code LOWTRAN 5. Air Force Geophysics Laboratory Report AFGL-TR-80-0067.

Kneizys, F. X., E. P. Shettle, W. O. Gallery, J. H. Chetwynd, Jr., L. W. Abreu, J. E. A. Selby, S. A. Clough, and R. W. Fenn (1983). Atmospheric Transmittance/

Radiance: Computer Code LOWTRAN 6. Air Force Geophysics Laboratory Report AFGL-TR-83-0187.

Scattering

Hansen, J. E., and L. D. Travis (1974). *Light Scattering in Planetary Atmospheres. Space Sci. Rev.* **16,** 527–610.

Bohren, C. F., and D. R. Huffman (1983). *Absorption and Scattering of Light by Small Particles*. Wiley, New York.

Van de Hulst, H. C. (1957). *Light Scattering by Small Particles*. Wiley, New York.

Remote Sensing

Colwell, R. N. (ed.) (1983). *Manual of Remote Sensing* (2nd ed). American Society of Photogrammetry, Falls Church, VA.

Lund, I. A., D. D. Grantham, and C. B. Elam, Jr. (1977). Atlas of Cloud-Free Line-of-Sight Probabilities, Part 3: United States of America. Air Force Geophysics Laboratory Report AFGL-TR-77-0188. Note: 5 reports cover Germany, U.S.S.R., U.S.A., Europe and North Africa, and Middle East.

3DNEPH, USAF Environmental Technical Applications Center, Bldg. 159, Navy Yard Annex, Washington, D.C. 20333.

Meteorology

List, R. J. (ed.) (1958). Smithsonian Meteorological Tables (6th ed.). Smithsonian Institution, Washington, D.C.

Weast, R. C. (ed.) (1982). *CRC Handbook of Chemistry and Physics* (63rd ed.). CRC Press, Boca Raton, FL. Section E.

Solar Atlases

Kurucz, R. L., I. Furenlid, J. Brault, and L. Testerman (1984). Solar Flux Atlas from 296 to 1300 nm, National Solar Observatory, Sunspot, NM.

Debouille, L., G. Roland, J. Brault, and L. Testerman (1981). Photometric Atlas of the Solar Spectrum from 1,850 to 10,000 cm^{-1}. Kitt Peak National Observatory, Tucson, AZ.

Goldman, A., J. W. Vanallen, R. D. Blatherwick, C. M. Bradford, D. G. Murcray, F. H. Murcray, and G. R. Cook (1980). *New Atlas of IR Solar Spectra*. University of Denver, Denver, CO.

Goldman, A., J. W. Vanallen, R. D. Blatherwick, F. H. Murcray, F. J. Murcray, and D. G. Murcray (1982). *New Atlas of Stratospheric IR Absorption Spectra*. University of Denver, Denver, CO.

Blatherwick, R. D., F. H. Murcray, F. J. Murcray, A. Goldman, and D. G. Murcray (1982). *Atlas of South Pole IR Solar Spectra*. University of Denver, Denver, CO.

Hall, D. N. B. (1974). *An Atlas of Infrared Spectra of the Solar Protosphere and of Sunspot Umbrae in the Spectral Intervals 4040-5095, 5550-6700, 7400-8790 cm^{-1}*. Kitt Peak National Observatory, Tucson, AZ.

Delbouille, L., and G. Roland (1963). Photometric Atlas of the Solar Spectrum from λ 7498 to λ 12016. *Mem. Soc. Roy. Sc. de Liege* **4.**

Migeotte, M., L. Neven, J. Swensson (1956). The Solar Spectrum from 2.8 to 23.7 microns. *Mem. Soc. Roy. Sc. de Liege* **1.**

Mohler, O. C., A. K. Pierce, R. R. McMath, and L. Goldberg (1950). *Photometric Atlas of the near Infra-red Solar Spectrum from λ 8465 to λ 25242*. University of Michigan, Ann Arbor, MI.

Farmer, C. B., and P. J. Key (1965). A Study of the Solar Spectrum from 7 to 400 microns. *Appl. Optics* **4,** 1051–1068.

Computed Spectra

Kyle, T. G., and A. Goldman (1975). *Atlas of Computed Infrared Atmospheric Absorption Spectra*. Atmospheric Transmissions in the Wave-Number Region from 1 to 2600 cm^{-1} for Altitudes above 54, 45, 40, 30, 14 and 4 km. National Center for Atmospheric Research, Boulder, CO.

Traub, W. A., and M. T. Stier (1976). Theoretical Atmospheric Transmission in the Mid- and Far-infrared at Four Altitudes. *Appl. Optics* **15,** 364–377. A more complete set of plots is available from the same authors (same title) as *Center for Astrophysics Preprint No. 369*, Cambridge, MA, (1975).

REFERENCES

Allen, C. W. (1973). *Astrophysical Quantities* (3rd ed.). University of London–The Athlone Press, London, England

Ben-Shalom, A., A. D. Devir, S. G. Lipson, U. P. Oppenheim, and E. Ribak (1985). Paper TuC21-1, Topical Meeting on Optical Remote Sensing of the Atmosphere, Optical Society of America, Lake Tahoe, NV.

Bohren, C. F., and D. R. Huffman (1983). *Absorption and Scattering of Light by Small Particles*. Wiley, New York.

Burch, D. E., and D. A. Gryvnak (1980). Continuum Absorption by H_2O Vapor in the Infrared and Millimeter Regions, in *Atmospheric Water Vapor*, A. Deepak, T. D. Wilkerson, and L. H. Ruhnke (eds.). Academic Press, New York.

Chang, Y. S., and J. H. Shaw (1977). A Nonlinear Least Squares Method of Determining Line Intensities and Half-widths. *Appl. Spec.* **31,** 213–220.

Clough, S. A., F. X. Kneizys, L. S. Rothman, G. P. Anderson, and E. P. Shettle (1986). Current Issues in Atmospheric Transparency. *International Meeting on Atmospheric transparency for Satellite Applications*. Capri, Italy (Sponsor: U. Naples)

Dianov-Klokov, V. I., V. M. Ivanov, V. N. Arefev, and N. I. Sizov (1981). Water Vapour Continuum Absorption at 8–13 μ. *J. Quant. Spectrosc. Radiat. Transfer* **25,** 83–92.

Diner, D. J., and J. V. Martonchik (1984). Atmospheric Transfer of Radiation above an Inhomogeneous Non-Lambertian Reflective Ground—I. Theory. *J. Quant. Spectrosc. Radiat. Transfer* **31,** 97–125.

Edlen, K. (1966). The Refractive Index of Air. *Metrologia* **2,** 12.

Gordon, J. E. (1985). A Simple Method for Approximating Mie Scattering. *J. Opt. Soc. Am. A* **2,** 156–159.

Grant, W. B. (1982). Helium-Neon Laser Remote Sensing of Methane, in *Resource Recovery from Solid Wastes*, S. Sengupta and K-F. V. Wong (eds.). Pergammon Press, New York.

Hansen, J. E. and L. D. Travis (1974). Light Scattering in Planetary Atmospheres. *Space Sci. Rev.* **16,** 527–610.

Hudson, R. D. (1971). Critical Review of Ultraviolet Photoabsorption Cross Sections for Molecules of Astrophysical and Aeronomic Interest. *Rev. Geophys. Space Phys.* **9,** 305–406.

Kneizys, F. X., E. P. Shettle, W. O. Gallery, J. H. Chetwynd, L. W. Abreu, J. E. A. Selby, S. A. Clough, and R. W. Fenn (1983). Atmospheric Transmittance/Radiance: Computer Code LOWTRAN 6. Air Force Geophysics Laboratory Report AFG-LTR-83-0187.

Labs, D., and H. Neckel (1968). The Radiation of the Solar Photosphere from 2000 Å to 100 μ. *Z. Astrophys.* **69,** 1–73.

Labs, D., and H. Neckel (1970). Transformation of the Absolute Solar Radiation Data into the International Practical Temperature Scale of 1968. *Solar Phys.* **15,** 79–87.

Le Doucen, R., C. Cousin, C. Boulet, and A. Henry (1985). Temperature Dependence of the Absorption in the Region Beyond the 4.3-μm Band Head of CO_2. 1: Pure CO_2 Case. *Appl. Opt.* **24,** 897–906.

Loper, G. L., M. A. O'Neill, and J. A. Gelbwachs (1983). Water-Vapor Continuum CO_2 Laser Absorption Spectra Between 27°C and −10°C. *Appl. Opt.* **22,** 3701–3710.

Lund, I. A., D. D. Grantham, and C. B. Elam Jr. (1977). Atlas of Cloud-Free Line-of-Sight Probabilities, Part 3: United States of America. Air Force Geophysics Laboratory Report AFGL-TR-77-0188.

Malick, J. D., J. H. Allen, and S. Zakanycz (1979). Calibrated Analytical Modeling of Cloud-Free Intervals, in SPIE Vol. 195, *Atmospheric Effects on Radiative Transfer* pp. 142–147.

Menzies, R. T., and M. J. Kavaya (1985). Coherent CO_2 Lidar Measures of Aerosol Backscatter Coefficients: Wavelength Dependence in the 9-11 Micron Region. Paper IX.3 in *Proceedings of the 3rd Topical Conference on Coherent Laser Radar: Technology and Applications*. July 7–11, Gt. Malvern, England. Sponsored by UK Institute of Physics Quantum Electronics Group and others.

Menzies, R. T., M. J. Kavaya, P. H. Flamant, and D. A. Haner (1984). Atmospheric Aerosol Backscatter Measurements using a Tunable Coherent CO_2 Lidar. *Appl. Opt.* **23,** 2510–2516.

Niple, E. (1980). Nonlinear Least Squares Analysis of Atmospheric Absorption Spectra. *App. Opt.* **19,** 3481–3490.

Norton, R. H., and R. Beer (1976). New Apodizing Functions for Fourier Spectrometry. *J. Opt. Soc. Am.* **66,** 259–264.

Pollack, J. B., and J. N. Cuzzi (1980). Scattering by Non-Spherical Particles of Size Comparable to a Wavelength: A New Semi-Empirical Theory and its Application to Tropospheric Aerosols. *J. Atmos. Sci.* **37,** 868–881.

Ralston, M. L., and R. I. Jennrich (1978). DUD, a Derivative-Free Algorithm for Nonlinear Least Squares. *Technometrics* **20,** 7–14.

Rothman, L. S., R. R. Gamache, A. Barbe, A. Goldman, J. R. Gillis, L. R. Brown, R. A. Toth, J-M Flaud, and C. Camy-Peyret (1983). *Appl. Opt.* **83,** 2247–2256.

Scott, N. A. (1974). A Direct Method of Computation of the Transmission Function of an Inhomogeneous Gaseous Medium—I: Description of the Method. *J. Quant. Spectrosc. Radiat. Transfer* **14,** 691–704.

Traub, W. A., and M. T. Stier (1976). Theoretical Atmospheric Transmission in the Mid- and Far-Infrared at Four Altitudes. *Appl. Opt.* **15,** 364–377.

Van de Hulst, H. C. (1957). *Light Scattering by Small Particles*. Wiley, New York.

Vernazza, J. E., E. H. Avrett, and R. Loeser (1976). Structure of the Solar Chromosphere II. The Underlying Photosphere and Temperature Minimum Region. *Astrophys. J. (Supplement Series)* **30,** 1–60.

Young, A. T. (1980). Revised Depolarization Corrections for Atmospheric Extinction. *Appl. Opt.* **19,** 3427–3428.

CHAPTER

3

INFRARED LASER ABSORPTION: THEORY AND APPLICATIONS

CHRISTOPHER R. WEBSTER, ROBERT T. MENZIES, and E. DAVID HINKLEY*

Jet Propulsion Laboratory
California Institute of Technology
Pasadena, California

**Present address:* Lockheed Advanced Aeronautics Company, 25115 Avenue Stanford, Valencia, CA 91355.

3.1 INTRODUCTION

This chapter describes the theory and applications of IR laser absorption measurements with particular emphasis on the use of tunable lasers, where the full molecular lineshapes are recorded. It is not intended as a review rich in references of all publications in the field. Rather, we have here combined a description of classical absorption and spectroscopic techniques with that of modern laser detection methods to produce a chapter that should prove useful to researchers actually engaged in making laboratory and field measurements, such as the analytical chemist using tunable diode laser harmonic spectroscopy to measure the concentration of a gas phase component in a mixture.

Several excellent introductory texts and review articles covering some of the material included in this chapter are available, including those of Allen and Cross (1963), Corney (1977), Demtroder (1981), Herzberg (1945), Hinkley (1976), Hollas (1982), Measures (1984), Penner (1959), Pugh and Rao (1976), Shimoda (1976), Siegman (1971), Steinfeld (1974), Townes and Schawlow (1975), and Wolfe and Zissis (1978).

In this chapter the symbol ν is used to denote frequency in units of hertz while $\tilde{\nu}$ is used for wavenumber in units of reciprocal centimeters.

3.2 INFRARED ABSORPTION AND SPECTRAL LINESHAPES

3.2.1 Absorption Coefficient and Beer–Lambert Law

A plane electromagnetic wave of angular frequency ω propagating in the Z direction through a medium of refractive index $n(\omega)$ will suffer both absorption (a decrease in amplitude) and dispersion [a change in phase velocity from its value c in vacuum to $v(\omega) = c/n(\omega)$]. These two properties of the interaction of electromagnetic radiation with an absorbing medium can be described by a classical model in which the atomic electrons are treated as damped harmonic oscillators that are forced to oscillate by the electric field of the incident wave. The resulting dispersion equation suggests a complex refractive index $n(\omega)$ that can be divided into two parts (see Measures, 1984):

$$n(\omega) = \eta(\omega) + \tfrac{1}{2}i\chi(\omega) \qquad (3.1)$$

The classical model shows that the real part $\eta(\omega)$, which is the experimentally measured, or absolute, refractive index, represents the dispersion of the wave, that is, the dependence of the phase velocity on frequency. The imaginary part of the complex refractive index represents the absorption of electromagnetic radiation and, as will be shown later, is proportional to the absorption coefficient of the medium.

The Kramers–Kronig dispersion relations relate absorption and dispersion using the complex refractive index and may be evaluated to describe the frequency dependence of $\eta(\omega)$ and $\chi(\omega)$ close to an atomic transition frequency at ω_0:

$$\eta(\omega) = 1 + Ne^2 f_0(\omega_0 - \omega)/4\epsilon_0 m_e \omega[(\omega_0 - \omega)^2 + (\gamma/2)^2] \qquad (3.2)$$

$$\chi(\omega) = Ne^2 f_0 \gamma/4\epsilon_0 m_e \omega[(\omega_0 - \omega)^2 + (\gamma/2)^2] \qquad (3.3)$$

Here N is the number of oscillators (atoms per unit volume), e the electron charge, ϵ_0 the vacuum permittivity (8.854×10^{-12} C^2 N^{-1} m^{-2}), m_e the electron's mass, and γ the damping constant for the electron motion. A plot of $\eta(\omega)$ and $\chi(\omega)$ against frequency near ω_0 is given in Figure 3.1. The value of f_0 ($\leqslant 1$) in the numerators of Equations (3.2) and (3.3) gives the fraction of a single classical harmonic oscillator represented by the transition. This will later be defined as the oscillator strength for the transition. If the irradiance of the incident plane wave is $I(Z)$, its attenuation along a distance dZ is

$$dI = -\alpha(\omega) I \, dZ \qquad (3.4)$$

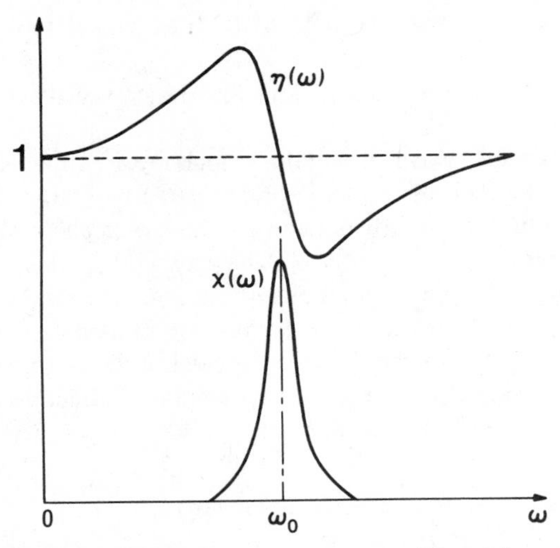

Figure 3.1. Variation of $\eta(\omega)$ and $\chi(\omega)$ near ω_0. [From Measures (1984).]

which on integration can be written in the form of the Beer–Lambert law:

$$I(\omega, Z) = I_0 \exp[-\alpha(\omega) Z] \tag{3.5}$$

The constant I_0 is the irradiance of the wave at $Z = 0$. The frequency-dependent term $\alpha(\omega)$, defined as the linear attenuation coefficient of the medium, is directly proportional to the imaginary part of the refractive index. That is,

$$\alpha(\omega) = \omega\chi(\omega)/c \tag{3.6}$$

From Equation (3.3) the attenuation coefficient becomes

$$\alpha(\omega) = (Ne^2 f_0 / 2m_e \epsilon_0 c)[(\gamma/2)/((\omega_0 - \omega)^2 + (\gamma/2)^2)] \tag{3.7}$$

It is customary to define the attenuation coefficient in units of reciprocal centimeters (linear attenuation coefficient) or squared centimeters per cubic centimeters (volume extinction coefficient). For applications of the Beer–Lambert law to spectroscopic measurements in the IR, we write

$$I(\tilde{\nu}, z) = I_0 \exp[-\kappa(\tilde{\nu}) z_D], \tag{3.8}$$

where $\tilde{\nu}$ is the vacuum wavenumber of the radiation in units of reciprocal centimeters related to the vacuum wavelength and frequency by

$$\tilde{\nu}\ (\mathrm{cm}^{-1}) \equiv 1/\lambda\ (\mathrm{cm}) \equiv \nu\ (\mathrm{Hz})/c\ (\mathrm{cm\ s}^{-1}) \equiv \omega\ (\mathrm{Hz})/2\pi c\ (\mathrm{cm\ s}^{-1}) \tag{3.9}$$

The dimensionless exponent $\kappa(\tilde{\nu})z_D$ is called the optical thickness of the medium, where $\kappa(\tilde{\nu})$ is the absorption coefficient at $\tilde{\nu}$. The parameter z_D is given by a concentration unit multiplied by a length unit and is sometimes referred to as the optical density, which is the number per unit area of absorbing molecules within the sample along the direction of propagation of the radiation (see Pugh and Rao, 1976). The units of the irradiance, I and I_0 are usually watts per squared centimeters.

The Beer–Lambert law describes the exponential attenuation of radiation traveling through an absorbing medium but is valid only for the linear absorption regime, that is, when the absorbed irradiance is strictly proportional to the incident irradiance, and any change of molecular energy level populations due to the radiation field can be neglected. When IR laser sources are employed, it is possible to observe a breakdown of the Beer–Lambert law as the populations of excited and ground energy levels change significantly, and dI is no longer proportional to I. For high-intensity radiation the absorption becomes nonlinear as the populations of levels coupled by the laser field tend to become equal (assuming their degeneracies are the same; Measures, 1984) and the absorption coefficient approaches zero. Under these circumstances the sample becomes transparent at the frequency of the incident radiation.

In the field of analytical chemistry, the Beer–Lambert law is often expressed in logarithmic form:

$$\kappa(\tilde{\nu})z = \log_e(I_0/I) = 2.303\ \log_{10}(I_0/I) \tag{3.10}$$

and the dimensionless quantity $\log_{10}(I_0/I)$ is defined as the absorbance, which is itself a function of $\tilde{\nu}$. The absorbance A is then written

$$A = \log_{10}(I_0/I) = \epsilon(\tilde{\nu})c'l \tag{3.11}$$

When the pathlength is in units of centimeters and the concentration is expressed in moles per liter, $\epsilon(\tilde{\nu})$ is the molar absorption (extinction) coefficient or molar absorptivity, and has dimensions $1/[\text{concentration } (c) \times \text{length } (l)]$. The following equations define spectral absorption (absorptance) $\mathrm{Ab}(\tilde{\nu})$ and spectral transmission (transmittance) $T(\tilde{\nu})$:

$$\mathrm{Ab}(\tilde{\nu}) = [I_0(\tilde{\nu}) - I(\tilde{\nu})]/I_0(\tilde{\nu}) \tag{3.12}$$

$$T(\tilde{\nu}) = I(\tilde{\nu})/I_0(\tilde{\nu}) \tag{3.13}$$

3.2.2 Molecular Absorption Lineshapes and Integrated Absorption Coefficient

Specifying the maximum value of the molar absorption (extinction) coefficient is not a satisfactory measure of the total absorption intensity since no inclusion of the width of the spectral feature is made. The intensity of a spectral feature must therefore be specified by the area under the absorption curve over the wavelength region of interest. For absorption by a separate line characterized by a definite shape, line center position, and width, the line intensity, or line-strength, is defined as the integrated absorption coefficient

$$S = \int_{-\infty}^{\infty} \kappa(\tilde{\nu})\, d\tilde{\nu} \tag{3.14}$$

The absorption coefficient $\kappa(\tilde{\nu})$ can then be written in the useful form

$$\kappa(\tilde{\nu}) = Sg(\tilde{\nu} - \tilde{\nu}_0) \tag{3.15}$$

where $g(\tilde{\nu} - \tilde{\nu}_0)$ is the lineshape function or ''shape function'' normalized so that

$$\int_{-\infty}^{\infty} g(\tilde{\nu} - \tilde{\nu}_0)\, d\nu = 1 \tag{3.16}$$

The normalized intensity profile $g(\tilde{\nu} - \tilde{\nu}_0) = C\Pi(\tilde{\nu})$ then allows different line profiles described by $\Pi(\tilde{\nu})$ to be compared directly.

The units of the shape function $g(\tilde{\nu} - \tilde{\nu}_0)$ are inverse linewidth, or centimeters. We will, in Section 3.2.11, express the line intensities S in units of centimeters per molecules, so that $\kappa(\tilde{\nu})$ will be in units of centimeters squared per molecules and z_D in the concentration times length units of molecules per cubic centimeters per centimeters. The units of $\kappa(\tilde{\nu})$ are in this case more in accord with its definition as an absorption cross section rather than an absorption coefficient.

Even in the absence of instrumental broadening, isolated spectral lines are observed to have a nonzero width and a definite, reproducible lineshape characterized by the shape function $g(\tilde{\nu} - \tilde{\nu}_0)$. The observed lineshape and its associated width depends on the relative importance of each of several broadening mechanisms; namely, natural line broadening, Doppler broadening, collisional (pressure) broadening (or narrowing), saturation broadening, and modulation broadening. These mechanisms will be described in more detail in the next three sections.

The spectral distribution of the absorbed or emitted intensity in the vicinity of the central frequency $\tilde{\nu}_0$ is described by the line profile function $\Pi(\tilde{\nu})$, shown in Fig. 3.2. The full width at half maximum (FWHM) of the spectral line is defined as the frequency interval $\delta\tilde{\nu} = (\tilde{\nu}_b - \tilde{\nu}_a)$ between frequencies $\tilde{\nu}_a$ and $\tilde{\nu}_b$ for which

$$\Pi(\tilde{\nu}_a) = \Pi(\tilde{\nu}_b) = \tfrac{1}{2}\Pi(\tilde{\nu}_0). \tag{3.17}$$

In this chapter we will use the term *half-width* to refer to the half-width at half maximum (HWHM) denoted by the symbol γ_s when in units of reciprocal centimeters, and the subscript s will describe the broadening mechanism. In terms of a full width as a spread in wavelengths, differentiating Equation (3.9) gives the identity

$$|\delta\lambda| = (c/\nu^2)\,\delta\nu = (1/\tilde{\nu}^2)\,\delta\tilde{\nu} \tag{3.18}$$

where the relative widths $|\delta\nu/\nu| = |\delta\tilde{\nu}/\tilde{\nu}| = |\delta\lambda/\lambda|$ are identified. As will be described in Section 3.2.12, the term *equivalent width* is a parameter useful in deriving linestrengths and linewidths from experimentally recorded spectra. The equivalent width W is the width (in cm^{-1}) of the rectangle of height I_0 that has the same area as that under the spectral line.

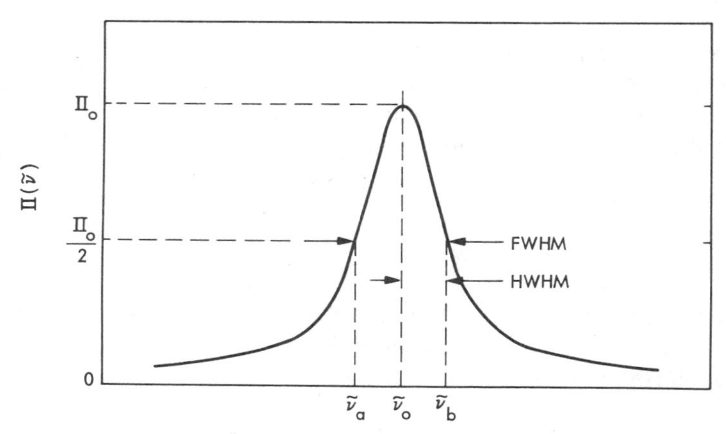

Figure 3.2. Plot of line profile function II $(\tilde{\nu})$ against $\tilde{\nu}$ showing HWHM and FWHM.

In this chapter we will quote linewidths in units of either reciprocal centimeters or hertz (kHz, MHz), but it should be pointed out that spectroscopists often quote linewidths in units of millikaysers, where

$$1 \text{ mK} = 10^{-3} \text{ cm}^{-1} \equiv 29.979 \text{ MHz}$$

In describing the various broadening mechanisms of atomic or molecular lineshapes, distinction must be made between homogeneous and inhomogeneous line broadening. Radiative and collisional processes produce homogeneously broadened lines because the probability of absorption or emission of radiation is the same for all atoms or molecules in the same energy level. Inhomogeneous broadening includes Doppler broadening, where the probability of absorption at a particular frequency ω is not the same for all atoms or molecules, but depends, in this case, on their velocity. As we will see in a discussion of Voigt profiles, the inhomogeneously broadened transition must be described by both a homogeneous linewidth, the width of the response of a single atom or molecule, and an inhomogeneous linewidth, determined by the spread of the individual atomic center frequencies by inhomogeneous mechanisms like the Doppler effect.

3.2.3 Natural or Radiative Line Broadening

When excited energy states are populated, they decay as a function of time with a relaxation time (lifetime) τ related to the Einstein coefficient for spontaneous emission, A_{nm}, by

$$\tau = 1/A_{nm} = (4\pi\epsilon_0)3hc^3/[|R^{nm}|^2(64\pi^4\nu^3)] \tag{3.19}$$

where R^{nm} is the transition moment for the transition between the states m and n. This finite lifetime of the energy states involved in the transition produces, according to the Heisenberg uncertainity principle ($\tau\,\Delta E \geqslant h/2\pi$), a spread, $\Delta\nu_n$, in the emission (absorption) frequencies called natural or radiative line broadening:

$$\Delta\nu_n \geqslant 1/2\pi\tau \geqslant 32\pi^3\nu^3|R^{nm}|^2/[(4\pi\epsilon_0)(3hc^3)] \tag{3.20}$$

The spectral distribution of a lifetime-limited emission line is the same as that of the classical absorber described by Equation (3.7) where $1/\tau = \gamma/2$. The distribution is Lorentzian and takes the form

$$\Pi(\omega - \omega_0) = \Pi_0/[(\omega - \omega_0)^2 + (\gamma/2)^2] \tag{3.21}$$

so that the shape function of the normalized Lorentzian profile may be written as

$$g_n(\tilde{\nu} - \tilde{\nu}_0) = (\gamma_n/\pi)/[(\tilde{\nu} - \tilde{\nu}_0)^2 + \gamma_n^2] \tag{3.22}$$

where γ_n represents the HWHM (here in cm^{-1}). The FWHM is given by

$$\begin{aligned} \delta\omega_n\,(\text{rad s}^{-1}) &= 2\gamma \\ \delta\omega_n\,(\text{Hz}) &= 2\gamma/2\pi \\ \delta\omega_n\,(\text{cm}^{-1}) &= 2\gamma/2\pi c = 2\gamma_n \end{aligned} \tag{3.23}$$

The natural linewidth of a spectral line will actually depend on the energy spread and therefore the lifetime of both energy states coupled by the radiation field. According to Equation (3.20), the natural or radiative linewidth has a ν^3 dependence on frequency. This linewidth can be a few tens of megahertz for transitions from excited electronic states, a few tens of the kilohertz from excited vibrational states, and only $\sim 10^{-4}$–10^{-5} Hz for transitions coupling rotational energy states. Expressed in an alternative manner, excited electronic states have short lifetimes, typically on the order of 10^{-8} s, excited vibrational states lifetimes on the order of 10^{-3} s, and excited rotational energy states long lifetimes on the order of 10^{-1} s. Even electronic transitions with small transition probabilities ("forbidden transitions") can be associated with excited states of long radiative lifetime (e.g., ~ 1 s) and therefore very narrow (~ 0.1 Hz) natural linewidths. These lines are often very weak and hard to observe (but not always), and the linewidths are usually masked by other effects such as Doppler or pressure broadening.

For a Lorentzian lineshape such as that resulting from natural or radiative line broadening, the absorption coefficient can be written as

$$\kappa(\tilde{\nu}) = (S/\pi)\gamma_n/[(\tilde{\nu} - \tilde{\nu}_0)^2 + \gamma_n^2] \tag{3.24}$$

At line center, where $\tilde{\nu} = \tilde{\nu}_0$, the absorption coefficient is

$$\kappa(\tilde{\nu}_0) = S/\pi\gamma_n \tag{3.25}$$

and when $\tilde{\nu} = \tilde{\nu}_0 \pm \gamma_n$,

$$\kappa(\tilde{\nu}_0 \pm \nu_n) = S/2\pi\gamma_n \tag{3.26}$$

3.2.4 Doppler Line Broadening

Doppler line broadening results from the motion of absorbing or emitting atoms or molecules relative to the source or detector. When an isolated molecule absorbs a photon, it is required to conserve both energy and momentum. This is done by a change in the molecule's translational energy by an amount $h\nu_0(v/c)\cos\theta$, where θ is the angle between the molecular velocity vector v and the photon direction, and $h\nu_0$ the transition energy producing the line of central frequency ν_0. A range of photon energies, determined by the molecular velocity distribution, can therefore be absorbed. For a given transition of resonant frequency ν_0 for stationary molecules, the actual frequency of light absorbed by molecules moving with a velocity v_z measured positively away from the source is

$$\nu = \nu_0(1 - v_z/c) \tag{3.27}$$

where c is the velocity of light. For typical average molecular velocities of 500 ms^{-1} at room temperature, we see that Doppler shifts on the order of $2 \times 10^{-6}\nu_0$ may be encountered. In the mid-IR region, this corresponds to shifts on the order of 100 MHz ($\simeq 3 \times 10^{-3}$ cm^{-1}), which gives us an estimate of the linewidths involved in Doppler broadening.

The shape of the Doppler-broadened line is determined by the proportion of molecules having each velocity around the most probable molecular velocity. The distribution of molecular speeds at thermal equilibrium is given by the Maxwell–Boltzmann distribution for translational energy states. For a given temperature, this distribution is a bell-shaped Gaussian curve, as is, therefore, the Doppler lineshape (see Figure 3.3). According to the kinetic theory for an ideal gas (see Atkins, 1978), the most probable velocity, v_p, which is the velocity at which the distribution passes through its maximum, is given by

$$v_p = (2kT/m)^{1/2} \tag{3.28}$$

where k is the Boltzmann constant, T the gas temperature in kelvins, and m the molecular mass. For NO_2 at 300 K, for example, the most probable velocity turns out to be $v_p \simeq 270$ ms^{-1}.

The Maxwellian distribution gives the probability $P(v_z)\,dv_z$ of an atom having a velocity between v_z and $v_z + dv_z$ as

$$P(v_z)\,dv_z = (m/2\pi kT)^{1/2}\exp(-mv_z^2/2kT)\,dv_z \tag{3.29}$$

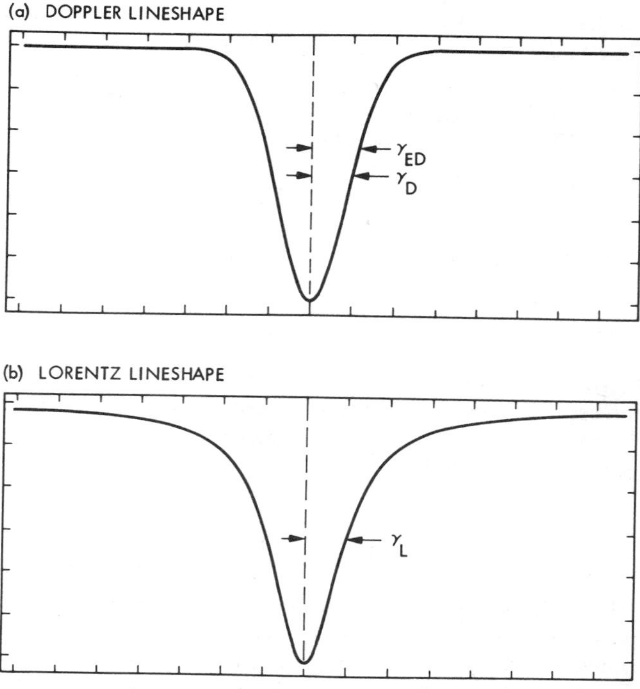

Figure 3.3. Comparison between (*a*) Doppler and (*b*) Lorentz lineshapes at same peak amplitude and half-width.

For a linewidth parameter

$$\gamma_D = (\tilde{\nu}_0/c)(2kT \ln 2/m)^{1/2} \tag{3.30}$$

we can use Equation (3.27) to write the probability of detecting radiation of wavenumber between $\tilde{\nu}$ and $\tilde{\nu} + d\tilde{\nu}$,

$$P(\tilde{\nu})\, d\tilde{\nu} = (1/\gamma_D)(\ln 2/\pi)^{1/2} \exp[-(\tilde{\nu} - \tilde{\nu}_0)^2 \ln 2/\gamma_D^2]\, d\tilde{\nu} \tag{3.31}$$

The normalized shape function for the Doppler-broadened line can therefore be written as

$$g_D(\tilde{\nu} - \tilde{\nu}_0) = (1/\gamma_D)(\ln 2/\pi)^{1/2} \exp[-(\tilde{\nu} - \tilde{\nu}_0)^2 \ln 2/\gamma_D^2] \tag{3.32}$$

where the Doppler HWHM is given by Equation (3.30). To avoid the clumsiness of the ln 2 factor, a width parameter γ_{ED}, known as the $1/e$ Doppler width, may be introduced such that

$$\begin{aligned} \gamma_{ED} &= \gamma_D/(\ln 2)^{1/2} \\ &= (\tilde{\nu}_0/c)(2kT/m)^{1/2} \end{aligned} \tag{3.33}$$

The shape function becomes

$$g_D(\tilde{\nu} - \tilde{\nu}_0) = (1/\gamma_{ED}\pi^{1/2})\exp[-(\tilde{\nu} - \tilde{\nu}_0)^2/\gamma_{ED}^2] \tag{3.34}$$

The Doppler width generally does not depend on gas pressure (see Section 3.2.6) but does increase with the square root of the temperature of the gas. If we write M to signify the molecular weight, the Doppler HWHM can be calculated from (Townes and Schawlow, 1975)

$$\gamma_D\,(\mathrm{cm}^{-1}) = 3.581 \times 10^{-7}\tilde{\nu}_0(T/M)^{1/2} \tag{3.35}$$

For NO_2 ($M = 46$) at 300 K absorbing IR light as $\tilde{\nu}_0 = 1600\ \mathrm{cm}^{-1}$, the value of γ_D would be 0.00146 cm^{-1} ($\cong$44 MHz). Large Doppler widths result for light molecules absorbing or emitting short wavelengths at high temperatures. For example, OH ($M = 17$) lines resulting from UV emission at 308 nm from a flame whose temperature is 1200 K have a Doppler-broadened HWHM close to 0.1 cm^{-1} (3 GHz).

For a Gaussian lineshape resulting from Doppler broadening, the absorption coefficient can be written as

$$\kappa(\tilde{\nu}) = (S/\gamma_D)(\ln 2/\pi)^{1/2}\exp[-(\tilde{\nu} - \tilde{\nu}_0)\ln 2/\gamma_D^2] \tag{3.36}$$

At line center, where $\tilde{\nu} = \tilde{\nu}_0$, the absorption coefficient is

$$\kappa(\tilde{\nu}_0) = (S/\gamma_D)(\ln 2/\pi)^{1/2} \tag{3.37}$$

3.2.5 Collisional Line Broadening

In describing natural or radiative broadening in Section 3.2.3, we saw that the finite lifetime of isolated absorbing or radiating states resulted in a spread in their energies according to the uncertainty principle, which determined the natural linewidth of the transition involved. Atoms and molecules, however, are subject to interaction forces of neighboring atoms, molecules, ions, or electrons that perturb the energy states and produce additional line broadening that is often greater than the radiative linewidth. This line broadening is of interest to theoreticians and experimentalists alike because its observation, measurement, and interpretation provides direct information on fundamental interatomic and intermolecular forces and mechanisms. The prime interest, however, in line broadening and shifts is to allow the quantitative interpretation of atmospheric and laboratory spectra. Except for collisions involving hydrogen and helium, short-range interatomic forces are less important to collisional line broadening than long-range attractive van der Waals forces. These interatomic forces are often represented for binary systems (excited atom plus single perturber) that for multiperturbers must be averaged using either the quasi-static approximation or the impact approximation, the latter being also known as the phase shift approximation.

Following the treatment of Corney (1977), at a frequency $\Delta\nu = \nu_0 - \nu$ from line center the shape of the spectral line profile is determined by the wavetrain emitted during the time interval Δt where $\Delta t = 1/2\pi\ \Delta\nu$. The time interval Δt, which is therefore determined by the frequency displacement from line center, will be the mean time between collisions, τ_c, if the atom emits a wavetrain until perturbed by a collision. The quasi-static approximation applies when the duration of a collision (distance of closest approach divided by the mean velocity at this distance) is greater than τ_c, and the perturber motion may be ignored. This approximation is good at lower temperatures when the mean velocities are smaller and at high pressures when the mean time between collisions, τ_c, is short. The quasi-static theory of line broadening is applicable to the far wings of the line, where the profile is non-Lorentzian and shows strong temperature dependence. The impact or phase shift approximation best represents the case where the duration of the collision is short compared to the mean time between collisions, τ_c. This approximation is very good at higher temperatures when the mean velocities are larger and at low pressures (up to several hundred Torr) where τ_c is long. The full lineshape may therefore be accurately represented using the impact approximation near line center and the quasi-static approximation for the wings. In the IR wavelength region where typical pressure-broadened linewidths are much less than the transition frequency, the lineshape is accurately approximated by the impact theory up to a few linewidths from line center. The complexity of both the molecular interactions and the required mathematics has so far prohibited the construction of a single analytical lineshape applicable at all frequencies.

In its simplest form due to Lorentz (1906), the impact approximation assumes that on collision the classically described atomic electron oscillation halts and then restarts following the collision with a phase completely unrelated to that before the event. A later approximation of Van Vleck and Weisskopf (1945) assumes that there is in fact some phase relation between the pre- and postcollision oscillations. Incoherence in the wavetrains was arbitrarily assumed to result only from collisions that produced a phase change of unity. Phase changes less than unity are included in the more recent adiabatic impact theories that, unlike the Van Vleck and Weisskopf theory, also account for the lineshifts observed in collisionally broadened lines.

According to the simple form due to Lorentz, the collisionally broadened normalized lineshape is written:

$$g_L(\tilde{\nu} - \tilde{\nu}_0) = (\gamma_L/\pi)/[(\tilde{\nu} - \tilde{\nu}_0)^2 + \gamma_L^2] \tag{3.38}$$

where γ_L is the HWHM in reciprocal centimeters. This shape function is identical to that given in Equation (3.22) for a radiatively broadened lineshape. The parameter γ_L may be identified, in the impact approximation, with the mean time between collisions. From kinetic theory the linewidth for a two-component mixture may be written (Corney, 1977):

$$\gamma_L = 1/\left\{8\pi kT\left[N_a D_{aa}^2(2/m_a)^{1/2} + N_b D_{ab}^2(1/m_a + 1/m_b)^{1/2}\right]\right\}^{1/2} \tag{3.39}$$

Where N_a and N_b are the number densities of absorbing molecules, a, and non-absorbing molecules, b; D_{aa} and D_{ab} are the sum diameters of collisions a to a and a to b between molecules; and m_a and m_b are the molecular masses.

The pressure and temperature dependence of the linewidth is given by

$$\gamma_L = \gamma_L^0(P/P_0)(T_0/T)^s \tag{3.40}$$

When P is the gas pressure and γ_L^0 is the HWHM at standard conditions of temperature (T_0) and pressure (P_0). Although the temperature dependence is not necessarily a power law, it is often represented in this form. The effective power s is often $\frac{1}{2}$, but this is not always true and depends on the perturber and parent molecule interaction and may depend on the rotational state. When both self-broadening and foreign gas broadening are included and are described by linewidths γ_a and γ_b, the Lorentzian HWHM may be written as a sum of two components, which for $s = \frac{1}{2}$ becomes

$$\gamma_L = [\gamma_a(P_a/P_0) + \gamma_b(P_b/P_0)](T_0/T)^{1/2} \tag{3.41}$$

As is evident from Equation (3.40), the pressure-broadened linewidth is inversely proportional to the square root (for $s = \frac{1}{2}$) of the temperature, unlike the Doppler-broadened linewidth, which is directly proportional to the square root of temperature. The linewidth γ_L is often denoted b_c by spectroscopists and termed the *pressure-broadening coefficient*. Self-broadening coefficients are generally higher than those due to foreign gas broadening. Highly polar molecules like NH_3 or HNO_3 have self-broadening coefficients as high as 1 cm^{-1} atm^{-1} and with wide variation observed line by line. The self-broadening coefficient of CH_4, on the other hand, is much lower ($\simeq 0.08$ cm^{-1} atm^{-1}) and shows little line-by-line variation. H_2–H_2 broadening is extremely low (<0.01 cm^{-1} atm^{-1}). Foreign gas-broadening coefficients can be as high as 0.15 cm^{-1} atm^{-1} for N_2 on NH_3 or H_2O and are extremely low where hydrogen or helium provide the foreign gas.

For a collisionally broadened line, the absorption coefficient can be written:

$$\kappa(\tilde{\nu}) = (S/\pi)\gamma_L/[(\tilde{\nu} - \tilde{\nu}_0)^2 + \gamma_L^2] \qquad (3.42)$$

which at line center, where $\tilde{\nu} = \tilde{\nu}_0$, becomes

$$\kappa(\tilde{\nu}_0) = S/\pi\gamma_L \qquad (3.43)$$

and when $\tilde{\nu} = \tilde{\nu}_0 \pm \gamma_L$,

$$\kappa(\tilde{\nu}_0 \pm \gamma_L) = S/2\pi\gamma_L \qquad (3.44)$$

It should be noted that the absorption coefficient is proportional both to the lineshape function $g_L(\tilde{\nu} - \tilde{\nu}_0)$ and to the number of absorbers. Since the Lorentzian shape function of a pressure-broadened line is inversely proportional to the linewidth γ_L, and therefore the collision rate, the peak absorption coefficient is independent of pressure in the purely self-broadened case. This may be seen in Figure 3.4 for gas pressures above ~ 10 Torr where the collision-broadened Lorentzian shape dominates the line profile.

3.2.6 Dicke Narrowing

In a discussion of collisional broadening of spectral lines we must also include the phenomenon of collisional, or Dicke, narrowing, whereby with increasing pressure the Doppler width is reduced by the effect of velocity-averaging elastic collisions. Collisional narrowing of this kind was first predicted by Dicke and later observed in magnetic dipole transitions in the microwave region. Several observations and measurements have been made for vibrational–rotational transitions, but to date none for electronic transitions. Dicke (1953) has described the effect, whose physics is the same as motional narrowing in magnetic resonance spectroscopy, by considering a particle in a one-dimensional box of length L whose allowed transitional energy states are proportional to $1/L^2$. As L decreases, the allowed transitional states move apart until adjacent states are separated by an energy greater than $h\nu_0(v/c)$. This occurs when L is approxi-

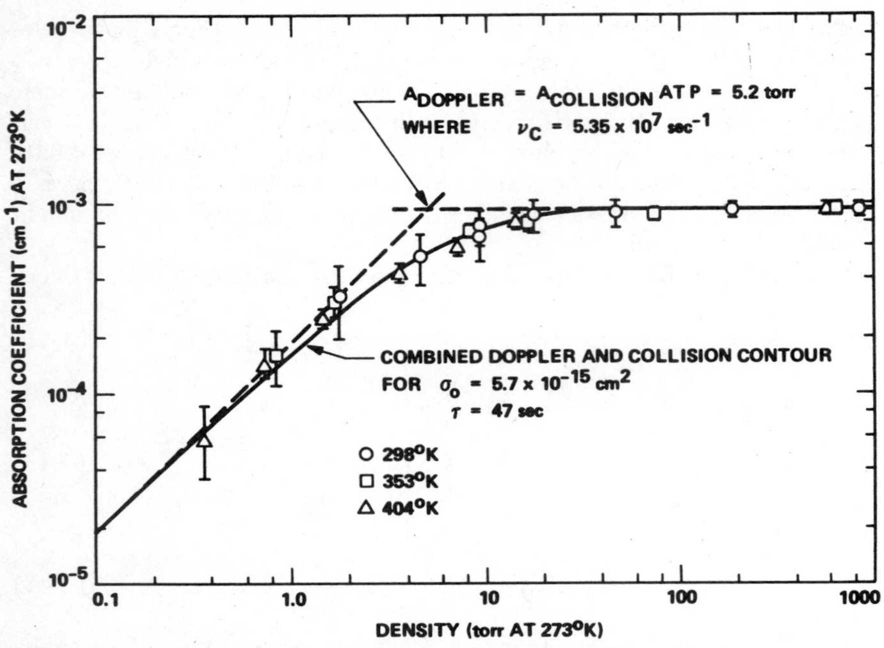

Figure 3.4. Absorption coefficient in CO_2 at 10.6 μm as function of CO_2 pressure. [After Gerry and Leonard (1966).]

mately equal to the wavelength ($\lambda/2\pi$) of the light and leads to an increase in the probability of absorbing a photon without the necessity of a change in the particle's translational energy (see discussion of Doppler effect, Section 3.2.4), since the walls (collisional partners) take up the excess energy. Collisional narrowing therefore occurs when the mean free path of the particle (molecule) is smaller than the absorbed or emitted wavelength. An alternative requirement is that the excited-state lifetime (considering an emission process) is long compared to the mean time between collisions. During this time the mean velocity of the molecule approaches zero, as the collisional averaging causes the molecule to take on all possible velocities. Under collisional narrowing the Gaussian shape of the Doppler-broadened line becomes Lorentzian-like with a width narrower than the Doppler width and inversely proportional to gas pressure.

Recent observations of collisional line narrowing include that observed in H_2O by Eng and co-workers (1972), in HF by Pine (1980), in N_2, CO, and NO by Hurst and co-workers (1984), and in NO by Maki and co-workers (1985). Because collisional narrowing can only be observed if the Doppler width is larger than the pressure-broadened linewidth, the effect is usually overwhelmed by collisional broadening. In the field of the observation and modeling of planetary atmospheres, collisional narrowing effects can be very important. For example, there have been several studies of Dicke narrowing of the hydrogen quadrupole lines belonging to vibrational–rotational bands observed in absorption from Jupiter (e.g., Fink and Belton, 1969). Figure 3.5 shows a plot of the line HWHM against hydrogen pressure for the 2-0 $S(1)$ line near 8604 cm^{-1} (Bragg, 1981). The effect is apparently observed because the Jovian atmosphere is primarily hydrogen (a) whose Doppler width is large and (b) whose pressure-broadening coefficient of about 5×10^{-4} cm^{-1} at STP is extremely small.

3.2.7 Mixed Lineshapes and Voigt Profile

In both laboratory and atmospheric studies, observed lineshapes are rarely due solely to one of the broadening mechanisms described in the previous section but are often mixed in shape. Temperature and pressure regions of interest often cover a transition from one lineshape to another. For example, the laboratory recording of an N_2O line near 612 cm^{-1} by Kim and co-workers (1978) seen in Figure 3.6 is a beautiful example of the transition between the Doppler regime, where γ_D is constant and the peak absorption varies with pressure, and the pressure-broadened Lorentzian regime, where the latter quantity remains constant while γ_L increases with increasing pressure.

The change from Gaussian to Lorentzian lineshape is also seen in spectral measurements of Earth's atmosphere due to the decreasing pressure with alti-

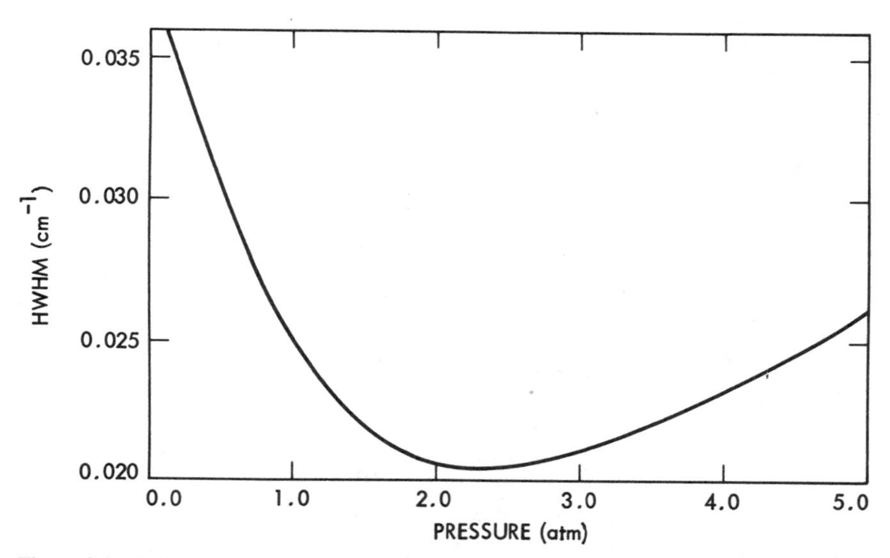

Figure 3.5. Half-width at half maximum width of 2-0 $S(1)$ line as function of hydrogen pressure. Minimum at 2.2 is region of greatest collisional narrowing. [Adapted from Bragg (1981).]

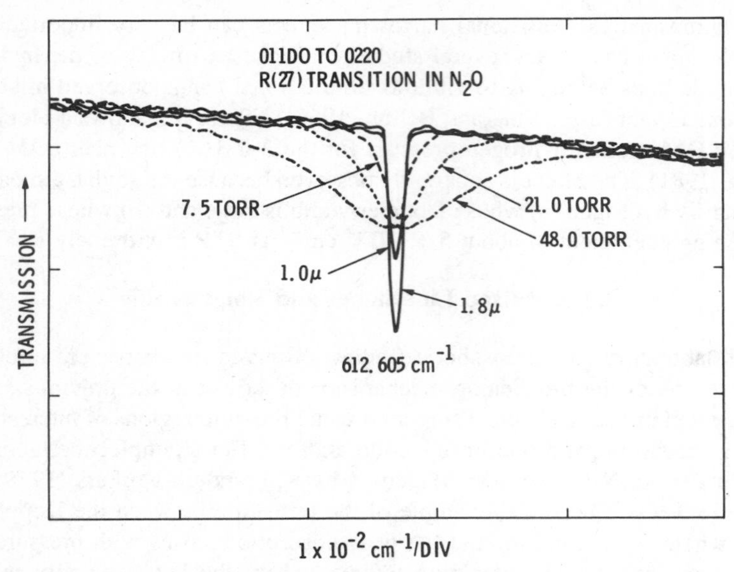

Figure 3.6. The $R(27)$ line of $01^{1d}0$ to $02^{2}0$ hot-band transition in N_2O measured at room temperature in 7-cm cell (dotted and dashed curves) and in long-path cell at 200 m (solid curves). [From Kim and co-workers (1978).]

tude and, to a smaller extent, to the effect of changing temperature. This can be seen in Table 3.1 for stratospheric altitudes between 20 and 40 km where, for example, a water vapor line at 1609 cm^{-1} changes from a shape that is $\simeq 7.8\%$ Lorentzian at 20 km to one that is $\simeq 90\%$ Gaussian at 40 km. The quantity tabulated is the Doppler content, given by $\gamma_D/(\gamma_D + \gamma_L)$. For atmo-

Table 3.1. Change in Lineshape as Function of Altitude for H_2O, NO, NO_2, and O_3 in Earth's Atmosphere

Geometric Altitude (km)	Pressure (mb)	Doppler Content $\gamma_D/(\gamma_D + \gamma_L)$			
		H_2O	NO	NO_2	O_3
20	58.5	0.22	0.30	0.20	0.21
25	26.7	0.40	0.50	0.37	0.38
30	12.7	0.64	0.69	0.56	0.57
35	6.2	0.79	0.82	0.73	0.74
40	3.2	0.89	0.91	0.85	0.86
$\tilde{\nu}$, cm^{-1}		1609	1880	1600	1880
b_L, cm^{-1} atm^{-1}, 300 K		0.094	0.06	0.07	0.077
Molecular weight		20	30	46	48

spheric measurements over large altitude ranges, as attainable, for example, using differential absorption lidar, it is very important to take into account a further variation, that of the pressure shift of the line centers. In fact, all atomic and molecular transitions exhibit homogeneous broadening of their transitions characterized by Lorentzian lineshapes due to radiative broadening. Although additional inhomogeneous broadening or homogeneous pressure broadening may take over as the dominant mechanisms, at any set of conditions the lineshape is truly a convolution of independent Gaussian and Lorentzian shapes, known as the Voigt lineshape or Voigt profile.

Consider a moving atom, whose rest frame absorption line center wavenumber is $\tilde{\nu}_0$, moving with velocity v_z in the z direction, which therefore absorbs light of wavenumber $\tilde{\nu}^*$ such that

$$\tilde{\nu}^* = \tilde{\nu}_0(1 - v_z/c) \tag{3.45}$$

Due to the finite radiative lifetime or to collisions, the absorbed radiation is made up of a distribution of wavenumber $\tilde{\nu}$ given by the Lorentzian shape function

$$g_L(\tilde{\nu} - \tilde{\nu}^*) = (\gamma_L/\pi)/[(\tilde{\nu} - \tilde{\nu}^*)^2 + \gamma_L^2] \tag{3.46}$$

The spectral profile of the ensemble of moving atoms is obtained by averaging the above equation over the thermal distribution given by Equation (3.31). The probability of detecting a wave of wavenumber frequency between $\tilde{\nu}^*$ and $\tilde{\nu}^* + d\tilde{\nu}^*$ is described by the shape function

$$g_D(\tilde{\nu}^* - \tilde{\nu}_0) = (1/\gamma_D)(\ln 2/\pi)^{1/2}\exp[-(\tilde{\nu}^* - \tilde{\nu}_0)^2\ln 2/\gamma_D^2] \tag{3.47}$$

so that the resulting spectral profile may be written as the convolution integral

$$g_V(\tilde{\nu}) = \int_{-\infty}^{\infty} g_D(\tilde{\nu}^* - \tilde{\nu}_0)g_L(\tilde{\nu} - \tilde{\nu}^*)\, d\tilde{\nu}^* \tag{3.48}$$

which is known as the Voigt profile (see Figure 3.7). If we define

$$y = \gamma_L(\ln 2)^{1/2}/\gamma_D = \gamma_L/\gamma_{\mathrm{ED}} \qquad t = (\tilde{\nu}^* - \tilde{\nu}_0)/\gamma_{\mathrm{ED}} \tag{3.49}$$

and

$$x = (\tilde{\nu} - \tilde{\nu}_0)(\ln 2)^{1/2}(1/\gamma_D) = (\tilde{\nu} - \tilde{\nu}_0)/\gamma_{\mathrm{ED}} \tag{3.50}$$

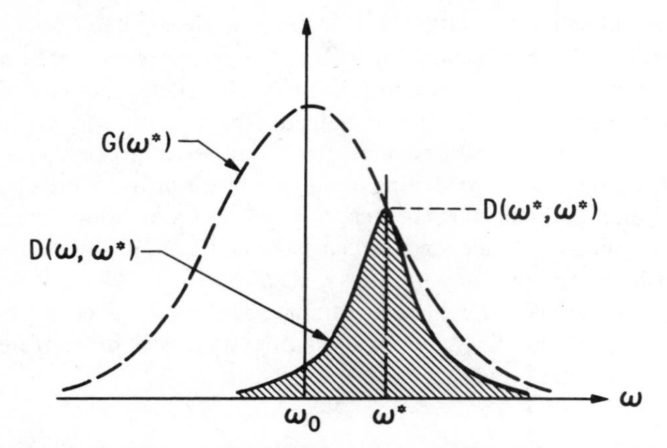

Figure 3.7. Convolution of Gaussian and Lorentzian profiles. [From Measures (1984).]

the Voigt profile shape function can be written as

$$g_V(\tilde{\nu}) = (1/\pi^{3/2}\gamma_{ED}) \int_{-\infty}^{\infty} y \exp(-t^2)/\left(y^2 + (x - t)^2\right) dt \quad (3.51)$$

The absorption coefficient is then written according to

$$\kappa(\tilde{\nu}) = S g_V(\tilde{\nu}) \quad (3.52)$$

It is customary to define the Voigt function $V(x, y)$ as the ratio of the absorption coefficient $\kappa(\tilde{\nu})$ to that at line center for a Doppler profile, κ_0, where

$$\kappa_0 = (S/\gamma_D)(\ln 2/\pi)^{1/2} = S/\pi^{1/2}\gamma_{ED} \quad (3.53)$$

That is,

$$V(x, y) = \kappa(\tilde{\nu})/\kappa_0 = (1/\pi) \int_{-\infty}^{\infty} y \exp(-t^2)/\left(y^2 + (x - t)^2\right) dt \quad (3.54)$$

or

$$\kappa(\tilde{\nu}) = \kappa_0 V(x, y) \quad (3.55)$$

The Voigt profile cannot be expressed in analytical form but can be evaluated numerically since the integral may be written as

$$\int_{-\infty}^{\infty} y \exp(-t^2)/\left(y^2 + (x - t)^2\right) dt = \pi R\omega(z) = \pi R\omega(x + iy) \quad (3.56)$$

where $R\omega(z)$ is the real part of the error function for complex arguments $\omega(z)$, where

$$\omega(z) = \exp(-z^2)\text{erfc}(-iz) \quad \text{and} \quad z = x + iy \quad (3.57)$$

Numerical values of $R\omega(z)$ are given in tables such as in Abramowitz and Stegun (1970) for various values of x and y (see Table 3.2). Note that the π appearing in Equation (3.56) cancels with that in Equation (3.54) so that the appropriate tabulated values are those for the Voigt function $V(x, y)$.

Close to line center, the profile is close to Gaussian, while in the wings, the profile is essentially Lorentzian. The shape is uniquely determined by y, the ratio of the Lorentz to Doppler widths, and x provides the frequency dependence of the profile. If $y = 0$, Equation (3.49) tells us that the line is pure Gaussian. In this case, at line center, where $\tilde{\nu} = \tilde{\nu}_0$, x is also zero, and

$$V(x, y) = V(0, 0) = \pi\pi R\omega(x + iy) = 1 \quad (3.58)$$

from the tables, so that $\kappa(\tilde{\nu}) = \kappa_0$, the line center absorption coefficient for a pure Doppler profile.

Note that either x or y, or both, can be specified to decide whether a Voigt calculation must be done or, alternatively, a pure Doppler or pure Lorentz approximation would suffice. For most applications to IR spectroscopy, the Lorentz approximation is used if $x > 5$ or $y > 3$, that is, if the point on the profile is more than 5 Dopper widths (γ_{ED}) away or the Lorentzian width (γ_L) is more than 3 times this Doppler width.

At line center, where $\tilde{\nu} = \tilde{\nu}_0$, the Voigt parameter x goes to zero. Depending on the value of y, the line center absorption coefficient $\kappa(\tilde{\nu}_0)$ will lie between the values of that of pure Doppler, given by Equation (3.53), and that of pure Lorentz, given by Equation (3.43), that is, between $S/\pi\gamma_L$ and $S/\pi^{1/2}\gamma_{\text{ED}}$. The line center absorption coefficient can therefore be approximated by a weighted function

$$\kappa(\tilde{\nu}_0) \simeq (\beta/\pi^{1/2}\gamma_{\text{ED}}) + (1 - \beta)(S/\pi\gamma_L) \quad (3.59)$$

Table 3.2. Voigt Table of $V(x, y)$ for $0 \leqslant x < 3.9$ and $0 \leqslant y \leqslant 3$[a]

y	$x=0$	$x=0.1$	$x=0.2$	$x=0.3$	$x=0.4$	$x=0.5$	$x=0.6$	$x=0.7$	$x=0.8$	$x=0.9$
0.0	1.000000	0.990050	0.960789	0.913931	0.852144	0.778801	0.697676	0.612626	0.527292	0.444858
0.1	0.896457	0.888479	0.864983	0.827246	0.777267	0.717588	0.651076	0.580698	0.509299	0.439421
0.2	0.809020	0.802567	0.783538	0.752895	0.712146	0.663223	0.608322	0.549739	0.489710	0.430271
0.3	0.734599	0.729337	0.713801	0.688720	0.655244	0.614852	0.569238	0.520192	0.469480	0.418736
0.4	0.670788	0.666463	0.653680	0.632996	0.605295	0.571717	0.533581	0.492289	0.449244	0.405763
0.5	0.615690	0.612109	0.601513	0.584333	0.561252	0.533157	0.501079	0.466127	0.429418	0.392021
0.6	0.567805	0.564818	0.555974	0.541605	0.522246	0.498591	0.471453	0.441712	0.410264	0.377977
0.7	0.525930	0.523423	0.515991	0.503896	0.487556	0.467521	0.444434	0.418998	0.391936	0.363957
0.8	0.489101	0.486982	0.480697	0.470452	0.456579	0.439512	0.419766	0.397906	0.374518	0.350182
0.9	0.456532	0.454731	0.449383	0.440655	0.428808	0.414191	0.397216	0.378341	0.358043	0.336799
1.0	0.427584	0.426044	0.421468	0.413989	0.403818	0.391234	0.376571	0.360200	0.342511	0.323899
1.1	0.401730	0.400406	0.396470	0.390028	0.381250	0.370363	0.357637	0.343375	0.327900	0.311537
1.2	0.378537	0.377393	0.373989	0.368412	0.360799	0.351335	0.340241	0.327766	0.314176	0.299741
1.3	0.357643	0.356649	0.353691	0.348839	0.342206	0.333942	0.324229	0.313273	0.301294	0.288519
1.4	0.338744	0.337876	0.335294	0.331054	0.325248	0.318001	0.309463	0.299804	0.289208	0.277865
1.5	0.321585	0.320825	0.318561	0.314839	0.309736	0.303355	0.295820	0.287274	0.277869	0.267766
1.6	0.305953	0.305284	0.303290	0.300009	0.295506	0.289866	0.283192	0.275602	0.267228	0.258203
1.7	0.291663	0.291072	0.289309	0.286406	0.282417	0.277412	0.271479	0.264718	0.257237	0.249151
1.8	0.278560	0.278035	0.276470	0.273892	0.270346	0.265890	0.260598	0.254554	0.247851	0.240586
1.9	0.266509	0.266042	0.264648	0.262350	0.259186	0.255205	0.250469	0.245050	0.239027	0.232482
2.0	0.255396	0.254978	0.253732	0.251677	0.248844	0.245276	0.241025	0.236152	0.230724	0.224813
2.1	0.245119	0.244745	0.243628	0.241783	0.239239	0.236031	0.232204	0.227810	0.222905	0.217552
2.2	0.235593	0.235256	0.234251	0.232592	0.230300	0.227407	0.223952	0.219978	0.215535	0.210676
2.3	0.226742	0.226438	0.225531	0.224033	0.221963	0.219347	0.216219	0.212616	0.208581	0.204160
2.4	0.218499	0.218224	0.217404	0.216047	0.214172	0.211800	0.208961	0.205686	0.202013	0.197982
2.5	0.210806	0.210557	0.209813	0.208582	0.206879	0.204723	0.202139	0.199155	0.195804	0.192120
2.6	0.203613	0.203387	0.202710	0.201589	0.200039	0.198074	0.195717	0.192992	0.189928	0.186554
2.7	0.196874	0.196668	0.196050	0.195028	0.193613	0.191818	0.189664	0.187170	0.184362	0.181265
2.8	0.190549	0.190360	0.189796	0.188861	0.187566	0.185924	0.183950	0.181662	0.179084	0.176237
2.9	0.184602	0.184429	0.183912	0.183056	0.181868	0.180361	0.178549	0.176447	0.174074	0.171452
3.0	0.179001	0.178842	0.178368	0.177581	0.176491	0.175105	0.173437	0.171502	0.169315	0.166895

y	$x=1.0$	$x=1.1$	$x=1.2$	$x=1.3$	$x=1.4$	$x=1.5$	$x=1.6$	$x=1.7$	$x=1.8$	$x=1.9$
0.0	0.367879	0.298197	0.236928	0.184520	0.140858	0.105399	0.077305	0.055576	0.039164	0.027052
0.1	0.373170	0.312136	0.257374	0.209431	0.168407	0.134049	0.105843	0.083112	0.065099	0.051038
0.2	0.373153	0.319717	0.270928	0.227362	0.189247	0.156521	0.128895	0.105929	0.087090	0.071811
0.3	0.369386	0.322586	0.279199	0.239793	0.204662	0.173865	0.147272	0.124612	0.105522	0.089592
0.4	0.363020	0.321993	0.283443	0.247908	0.215711	0.186984	0.161702	0.139717	0.120793	0.104641
0.5	0.354900	0.318884	0.284638	0.252654	0.223262	0.196636	0.172820	0.151751	0.133288	0.117233
0.6	0.345649	0.313978	0.283540	0.254784	0.228026	0.203461	0.181177	0.161171	0.143369	0.127644
0.7	0.335721	0.307816	0.280740	0.254895	0.230578	0.207990	0.187245	0.168379	0.151366	0.136134
0.8	0.325446	0.300807	0.276693	0.253461	0.231385	0.210664	0.191423	0.173725	0.157578	0.142949
0.9	0.315064	0.293259	0.271752	0.250858	0.230826	0.211846	0.194049	0.177513	0.162268	0.148310
1.0	0.304744	0.285402	0.266189	0.247381	0.229205	0.211837	0.195407	0.180002	0.165667	0.152418
1.1	0.294606	0.277407	0.260213	0.243266	0.226767	0.210881	0.195734	0.181414	0.167977	0.155452
1.2	0.284731	0.269401	0.253985	0.238695	0.223710	0.209182	0.195228	0.181938	0.169373	0.157569
1.3	0.275174	0.261476	0.247628	0.233813	0.220192	0.206902	0.194053	0.181733	0.170003	0.158906
1.4	0.265967	0.253697	0.241233	0.228733	0.216340	0.204177	0.192347	0.180933	0.169997	0.159585
1.5	0.257128	0.246112	0.234870	0.223542	0.212253	0.201115	0.190222	0.179651	0.169465	0.159709
1.6	0.248665	0.238752	0.228592	0.218309	0.208014	0.197806	0.187772	0.177983	0.168500	0.159369
1.7	0.240578	0.231635	0.222436	0.213086	0.203684	0.194320	0.185073	0.176008	0.167183	0.158641
1.8	0.232861	0.224775	0.216428	0.207912	0.199315	0.190717	0.182189	0.173792	0.165579	0.157593
1.9	0.225503	0.218176	0.210587	0.202818	0.194947	0.187043	0.179172	0.171390	0.163746	0.156282
2.0	0.218493	0.211839	0.204926	0.197827	0.190608	0.183335	0.176064	0.168849	0.161733	0.154757
2.1	0.211816	0.205760	0.199452	0.192953	0.186324	0.179623	0.172901	0.166206	0.159580	0.153059
2.2	0.205457	0.199935	0.194166	0.188208	0.182112	0.175930	0.169710	0.163493	0.157320	0.151224
2.3	0.199402	0.194356	0.189072	0.183599	0.177985	0.172276	0.166513	0.160737	0.154982	0.149281
2.4	0.193634	0.189014	0.184165	0.179131	0.173954	0.168674	0.163330	0.157958	0.152591	0.147256
2.5	0.188139	0.183901	0.179444	0.174805	0.170024	0.165136	0.160175	0.155175	0.150165	0.145172
2.6	0.182903	0.179008	0.174903	0.170623	0.166201	0.161669	0.157060	0.152402	0.147722	0.143045
2.7	0.177910	0.174324	0.170538	0.166582	0.162487	0.158281	0.153993	0.149649	0.145274	0.140892
2.8	0.173147	0.169840	0.166342	0.162681	0.158883	0.154975	0.150981	0.146927	0.142834	0.138725
2.9	0.168602	0.165546	0.162310	0.158916	0.155389	0.151753	0.148030	0.144243	0.140411	0.136555
3.0	0.164261	0.161434	0.158435	0.155285	0.152005	0.148618	0.145144	0.141602	0.138012	0.134391

Table 3.2. (*Continued*)

y	x=2.0	x=2.1	x=2.2	x=2.3	x=2.4	x=2.5	x=2.6	x=2.7	x=2.8	x=2.9
0.0	0.018316	0.012155	0.007907	0.005042	0.003151	0.001930	0.001159	0.000682	0.000394	0.000223
0.1	0.040201	0.031936	0.025678	0.020958	0.017397	0.014698	0.012635	0.011037	0.009778	0.008769
0.2	0.059531	0.049726	0.041927	0.035728	0.030792	0.026841	0.023653	0.021057	0.018918	0.017134
0.3	0.076396	0.065521	0.056586	0.049248	0.043211	0.038226	0.034087	0.030626	0.027707	0.025225
0.4	0.090944	0.079385	0.069655	0.061473	0.054585	0.048773	0.043849	0.039656	0.036064	0.032967
0.5	0.103359	0.091422	0.081182	0.072408	0.064890	0.058437	0.052885	0.048090	0.043930	0.040304
0.6	0.113836	0.101765	0.091245	0.082092	0.074132	0.067205	0.061167	0.055890	0.051264	0.047194
0.7	0.122574	0.110558	0.099943	0.090585	0.082345	0.075088	0.068691	0.063043	0.058046	0.053611
0.8	0.129768	0.117948	0.107383	0.097963	0.089576	0.082112	0.075467	0.069548	0.064266	0.059543
0.9	0.135600	0.124081	0.113679	0.104309	0.095884	0.088317	0.081521	0.075416	0.069927	0.064986
1.0	0.140240	0.129097	0.118941	0.109709	0.101336	0.093751	0.086885	0.080670	0.075043	0.069944
1.1	0.143840	0.133125	0.123277	0.114251	0.105999	0.098466	0.091598	0.085338	0.079632	0.074431
1.2	0.146541	0.136286	0.126788	0.118019	0.109942	0.102518	0.095702	0.089451	0.083718	0.078462
1.3	0.148466	0.138689	0.129570	0.121092	0.113232	0.105960	0.099243	0.093044	0.087328	0.082059
1.4	0.149725	0.140432	0.131709	0.123548	0.115935	0.108848	0.102264	0.096155	0.090492	0.085245
1.5	0.150415	0.141604	0.133284	0.125454	0.118109	0.111233	0.104811	0.098820	0.093239	0.088044
1.6	0.150622	0.142283	0.134367	0.126877	0.119812	0.113165	0.106925	0.101076	0.095601	0.090482
1.7	0.150418	0.142540	0.135021	0.127873	0.121096	0.114690	0.108647	0.102957	0.097608	0.092584
1.8	0.149870	0.142434	0.135305	0.128495	0.122010	0.115851	0.110016	0.104498	0.099288	0.094376
1.9	0.149032	0.142021	0.135269	0.128792	0.122597	0.116689	0.111067	0.105730	0.100671	0.095882
2.0	0.147953	0.141347	0.134959	0.128805	0.122897	0.117239	0.111834	0.106683	0.101783	0.097127
2.1	0.146675	0.140453	0.134414	0.128574	0.122945	0.117534	0.112347	0.107386	0.102649	0.098133
2.2	0.145234	0.139375	0.133669	0.128130	0.122773	0.117606	0.112635	0.107864	0.103293	0.098922
2.3	0.143660	0.138145	0.132755	0.127506	0.122411	0.117481	0.112723	0.108140	0.103737	0.099513
2.4	0.141982	0.136789	0.131699	0.126726	0.121884	0.117184	0.112633	0.108238	0.104002	0.099925
2.5	0.140220	0.135331	0.130524	0.125814	0.121215	0.116737	0.112389	0.108177	0.104105	0.100177
2.6	0.138395	0.133791	0.129252	0.124792	0.120424	0.116160	0.112008	0.107975	0.104066	0.100284
2.7	0.136523	0.132187	0.127900	0.123676	0.119530	0.115471	0.111508	0.107648	0.103898	0.100261
2.8	0.134619	0.130533	0.126483	0.122484	0.118548	0.114685	0.110904	0.107213	0.103617	0.100122
2.9	0.132693	0.128842	0.125016	0.121229	0.117492	0.113816	0.110210	0.106682	0.103236	0.099879
3.0	0.130757	0.127125	0.123510	0.119922	0.116375	0.112878	0.109439	0.106067	0.102767	0.099544

y	x=3.0	x=3.1	x=3.2	x=3.3	x=3.4	x=3.5	x=3.6	x=3.7	x=3.8	x=3.9
0.0	0.000123	0.000067	0.000036	0.000019	0.000010	0.000005	0.000002	0.000001	0.000001	0.000000
0.1	0.007943	0.007254	0.006670	0.006167	0.005728	0.005340	0.004995	0.004685	0.004406	0.004153
0.2	0.015627	0.014338	0.013225	0.012252	0.011394	0.010633	0.009952	0.009339	0.008786	0.008282
0.3	0.023095	0.021250	0.019639	0.018222	0.016966	0.015846	0.014841	0.013935	0.013115	0.012368
0.4	0.030279	0.027929	0.025862	0.024032	0.022403	0.020944	0.019632	0.018446	0.017370	0.016389
0.5	0.037126	0.034328	0.031849	0.029643	0.027670	0.025897	0.024297	0.022847	0.021529	0.020326
0.6	0.043598	0.040407	0.037565	0.035022	0.032738	0.030677	0.028812	0.027118	0.025574	0.024162
0.7	0.049665	0.046141	0.042983	0.040144	0.037582	0.035263	0.033158	0.031239	0.029486	0.027880
0.8	0.055311	0.051509	0.048083	0.044989	0.042185	0.039637	0.037316	0.035195	0.033253	0.031469
0.9	0.060529	0.056501	0.052854	0.049544	0.046532	0.043785	0.041274	0.038974	0.036861	0.034916
1.0	0.065318	0.061114	0.057289	0.053801	0.050615	0.047698	0.045023	0.042565	0.040301	0.038212
1.1	0.069685	0.065350	0.061387	0.057757	0.054428	0.051370	0.048556	0.045962	0.043567	0.041352
1.2	0.073641	0.069216	0.065151	0.061413	0.057971	0.054798	0.051869	0.049161	0.046653	0.044328
1.3	0.077202	0.072722	0.068589	0.064773	0.061246	0.057984	0.054962	0.052159	0.049558	0.047139
1.4	0.080385	0.075883	0.071711	0.067844	0.064258	0.060928	0.057835	0.054958	0.052279	0.049783
1.5	0.083210	0.078712	0.074529	0.070636	0.067012	0.063637	0.060491	0.057557	0.054819	0.052260
1.6	0.085697	0.081229	0.077055	0.073158	0.069518	0.066116	0.062936	0.059962	0.057179	0.054572
1.7	0.087870	0.083450	0.079306	0.075423	0.071785	0.068374	0.065176	0.062177	0.059362	0.056720
1.8	0.089749	0.085394	0.081297	0.077445	0.073823	0.070419	0.067217	0.064206	0.061374	0.058708
1.9	0.091355	0.087080	0.083044	0.079236	0.075646	0.072260	0.069068	0.066058	0.063219	0.060540
2.0	0.092711	0.088525	0.084562	0.080811	0.077263	0.073908	0.070736	0.067738	0.064903	0.062222
2.1	0.093835	0.089749	0.085867	0.082182	0.078687	0.075373	0.072232	0.069254	0.066433	0.063759
2.2	0.094748	0.090767	0.086974	0.083364	0.079930	0.076666	0.073563	0.070615	0.067815	0.065156
2.3	0.095467	0.091597	0.087900	0.084370	0.081004	0.077796	0.074739	0.071829	0.069058	0.066420
2.4	0.096010	0.092255	0.088657	0.085213	0.081921	0.078774	0.075770	0.072902	0.070166	0.067556
2.5	0.096393	0.092754	0.089259	0.085905	0.082690	0.079611	0.076664	0.073845	0.071149	0.068572
2.6	0.096632	0.093110	0.089719	0.086458	0.083324	0.080316	0.077430	0.074663	0.072013	0.069474
2.7	0.096739	0.093336	0.090050	0.086883	0.083832	0.080898	0.078076	0.075366	0.072764	0.070267
2.8	0.096729	0.093442	0.090263	0.087190	0.084225	0.081366	0.078612	0.075961	0.073411	0.070959
2.9	0.096613	0.093442	0.090368	0.087391	0.084511	0.081730	0.079044	0.076455	0.073959	0.071555
3.0	0.096402	0.093345	0.090375	0.087493	0.084700	0.081996	0.079381	0.076855	0.074415	0.072061

[a]Adapted from Abramowitz and Stegun (1979).

where

$$\beta = \gamma_{ED}/(\gamma_L + \gamma_{ED}) \tag{3.60}$$

Empirical fits to the Voigt line width (HWHM), γ_V, can be made with accuracies up to 0.01%. Olivero and Longbothum (1977) modifed a simple expression due to Whiting that now attains an accuracy to 0.02%:

$$\gamma_V = 0.5346\gamma_L + (0.2166\gamma_L^2 + \gamma_D^2)^{1/2} \tag{3.61}$$

The use of a "look-up" table as described here is only one method of reproducing Voigt profiles. A number of techniques for calculating approximate values of the Voigt function are available, such as the fast method of Pierlussi et al. (1977), which uses three individual approximations depending on the values of x and y, or the Humlicek (1979) method.

3.2.8 Saturation and Modulation Broadening

In Section 3.2.1 we described how the Beer–Lambert law no longer applied for intense light sources that produced substantial changes in the populations of levels coupled by the field. In this regime of nonlinear absorption the transition becomes saturated, and dI is no longer proportional to I. The absorption coefficient itself then becomes dependent on the intensity I of the incident radiation. The approximation of an essentially constant initial-state population, which resulted in a time-independent transition probability, is no longer valid, and the strong-signal theory of Rabi leads to a time-dependent probability of the atom being in either the upper or lower level. In fact, the transition probability is a periodic function of time, the system oscillating with a characteristic frequency called the Rabi frequency, given by

$$\Omega_0 = R^{nm} 2\pi E_0/h \tag{3.62}$$

and depends on the electric dipole transition moment R^{nm}, the field strength E_0, and the detuning from line center.

It can be shown (Shimoda, 1976) that the lineshape associated with a saturated transition is, in the absence of inhomogeneous broadening, Lorentzian with a half-width (HWHM) given by

$$\gamma_{sat} \simeq \left(\gamma^2 + \gamma\tau|\Omega|^2\right)^{1/2} \tag{3.63}$$

where

$$\gamma = \tfrac{1}{2}(\gamma_1 + \gamma_2) \quad \text{and} \quad \tau = \tfrac{1}{2}(\gamma_1^{-1} + \gamma_2^{-1}) \tag{3.64}$$

for the two-level system. Thus, the linewidth increases with the optical field intensity, producing saturation broadening of the transition. The phenomenon is also known as power broadening because $|\Omega_0|^2$ is proportional to the incident intensity. The ability to saturate a transition is therefore the ability to approach photon flux densities high enough to compete with all relaxation processes from the excited level. This can often be achieved in the IR region using laser sources.

For a monochromatic laser source of intensity (irradiance, radiant flux density) I (in W m^{-2}) the time-averaged field E_0 is given by (Yariv, 1976)

$$E^2 = 2I/c\epsilon_0 \tag{3.65}$$

We calculate, for example, that $E \simeq 274$ V cm^{-1} for a source of intensity 100 W cm^{-2}, which might be a low-power CO_2 laser output. For $R^{nm} \simeq 1$ D (Debye) ($= 10^{-18}$ esu cm $= 3.3 \times 10^{-30}$ C m) the characteristic frequency $|\Omega_0|$ becomes $\sim 1 \times 10^8$ s^{-1}. Therefore, a power of 1 W mm^{-2} can saturate a transition with $R^{nm} = 1$ D if the relaxation time of the upper state, through spontaneous radiative, collisional, and transit time effects, is greater than $\simeq 100$ μs.

The HWHM can be written as (Shimoda, 1976)

$$\gamma_{\text{sat}} = \gamma(1 + I/I_{\text{sat}})^{1/2} \tag{3.66}$$

where

$$I = c\epsilon_0 E_0^2/2 = c\epsilon_0 h^2|\Omega_0|^2/(2|R^{nm}|^2) \tag{3.67}$$

and

$$I_{\text{sat}} = c\epsilon_0 h_\gamma^2/(2\tau|R^{nm}|^2) \tag{3.68}$$

The quantity I/I_{sat} is known as the degree of saturation, or saturation parameter, where

$$I/I_{\text{sat}} = |\Omega_0|^2\tau/\gamma \tag{3.69}$$

which tends to zero for low-intensity sources.

In the strong-field limit, where the degree of saturation is much greater than unity, a bleaching of the absorption results, and the light attenuation becomes linear in I (Measures, 1984). In the weak-field exponential (Beer–Lambert) attenuation regime and the strong-field linear attenuation regime, a quantitative description of the absorption is straightforward. It is the intermediate regime that proves very difficult to model.

Important differences exist between the saturation characteristics of homogeneously broadened transitions and those of inhomogeneously broadened transitions (Siegman, 1971). For a homogeneously broadened transition, applying a strong saturating monochromatic signal at a frequency off line center (e.g., in the wings of the profile at $\tilde{\nu}'$) will saturate all atoms equally, by an amount that will increase as the saturating signal is tuned toward the line center. That is, the entire atomic response saturates uniformly for a saturating signal applied anywhere within the linewidth. Because of this uniformity, the lineshape and linewidth are unchanged with saturation, in contrast with the case of saturation of inhomogeneously broadened transitions. For an inhomogeneously broadened transition, saturation of the population difference will occur only for those atoms probed by the saturating signal, and with increasing intensity a "hole" will be burnt in the line profile only at ν'.

However, we must distinguish this comparison from the case where a strong saturating signal is fixed at ν' and a second laser is scanned over a range of frequency to produce a recorded line profile. Then both homogeneously and inhomogeneously broadened transitions will produce line profiles where the shape and width is changed by saturation.

Modulation broadening is another broadening mechanism that can become significant in the IR region but is more commonly of concern to microwave spectroscopists, who employ Stark modulation techniques at high frequencies for sensitive absorption measurements and detection methods. Modulating the applied Stark field at wavenumber $\tilde{\nu}_m$ to produce lineshifts less than or equal to the linewidth $\Delta\tilde{\nu}$ and ramping the field over a transition is analogous to modulating the wavenumber $\tilde{\nu}_L$ of a laser at a wavenumber $\tilde{\nu}_m$ and scanning over that linewidth. Due to the modulation, the laser wavenumber seen by the absorbing molecules will comprise components at $\tilde{\nu}_L \pm \tilde{\nu}_m$. Thus, as the wavenumbers of $\tilde{\nu}_m$ approaches the linewidth $\Delta\tilde{\nu}$, a broadening will become observable, called modulation broadening. At higher frequencies of wavenumbers of $\tilde{\nu}_m$ the line observed will split into two ("sidebands") components that will move apart as the wavenumber of $\tilde{\nu}_m$ increases. For molecular linewidths on the order of 50–100 kHz (HWHM) observed in microwave spectroscopy, this sets a limit on the highest modulation frequency of the Stark field that can be used. In the mid-IR region, linewidths on the order of $\leqslant$100 MHz are encountered in high-resolution studies, and so modulation broadening is less often a limitation.

We have established that modulation frequencies comparable to the molecular linewidth cause modulation broadening to become significant. Since the molecular linewidth is a measure of the relaxation processes in the molecule, we can express this limit in another way. If the line is pressure broadened, for example, we conclude that modulation broadening becomes significant when the frequency of modulation is comparable to or greater than the collisional rate. That is, modulation broadening results when the relaxation process cannot relax the molecules fast enough to follow the modulation frequency (or electric field).

3.2.9 Boltzmann Distribution and Partition Functions

The Boltzman distribution law relates the populations N_n and N_m at temperature T of energy levels n and m separated by an energy ΔE, according to

$$N_n g_m / N_m g_n = \exp(-\Delta E/kT) \tag{3.70}$$

where k is Boltzmann's constant (1.38066×10^{-23} J K^{-1}), T is the temperature in kelvins, g_n is the degeneracy of the nth level, and $\Delta E = E_n - E_m$ is the energy separation of the two levels n and m. The collisional relaxation rate is assumed high enough to maintain the Boltzmann distribution and avoid saturation. The probability of finding an atom or molecule in any given level n is a multilevel system is then

$$\frac{N_m}{\Sigma N_j} = \frac{g_m \exp(-E_m/kT)}{\sum_i g_i \exp(-E_i/kT)} \tag{3.71}$$

The denominator of the above expression is called the partition function, Q, where

$$Q = \Sigma\, g_i \exp(-E_i/kT) \tag{3.72}$$

The fractional population of the mth state, F_m, described by Equation (3.71), can be split into its electronic, F_e, vibrational, F_v, rotational, F_r, and nuclear spin, F_{ns}, components according to

$$F_m = F_e F_v F_r F_{ns} \tag{3.73}$$

assuming that the Born–Openheimer approximation holds and allows a similar partitioning of the molecular wavefunction. The fractional population is an important quantity to evaluate for calculating or measuring individual line intensities (see Kroto, 1975). For electronic states, the fraction of molecules in the lowest state, F_e, is usually 1, although for some molecules which have low-lying excited electronic states or degenerate ground states, that is not true. For example (Atkins, 1978), the $^2\Pi_{3/2}$ first excited state of NO is only ~ 121 cm^{-1} higher in energy than the ground $^2\Pi_{1/2}$ electronic state. The electronic partition function for NO would be

$$Q_{elec} = \Sigma\, g_i \exp(-E_i/kT) = 2\exp(-E_1/kT) + 2\exp(-E_2/kT)$$

or

$$Q_{elec} = 2[1 + \exp(-\Delta E/kT)]$$

as $E_1 = 0$ so $\Delta E = E_2$. At room temperature, $kT \simeq 200\ \text{cm}^{-1}$, and so $Q_{\text{elec}} \simeq 2.8$. Therefore, the fraction of the molecules in the lower electronic state is $F_e = 2/2.8 = 0.71$.

The vibrational partition function Q_{vib} of a polyatomic molecule must be written as the product of those of the $3N - 6$ normal modes of vibration,

$$Q_{\text{vib}} = Q_{\text{vib}}(1)Q_{\text{vib}}(2)Q_{\text{vib}}(3) \cdots Q_{\text{vid}}(3N - 6) \tag{3.74}$$

where for each nondegenerate mode,

$$Q_{\text{vib}}(j) = \Sigma \exp(-E_v/kT) \tag{3.75}$$

If the harmonic oscillator approximation is used, the energies may be written as

$$E_v = (v + 1/2)h\nu \qquad v = 0, 1, 2, \ldots, 3N - 6 \tag{3.76}$$

where $\nu = (k'/\mu)^{1/2}/2\pi$, k' is the force constant, and μ is the reduced mass.

Measuring energies from the zero-point level, the vibrational partition function may be written as

$$Q_{\text{vib}}(j) = \sum_n \exp(-nhc\tilde{\nu}_j/kT) = \frac{1}{1 - \exp(-hc\tilde{\nu}_j/kT)} \tag{3.77}$$

where n is an integer and $\tilde{\nu}_j$ is the wavenumber corresponding to the jth vibrational mode. Thus, for a molecule with normal modes of wavenumber $\tilde{\nu}_1$, $\tilde{\nu}_2$, . . . , the total vibrational partition function is

$$Q_{\text{vib}} = \prod_j \frac{1}{1 - \exp(-hc\tilde{\nu}_j/kT)} \tag{3.78}$$

For example (Atkins, 1978), the three normal modes of the water molecule are $\tilde{\nu}_1 = 3656.7\ \text{cm}^{-1}$, $\tilde{\nu}_2 = 1594.8\ \text{cm}^{-1}$, and $\tilde{\nu}_3 = 3755.8\ \text{cm}^{-1}$. At 298 K, therefore, the total vibration partition function is

$$Q_{\text{vib}} = Q_{\text{vib}}(1)Q_{\text{vib}}(2)Q_{\text{vib}}(3) = 1.0000 \times 1.0005 \times 1.0000 = 1.0005$$

showing that the vibrations are all of high enough frequency that at room temperature virtually all molecules are in the ground state.

The fractional population of a given vibrational level is written as

$$F_v = (g_v/Q_{\text{vib}})\exp[-(E_v - E_0)/kT] \tag{3.79}$$

The rotational partition function, written as a sum over J and M, becomes

$$Q_{\text{rot}} = \sum_{J,M} \exp(-E_J/kT) = \sum_J (2J + 1)\exp(-E_j/kT) \tag{3.80}$$

since each J level has a degeneracy in M of $2J + 1$. The rotational energy levels of most molecules lie close to one another compared to kT, so that at room temperature a large number are populated and large values of Q_{rot} result. The summation in the above equation can then be replaced by an integral sign. Furthermore, a factor σ known as the symmetry number (Atkins, 1978) must be introduced to ensure correct counting of energy levels, so that the rotational partition function for a linear molecule can be approximated to

$$Q_{\text{rot}} = 2IkT/h^2\sigma = kT/B\sigma \tag{3.81}$$

where B is the rotational constant and I the moment of inertia. For all other molecules,

$$\begin{aligned} Q_{\text{rot}} &\simeq (\pi^{1/2}/\sigma)\left[(2I_A kT/h^2)(2I_B kT/h^2)(2I_c kT/h^2)\right]^{1/2} \\ &\simeq \left[\pi(kT)^3/(ABC)\right]^{1/2}\sigma \end{aligned} \tag{3.82}$$

where I_A, I_B, and I_C are the three moments of inertia and A, B, and C their respective rotational constants, neglecting the spin statistics and nuclear spin degeneracy. These approximations are valid only in the rigid rotor approximation and where no internal rotation exists and are not applicable at low temperatures. For further details see Herzberg (1945). We can now write the fractional population of the Jth rotational level

$$F_R = (g_R/Q_{\text{rot}})\exp[-(E_R - E_0)kT] \tag{3.83}$$

which for a linear molecule becomes

$$\begin{aligned} F_R &= \left\{(2J + 1)\exp[-BJ(J + 1)/kT]\right\}Q_{\text{rot}} \\ &= \left\{(2J + 1)\exp[-BJ(J + 1)/kT]\right\}/(kT/B\sigma) \end{aligned} \tag{3.84}$$

3.2.10 Oscillator Strengths, Einstein Coefficients, and Transition Probabilities

In Equation (3.7) we defined the absorption coefficient in terms of the classical oscillating electron and the oscillator strength. Since the linestrength intensity S is the spectrally integrated absorption coefficient [Eq. (3.14)], we can define the dimensionless oscillator strength f_{nm} by integrating Equation (3.7):

$$f_{nm} = [2m_e c\epsilon_0/Ne^2\pi]S \tag{3.85}$$

for S in units of reciprocal centimeters and c in centimeters per second. The oscillator strength is the ratio of the strength of the observed transition to that of an electric dipole transition between two states of an electron oscillating harmonically in three dimensions. For a single atom or molecule $\Sigma_m f_{nm} = 1$. Oscillator strengths, or f values, are usually employed for describing electronic transitions but retain their meaning for describing any atomic or molecular transition probability. Oscillator strengths provide an important link between theory and experiment since they may be theoretically calculated through the Einstein coefficients, which are directly related to the transition dipole moment, or they can be experimentally determined through a measurement of the integrated absorption coefficient as described above. In terms of the Einstein coefficients, the oscillator strength may be written (Demtroder, 1981):

$$\begin{aligned} f_{nm} &= \frac{2m_e\epsilon_0 h\nu_{nm}}{\pi e^2} B_{nm}(\nu) \\ &= \frac{m_e\epsilon_0 c^3}{A\pi^2\nu_{nm}^2 e^2} A_{nm}(\nu) \end{aligned} \tag{3.86}$$

In turn, the Einstein coefficients are related to the transition dipole moment R^{nm} by

$$A_{nm}(\nu) = (16\pi^3\nu_{nm}^3/3g_n\epsilon_0 hc^3)|R^{nm}|^2$$

and

$$B_{nm}(\nu) = (2\pi^2/3g_n\epsilon_0 h^2)|R^{nm}|^2 \tag{3.87}$$

Considering only the interaction of the electric vector of the radiation, $|R^{nm}|^2$ is zero for electric dipole forbidden transitions and nonzero for electric dipole allowed transitions, and thus sets the quantum-mechanical selection rules for the absorption process. The transition moment is usually resolved into three components about Cartesian axes according to those of the dipole moment, so that the selection rules can be specifically related to the molecular symmetry. Here R^{nm} may be defined in terms of the initial state ψ_n, final state ψ_m, and the dipole moment operator, μ:

$$R^{nm} = \int_{-\infty}^{\infty} \psi_m^* \mu \psi_n \, d\tau \tag{3.88}$$

It can also be related to the oscillator strength by

$$|R^{nm}|^2 = (3g_n he^2/8\pi^2 m_e \nu_{nm}) f_{nm}$$
$$= 7.09 \times 10^{-43} \times (f_{nm}/\nu_{nm}) \quad \mathrm{C^2\ m^2} \tag{3.89}$$

Thus, an electronic transition at 20,000 cm^{-1} ($\nu_{nm} \simeq 6 \times 10^{14}$ Hz) with an oscillator strength of $f = 1 \times 10^{-3}$ would have a transition dipole moment of $R^{nm} = 1.087 \times 10^{-30}$ C m $= 0.33$ D (Debye), which can be visualized as corresponding to moving a charge of 0.33 of an electron through a distance on the order of an atomic radius.

Electronic transitions in polyatomic molecules are usually accompanied by vibrational and rotational energy changes. If the Born–Oppenheimer approximation can be used, the wavefunctions in Equation (3.88) can be separated into products of electronic, vibrational, and rotational wavefunctions, but the integration is now over both electronic and nuclear coordinates (see Hollas, 1982). Splitting the dipole moment operator μ into the sum of an electronic part, $\mu_e = -e \Sigma r_i$, and a nuclear part, $\mu_N = +e \Sigma r_i$, the matrix element for a transition between two different electronic states can be written as

$$R^{nm}_{\mathrm{rovib}} = R^{a'a''}_{\mathrm{el}} \times R^{v'v''}_{\mathrm{vib}} \times R^{J'K',J''K''}_{\mathrm{rot}} \tag{3.90}$$

where the first factor is the electronic transition moment, and the second factor is proportional to the Franck–Condon factor

$$R^{v'v''}_{\mathrm{vib}} = \int \Psi'_{\mathrm{vib}}\, \psi''_{\mathrm{vib}}\, d\tau_{\mathrm{vib}} \tag{3.91}$$

which represents the overlap integral of the vibrational wave functions in each of the electronic states. The last factor is proportional to the Hönl–London factor

$$R^{J'K',J''K''}_{\mathrm{rot}} = \int \psi''(J'', K'')\psi'(J', K')\, d\tau_{\mathrm{rot}} \tag{3.92}$$

The factor contains the wavefunctions of the rigid rotor, which many include dependence on other rotational quantum numbers (e.g., K for symmetric top transitions), and may be represented by spherical harmonics. From a knowledge of the rotational quantum numbers J and K for each state, the band type, and branch, $P(\Delta J = -1)$, $Q(\Delta J = 0)$, or $R(\Delta J = +1)$, the Hönl–London factors may be evaluated from tables based on direction cosine matrix elements.

Vibrational transitions within a single electronic state can be represented by the vibrational transition moment

$$R^{v'v''} = \int \psi_v'^* \mu \psi_v'' \, dx, \tag{3.93}$$

where μ is the electronic dipole moment and x is the displacement of the internuclear distance from equilibrium. The variation of μ with x may be expressed in a Taylor series expansion,

$$\mu = \mu_e + \left(\frac{d\psi}{dx}\right)_e x + \frac{1}{2!}\left(\frac{d^2\mu}{dx}\right)_e x^2 + \cdots . \tag{3.94}$$

so that with the first integral equated to zero because of the orthogonality of the two wavefunctions ($v' \neq v''$), the transition moment becomes

$$R^{v'v''} = (d\mu/dx)_e \int \psi_v'^* x \psi_v'' \, dx + \cdots \tag{3.95}$$

The vibrational selection rule $\Delta v = \pm 1$ results from the first term if $R^{v'v''}$ must be nonzero. The transition intensities will depend on the square of $R^{v'v''}$ and therefore on $(d\mu/dx)_e^2$. For real molecules, the harmonic oscillator approximation does not hold, and terms in higher powers of x (x^2, x^3) must be included (see Hollas, 1982). This electrical anharmonicity will cause transitions with $\Delta v = \pm 2, \pm 3, \ldots$ to be observed, the intensity of the $\Delta v = \pm 2$ transition being dependent on the square of $(d^2\mu/dx^2)_e$. Mechanical anharmonicity, that is, the departure of the potential function, vibrational energy levels, and wavefunctions from a vibrational motion that obeys Hooke's law, also contributes to $\Delta v = \pm 2, \pm 3, \ldots$ transition intensity. Furthermore, mechanical anharmonicity modifies the vibrational term values and wavefunctions.

For many molecules, vibrational–rotational interactions cannot be ignored. The square of the rovibrational transition moment is then written as

$$|R_{v''J''}^{v'J'}|^2 = |M_{v''J''}^{v'J'}|^2 |R_{J''}^{J'}|^2 \tag{3.96}$$

where the last factor is related to the Hönl–London factor, which may include dependence on other rotational quantum numbers (e.g., K for symmetric top transitions). The square of the matrix element of the vibrational transition is now factored into a rotationally independent pure vibrational part, $|R_{v''}^{v'}|^2$, and a factor $F_{vR}(m)$ that takes into account any vibrational–rotational interaction. Thus,

$$|M_{v''J''}^{v'J'}|^2 = |R_{v''}^{v'}|^2 F_{vR}(m) \tag{3.97}$$

For linear molecules, the F factor, or Herman–Wallis factor (Herman and Wallis, 1955), can be expressed in terms of m as

$$F_{vR} = 1 + am + bm^2 + \cdots \tag{3.98}$$

where the constants a and b depend on the molecular constants and $m = -J''$ for a P-branch and $m = J'' + 1$ for an R-branch transition. The same form of the F factor has been applied to both diatomic and polyatomic linear molecules and to symmetric and asymmetric tops. This factor essentially represents the nonrigidity of the rotating molecule. Generally, departure of F-factor values from unity are more noticeably for high J or $|m|$ values and light molecules. Figure 3.8 shows plots of calculated F factors as a function of m for various bands of the HCl molecule (Herman and co-workers, 1958).

For the CO molecule, for example, much smaller values result. Toth and co-workers (1969) measured linear and quadratic Herman–Wallis coefficients of ~ 0.01 and 0.0002 for the ν_3 band. In many cases, the inability to observe, within experimental error, any variation in $|M_{v''J''}^{v'J'}|^2$ with rotational quantum number is sufficient to approximate $F_{vR}(m)$ to unity.

3.2.11 Vibrational–Rotational Linestrengths and Band Intensities

In this section we will write down specific expressions for the vibrational–rotational linestrengths by combining the expressions for level populations, par-

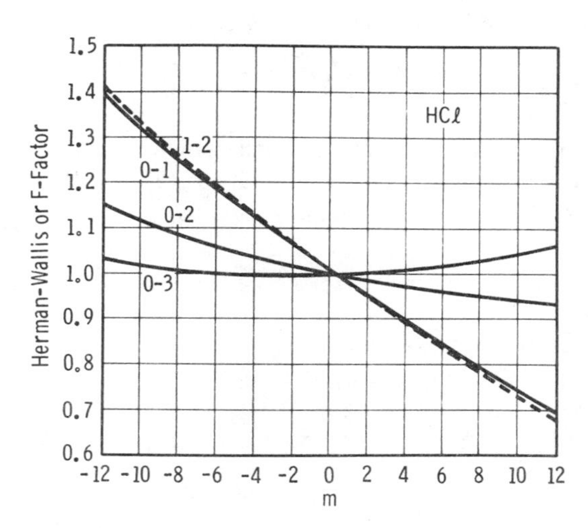

Figure 3.8. Herman–Wallis factor vs. m for 0-1, 1-2, 0-2, and 0-3 transitions of HCl molecule. [Adapted from Herman and co-workers (1958).]

tition functions, and the matrix elements of the transition dipole moment. The reader is referred to Penner (1959), who derived the quantum-mechanical expression for the strength of a spectral line arising from a transition from the state m to the state n:

$$S_{mn}\,(\mathrm{cm}^{-2}\,\mathrm{atm}^{-1}) = (8\pi^3/4\pi\epsilon_0 3hc)(N_m/g_m)\tilde{\nu}_{mn} \cdot [1 - \exp(-hc\tilde{\nu}_{mn}/kT)]|R^{mn}|^2 \tag{3.99}$$

Here N_m is the number of molecules of the absorbing gas per cubic centimeters per atmospheres in the lower state m, R^{mn} is in units of Coulombs per centimeter, and S_{mn} is the therefore in units of reciprocal square centimeters per atmospheres. The term $1 - \exp(-hc\tilde{\nu}_{mn}/kT)$ represents the effects of induced emission. For many applications in IR spectroscopy, since $hc\tilde{\nu}_{mn} >> kT$, this factor can be set equal to 1, but this is not always true, especially for longer wavelength transitions or high-temperature studies.

Using the Boltzmann distribution, we can write

$$N_m/g_m = (N/Q)\exp(-hc\tilde{\nu}_m/kT) \tag{3.100}$$

where N is the total number of molecules of absorbing gas (in $\mathrm{cm}^{-3}\,\mathrm{atm}^{-1}$), Q is the total partition function, and $\tilde{\nu}_m$ is the lower state energy (in cm^{-1}). Therefore, assuming $Q_e = 1$, and replacing the term $8\pi^3/4\pi\epsilon_0 3hc$ by C',

$$S_{mn}\,(\mathrm{cm}^{-2}\,\mathrm{atm}^{-1}) = (C'N\tilde{\nu}_{mn}/Q_{\mathrm{vib}}Q_{\mathrm{rot}})\exp(-hc\tilde{\nu}_m/kT) \cdot [1 - \exp(-hc\tilde{\nu}_{mn}/kT)]|R^{mn}|^2 \tag{3.101}$$

For an ideal gas the number of molecules per cubic centimeter in 1 atm of gas at 273.15 K is Loschmidt's number, $N_L = 2.68675 \times 10^{19}$ molecules cm^{-3}. At temperature T, this number becomes

$$N = N_L(273.15/T) \tag{3.102}$$

The units most commonly used for the linestrength are those used in the Air Force Geophysical Laboratory (AFGL) trace-gas compilation (see Rothman and co-workers, 1983), (i.e., cm $\mathrm{molecule}^{-1}$). The linestrength is converted to these units by division with Loschmidt's number at temperature T. Thus,

$$S_{mn}\,(\mathrm{cm\ molec}^{-1}) = (C'\tilde{\nu}_{mn}/Q_{\mathrm{vib}}Q_{\mathrm{rot}})\exp(-hc\tilde{\nu}_m/kT) \cdot [1 - \exp(-hc\tilde{\nu}_{mn})/kT]|R^{mn}|^2 \tag{3.103}$$

The partition functions are themselves temperature dependent as shown in Section 3.2.9. The linestrength at any temperature T may be normalized to that at T_0 (296 K for the AFGL listings) by

$$S(T_0) = S(T)[Q_{\text{vib}}(T)Q_{\text{rot}}(T)/Q_{\text{vib}}(T_0)Q_{\text{rot}}(T_0)] \cdot \exp[1.439\tilde{\nu}_m(T_0 - T)/T_0T] \quad (3.104)$$

The induced emission term is not included, since $hc\tilde{\nu}_{mn} >> kT$ in the IR region; for the temperature dependence of Q_{vib}, a similar argument holds. The rotational partition function depends on the molecular structure. For a linear molecule, the above equation simplifies to

$$S(T_0) = S(T)(T/T_0)\exp[1.439\tilde{\nu}_m(T_0 - T)/T_0T] \quad (3.105)$$

the rotational partition function dependence being the T/T_0 term. The ratio $S(T)/S(T_0)$ given by the above equation is plotted in Figure 3.9.

It is customary to express vibrational–rotational linestrengths in terms of the rotational quantum number m defined in the Herman–Wallis equation [Eq.

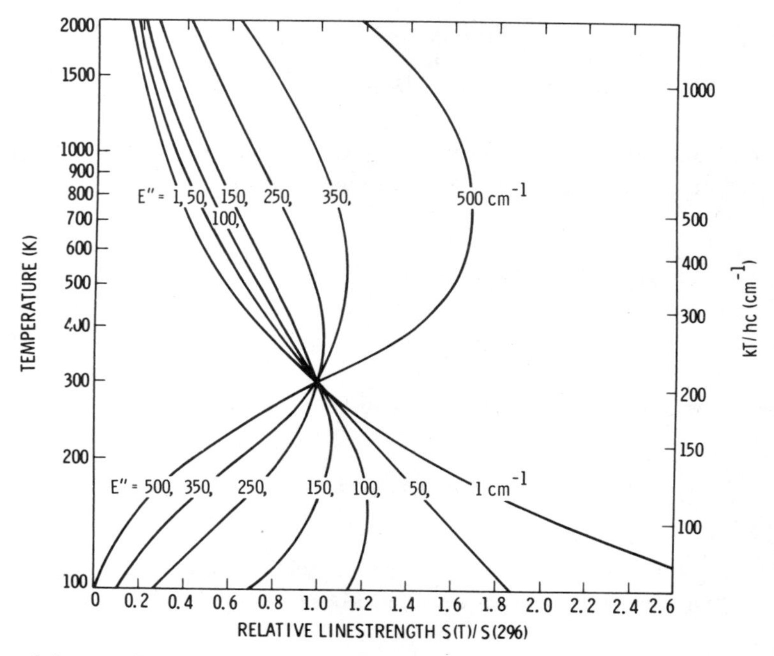

Figure 3.9. Plot of ratio $S(T)/S(T_0)$ for linear molecule, not including induced emission term.

(3.97)] with a magnitude of J'' for P- and Q-branch transitions and $J'' + 1$ for R-branch transitions. The linestrength (in cm molecule^{-1}) is then

$$S(m) = [C'\tilde{\nu}(m)/Q_{\text{vib}}Q_{\text{rot}}]\exp(-hc\tilde{\nu}_m/kT) \cdot [1 - \exp(-hc\tilde{\nu}(m)/kT)]|R^{mn}|^2 \quad (3.106)$$

The matrix elements of R^{mn} have been described in the previous section. We will write

$$|R^{mn}|^2 = |R^{J'}_{J''}|^2 |R(0)|^2 F(m) \quad (3.107)$$

where $|R^{J'}_{J''}|^2$ is the rotational factor containing the Hönl–London $U^{J'}_{J''}$ factor and the statistical weight factor

$$|R^{J'}_{J''}|^2 = g_{J'} U^{J'}_{J''} \quad (3.108)$$

where J refers to all rotational quantum numbers, and $g_{J'}$ is the statistical weight, including nuclear spin, of the upper level. If we now expand the energy term into a product of vibrational and rotational contributions, then in terms of the rotational quantum number m, the linestrength becomes

$$S(m) = [C'\tilde{\nu}(m)/Q_{\text{vib}}Q_{\text{rot}}]\exp[-hc\tilde{\nu}_{\text{rot}}(m)/kT]\exp(-hc\tilde{\nu}_{\text{vib}}/kT) \times [1 - \exp(-hc\tilde{\nu}/kT)]g_{J'}U^{J'}_{J''}|R(0)|^2 F(m) \quad (3.109)$$

The total band intensity, or bandstrength, can now be written by summing the above equation over all the rotational quantum numbers:

$$S_{\text{band}} = \sum_{m=-\infty}^{\infty} S(m) = (C'/Q_{\text{vib}})\exp(-hc\tilde{\nu}_{\text{vib}}/kT) \cdot [1 - \exp(-hc\tilde{\nu}_0/kT)]|R(0)|^2 \bar{\nu} \quad (3.110)$$

where $\bar{\nu}_0$, the effective band center wavenumber, is given by

$$\bar{\nu}_0 = (1/Q_{\text{rot}}) \sum_m F(m) g_{J'} U^{J'}_{J''} \tilde{\nu}(m)\exp(-hc\tilde{\nu}_{\text{rot}}/kT) \quad (3.111)$$

The linestrength $S(m)$ therefore bears a simple relationship to the bandstrength, which is found by division of Equation (3.109) by Equation (3.110):

$$S(m) = S_{\text{band}}[\tilde{\nu}(m)/Q_{\text{rot}}\bar{\nu}]\exp[-hc\tilde{\nu}_{\text{rot}}(m)/kT]g_{J'}U^{J'}_{J''}F(m) \quad (3.112)$$

or, alternatively,

$$S_{\text{band}} = S(m)Q_{\text{rot}}\bar{\nu}/[\tilde{\nu}(m)\exp[-hc\tilde{\nu}_{\text{rot}}(m)/kT]g_{J'}U_{J''}^{J'}F(m) \quad (3.113)$$

For linear molecules, including diatomics, $g_{J'}U_{J''}^{J'} = |m| = J''$ for the P branch and $J'' + 1$ for the R branch. For the ν_2 band of the linear CO_2 molecule (see Valero and co-workers, 1979), J must be replaced by J, l, where l is the quantum number associated with the degeneracy of the ν_2 vibration, so that $g_{J'}l' = 2(2J' + 1)$ for $l \neq 0$. For the $(\nu_4 + \nu_5)^0$ combination band of acetylene (Podolske and co-workers, 1984), the linestrength or line intensity given by Equation (3.112) must include a nuclear statistical weight factor because of the two identical hydrogen nuclei in $^{12}C_2H_2$, where $g_J = 3$ for odd J *and* $g_J = 1$ for even J.

As an example of a recent application to a slightly asymmetric top molecule, Zahniser and Stanton (1983) measured linestrengths for lines in the $\tilde{\nu}_3$ band of HO_2 near 1080 cm^{-1}. The band shows parallel band ($\Delta K_a = 0$) structure, the Q-branch intensity being suppressed due to the large ratio of major to minor moments of inertia. Each rotational level is split by the electronic spin of the unpaired electron into F_1 levels with $J = N + \frac{1}{2}$ and F_2 levels ($J = N - \frac{1}{2}$). The linestrength for this slightly asymmetric top may be related to the bandstrength by Equation (3.112) using the notation of Zahniser and Stanton:

$$S_{NK_aK_c} = S_{\text{band}}U_{K_aJ}(2J + 1)\exp[-F(N, K_a, K_c)/kT]/ \\ \cdot 2[(\pi/ABC)(kT/hc)^3]^{1/2} \quad (3.114)$$

Here K_a and K_c denote the angular momentum about the a and c principal axes, respectively, and $F(N, K_a, K_c)$ is the energy of the lower rotational level. The denominator contains the asymptotic expansion for the rotational partition function for an asymmetric top molecule [see Eq. (3.82)] and an additional factor of 2 to account for the electronic degeneracy of the ground state. The term U_{K_aJ} is the Hönl–London factor for a symmetric top molecule but is applicable in this case for HO_2 since the asymmetry splitting is so small. For the P branch ($\Delta J = -1$), this is $U_{K_aJ} = (J^2 - K_a^2)/J(2J + 1)$. The above equation gives a ratio of $S_{\text{band}}/S_{\text{line}} = 436$ for the $6_{15} = 7_{16}\ F_1$ transition. For R-branch ($\Delta J = +1$) transitions in this band, $U_{K_aJ} = [(J + 1)^2 - K_a^2]/(J + 1)(2J + 1)$, and for Q-branch ($\Delta J = 0$) transitions, $U_{K_aJ} = K_a^2/J(J + 1)$.

3.2.12 Experimental Measurement of Intensities from Infrared Spectra

In the preceding sections, describing the intensities and shapes of spectral lines, we have not included instrumental broadening nor have we discussed the experimental measurement of spectral parameters such as the absorption line-strength. Let us first assume that the spectral line density is low enough to allow individual lines to be resolved. When nonlaser instruments are used for IR absorption measurements, the instrumental widths are often greater than the molecular linewidth, especially in the low-pressure regime, where narrow, Doppler-limited widths would otherwise be identified with individual lines. In this case the line intensity and width is derived either from a nonlinear least-squares synthetic spectral fitting method or from the equivalent width method (Jansson and Korb, 1968) in which this parameter is measured under varying conditions. The equivalent width, W, is defined as the width (in cm^{-1}) of a rectangular spectral hole burned into the incident radiation spectrum and having the same area as that of the actual absorption line. That is

$$W = \int_{\text{line}} \left[1 - \exp(-\kappa(\tilde{\nu})z\right] d\tilde{\nu} \tag{3.115}$$

so that

$$W = (1/I_0) \int_{\text{line}} \left[I_0 - I(\tilde{\nu})\right] d\tilde{\nu} \tag{3.116}$$

for constant I_0. The integral in the above equation is the area under the spectral line, which may be measured from the experimental recording. The equivalent width is a very useful parameter because it is independent of the instrument function (resolving power) if either I_0 is constant or I_0 varies linearly with $\tilde{\nu}$ and the instrument response function is symmetric. In the optically thin case, where $\kappa(\tilde{\nu})z \ll 1$, the equivalent width may be approximated by $W \simeq \int_{\text{line}} \kappa(\tilde{\nu})z \, d\tilde{\nu}$, and the linestrength can be measured directly from a measurement of the equivalent width provided the gas density N is known. For optically thick samples, the complete expression of Equation (3.115) must be used (see Pugh and Rao, 1976). The frequency dependence of $\kappa(\tilde{\nu})$ is given by the shape function $g(\tilde{\nu} - \tilde{\nu}_0)$, which for a Voigt lineshape depends on the ratio $\gamma_L/\gamma_{\text{ED}}$, which is unknown. In order to extract the linestrength, values of log W against log N must be plotted as a curve of growth and compared to theoretical curves using $\gamma_L/\gamma_{\text{ED}}$ as a variable parameter. At low densities, W is proportional to N, but with increasing N, the curve of growth flattens to a limit proportional to $N^{1/2}$ (see Corney, 1977). These methods are used extensively by astrophysicists measuring oscillator strengths. The curve-of-growth method is subject to error due to wing absorption, and where pressure broadening contributes to the observed lineshape, wing and base corrections (see Korb and co-workers, 1968) must be applied to the determined equivalent width. Tables of equivalent widths of isolated lines with combined Doppler- and collision-broadened profiles are available (Jansson and Korb, 1968).

When data reduction techniques such as the equivalent width method are employed, uncertainties in the laboratory measurement of linestrengths are typically 5–25%, these uncertainties often representing the largest contribution to uncertainties in the remote measurements of species of atmospheric interest.

For the measurement of bandstrengths, the principle of summing the individually measured linestrengths is complicated by the blending or weakness in intensity of some lines. According to Pugh and Rao (1976), the method of Wilson and Wells (1946) is the most popular technique for measuring bandstrengths. If the bandstrength is written as

$$S = \int_0^\infty (1/z)\ln[I_0/I(\tilde{\nu})]\, d\tilde{\nu} \tag{3.117}$$

this method is based on the observation that the measured value of S approaches the required value of S (in the absence of instrumental broadening) as z approaches zero. However, this method is not suitable for spectra comprising overlapping bands and suffers from uncertainties both in the extrapolation to small z values and in the location of the baseline due to wing absorption. Until the arrival of tunable diode lasers and difference frequency lasers with output bandwidths typically <10 MHz ($<0.0003\ \mathrm{cm}^{-1}$), and indeed modern Fourier transform infrared (FTIR) spectrometers with a resolution of $<0.0005\ \mathrm{cm}^{-1}$, linestrength or bandstrength determinations concentrated on devising methods that were independent of the instrumental function since this was often greater than the molecular widths studied. Tunable diode laser linewidths are often sufficiently narrow that to a first approximation their contribution to a Doppler- or collision-broadened lineshape is negligible. In this case, the linestrength at low optical depth may be written as

$$\begin{aligned} S_{\text{line}} &= \int_0^\infty \kappa(\tilde{\nu})\, d\tilde{\nu} = (1/z)\int_0^\infty (\Delta I/I_0)\, d\tilde{\nu} \\ &= (1/cl)(\Delta I/I_0)(a/h) \end{aligned} \tag{3.118}$$

where c is the gas concentration, l the pathlength, and a/h the area-to-height ratio of the line profile, which may be determined, for example, by tracing the line with a polar planimeter. The theoretical values of a/h for normalized pure Doppler and pure Lorentz profiles are

$$(a/h)_D = \gamma_D/(\ln 2/\pi)^{1/2} = 2.1289\gamma_D \tag{3.119}$$

and

$$(a/h)_L = \pi\gamma_L = 3.1416\gamma_L \tag{3.120}$$

Therefore, where a pure lineshape (e.g., Doppler) exists, comparison of the measured a/h to the theoretically predicted a/h can provide a measurement of the laser bandwidth. Usually, this comparison is made only to confirm the negligibility of the bandwidth.

Instrumental broadening due to finite laser bandwidth must often be included. The observed profile is then a convolution of the instrumental and molecular lineshapes. If the laser output frequency profile itself can be approximated by a pure Gaussian or Lorentzian shape, the following two theorems (Corney, 1977) can be applied:

1. The convolution of two Gaussian profiles characterized by HWHM widths β_1 and β_2 is also a Gaussian with a HWHM width

$$\beta = (\beta_1^2 + \beta_2^2)^{1/2} \tag{3.121}$$

2. The convolution of two Lorentzian profiles characterized by HWHM widths γ_1 and γ_2 is also a Lorentzian with a HWHM width

$$\gamma = \gamma_1 + \gamma_2 \tag{3.122}$$

The output bandwidth of any given lead–salt tunable diode laser can vary widely with junction temperature, current through the diode, and environment (acoustical broadening is often troublesome in closed-cycle refrigerators) and must be measured for each spectral region studied (see Section 3.3.7). The response function of a tunable diode laser spectrometer has recently been determined by May (1987) using a deconvolution procedure which does not require specific knowledge of the origin of the excess laser linewidth. This method has been successfully applied to the spectra of CH_4, HNO_3, and HO_2NO_2. When high-resolution FTIR spectrometers are used, the instrumental width is well understood, and then line-fitting procedures are applicable for molecular bands of low line density. Brown and co-workers (1983) recently fitted simultaneously absorption line positions, strengths, linewidths, and continuum parameters for the ν_4 and $\nu_1 + \nu_4$ bands of CH_4 to allow relative linestrengths to be measured with accuracies of 2% or better. Lineshape-fitting procedures are also preferred for the measurement of line-broadening coefficients using tunable diode laser spectrometers [see Section 3.3.7 and Lundqvist and co-workers (1982)].

3.2.13 Intensity Units and Conversion Tables

In defining the absorption coefficient $\kappa(\tilde{\nu})$ in the Beer–Lambert law expressed in Equation (3.8), we note that its units are $1/z$, where z is concentration times length. Since the lineshape function $g(\tilde{\nu} - \tilde{\nu}_0)$ has units of centimeters, and $\kappa(\tilde{\nu}) = Sg(\tilde{\nu} - \tilde{\nu}_0)$, then the linestrength or line intensity S must have units of cm^{-1}/z or cm^{-2}/concentration. When the concentration is given in units of molecules cm^{-3}, S is in units of cm molecule^{-1}; when the concentration is given as a pressure in atmospheres, S is in units of $\text{cm}^{-2}\ \text{atm}^{-1}$. The former concentration units are preferred because they are independent of temperature. When pressure is the concentration unit, the temperature of the gas and the equation of state must also be specified, as pointed out by Pugh and Rao (1976).

As an example, we now will calculate the line center absorptance at 310 K for a Doppler-broadened ($\gamma_D = 0.001\ \text{cm}^{-1}$ at 300 K) line of a linear molecule transition with $\tilde{\nu}'' = 373\ \text{cm}^{-1}$ and a linestrength of 2.2×10^{-20} cm molecule^{-1} at 296 K assuming a 10-cm cell length containing the pure absorbing gas at a pressure of 0.01 Torr.

First, we convert the Doppler width to its value of $0.00102\ \text{cm}^{-1}$ at 310 K according to Equation (3.35) and also convert the linestrength to its value of 2.123×10^{-20} cm molecules^{-1} at 310 K according to Equation (3.105). The line center absorption coefficient at 310 K is then

$$\begin{aligned}\kappa(\tilde{\nu}_0) &= [S(310)/\gamma_D(310)](\ln 2/\pi)^{1/2} \\ &= 9.7766 \times 10^{-18}\ \text{cm}^2\ \text{molecule}^{-1}\end{aligned}$$

so that the exponent is

$$\begin{aligned}\kappa(\tilde{\nu}_0)z = {} & 9.7766 \times 10^{-18}\ \text{cm}^2\ \text{molecule}^{-1}\ (0.01\ \text{Torr}/760\ \text{Torr atm}^{-1}) \\ & \times (310/273.15) \times 2.68675 \times 10^{19}\ \text{molecule cm}^{-3}\ \text{atm}^{-1} \\ & \times 10\ \text{cm}\end{aligned}$$

which is equal to 0.0393, making the transmission at line center 0.966. The absorption at line center is therefore equal to 0.0385, or 3.8%. At low optical depths, as in this case, the absorptance is nearly equal to the exponent value.

In Table 3.3 the bandstrength conversion factor table of Pugh and Rao (1976) is reproduced. In this table, the conversion from one set of intensity units to another was made by the authors, assuming the ideal gas law and employing the following physical constants:

Table 3.3. Bandstrength Conversion Factors[a]

	$cm^{-2}\ atm^{-1}$ at 300°K	$cm^{-1}\ sec^{-1}\ atm^{-1}$ at 300°K	$cm^{-2}\ atm^{-1}$ at T	$cm^{-1}\ sec^{-1}\ atm^{-1}$ at T	$cm^{-2}\ atm^{-1}$ at STP	$cm^{-1}\ sec^{-1}\ atm^{-1}$ at STP
$cm^{-2}\ atm^{-1}$ at 300 K	1.0	3.3356×10^{-11}	$3.3333T \times 10^{-3}$	$1.1119T \times 10^{-13}$	9.1053×10^{-1}	3.0372×10^{-11}
$cm^{-1}\ sec^{-1}\ atm^{-1}$ at 300 K	2.997925×10^{10}	1.0	$9.9931T \times 10^{7}$	$3.3333T \times 10^{-3}$	2.7297×10^{10}	9.1053×10^{-1}
$cm^{-2}\ atm^{-1}$ at T	$\frac{300}{T}$	$\frac{1.0007}{T} \times 10^{-8}$	1.0	3.3356×10^{-11}	$\frac{273.16}{T}$	$\frac{9.1116}{T} \times 10^{-9}$
$cm^{-1}\ sec^{-1}\ atm^{-1}$ at T	$\frac{8.9938}{T} \times 10^{12}$	$\frac{300}{T}$	2.997925×10^{10}	1.0	$\frac{8.1891}{T} \times 10^{12}$	$\frac{273.16}{T}$
$cm^{-2}\ atm^{-1}$ at STP	1.0983	3.6634×10^{-11}	$3.6609T \times 10^{-3}$	$1.2211T \times 10^{-13}$	1.0	3.3356×10^{-11}
$cm^{-1}\ sec^{-1}\ atm^{-1}$ at STP	3.2925×10^{10}	1.0983	$1.0975T \times 10^{8}$	$3.6609T \times 10^{-3}$	2.997925×10^{10}	1.0
$cm\ mole^{-1}$	2.4617×10^{4}	8.2113×10^{-7}	$8.2056T \times 10^{1}$	$2.7371T \times 10^{-9}$	2.2414×10^{4}	7.4766×10^{-7}
$cm^{2}\ sec^{-1}\ mole^{-1}$	7.3799×10^{14}	2.4617×10^{4}	$2.4600T \times 10^{12}$	$8.2056T \times 10^{1}$	6.7197×10^{14}	2.2414×10^{4}
$cm\ millimole^{-1}$ (also called dark)	2.4617×10^{1}	8.2113×10^{-10}	$8.2056T \times 10^{-2}$	$2.7371T \times 10^{-12}$	2.2414×10^{1}	7.4766×10^{-10}
$cm^{2}\ sec^{-1}\ millimole^{-1}$	7.3799×10^{11}	2.4617×10^{1}	$2.4600T \times 10^{9}$	$8.2056T \times 10^{-2}$	6.7197×10^{11}	2.2414×10^{1}
$cm^{-2}\ liter\ mole^{-1}$	2.4617×10^{1}	8.2113×10^{-10}	$8.2056T \times 10^{-2}$	$2.7371T \times 10^{-12}$	2.2414×10^{1}	7.4766×10^{-10}
$cm^{-1}\ sec^{-1}\ liter\ mole^{-1}$	7.3799×10^{11}	2.4617×10^{1}	$2.4600T \times 10^{9}$	$8.2056T \times 10^{-2}$	6.7197×10^{11}	2.2414×10^{1}
$cm\ molecule^{-1}$	4.0877×10^{-20}	1.3635×10^{-30}	$1.3626T \times 10^{-22}$	$4.5450T \times 10^{-33}$	3.7220×10^{-20}	1.2415×10^{-30}
$cm^{2}\ sec^{-1}\ molecule^{-1}$	1.2255×10^{-9}	4.0877×10^{-20}	$4.0849T \times 10^{-12}$	$1.3626T \times 10^{-22}$	1.1158×10^{-9}	3.7220×10^{-20}

[a] In converting from the units labeled in the top horizontal row to the units labeled in the left vertical column, move along the top row to the appropriate unit and down the column to the number which appears against the unit to be converted to. This numerical factor is the multiplicative factor. For example, S in units of cm $mole^{-1}$ can be converted

Table 3.3. (*Continued*)

cm $mole^{-1}$	cm^2 sec^{-1} $mole^{-1}$	cm $millimole^{-1}$ (also called dark)	cm^2 sec^{-1} $millimole^{-1}$	cm^{-2} liter $mole^{-1}$	cm^{-1} sec^{-1} liter $mole^{-1}$	cm $molecule^{-1}$	cm^2 sec^{-1} $molecule^{-1}$
4.0623×10^{-5}	1.3550×10^{-15}	4.0623×10^{-2}	1.3550×10^{-12}	4.0623×10^{-2}	1.3550×10^{-12}	2.4464×10^{19}	8.1602×10^{8}
1.2178×10^{6}	4.0623×10^{-5}	1.2178×10^{9}	4.0623×10^{-2}	1.2178×10^{9}	4.0623×10^{-2}	7.3340×10^{29}	2.4464×10^{19}
$\frac{1.2187}{T} \times 10^{-2}$	$\frac{4.0651}{T} \times 10^{-13}$	$\frac{1.2187}{T} \times 10^{1}$	$\frac{4.0651}{T} \times 10^{-10}$	$\frac{1.2178}{T} \times 10^{1}$	$\frac{4.0651}{T} \times 10^{-10}$	$\frac{7.3391}{T} \times 10^{21}$	$\frac{2.4481}{T} \times 10^{11}$
$\frac{3.6535}{T} \times 10^{8}$	$\frac{1.2187}{T} \times 10^{-2}$	$\frac{3.6535}{T} \times 10^{11}$	$\frac{1.2187}{T} \times 10^{1}$	$\frac{3.6535}{T} \times 10^{11}$	$\frac{1.2187}{T} \times 10^{1}$	$\frac{2.2002}{T} \times 10^{32}$	$\frac{7.3391}{T} \times 10^{21}$
4.4614×10^{-5}	1.4882×10^{-15}	4.4614×10^{-2}	1.4882×10^{-12}	4.4614×10^{-2}	1.4882×10^{-12}	2.6867×10^{19}	8.9620×10^{8}
1.3375×10^{6}	4.4614×10^{-5}	1.3375×10^{9}	4.4614×10^{-2}	1.3375×10^{9}	4.4614×10^{-2}	8.0546×10^{29}	2.6867×10^{19}
1.0	3.3356×10^{-11}	1.0×10^{3}	3.3356×10^{-8}	1.0×10^{3}	3.3356×10^{-8}	6.0221×10^{23}	2.0088×10^{13}
2.997925×10^{10}	1.0	2.997925×10^{13}	1.0×10^{3}	2.997925×10^{13}	1.0×10^{3}	1.8054×10^{34}	6.0221×10^{23}
1.0×10^{-3}	3.3356×10^{-14}	1.0	3.3356×10^{-11}	1.0	3.3356×10^{-11}	6.0221×10^{20}	2.0088×10^{10}
2.997925×10^{7}	1.0×10^{-3}	2.997925×10^{10}	1.0	2.997925×10^{10}	1.0	1.8054×10^{31}	6.0221×10^{20}
1.0×10^{-3}	3.3356×10^{-14}	1.0	3.3356×10^{-11}	1.0	3.3356×10^{-11}	6.0221×10^{20}	2.0088×10^{10}
2.997925×10^{7}	1.0×10^{-3}	2.997925×10^{10}	1.0	2.997925×10^{10}	1.0	1.8054×10^{31}	6.0221×10^{20}
1.6605×10^{-24}	5.5390×10^{-35}	1.6605×10^{-21}	5.5390×10^{-32}	1.6605×10^{-21}	5.5390×10^{-32}	1.0	3.3356×10^{-11}
4.9782×10^{-14}	1.6605×10^{-24}	4.9782×10^{-11}	1.6605×10^{-21}	4.9782×10^{-11}	1.6605×10^{-21}	2.997925×10^{10}	1.0

to S in cm^{-2} atm^{-1} at 300°K by using S(cm^{-2} atm^{-1} at 300°K) = (4.0623×10^{-5}) S(cm $mole^{-1}$). Also S in units of cm $mole^{-1}$ can be converted to S in cm $molecule^{-1}$ by using S(cm $molecule^{-1}$) = (1.6605×10^{-24}) · S(cm $mole^{-1}$)

Source: Pugh and Rao (1976).

$$R = 8.3143 \times 10^7 \text{ erg/(mol K)}$$

$$k = 1.38062 \times 10^{-16} \text{ erg/K}$$

$$c = 2.997925 \times 10^{10} \text{ cm/s}$$

In addition, 0 °C = 273.16 K and 1 atm = 1.01325×10^6 dyn/cm^2 were used. The footnote to this table gives examples of how to make use of these conversion factors.

3.3 HARMONIC DETECTION TECHNIQUES FOR TUNABLE DIODE LASER SPECTROSCOPY

3.3.1 Tunable Diode Lasers (TDLs)

Lead–salt diode lasers provide tunable IR sources in the 2.5–30-μm wavelength region, with output powers typically 0.1–1 mW in a single-mode band width of only $\simeq$0.0002 cm^{-1}. They can individually be tuned over several hundred reciprocal centimeters by overlapping the tuning ranges of a series of modes or a few reciprocal centimeters for a given longitudinal mode. Until recently, commercially available TDLs operated only at temperatures <77 K, and liquid helium cryogenics or refrigeration was required.

TDLs with single-mode operation and output powers in excess of 50 μW that operate at temperatures >100 K are now available using short-cavity mesa-stripe, cleaved-coupled-cavity, or quantum-well technology. The growth of a new lead–salt semiconductor, $Pb_{1-x}Eu_xSe_yTe_{1-y}$, has now allowed much higher operating temperatures to be achieved. In the 2.7–6.6-μm range, double heterojunction lasers have operated CW up to 147 K. Single quantum-well devices push the operating temperature still higher, so that TDLs in the 4-μm region have been operated CW up to 174 K and pulsed up to 241 K (Partin and coworkers, 1984).

The output wavelength of a TDL depends on the chemical composition of the diode and its temperature, once sufficient current is passed through the diode to reach the threshold for laser action. Increasing the TDL current increases the power output but also increases the temperature of the junction, thereby providing a mechanism for fine tuning the laser output frequency or wavelength. The quantum-well devices operate at low threshold currents, and input power requirements of only $\simeq$10 mW are necessary for 50-μW laser light output levels at operating temperatures in the 120-K region.

Tunable diode lasers, with their sub-Doppler resolution capability, high spectral purity and energy flux, and amenability to sensitive modulation techniques, have been used extensively for various kinds of spectroscopic applications. Multispecies absorption spectrometers are now routinely used in research and industrial laboratories and are finding increased use in airborne and balloon-borne atmospheric monitoring instruments.

For recording absorption spectra, TDLs may be tuned by one of several methods, the most popular being by increasing the injected current across the junction. Current changes of 1 mA will typically tune a TDL 0.005 cm^{-1}. Figure 3.10 shows a 40-mA scan over NO features near 5.2 μm at Doppler-limited resolution. In this direct-absorption mode, however, only peak absorptances of $>1\%$ can be measured (excluding sweep integration methods). For sensitive measurements of weak absorptions due to low concentrations or low absorption cross sections, harmonic detection methods are used, as described below.

3.3.2 Principles of Harmonic Detection

Harmonic detection is an experimental technique widely used to monitor weak signals in Stark, Zeeman, and magnetic resonance spectroscopy. This technique has more recently been used in conjunction with tunable diode lasers to dem-

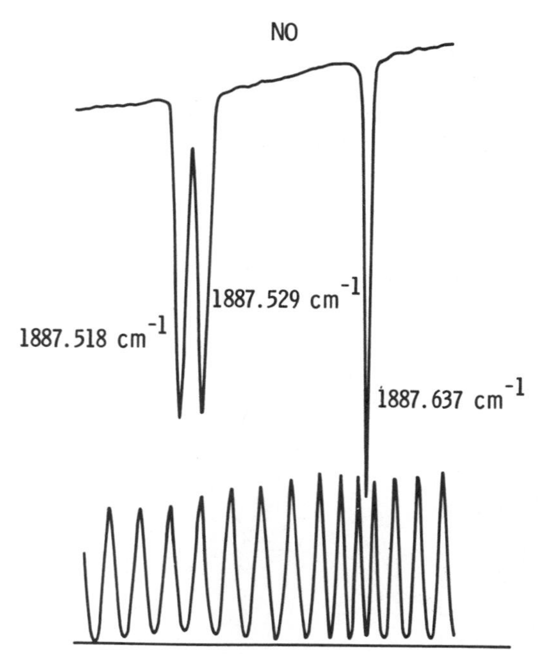

Figure 3.10. TDL scan over NO features near 5.2 μm at Doppler-limited resolution.

onstrate a spectroscopic means of measuring IR absorption coefficients as low as 10^{-7} m^{-1}. To achieve this high sensitivity, the wavelength of the TDL is modulated at a high frequency (kHz) by an amount typically several times smaller than the width of the absorption line of the species studied. Scanning the laser wavelength and using ac detection at the modulation frequency (f), or twice the modulation frequency ($2f$), then produces a detected signal proportional (for small modulation amplitude) to the first derivative and second derivative, respectively, of the absorption line. (See Figure 3.11). The increased sensitivity of harmonic detection results from (i) using high-modulation frequencies (kHz), which removes detection from the $1/f$ region; (ii) the large discrimination of the technique against signals that do not show wavelength-dependent changes; and (iii) the removal of the sloping background level.

Harmonic detection provides a significant improvement in sensitivity over direct absorption and allows absorptances in the 10^{-4}–10^{-5} range to be measured. For atmospheric monitoring this is done from a knowledge of the Voigt lineshape of the species monitored and the measured stratospheric temperature. This requires a knowledge of the linestrength and line-broadening coefficients. Relating the harmonic signal to the total returned signal to yield the mixing ratio is then possible using the measured frequency modulation and the TDL linewidth.

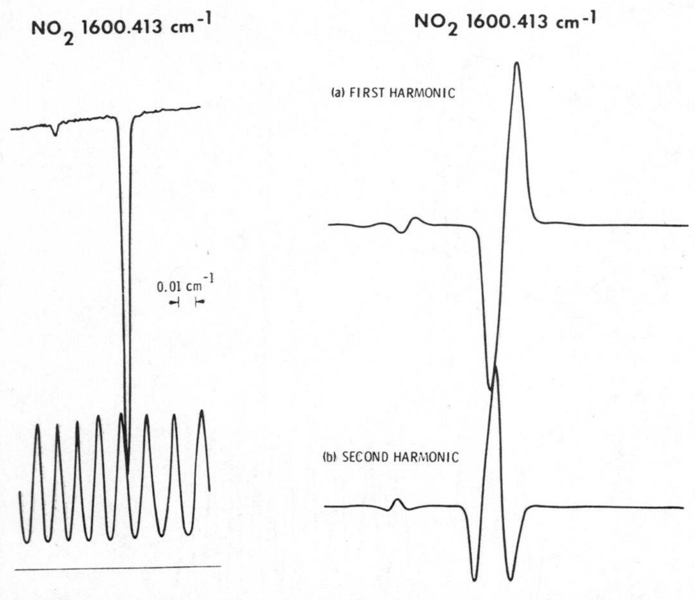

Figure 3.11. Direct, first- and second-harmonic lineshapes recorded experimentally for NO_2 features near 1600.4 cm^{-1}.

Consider a TDL with an IR output radiation of frequency ν_{TDL} that typically lies in the range 0.3×10^{14} Hz (10 μm) to 1×10^{14} Hz (3 μm). If this output frequency is modulated by applying small sinusoidal variation to the TDL current, the laser radiation has a frequency

$$\nu(t) = \nu_{TDL} + a \cos \omega t \tag{3.123}$$

where ω is the angular modulation frequency, and a is half the peak-to-peak amplitude of the modulation. If the TDL frequency is now tuned over an absorption lineshape characterized by an absorption coefficient $\kappa(\nu)$ with a line center such that

$$\nu_{TDL} = \nu_0 + \nu_1 \tag{3.124}$$

and a dimensionless parameter called the modulation coefficient is defined as

$$m = a/\gamma \tag{3.125}$$

where γ is the HWHM of the absorption line, the laser radiation frequency may be written as

$$\nu(t) = \nu_0 + \nu_1 + m\gamma \cos \omega t \tag{3.126}$$

Scanning this frequency over an absorption line will produce a signal whose waveform will depend on the lineshape "sampled" and the modulation coefficient. In the limiting case in which $m \ll 1$ and the laser samples over only a small part of the side of a lineshape which is assumed linear, the signal $\kappa(\nu_1, m, t)$ would be a sine wave of frequency ω and an amplitude proportional to m. However, molecular lineshapes do not show linearity with frequency interval, and the signal produced $\kappa(\nu_1, m, t)$ is not sinusoidal, although still a function periodic in ω (see Figure 3.12). According to Fourier's theorem, such a periodic wave can be represented by a series of sinusoidal components of frequencies 2, 3, . . . , n. The relative contributions of these "harmonics" (ω = first harmonic) depends primarily on the lineshape function, whether it be a Gaussian (Doppler), Lorentz, or Voigt profile. We should note that in the limit that the modulation amplitude (and therefore m) approaches zero and the sampled region of the lineshape therefore approaches linearity, the signal components at higher harmonics (2ω, 3ω, . . . , $n\omega$) tend to zero. Thus, the relative contribution of the harmonic components to the signal shows a definite m dependence, which will be given later.

The application of phase-sensitive discrimination is illustrated in Figure 3.13 for (a) ν_{TDL} off line center and (b) $\nu_{TDL} = \nu_0$. We can see how frequency-modulated lasers scanning over absorption lineshapes produce dispersive-shaped lines for detection at the modulation frequency: exactly at line center, quadra-

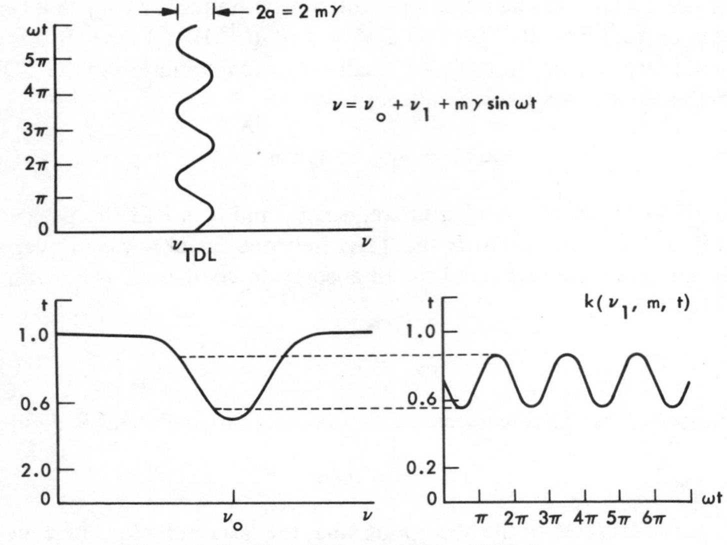

Figure 3.12. Sinusoidal modulation of TDL frequency.

ture occurs and the mean dc output of the lock-in amplifier is zero, producing the "crossing point" of the first harmonic signal.

Because the wavelength-dependent signal has components at 2ω, 3ω, and 4ω, and so on, as discussed in the previous section, these signals may also be extracted by the selective amplifier-demodulator using the same reference frequency. The labeling of detection at 2ω as second-derivative detection refers only to cases of very small modulation amplitude relative to the absorption linewidth, that is, to cases in which $m\ (= a/\gamma) \ll 1$. The lineshapes produced at ω, 2ω, . . . , are then directly proportional to the first, second, and so on, derivatives, respectively, of the lineshape concerned. For modulation amplitudes that produce values of m greater than unity, this proportionality no longer holds, and detection at 2ω must be labeled with the more general term of second-harmonic detection.

3.3.3 Low-Modulation Derivative Spectroscopy

3.3.3.1 Lorentzian Lineshape

Inserting Equation (3.126) into the Lorentzian expression of Equation (3.42) yields

$$\kappa(\nu_1, m, t) = S\gamma_L/\left[\pi(\nu_1 + m\gamma_L \cos \omega t)^2 + \pi\gamma_L^2\right] \qquad (3.127)$$

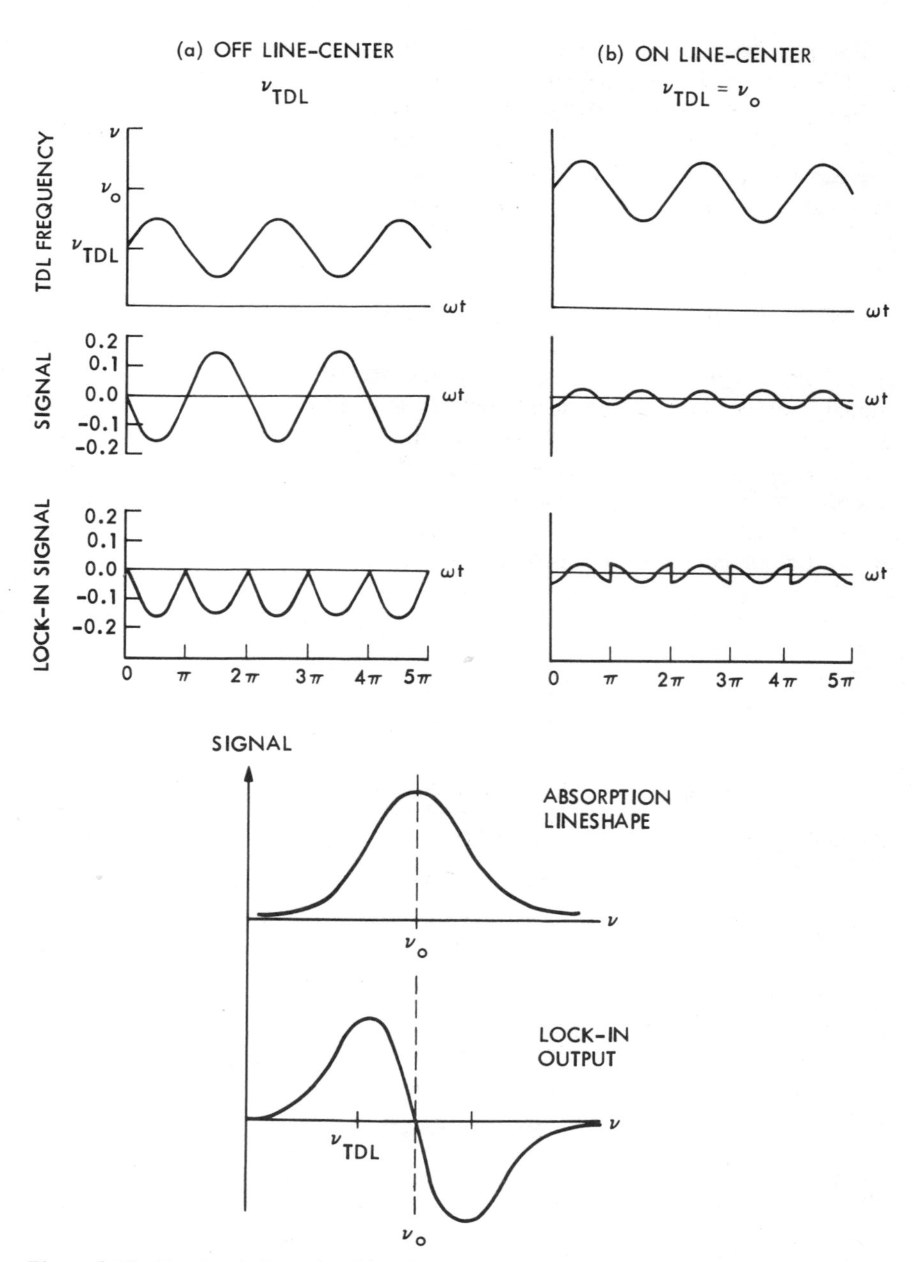

Figure 3.13. Signal waveforms resulting from sinusoidal modulation of laser frequency with detection at fundamental frequency.

which then expanded in the form $1/(1 + x) = 1 - x + x^2 - x^3, \ldots$, for $m < 1$, reveals that the signal consists of a dc component, $S\gamma_L/\pi(\nu_1^2 + \gamma_L^2)$, a component at the modulation frequency ω, and components at all the harmonics of ω. For $m \ll 1$ the high powers of m may be neglected. The signals at ω and 2ω may then be compared to the real derivatives of the Lorentzian absorption coefficient (see Table 3.4), and it is found (Kavaya, 1982) that

$$\text{Signal at } \omega = m\gamma_L \times [\text{first derivative of } \kappa(\nu)]$$
$$\text{Signal at } 2\omega = (m\gamma_L/2)^2 \times [\text{second derivative of } \kappa(\nu)] \quad (3.128)$$

Thus, the ω signal is proportional to the first derivative and to $m\gamma_L$, and the 2ω signal is proportional to the second derivative and to $(m\gamma_L)^2$. Each signal may be increased by increasing m. However, this will decrease the resolution in ν_1 and eventually violate the assumption $m \ll 1$. The effect of high-modulation coefficients, that is, $m > 1$, will be discussed in the next section.

From Table 3.4 (Kavaya, 1982) it is seen that the peak-to-peak excursions of the Lorentzian curve and its first two derivatives are

$$K_{\text{p-p}} = 0.318S/\gamma_L$$
$$K'_{\text{p-p}} = 0.413S/\gamma_L^2$$
$$K''_{\text{p-p}} = 0.796S/\gamma_L^3 \quad (3.129)$$

Table 3.4. Lorentzian Lineshape and Its Derivatives[a]

Symbol	Function	Maxima
	Lorentzian Lineshape	
$k(\nu)$	$\dfrac{S}{\pi}\dfrac{\gamma_L}{(\nu-\nu_0)^2+\gamma_L^2}$	$\dfrac{S}{\pi\gamma_L}$ at $\nu - \nu_0 = 0$
	First Derivative	
$\dfrac{d}{d\nu}k(\nu)$	$\dfrac{-2(\nu-\nu_0)}{(\nu-\nu_0)^2+\gamma_L^2}k(\nu)$	$-\dfrac{\pm 9S}{8\sqrt{3}\pi\gamma_L^2}$ at $\nu - \nu_0 = \dfrac{\pm\gamma_L}{\sqrt{3}}$
	Second Derivative	
$\dfrac{d^2}{d\nu^2}k(\nu)$	$\dfrac{2[3(\nu-\nu_0)^2-\gamma_L^2]}{[(\nu-\nu_0)^2+\gamma_L^2]^2}k(\nu)$	$\dfrac{-2S}{\pi\gamma_L^3}$ at $\nu - \nu_0 = 0$ $\dfrac{S}{2\pi\gamma_L^3}$ at $\nu - \nu_0 = \pm\gamma_L$

[a]From Kavaya (1982).

The peak-to-peak size of the signals at ω and 2ω will, of course, depend on the modulation parameter and are given by relating Equations (3.128) with Equations (3.129). The ratios of the first- and second-harmonic signal excursions to the chopped-signal excursion are 1.30 m and 0.63 m^2, respectively, and the ratio of the second harmonic to the first harmonic is $0.48m$.

3.3.3.2 Doppler Lineshape

Equations (3.128) apply also to the Doppler case with the same restrictions on m. Table 3.5 (Kavaya, 1982) provides the corresponding results for the Doppler case. Note that the peaks of the first derivative (zero crossings of the second derivative) occur at $\nu - \nu_0 = \pm\gamma_D/(2 \ln 2)^{1/2} = \pm 0.85\gamma_D$. Note also that the Doppler signal amplitudes are larger than the corresponding Lorentzian signal amplitudes.

Table 3.5. Doppler Lineshape and Its Derivatives[a]

Symbol	Function	Maxima
$k(\nu)$	$\frac{S}{\gamma_D}\left(\frac{\ln 2}{\pi}\right)^{1/2} \exp\left(-\frac{(\nu-\nu_0)^2 \ln 2}{\gamma_D^2}\right)$	$\frac{S}{\gamma_D}\left(\frac{\ln 2}{\pi}\right)^{1/2}$ at $\nu - \nu_0 = 0$
	First Derivative	
$\frac{d}{d\nu}k(\nu)$	$\frac{-2(\ln 2)(\nu-\nu_0)}{\gamma_D^2} k(\nu)$	$-\frac{\pm S \ln 2}{\gamma_D^2}\left(\frac{2}{\pi e}\right)^{1/2}$ at $\nu - \nu_0 = \frac{\pm\gamma_D}{(2 \ln 2)^{1/2}}$
	Second Derivative	
$\frac{d^2}{d\nu^2}k(\nu)$	$\left(\frac{2 \ln 2}{\gamma_D^2}\right)\left(\frac{2 \ln 2(\nu-\nu_0)^2}{\gamma_D^2} - 1\right)k(\nu)$	$\frac{-2S(\ln 2)^{1/2}}{\gamma_D^3 \pi^{1/2}}$ at $\nu - \nu_0 = 0$ and $\frac{4S(\ln 2)^{3/2}}{\gamma_D^3 \pi^{1/2} e^{3/2}}$ at $\nu - \nu_0 = \pm\left(\frac{3}{2 \ln 2}\right)^{1/2}\gamma_D$

[a]From Kavaya (1982).

The peak-to-peak excursions of the Doppler curve and its derivatives are

$$K_{\text{p-p}} = 0.47S/\gamma_D$$
$$K'_{\text{p-p}} = 0.671S/\gamma_D^2$$
$$K''_{\text{p-p}} = 0.942S/\gamma_D^3 \quad (3.130)$$

The ratios of the m-dependent first- and second-harmonic signal excursions to the chopped signal excursion are 1.43 m and 0.50 m^2, respectively, and the ratio of the second harmonic to the first harmonic is 0.35 m.

3.3.4 High-Modulation Harmonic Spectroscopy

In the previous section, approximations were made based on values of m that do not hold for $m > 1$. For the detection of weak absorption signals using second-harmonic detection, the SNR must be optimized by increasing m above unity to maximize the size of the signal at 2ω. The exact expression for a Lorentzian and for a Doppler lineshape may be Fourier analyzed to find the Fourier coefficients of the ω and 2ω signal components. For sinusoidal modulation the time variation of the diode frequency may be written:

$$\nu(t) = \nu_{\text{TDL}} + a \cos \omega t \quad \text{or} \quad \nu(t) = \nu_0 + \nu_1 + m\gamma \cos \omega t \quad (3.131)$$

The time-dependent term of the transmitted intensity is $\kappa(\nu_0 + \nu_1 + m\gamma \cos \omega t)$, which is a periodic and an even function of ωt (time) and may therefore be expanded in a cosine Fourier series

$$\kappa(\nu_0 + \nu_1 + m\gamma \cos \omega t) = \sum_{n=0} H_n(\nu_0 + \nu_1) \cos n\omega t \quad (3.132)$$

where $\nu_0 + \nu_1$ is considered a constant over the modulation period. Then $H_n(\nu_0 + \nu_1)$ is the nth Fourier component of the modulated absorption coefficient. These dimensionless coefficients give the relative magnitudes of the signals that would be obtained with detection at the different harmonics of the modulation frequency.

These results may be rewritten using the convenient parameter x, where

$$x = \nu_1/\gamma \quad (3.133)$$

and

$$\kappa(x, m) = \sum_{n=0} H_n(x, m) \cos n\omega t \quad (3.134)$$

The Fourier coefficients are then given by

$$H_n(x, t) = \frac{2}{\pi} \int_0^{\pi} \kappa(x + m \cos \theta) \cos n\theta \, d\theta \tag{3.135}$$

The first and second coefficients are listed in Table 3.6, together with the normalized absorption ceofficients, for Doppler, Lorentz, and Voigt lineshapes.

For the absorption coefficients given here, the absorption coefficient at line center is normalized to unity. For example, for the Lorentzian lineshape, $S/\pi\gamma_L = 1$, and so the absorption coefficient becomes

$$\kappa(\nu) = 1/\left[1 + ((\nu - \nu_0)/\gamma_L)^2\right] \tag{3.136}$$

Table 3.6. Fourier Coefficients for Gaussian, Lorentzian, and Voigt Lineshapes

	Gaussian Lineshape
Normalized absorption coefficient	$k^G(x, m) = \exp\left[-\ln 2(x + m \cos \omega t)^2\right]$
First Fourier coefficient	$H_1^G(x, m) = \frac{2}{\pi} \int_0^{\pi} \exp\left[-\ln 2(x + m \cos \theta)^2\right] \cos \theta \, d\theta$
Second Fourier coefficient	$H_2^G(x, m) = \frac{2}{\pi} \int_0^{\pi} \exp\left[-\ln 2(x + m \cos \theta)^2\right] \cos 2\theta \, d\theta$
	Lorentz Lineshape
Normalized absorption coefficient	$k^L(x, m) = 1/\left[1 + (x + m \cos \omega t)^2\right]$
First Fourier coefficient	$H_1^L(x, m) = \frac{2}{\pi} \int_0^{\pi} \frac{\cos \theta}{1 + (x + m \cos \theta)^2} \, d\theta$
Second Fourier coefficient	$H_2^L(x, m) = \frac{2}{\pi} \int_0^{\pi} \frac{\cos 2\theta}{1 + (x + m \cos \theta)^2} \, d\theta$
	Voigt Lineshape
Normalized absorption coefficient	$k^V(y, z) = \frac{V(y, z)}{V(0, z)}$ where $z = \gamma_L/\gamma_D(\ln 2)^2$
First Fourier coefficient	$H_1^V(x, m) = \frac{2}{\pi} \frac{1}{V(0, z)} \int_0^{\pi} V[(x + m \cos \theta) y_{1/2}, z) \cos \theta \, d\theta$
Second Fourier coefficient	$H_2^V(x, m) = \frac{2}{\pi} \frac{1}{V(0, z)} \int_0^{\pi} V[(x + m \cos \theta) y_{1/2}, z) \cos \theta \, d\theta$
	where $y = z(\nu - \nu_0)/\gamma_L$ and $V(y_{1/2}, z) = \frac{1}{2} V(0, z)$

or

$$\kappa^L(x, m) = 1/\left[1 + (x + m \cos \omega t)^2\right] \tag{3.137}$$

as in Table 3.6. The integrals in this table may be evaluated numerically.

Figure 3.14 (Kavaya, 1982) shows plots of $H^L(x, m)$ against x for various values of the modulation coefficient m. In addition to an increasing signal with larger values of m, it is evident that the curves widen in the horizontal or frequency offset dimension. From Table 3.4, it is seen that the peak values of the first derivative (zero crossings of the second derivative) occur at $\nu - \nu_0 = \pm\gamma_L/3^{1/2} = \pm 0.577\gamma_L$. These frequency offset coordinates are marked on the figure.

Numerical evaluation of the integrals appropriate to the *Doppler* case [i.e., $H^G(x, m)$] yields the curves shown in Figure 3.15, where $H^G(x, m)$ is plotted against values of the modulation coefficient m (Kavaya, 1982). From Table 3.5 the peak values of the first derivative (zero crossings of the second derivative) occur at $\nu - \nu_0 = \pm\gamma_D/(2 \ln 2)^{1/2}$, these points being marked on the figure.

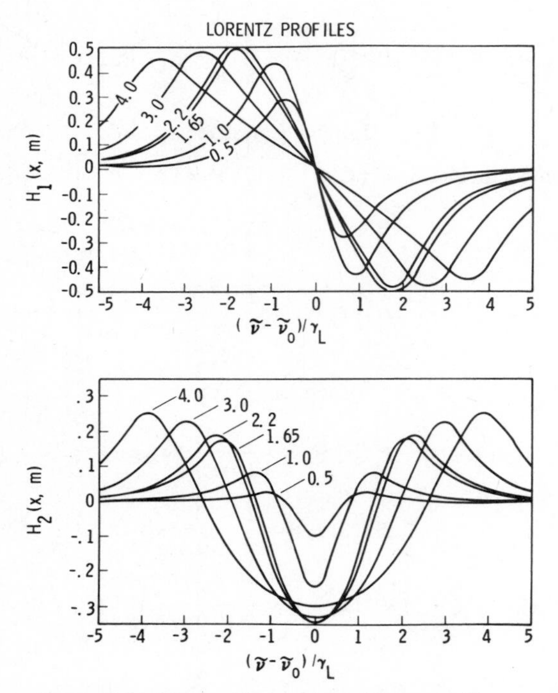

Figure 3.14. Plots of $H_n(x, m)$ for Lorentz profiles.

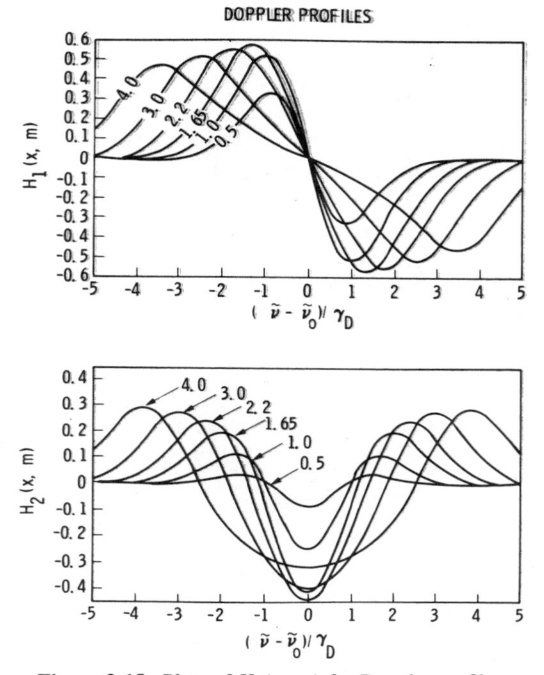

Figure 3.15. Plots of $H_n(x, m)$ for Doppler profiles.

3.3.5 Analytical Expressions

We will now investigate in detail the dependence of the lineshape and signal size on m. Following the treatment of Arndt (1965), the signal at the detector output for a *Lorentzian lineshape* is, for $n \geqslant 1$,

$$S_n(x) = (i^n/m^n)\left\{\left[(1 - ix)^2 + m^2\right]^{1/2} - (1 - ix)\right\}/\left[(1 - ix)^2 + m^2\right]^{1/2} + cc \qquad (3.138)$$

For $n = 1$, the *first-harmonic signal* is

$$S_1(x) = \left[2^{1/2}/m(M^2 + 4x^2)^{1/2}\right]\left\{\left[(M + 4x^2)^{1/2} + M\right]^{1/2} - \left[(M^2 + 4x^2)^{1/2} - M\right]^{1/2}\right\} \qquad (3.139)$$

where

$$M = 1 - x^2 + m^2. \qquad (3.140)$$

The first-order approximation yields

$$S_1(x) = -2xm/(1 + x^2)^2 \tag{3.141}$$

which is identical with the first derivative, as expected for a low m value. The points of maximum amplitude for $S_1(x)$ occur at

$$x_{\text{max}} = \pm 3^{1/2}[(4 + 3m^2)^{1/2} - 1] \tag{3.142}$$

which is $\pm 3^{-1/2}$ for small modulation, as given in Table 3.4. Then

$$S_1(x_{\text{max}}) = 3^{1/2}[(4 + 3m^2)^{1/2} - 2]/2m[(4 + 3m^2)^{1/2} - 1]^{1/2} \tag{3.143}$$

An important result is that $S_1(x_{\text{max}})$ is maximal at

$$m_{\text{max}} = 2 \tag{3.144}$$

with the amplitude

$$S_1(x_{\text{max}})_{\text{max}} = \pm 0.5 \tag{3.145}$$

and the maximal amplitudes occur at

$$(x_{\text{max}})_{\text{max}} = \pm 3^{1/2} \tag{3.146}$$

The *second-harmonic signal* is obtained by putting $n = 2$ into Equation (3.138) and eliminating the imaginary part to yield

$$S_2(x) = (8^{1/2}/m^2) - (2^{1/2}/M'm^2) \cdot [(M + 1 - x^2)(M' + M)^{1/2} + 4x(M' - M)^{1/2}] \tag{3.147}$$

where

$$M' = (M^2 + 4x^2)^{1/2} \tag{3.148}$$

For very large x $(=\nu_1/\gamma)$, $S_2(x)$ approaches zero. When $x = 0$, at line center,

$$S_2(0) = (2/m^2)[2 - (2 + m^2)/(1 + m^2)^{1/2}] \tag{3.149}$$

the value of $S_2(0)$ reaching a maximum for

$$m_{\text{max}} = (2 + 8^{1/2})^{1/2} = 2.197 \tag{3.150}$$

where the largest signal amplitude becomes

$$S_2(0)_{\text{max}} = -2/(3 + 8^{1/2}) = -0.343 \tag{3.151}$$

The results of Equations (3.143) and (3.149) are identified as H_1 and H_2 in Figure 3.16, adapted from a figure published by Wilson (1963).

For a *Doppler-broadened line*, Wilson has evaluated the integrals given in Table 3.6. His results for the signal amplitudes as a function of m are given in Figure 3.17. The first-harmonic signal reaches a maximum of

$$H_1^G(x, 1.65) = 0.57 \tag{3.152}$$

at $m = 1.65$, while the second-harmonic signal reaches a maximum of

$$H_2^G(0, 2.2) = 0.438 \tag{3.153}$$

at $m = 2.2$. Results for the Voigt signals will lie between the Lorentz and Doppler values. Table 3.7 lists the maximum signal sizes and required modulation coefficients. A result immediately evident from the table is that the max-

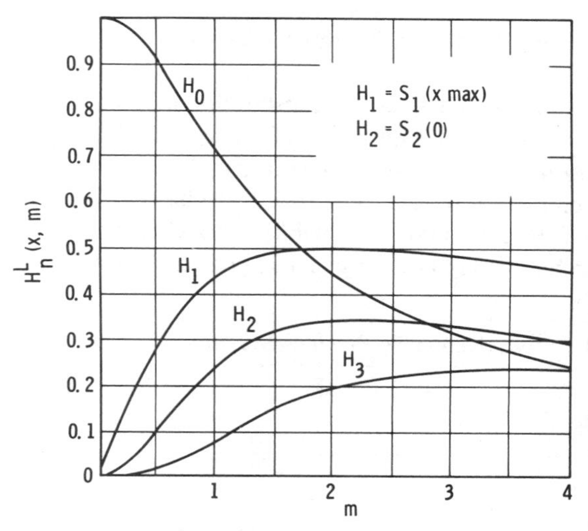

Figure 3.16. Plots of $H_n^L(x, m)$ against m. [Adapted from Wilson (1963).]

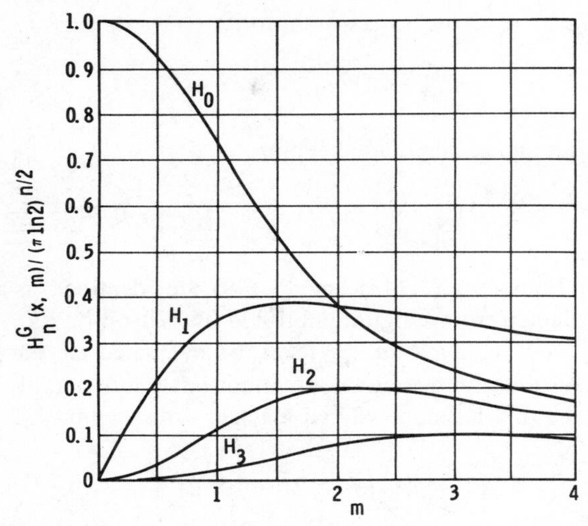

Figure 3.17. Plots of $H_n^G(x, m)/(\ln 2)^{n/2}$ against m. [Adapted from Wilson (1963).]

imum values of the Lorentz, Gaussian, and Voigt second-harmonic signals are all obtained for a modulation coefficient of $m = 2.2$. This is illustrated in Figure 3.18 from Reid and Labrie (1981), where

$$P = H_2(x, m)_{\max} \tag{3.154}$$

3.3.6 Sensitivities of Tunable Diode Laser Spectrometers

The sensitivity of a tunable diode laser absorption measurement using harmonic detection techniques depends on the experimental configuration (e.g., whether short-path, long-path, or multipass cell), the molecular environment (e.g., gas pressure and temperature), and the individual characteristics of key elements used (e.g., the TDL output power, tunability, stability, bandwidth, and detector or preamplifier characteristics). In general, peak absorptances of 0.001–0.01%

Table 3.7. Required m To Produce Maximum Values of $H(x, m)_{\max}$

	First Harmonic		Second Harmonic	
Signal	$m_{\max}$	$H_1(x, m)_{\max}$	$m_{\max}$	$H_2(x, m)_{\max}$
Lorentz	2.0	0.5	2.2	0.343
Gaussian	1.65	0.57	2.2	0.438
Voigt	1.65–2.0	0.5–0.57	2.2	0.34–0.44

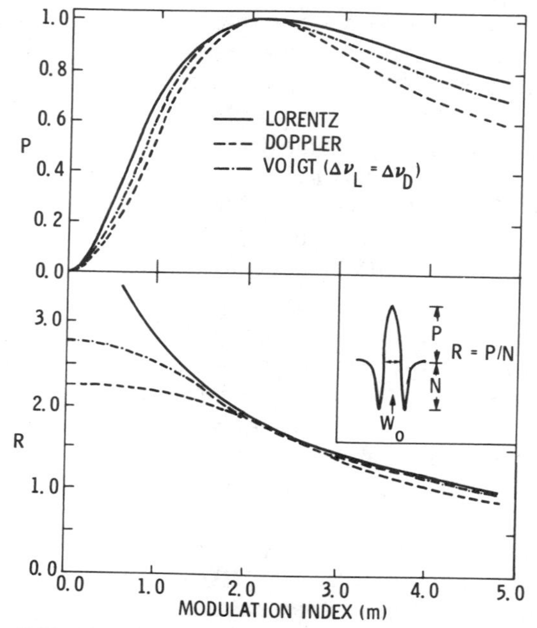

Figure 3.18. Plots of P and R vs. m. [From Reid and Labrie (1981).]

may be detected for isolated lines recorded at low pressures, near the Doppler-limited region. For atmospherically broadened lines, where linewidths are typically 0.05–0.2 cm^{-1} (HWHM), this detection limit is reduced to the 0.01–0.02% level.

The limitations to the sensitivity of a given TDL absorption measurement usually depend on the relative magnitude of three basic contributions to the observed background signal seen in the absence of the absorbing gas: (i) detector–signal channel noise, (ii) optical beam noise, and (iii) Fabry–Perot interference fringes.

3.3.6.1 Detector–Signal Channel Noise

For TDL measurements in the mid-IR region, HgCdTe detectors at 77 K in the photoconductive mode are usually employed for signal detection. The output voltage from the detector may be represented by the sum of two contributions: a signal voltage that is coherent with the signal radiation power and a noise voltage that is not. The magnitude of the signal voltage will depend on the bias voltage, the modulation frequency, the wavelength, the incident irradiance, and the detector area. The rms voltage of the detector noise is defined as the square root of the time average of the square of the difference between the instantaneous voltage and the time-averaged voltage (see Wolfe and Zissis, 1978). For ac-coupled amplifiers where the average value of $v(t)$ is zero, this becomes v_n (rms) $= \{\overline{[v_n(t)]^2}\}^{1/2}$. The random detector noise depends on the bias voltage, the modulation frequency, and the detector area. The signal-to-noise ratio is then usually expressed as $S/N =$ average signal$/v_n$ (rms). For constant-bias voltage and modulation frequency, the signal-to-noise ratio varies approximately directly as the square root of the detector area.

Commercially available HgCdTe detectors operating at 77 K have high detectivity D_λ^* ($>10^{10}$ cm Hz$^{1/2}$ W^{-1}) and high responsivity R (>250 V W^{-1}) at mid-IR wavelengths, the rms noise voltage over a few kilohertz bandwidth being typically <1 μV. Noise equivalent powers (NEP) of $\leqslant 1$ nW are typical. In order to see an absorptance of 0.01% with a signal-to-noise ratio of 1, a few microwatts of laser power at the detector is therefore usually required. For some absorption experiments, such as those using high multipass geometries, topographic targets, or very long-path retroreflector arrangements, these power levels may not be achieved. The signal-to-noise ratio will then be determined by the detector–signal channel noise characteristics. Photoconductive HgCdTe detectors are generally of low impedance ($\leqslant 100$ Ω) so that the preamplifier input noise level's contribution will be nonnegligible in some cases. The "detector noise" for photoconductive detectors will comprise contributions from the generation-recombination noise due to the background photon flux and to thermal equilibrium carriers in the semiconductor and from the Johnson noise. It is often necessary to fabricate the detector so that the background noise-limited (RLIP) D_λ^* is approached; then the only detector parameter on which D_λ^* depends is the photon quantum efficiency.

For detector- or signal-channel-noise-limited measurements, the signal-to-noise ratio improves with integration time, which may be done by integration at each discrete point of the spectral scan or by the coaddition of multiple scans.

3.3.6.2 Optical Beam Noise

When the diode laser power is allowed to fall onto the receiving detector but the frequency modulation is removed, random fluctuations in excess of the detector noise, which increase with increasing laser power, will be observed. This noise is described by diode laser spectroscopists as optical beam noise. Its magnitude depends on the power fluctuations (through the bandwidth of interest) at the particular laser used and often depends on the temperature and current operating parameters. Mechanical vibrations of the diode and the resulting beam wander from the optical path are often the major contributions. Indeed, where closed-cycle refrigeration is employed, it is not uncommon to observe a periodic acoustic component to the beam noise caused by the piston cycling in the coldhead.

Many of the components of optical beam noise are generated within the laser itself and may be labeled source noise. This would include, for example, the laser specific component dependent on junction temperature and current and acoustic components generated by the coldhead refrigeration. The large increase in sensitivity using harmonic detection results in part from the fact that source noise in TDLs at, for example, 2 kHz is typically $\sim$20 dB lower than at the few hundred hertz frequencies of chopper-modulated experiments. Recently, Laguna (1984) demonstrated a novel technique for source noise reduction in TDL spectroscopy using polarization rotation via the Faraday effect. A reduction by about an order of magnitude in source noise is possible with this technique, but it is only applicable to paramagnetic molecules with permanent dipole moments.

For many applications of TDL instruments to atmospheric monitoring, open-path remote sensing at atmospheric pressure is required. At this higher pressure, increased linewidths ($\simeq 0.1\ cm^{-1}$), spectral interferences, and atmospheric turbulence all reduce the measurement sensitivity. Distinguishing the absorption signal from the ever-present background signal becomes particularly difficult because the absorbing gas cannot be removed from the path. The background signal becomes more significant when making atmospheric pressure measurements because the laser must be tuned over a larger spectral range, which means that both the change in laser power over the line profile and the time taken over the scan increase. Using two detectors to difference the signal with that in a reference, absorber-free path produces little improvement, since the correlation is poor and depends critically on the alignment. Reid and co-workers (1985) recognized that the absorption signal is related to the oscillation of the TDL wavelength with current, while the background is related to the oscillation of the TDL power with current. Therefore, because these two oscillators were of different phase, they could be successfully separated using a phase-sensitive detector.

3.3.6.3 Fabry–Perot Interference Fringes

The sensitivity of diode laser absorption measurements is often limited by the production of Fabry–Perot interference fringes, which are generated through multiple-beam interferences between optical surfaces located within the source-to-detector path. The observed fringes may result from laser transmission through individual optical elements such as windows or lenses. For both laboratory and field studies, even very careful optical alignment procedures cannot in practice completely remove interference fringes. Fringes of peak-to-peak optical depth of a few parts in 10^4, for example, are produced by surface reflectivities as low as 0.01%. For multipass absorption cells, the problem is more serious due to the increased beam overlap and therefore interference, which results from trying to maximize the effective pathlength by increasing the number of passes.

Tunable diode laser spectrometers are particularly suited to high-sensitivity studies, in part because they may be readily frequency modulated, and harmonic techniques are used to measure molecular absorptions smaller than 0.01%. At this level, however, even though optical beam noise may be reduced by increased integration time, the measurement sensitivity is usually limited by the presence of optical interference fringes.

Reid and co-workers successfully used a jitter-modulation technique and later developed a two-tone modulation technique (Cassidy and Reid, 1982) to reduce optical fringing. A related approach was taken by Koga and co-workers (1981), who proposed modulating the TDL current not with a sine wave but with an exponential function approximating an inverse integrated raised-cosine profile. All three of these techniques attack the fringing problem through laser modulation. They use complex modulations that produce harmonic signal sizes and linewidths different from the normal single sinusoidal modulation method, which makes transformation back to gas concentration measurements unnecessarily difficult.

At the Jet Propulsion Laboratory (JPL) a novel optical technique has recently been developed for improving the sensitivity of TDL and other tunable laser spectrometers by an order of magnitude (Webster, 1985). This very simple technique substantially reduces the amplitude of unwanted interference fringes without distortion of the molecular lineshape. The technique of the Brewster-plate spoiler is applicable to tunable lasers of any wavelength, to measurements using short pathlengths, long pathlengths with retroreflectors, or multipass cells. This method relies on continuously changing the effective cavity length so that the interference effects producing the fringes are averaged during the laser wavelength scan. Placing a Brewster plate into the effective optical cavity and oscillating it by a minimum amount of one-half of the cavity free-spectral range then spoils the interference conditions so that the fringes are "washed out." An example of the ability of this technique to clean up a high-sensitivity spectrum (here of N_2O) is shown in Figure 3.19 (Webster, 1985). For TDL spectroscopy in the mid-IR region, the Brewster plate spoiler should allow the harmonic detection of absorptances of $\leqslant 10^{-5}$ in a single laser scan without subtraction techniques, complex laser frequency modulation, and distortion of the molecular lineshape signal. The absorption sensitivity will then be limited by the beam noise or the detector–signal channel noise, both of which can be improved by longer integration times.

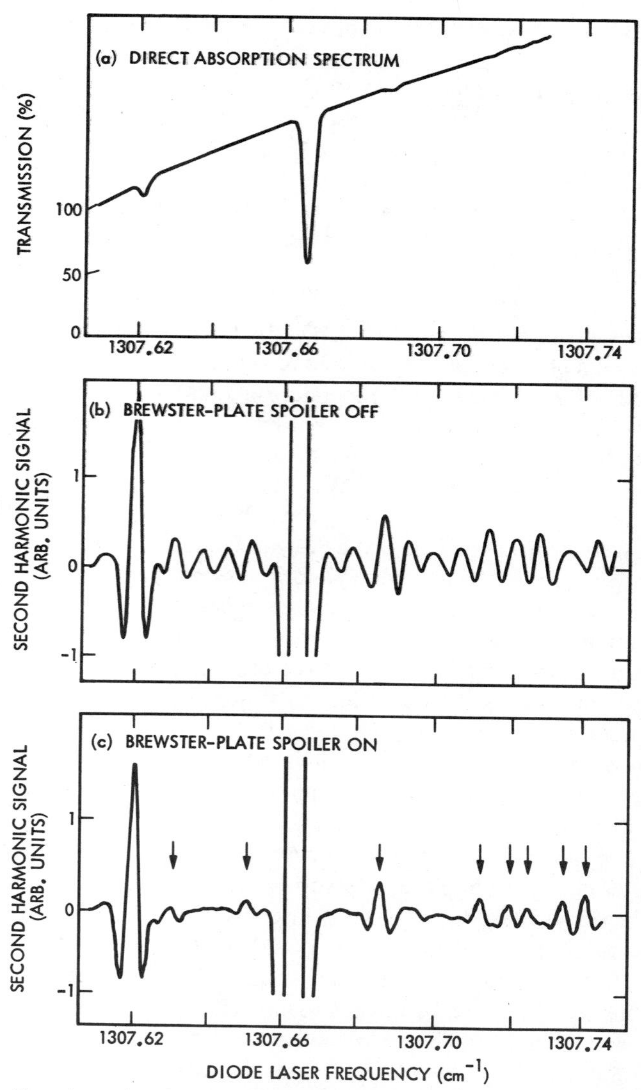

Figure 3.19. Experimental application of Brewster plate spoiler technique for reducing amplitude of unwanted fringes. [From Webster (1985).]

3.3.7 Measurement of Spectroscopic Parameters

Tunable diode lasers have been successfully used for several years to measure with high precision spectroscopic parameters such as line positions, linestrengths, broadening coefficients, and the temperature dependences of the last two parameters. For some experiments, the diode laser linewidth is negligible. However, this is not always true, for example, when narrow, Doppler-limited lines are used for linestrength measurements or when measurements of collisional narrowing are made.

Early heterodyne experiments on TDLs mounted in liquid helium Dewars reported linewidths as narrow as 27 kHz HWHM, but more recent results using commercial TDLs mounted in closed-cycle refrigerators quote linewidths in the range of 5–20 MHz HWHM and temperature fluctuations associated with the environment of the closed-cycle coolers, but experiments to improve the vibration isolation of the TDL do not always result in significant narrowing of the linewidth. At JPL the linewidths of TDLs have been studied by two different techniques (Reid and co-workers, 1982). Four diode lasers that operated in the 9-μm region were heterodyned against a 9.4-μm CO_2 laser. For these TDLs the effect of a new commercially available vibration isolation system was evaluated by comparing the linewidths with the closed-cycle cooler on and off. These TDLs had a minimum linewidth in the 1–10-MHz range, but this linewidth was often more than doubled by the action of the refrigeration system. In addition, the observed linewidth was found to depend strongly on operating temperature and injection current.

For many experiments using TDLs it is advantageous to operate with as narrow a linewidth as possible, but it is not possible to a priori know the operating conditions that lead to minimum linewidth. Heterodyning the TDL with a CO_2 laser is a relatively complex procedure requiring specialized equipment. Also, the heterodyne technique of measuring linewidths cannot be applied in the spectral regions where CO_2 (or CO) lasers do not operate. Consequently, we have devised a simple technique (Reid and co-workers, 1982) using a Fabry–Perot etalon that allows one to make a quantitative estimate of the TDL linewidth. This technique can be applied to any TDL and allows the modes that exhibit narrow linewidths to be quickly and easily identified. Part of this excess linewidth is undoubtedly caused by pressure. We will now briefly discuss the measurement of line positions, widths, and strengths using TDL sources.

One of the most attractive features of diode lasers is their tunability. However, just as with tunable dye lasers, this same characteristic means that accurate knowledge of the absolute wavelength cannot be obtained without reference to other standards. Even for a single-mode TDL operating at a given junction temperature and current, the absolute wavelength reproducibility following a thermal cycle is several orders of magnitude greater than the linewidth. The problem of wavelength calibration is further aggravated by the nonlinear tuning of any given TDL with injected current, as is evident from the etalon fringe recording in Figure 3.10.

Wavelength calibration must therefore by done by using simultaneously recorded reference lines of a molecular gas whose line positions are accurately known and using the fringes from a transmission spectrum of a Fabry–Perot etalon for interpolation between the reference lines. The absolute wavelength accuracy of a feature measured in the TDL spectrum will therefore depend on the signal-to-noise ratios obtained in both this and the reference spectrum, the absolute accuracy of the known reference line positions, and any changes in the free spectral range of the etalon during the spectral scan. The use of an appropriate Fabry–Perot etalon allows relative wavelength measurements to be accurately and reliably made with a TDL system. However, for multiline large spectral coverage, high-resolution Fourier transform instruments are better suited to absolute line position measurements, which can be made with uncertainties of smaller than one-tenth of the typical room temperature Doppler widths. When fixed-frequency laser sources near the features of interest are available, the method of measuring beat frequencies may be used with exceptionally high precision.

Once the TDL linewidth has been either measured using methods already discussed or deemed negligible compared to the molecular linewidths studied, the TDL can be very successfully applied to studying lineshapes and extracting broadening parameters. Where pressure-broadening coefficients are to be determined, a Doppler width measured from low-pressure scans is compared with theory to assess the contribution of other sources of broadening, such as laser linewidth. With increasing gas pressure, the whole lineshape may be fitted to extract the pressure-broadening coefficients.

Often, several spectroscopic parameters may be extracted simultaneously from digitization of the spectral data and a full lineshape analysis on a computer using least-squares fitting procedures. This makes it possible to determine the continuum level by making it one of the least-squares fit parameters. Using this approach, broadening coefficients may be measured with uncertainties of less than 5%.

The high resolution obtainable from TDL spectrometers is of considerable advantage in linestrength determinations, when individual linestrengths can be measured at low pressures where the line is predominantly Gaussian in shape. For lines of this profile, the error in locating the baseline is greatly reduced because the wings of the profile die out very quickly. Experimental measurements of intensities from infrared spectra have been discussed in Section 3.2.12. Although line intensities (or band intensities) can be experimentally measured to $\leqslant 2\%$ in favorable cases, it should be noted that field measurements of the concentrations of species of atmospheric interest are often limited in accuracy by factors other than the laboratory-measured linestrengths.

3.4 INFRARED LASER SIGNAL DETECTION AND PROCESSING TECHNIQUES

In this section we address the topic introduced in Section 3.3: the detection of weak laser or laser-induced signals, or the detection of signals resulting from small amplitude or phase modulation of CW laser signals. The sensitive harmonic detection techniques discussed in Section 3.3 rely on direct photovoltaic or photoconductive detection of signals by absorption of modulated CW laser radiation. The objective now is to present a brief summary of other spectroscopic techniques that have been employed for sensitive detection and quantitative measurement of trace species. Since these techniques involve a variety of physical interaction mechanisms, many of them have served to elucidate the energy level structures and the reaction kinetics of atomic, molecular, and free radical species when used in complement with direct absorption spectroscopic studies. It is appropriate to include a caveat at this point. There is a large body of literature associated with each signal detection and processing techniques discussed in this section; however, only a relatively few of the many excellent publications are cited as examples. The interested reader is encouraged to broaden the scope of investigation by consulting the references contained in these examples.

3.4.1 Signal Processing for CW and Pulsed Lasers

We begin by discussing certain generic processing techniques that are ubiquitous among laser spectroscopy laboratories and provide the ability to detect the (usually) weak signals that are generated by the variety of interaction mechanisms described. These are the phase-synchronous detection techniques commonly used in CW laser spectroscopy and the time-gated detection techniques commonly used in pulsed laser spectroscopy.

The phase-synchronous detection technique permits time averaging of a repetitive signal without the troublesome aspects associated with simple low-pass filtering. The objective is to enhance the repetitive (periodic) signal relative to the broadband noise. This technique is really a bandwidth reduction technique, a common property among all signal-averaging schemes. If the signal were steady, and if the detection and amplification electronics were very stable, the use of low-pass filtering would result in a definite improvement. In practice, however, the noise has a $1/f$ component in addition to a broadband "white-noise'' component, that is, the noise density is larger near dc than at higher frequencies. Slow drifts in the electronics also wreak havoc under these circumstances. The phase-synchronous detection technique greatly alleviates these practical problems if the signal is periodic or can be ''forced'' to be periodic, for example, by using a periodic chopper or switch or by modulating the source of radiation, as discussed in Section 3.3. This technique was suggested and applied in the early days of modern radio astronomy by periodically switching the antenna signal receiver between sky-viewing and reference-load-viewing modes (Dicke, 1946), and the term *Dicke switch* was born.

In phase-synchronous detection one considers a periodic signal passing through a linear amplifier whose gain is reversed in sign by a square-wave reference signal controlling, for example, an FET switch (Horowitz and Hill, 1980). (In the early days of audio frequency phase-synchronous detection, before FETs and linear op amps, coils were alternately energized and deenergized.) Then the output of this phase detection is passed through a low-pass filter, usually one or two stages of *RC* filtering. Assume we apply a signal

$$E_s \sin(\omega T + \Phi)$$

to a phase-synchronous detector whose reference signal is a square wave of frequency ω with transitions at $T = 0, \pi/\omega, 2\pi/\omega$, and so on. Let us assume further that the low-pass filter time constant $\tau \gg 2\pi/\omega$. Then the low-pass filtered output is the time average $\langle E_s \sin(\omega T + \phi) \rangle$ over the time period 0–π/ω minus the time average $\langle E_s \sin(\omega T + \phi) \rangle$ over the period π/ω–$2\pi/\omega$. Thus, the output voltage is

$$\langle V_0 \rangle = (4E_s/\pi)\cos \phi$$

If there exists an unwanted (noise) component at frequency $\omega + \Delta\omega$ in addition to the signal, the output voltage due to this component will be $\langle V_0 \rangle = (4E_N/\pi)\cos(\Delta\omega)T$, where we have replaced a fixed phase ϕ with a time-varying phase shift $\phi = T\,\Delta\omega$. This output voltage is a slow periodic signal that will be attenuated if the low-pass filter time constant $\tau > 1/\Delta\omega$. Thus, the phase-synchronous detection technique achieves bandwidth narrowing about the desired signal frequency while avoiding the difficulties of operating in the region near dc where slow drifts and $1/f$ noise dominate. Suppliers of phase-synchronous detection equipment (e.g., EG&G Princeton Applied Research Corp., Ithaco Corp.) should be consulted for much more extensive descriptions of their products.

Time-gated detection is another signal-averaging technique that is suitable for detecting transient signals arising from the interaction of a laser pulse with the medium under investigation. The enhancement of signal over noise occurs at two levels. First, because one knows the time of the pulse and its duration, the detection circuit can be triggered to display the result only during this specific interval, and the time-independent component of the noise power is discriminated against. Because one can often achieve much higher peak pulse output powers from the laser sources when they are pulsed than the corresponding average CW output powers, the detected signal voltage level is higher. Unfortunately, the noise voltage is also higher because a larger bandwidth is necessary in order to respond to the transient signal than when detecting a periodically modulated CW signal. However, repetitive pulsing allows one to form a cumulative sum of signal voltages that normally increase in direct proportion with time. Remembering that the "noise" is really the *fluctuation* level of the background sources of voltage at the output, if we assume that the background components at successive time gates are uncorrelated and Poisson-distributed events, the noise level increases as $T^{1/2}$. Thus, the signal-to-noise ratio increases as $T^{1/2}$, or as the square root of the number of pulses in the average.

Various techniques are used to detect repetitive transient signals. Single-photon counting with multichannel scalers is a well-established technique that has been used at visible and UV wavelengths in various scattering and fluorescence studies (McDermid, 1983). This is not feasible in the IR at wavelengths greater than 1 μm because of the poor performance of photoemissive devices. However, transient recorders that accept analog signals are commonly used in IR pulsed laser spectroscopy. The term *transient recorder* is usually applied to an instrument that writes the signal into a memory where it is stored for viewing or processing. The storage oscilloscope is in this category. The fastest storage scopes can display single transient signals with rise times as short as 2 ns. If the signal is repetitive, maintaining the same shape from pulse to pulse, sampling techniques can be applied to increase the bandwidth of digital storage scopes to several gigahertz. The proliferation of fast digital storage oscilloscopes has blurred the distinction between the storage scope and the transient digitizer, a device that has a high-speed analog preamplifier at the front end and a means of digitizing the information and storing it in memory, normally with computer interface capability. Transient digitizer manufacturers usually use an analog-to-digital converter (ADC), which samples the analog voltage at successive present sample intervals after the trigger, converts each sample voltage to a digital value, and allows the digital number to be stored in a channel of memory. The details of the ADC properties are often important in deciding which transient digitizer is adequate for a specific application. There is always a compromise between speed and resolution of the analog-to-digital (A/D) conversion. It is important to recognize the difference between the ultimate resolution capability defined by the ADC hardware design, which is usually the advertised quantity (e.g., an 8-bit digitizer is often assumed to provide 8 bits of resolution and corresponding accuracy) and the dynamic accuracy, which tells the user how many effective bits of accuracy are available at a specific frequency. Digitizers tend to exhibit fewer effective bits at higher frequencies. Many factors influence the dynamic accuracy, including, of course, the quality of the analog input preamplifier and signal conditioner, and the common tests for dynamic accuracy are described in at least one supplier's literature (Ramirez and DeWitt, 1985).

3.4.2 Heterodyne Detection

Heterodyne detection at optical and IR wavelengths is an extension of the superheterodyne radio frequency receiver concept, which attained feasibility only after single-frequency (single-mode) laser sources became available. The development of stable, single-mode laser local oscillator radiation with milliwatt-level CW output power, along with high-speed photodetectors that could be used as efficient mixers at optical frequencies for local oscillator and signal frequencies within 1–2 GHz of each other, stimulated research in several applications areas beginning in the late 1960s.

Figure 3.20 depicts the salient features that distinguish heterodyne detection from direct detection at optical and IR wavelengths. The detection process is fundamentally a "square-law" process when one uses a photodetector that absorbs photons and releases charge carriers as a direct result of photon absorption. Then the photocurrent is proportional to the square of the total electric

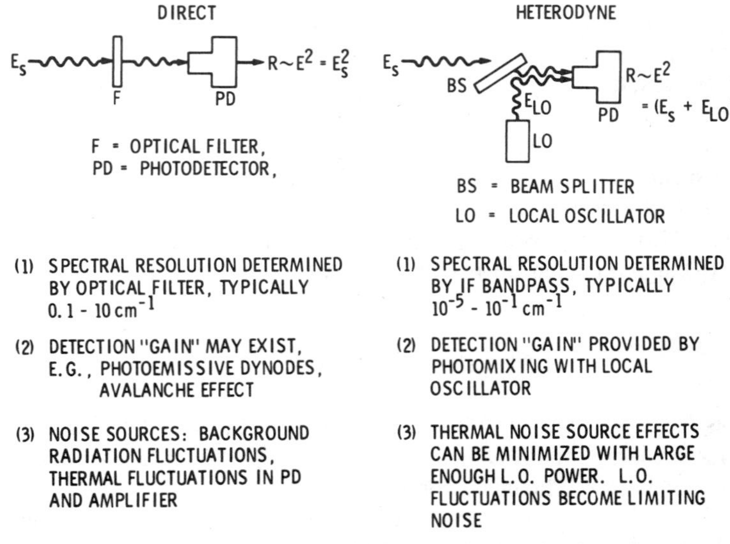

Figure 3.20. Major features that distinguish heterodyne detection from direct detection.

field of the incident radiation. In direct detection the photocurrent is proportional to the square of the electric field of the signal radiation, that is, proportional to the power of the optical or IR signal. In heterodyne detection the addition of a local oscillator produces a photocurrent response from the photodetector that contains terms proportional to the signal power, the local oscillator power, and a mixing term that is proportional to $E_s E_{LO}$ as long as the difference frequency is within the frequency response capability of the photodetector. It is this mixing term that is treated as the new signal, and one can view this process as a downconversion of the signal frequency from the optical region to the radio frequency region. Thus, the heterodyne receiver can detect radiation in a narrow spectral band (by optical standards) near the local oscillator frequency, and the bandwidth is determined by electronic filters instead of optical filters or dispersive optics.

Heterodyne detection had been recognized by 1970 as a valuable technique for optical communication and radar systems (Arams et al., 1967; Goodwin and Pedinoff, 1966; Teich, 1968), particularly at IR wavelengths, for velocimetry (Yeh and Cummins, 1964; Foreman et al., 1966; Huffaker et al., 1970), and for astronomical observations of, for example, stellar positions, high spectral resolution flux, and velocities (Townes, 1970; Nieuwenhuijzen, 1970; Van Buren, 1975; Sutton et al., 1982).

In the 1970–1971 time period the use of heterodyne detection in spectroscopic applications to trace-gas measurements were first enunciated (Menzies, 1971; Hinkley and Kelley, 1971). Although it was known that the etendue of a heterodyne receiver, according to the antenna theorem, was quite small for an optical instrument viewing thermal radiation (Siegman, 1966), the high spectral resolution and low noise properties intrinsic to an optimized heterodyne detection process stimulated investigations in the use of laser heterodyne receivers for trace molecular species measurements at wavelengths in the mid-IR.

Several reviews of coherent, or heterodyne, detection exist that emphasize its use in spectroscopic measurements, and extensive derivations of signal-to-noise ratio (SNR) concepts are contained in them (Blaney, 1975; Menzies, 1976; Kingston, 1978). The power SNR [ratio of the square of the intermediate frequency (IF) signal current to the mean-square noise current] of a heterodyne receiver is

$$\mathrm{SNR} = \eta P_s / h\nu B \tag{3.155}$$

where η is the heterodyne quantum efficiency, P_s is the signal power, $h\nu$ is the signal photon energy, and B is the bandwidth of the receiver IF channel following the photodetector. The usual configuration for the detection of IR radiation in absorption or emission spectroscopic measurements is the Dicke radiometer configuration, in which the receiver alternately views the signal source and a reference source, and the difference signal is phase synchronously detected and integrated over many cycles of alternation. In this case the noise power is effectively reduced by a factor of $(2B\tau)^{1/2}/\pi$, where τ is the integration time (Dicke, 1946; Tiuri, 1966). This factor is ordinarily very large because spectral linewidths in the mid-IR, as discussed in Section 3.2, vary between 20 MHz (Doppler-broadened region) and 3 GHz (pressure-broadened region, at 1 atm). Thus, values of B, which dictate the spectral resolution of the instrument, may range between 10^7 and 10^9 Hz, while τ might vary between 10^{-1} and 10^3 s.

The sensitivity of a heterodyne radiometer increases as the wavelength increases, and this characteristic, when combined with the blackbody power spectral distribution for a range of temperatures from cold regions of Earth's atmosphere (e.g., a tropopause region at 200 K) to the solar disk ($\sim$5500 K), results in the effectiveness of the heterodyne radiometer increasing wavelength throughout the IR. If the heterodyne radiometer were detecting a weak single-mode laser signal of power P_s, then according to Equation (3.155) the signal-to-noise ratio would increase linearly with wavelength. If the heterodyne radiometer were detecting thermal radiation, then the dependence of the signal-to-noise ratio on wavelength can be obtained using the following considerations. The thermal radiation energy density in a spectral interval $d\nu$ is the product of the photon (mode) occupation probability factor (Bose–Einstein statistics).

$$F(\nu, T) = \left[\exp(h\nu/kT) - 1\right]^{-1} \tag{3.156}$$

times the volume density of spatial modes (single polarization) in the frequency interval $d\nu$,

$$N(\nu) = (4\pi\nu^2/c^3)\, d\nu \tag{3.157}$$

times the photon energy, $h\nu$ (Menzies, 1976). The etendue of the heterodyne radiometer, according to the antenna theorem (Siegman, 1966), is

$$A_R\Omega_R = \lambda^2 \tag{3.158}$$

where A_R, Ω_R are the receiver collector area and field of view, respectively. If we convert the thermal radiation energy density (J m^{-3}),

$$\rho(\nu, d\nu) = h\nu N(\nu)F(\nu, T) \tag{3.159}$$

to intensity, or brightness (W m^{-2} Hz^{-1} sr^{-1}),

$$B(\nu, T) = (2h\nu^3/c^3)F(\nu, T) \tag{3.160}$$

and then use the etendue expression (3.158) to calculate the fraction of the total incident thermal radiation power that is effective in photomixing with the local oscillator, we derive

$$\mathrm{SNR} = (\sqrt{2}/\pi)(\tau B)^{1/2} \eta F(\nu, T) \tag{3.161}$$

when using (3.155) for the signal-to-noise ratio expression. This result makes evident the fact that the heterodyne receiver responds only to a single spatial mode and single polarization of the thermal radiation field. The derivation of Equation (3.155), which is the expression most commonly found in heterodyne detection references, considers the local-oscillator-induced shot noise as the predominant noise source. This depends on the assumption that the laser local oscillator radiation is that of a coherent state for which the photon fluctuation variance is the same as that of the Poisson distribution. This is a good approximation when stable, well-balanced lasers are used. The expression (3.155) is actually the ratio of the average signal power to the average noise power measured in the absence of a signal. A more generally applicable signal-to-noise ratio for stochastic signal and noise fields is the following expression (Elbaum and Teich, 1978):

$$\mathrm{SNR}_H = \frac{\langle i^2(t) \rangle - \langle i^2(t)_0 \rangle}{\left\{ \mathrm{Var}\left[i^2(t) \right] \right\}^{1/2}} \tag{3.162}$$

where $\langle i^2(t) \rangle$ and $\langle i^2(t)_0 \rangle$ represent average values of the IF electric power associated with the photocurrent in the presence and in the absence of the incident signal radiation, respectively. (The denominator is the square root of the variance when the signal radiation is incident.) Using this expression for the signal-to-noise when using a heterodyne radiometer to view a thermal radiation source, the expression given by Equation (3.161) is modified to

$$\mathrm{SNR}_H = (\sqrt{2}/\pi)(\tau B)^{1/2} \eta F(\nu, T) / [1 + \eta F(\nu, T)] \tag{3.163}$$

The additional factor $1 + \eta F(\nu, T)$ arises due to the photon correlation properties of a thermal radiation field, which obeys Bose–Einstein statistics. [This is treated in more detailed in, e.g., Menzies (1976).] The photon correlations add to the variance in photocounts that would exist if the photons were truly independent, obeying Poisson statistics. This additional factor is important if one is measuring spectral absorption features due to terrestrial atmospheric species in the solar background continuation at mid-IR wavelengths. For observation at 10 μm wavelength using the solar disk as the source, the expression $h\nu/kT \approx 0.27$ in Equation (3.156), and consequently $F(\tilde{\nu} = 1000\ \mathrm{cm}^{-1}, T = 5500\ \mathrm{K}) = 3.2$. The factor $F(\nu, T)$ is plotted in Figure 3.21, as a function of $\lambda = c/\nu$ for various temperatures of thermal radiation sources.

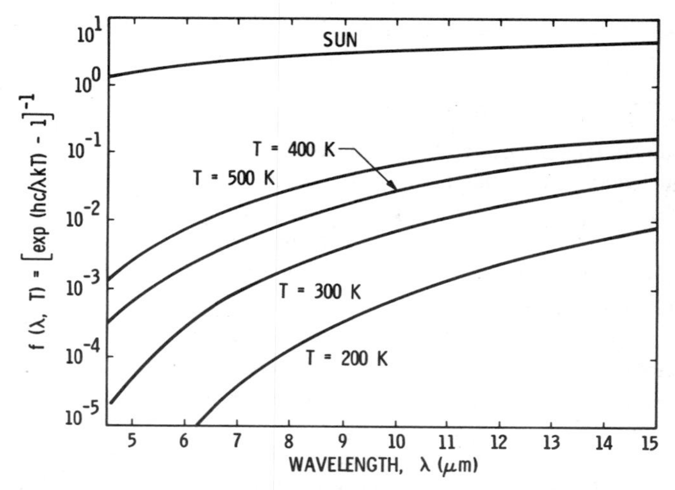

Figure 3.21. Wavelength dependence of heterodyne sensitivity to thermal radiation.

Based on early demonstrations of atmospheric trace-gas measurements using laboratory heterodyne spectroradiometers (e.g., Menzies and Seals, 1977; Frerking and Muehlner, 1977) and spectroscopic observations of planetary atmospheres (e.g., Johnson et al., 1976), flight instruments were developed to measure stratospheric species in the late 1970s. A balloon-borne heterodyne spectroradiometer, which has been flown several times for stratospheric observations, is described in Section 3.5.1. This instrument was designed with a grating-tunable carbon dioxide laser serving as the local oscillator. Several observations of molecular spectra in planetary atmospheres using discretely tunable carbon dioxide local oscillators have been reported also (Kostick and Mumma, 1983). The concept of a continuously tunable local oscillator spurred the development of an airborne laser heterodyne spectroradiometer using lead-salt semiconductor TDLs as local oscillators (Allario et al., 1980). The effectiveness of the approach is greatly dependent on the quality of the TDLs. Single-mode TDLs are in a state of technological adolescence, and the prospect of the maturation of this technology gives hope that future infrared heterodyne spectroradiometers designs will be compact, employing selectable TDL local oscillators, each continuously tunable over a portion of the mid-IR region.

3.4.3 FM Spectroscopy

The sensitive harmonic detection techniques discussed in Section 3.3 utilize modulation frequencies that are low compared with spectral widths of the spectral lines probed. As discussed earlier, values of the HWHM in the mid-IR extend from Doppler-limited values of 20 MHz for low-pressure ($\leqslant$5-Torr) samples to pressure-broadened values of approximately 3 GHz at 1 atm. In the visible, Doppler-broadened HWHM values are typically $\geqslant$500 MHz. The use of relatively low modulation frequencies produces spectra that have a resemblance to derivatives of the actual absorption lineshape. Recently, the technique of frequency modulation spectroscopy was applied to the measurement of weak absorption and dispersion spectra (Bjorklund, 1980). This approach utilizes modulation frequencies that exceed the spectral linewidth, and the widely separated sidebands produced can then individually interact with the absorption line of interest as the laser source is tuned through the spectral region.

The techniques used by Bjorklund and colleagues involved external phase modulation of a narrow-width CW tunable dye laser. Consider a single-frequency laser source of radiation at frequency ν_c passing through a phase modulator driven by a sinusoidally varying rf voltage at frequency ν_m. Then the input electric field to the phase modulator is given by $E_1(t) = \frac{1}{2}\tilde{E}_1(t) + cc$, where $\tilde{E}_1(t) = E_0 \exp(i\omega_c t)$. The phase-modulated output electric field has a pure FM spectrum given by

$$E_2(t) = \tfrac{1}{2}E_0 \sum_{n=-\infty}^{\infty} J_n(M)\exp\left[i(\omega_c + n\omega_m)t\right] + cc \tag{3.164}$$

where M is the modulation index and the J_n are the Bessel functions of order n. Assume $M \ll 1$, so that $J_0(M) \simeq 1$, $J_{\pm 1}(M) = \pm M/2$, and the higher terms become negligible. Then the frequency-modulated laser radiation is described by a carrier at frequency ν_c ($= \omega_c/2\pi$) and two weak sidebands at frequencies $\nu_c \pm \nu_m$. When this beam is passed through the gas whose spectral feature is being probed, with absorption coefficient $\alpha(\nu)$ and refractive index $n(\nu)$, the relative amplitudes and phases of the two sidebands are affected. Following the notation used by Bjorklund, we define the complex transmissivity $T_j = \exp(-\delta_j - i\phi_j)$, where $\delta_j = \alpha_j/2$, $\phi_j = n_j(\omega_c + \omega_m)/c$, where $j = 0$, ± 1 denote the components at ν_c and at $\nu_c \pm \nu_m$, respectively. Thus, δ_j describes the amplitude attenuation and ϕ_j describes the optical phase shift produced by the absorber. Then the (complex) transmitted electric field is

$$\begin{aligned} E_3(t) = \tfrac{1}{2}E_0\Big\{ &-T_{-1}M \exp\left[i(\omega_c - \omega_m)t\right] + 2T_0 \exp\left[i\omega_c t\right] \\ &+ T_1 M \exp\left[i(\omega_c + \omega_m)t\right]\Big\} \end{aligned} \tag{3.165}$$

The slowly varying envelope of the intensity, $I_3(t)$, that is photodetected is then given by $I_3(t) = c\,|E_3(t)|^2/8\pi$. Using Equation (3.165) for $E_3(t)$, dropping terms of order M^2, and assuming $|\delta_0 - \delta_{-1}|$, $|\delta_0 - \delta_1|$, $|\phi_0 - \phi_{-1}|$, and $|\phi_0 - \phi_1| << 1$, we obtain

$$I_3(t) = \frac{cE_0^2}{8\pi}\exp(-2\delta_0)\{1 + (\delta_{-1} - \delta_1)M\cos\omega_m t + (\phi_1 + \phi_{-1} - 2\phi_0)M\sin\omega_m t\} \quad (3.166)$$

Thus, the photodetector current will contain a beat signal at the rf modulation frequency if there is a spectral feature that produces an unbalanced loss ($\delta_{-1} - \delta_1 \neq 0$) or carrier frequency phase shift different from the average of the phase shifts experienced by the sidebands ($\phi_1 + \phi_{-1} - 2\phi_0 \neq 0$). If ν_m is large enough that the spectral feature is being probed by only a single sideband at a time, the in-phase and quadrative components of the "beat signal" [Eq. (3.166)] are proportional, respectively, to the absorption and dispersion induced by the spectral line. The null signal at frequency ν_m when the frequency-modulated radiation is not distorted is due to the cancellation of the rf signal arising from the upper sideband beating against the carrier with that arising from the lower sideband beating against the carrier.

If the experimental arrangement is such that all external sources of imbalance can be maintained at a low level that does not change as the laser frequency is tuned, then this technique is potentially a very sensitive means of probing a weak spectral feature, making it applicable to trace-gas detection. In practice, the best results have been obtained when narrow-linewidth CW tunable dye lasers have been used (Bjorklund et al., 1981). The application of this technique in the IR has been demonstrated by tuning GaAlAs diode lasers (Lenth, 1983), modulating the frequency by modulating the injection current. The modulation of the injection current in a semiconductor diode laser produces amplitude modulation in addition to frequency modulation. The result is that there are contributions from spectral absorption and dispersion to both the in-phase and quadrature components of the heterodyne beat signal. The signals observed when tuning the laser through the spectral region of interest are more complex, but if the (AM and FM) modulation indices are accurately known, an analysis of the lineshapes and calibration of the system with a known absorbing linestrength make it possible to perform quantitative high-resolution spectroscopy. Attempts to use this technique with fixed frequency or discretely tunable laser sources in the IR (e.g., CO_2 lasers) by tuning the modulator rf drive frequency have met with mixed results (Molina and Grant, 1984). The coupling efficiency to the phase modulator crystal depends somewhat on rf frequency, which results in a residual beat frequency signal that is not constant as the sideband is tuned through an absorption line. This characteristic, along with the difficulty of reducing unavoidable dispersion effects and low-level spectral interference effects ("channeling" in the spectra) when tuning either the modulator or the (tunable) laser, make it difficult to achieve theoretical sensitivity performance. The frequency-modulated spectroscopic technique and its variants have proven to be useful, however, for sub-Doppler nonlinear spectroscopic studies (Hall et al., 1981; Bjorklund and Levenson, 1982) in the visible wavelength region.

3.4.4 Photoacoustic and Photothermal Detection Spectroscopy

When tunable IR laser radiation is absorbed by a gas in a cell or chamber, producing molecular or atomic excitation, the acquired energy may be converted to kinetic energy through various collision processes, for example, electronic to translational or, more properly, vibrational to translational energy transfer. An increase in the translational (thermal) energy will produce an increase in pressure in a fixed-volume cell or in an adiabatic process. If the radiation is modulated in amplitude at an audio frequency, a sound wave can be generated in the cell, and a microphone can be used to convert this to a phase-synchronous voltage signal. This technique was explored by Bell long before the existence of lasers (Bell, 1881); however, the optoacoustic effect was seldom used as a spectroscopic tool until the advent of the laser, which contributed a remarkable impetus to its value in trace-gas detection and spectroscopy (Kreuzer, 1971) and in absorption spectroscopy of solid materials (Rosencwaig and Gersho, 1976). The decade of the 1970s saw a large number of technological developments and novel applications of photoacoustics (Pao, 1977).

If a laser at a suitable absorbing wavelength is intensity modulated with an optical chopper, as in Figure 3.22, producing a good approximation to a square-wave power input to the spectrophone cell (i.e., $t_r \lll T$, where t_r is the finite rise time of the power due to the movement of the chopper blade through the beam, and T is the modulation period), then the rms value of the first harmonic of the pressure signal, $\rho(\omega)$ (where $\omega = 2\pi/T$), is (Rosengren, 1975)

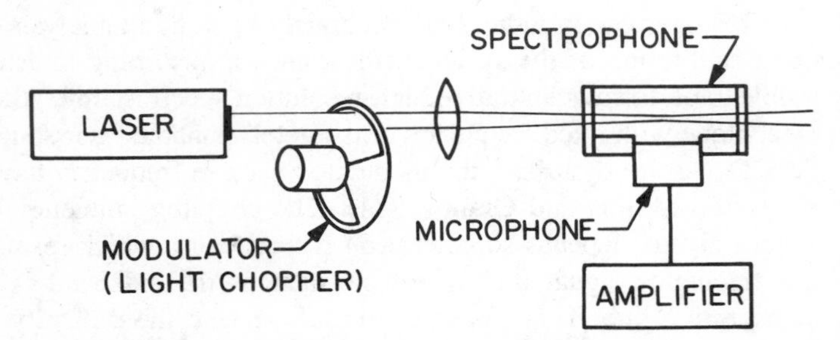

Figure 3.22. Block diagram of typical photoacoustic experimental arrangement (Kavaya, 1982).

$$P(\omega) = \frac{2\sqrt{2}\beta\overline{U}\sigma N l Q(\omega)\tau_t}{3\pi V[1 + (\omega\tau_t)^2]^{1/2}[1 + (\omega\tau_c)^2]^{1/2}} \tag{3.167}$$

where $\overline{U}$ is the mean laser radiation power integrated over the volume of the cell when the chopper passes the radiation, β is the fraction of the total absorbed energy converted to molecular translational energy via V–T collisions; N is the number of absorbing molecules, with absorption cross section σ; $Q(\omega)$ is the acoustical quality factor of the cell, which is equal to unity if there is no resonance at the chopper modulation frequency; l and V are the length and internal volume of the cell; τ_t is the thermal relaxation time, which is usually governed by the thermal conduction of the excess translational energy to the cell surfaces; and τ_c is the V–T collisional relaxation time. We have assumed that the collisional relaxation time is much shorter than the radiative relaxation time of the energy in the (vibrationally) excited molecules after absorption of the laser radiation. From Equation (3.167) it is apparent that small optoacoustic cell volume is desired for maximum pressure signal response. Very high sensitivities to weakly absorbing media can be obtained in this way. In certain applications this presents problems, for example, if the sample gas interacts with surfaces and is absorbed or slowly destroyed. In cases such as these, a larger volume cell is mandatory, and the use of acoustic resonances is advantageous. A slow gas flow is also more feasible in a large-volume resonant cell if the inlet and outlet ports are properly located. Values of acoustic quality factor greater than 100 were obtained in early demonstrations of the resonant cell technique (Dewey et al., 1973; Bruce and Pinnick, 1977). Measurements of water vapor absorption at CO_2 laser wavelengths were made with a large resonant cell (Shumate et al., 1976) due to the difficulty of maintaining stable mixtures of various specific humidities when enclosed in small volumes.

A synchronous "background" signal exists in photoacoustic spectroscopy whenever a spectrophone cell with windows is used due to the phase-synchronous signal from cell window absorption and consequent heating of the internal gas sample. The *differential cell* approach, first reported by Deaton et al. (1975), made use of a tandem chamber arrangement with a common window connecting them. The window absorption signals canceled to a large extent, while the absorption gas signal, coming from only one cell, produced a differential pressure signal at an acoustic frequency. Their reported equivalent absorption background signal was reduced to only $3.3 \times 10^{-7}\ \text{m}^{-1}$ when using a 1-W laser. A more successful approach was taken by Kavaya et al. (1979), who used Stark modulation optoacoustic detection.

The background signals due to window absorption and other sources are not as vexing when a narrow-linewidth tunable laser can be used, and very high sensitivity to nitric oxide was achieved using a spin-flip Raman laser to tune through selected NO absorption line frequencies (Patel and Kerl, 1977). This technique was used in a balloon-borne measurement of stratospheric NO (Burkhardt et al., 1975), an impressive achievement when considering the sophisticated technology involved.

The localized heating of gases, liquids, and solid surfaces that interact with spectrally resonant laser radiation produces several other possibilities for spectral detection in addition to the measurement of a pressure change in a cell. Two major detection schemes are based on photothermal lensing and photothermal probe beam deflection. The thermal lens technique has been applied to organic liquid spectroscopy (Swofford and Morrell, 1978), in which a spectrally resonant tunable dye laser creates an azimuthally symmetric refractive index perturbation that is parabolic in the radial coordinate, changing the focal properties of the laser beam. The existence of this thermal lens can be sensitively detected, and a spectrum is produced as the tunable laser sweeps through the liquid absorption resonances. The photothermal deflection technique takes advantage of the fact that the modulated optically induced heating by the spectrally resonant "pump" laser radiation will cause a refractive index perturbation that has strong spatial gradients, and the gradient of the modulated refractive index is used to periodically deflect a weak laser probe beam propagating through the material (Gerlach and Amer, 1980). Various geometries can be used to optimize the probe beam deflection for particular applications (Jackson et al., 1981). The probe beam is not resonant with the material's absorption features; the requirement is only that the refractive index perturbation exist at the probe laser wavelength. HeNe lasers have been commonly used as the probe lasers in photothermal deflection experiments.

A major application area for photoacoustic and photothermal spectroscopies is in flame studies. A variety of acronyms have been created to distinguish the various techniques. In photoacoustic spectroscopy (PAS) the pressure change induced by resonant absorption by a flame species is detected acoustically; however, the photoacoustic deflection spectroscopy (PADS) technique relies on detecting this pressure change optically, as in the photothermal deflection technique. The PADS technique has advantages in flame studies because the optical detection can be placed as far away from the flame as necessary without adversely affecting the sensitivity. The sensitivity depends on the separation between the pump and probe beams inside the flame, and not on the position of the detector. The PADS technique is to be distinguished from the photothermal deflection spectroscopy (PTDS) techniques in that the probe beam may be at a distance away from the pump beam and respond, with a time delay, to an acoustic wave generated by the pump beam. An alternate application for PADS is optoacoustic laser beam deflection (OLD). Recently, measurements of OH concentration maps and temperature profiles in a flame were conducted using a combination of these techniques (Kizirnis et al., 1984). The thermal lensing

(TL) and OLD techniques have also been used to study the phenomenon of photochemical nucleation of particles ("laser snow") in organic vapors (Tam et al., 1985).

3.4.5 Optogalvanic Detection Spectroscopy

Laser optogalvanic (LOG) spectroscopy is a simple, sensitive technique (Green et al., 1976) with applications in laser physics and chemistry. Discovered over 50 years ago (Penning, 1928), the optogalvanic effect has awaited the development of tunable lasers to demonstrate its sensitivity and wide applicability. Many areas of research have since benefited from this powerful technique; these include plasma physics, atomic and molecular spectroscopy, combustion diagnostics, trace element detection, and chemical analysis. In addition, the effect has been utilized in numerous schemes for laser stabilization and calibration.

In LOG spectroscopy, a tunable laser is used to probe the spectral characteristics of atomic or molecular species generated within an electrical discharge in a low-pressure gas. Optogalvanic signals arise when the impedance of the discharge changes in response to the absorption of laser radiation. This change in impedance may be conveniently monitored as a change in the voltage across the discharge tube. Figure 3.23 shows a simple experimental setup. In this arrangement, LOG spectra are recorded by scanning the wavelength of a chopped CW laser while monitoring the discharge voltage with a lock-in amplifier. A coupling capacitor blocks the high-dc voltages and allows the lock-in amplifier to record the laser-induced ac voltage changes, which are detected at the chopping frequency (typically a few hundred hertz). To take advantage of the laser collimation, hollow metal electrodes are mounted in a narrow discharge tube that is about $\frac{1}{4}$ in. in diameter and 3 in. long. Typical operating conditions include a potential of 300–500 V across the tube, a discharge current of about 1 mA, and a gas pressure of 0.1–5 Torr. If a pulsed tunable laser is used, the lock-in amplifier is replaced by suitable gated electronics.

What is particularly attractive about this detection technique is its simplicity and the wide variation of experimental measurements that can be made. The discharge tube that is the detector is also the generator of the species studied, which may be neutral atomic, molecular and radical species, or atomic and molecular positive or negative ions. Furthermore, through the detection and spectral characterization of these species as a function of location within the discharge, the technique offers in principle the ability to investigate fundamental electron–molecule–ion interaction processes.

One novel application of this technique was the recording of the I^- photodetachment spectrum (Webster et al., 1983) by probing an I_2 discharge with a pulsed dye laser operating near 405 nm. The release through photodetachment of the more mobile electrons at the expense of the less mobile negative ions

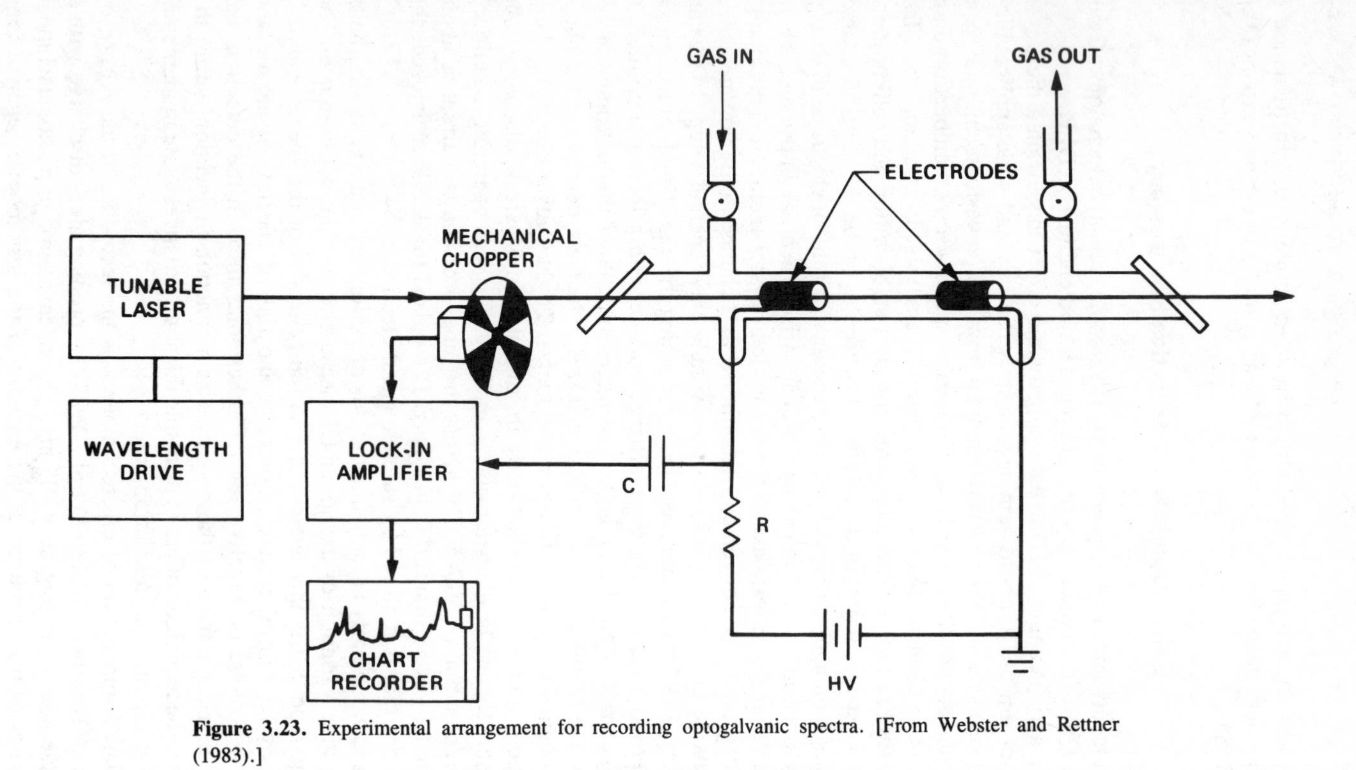

Figure 3.23. Experimental arrangement for recording optogalvanic spectra. [From Webster and Rettner (1983).]

produced a large increase in discharge current when the laser was tuned over the detachment threshold. Measurement of the threshold wavelength allowed the electron affinity of iodine to be measured with very high precision.

Only recently has LOG spectroscopy been applied to study molecular species (see the review of Webster and Rettner, 1983) where visible wavelength sources provided by CW and pulsed dye lasers have been employed. These same sources were used in the recently reported first detection of molecular ions (Walkup et al., 1983) using the optogalvanic technique. In this work transitions originating from the ground electronic states of N_2^+ and CO^+ were detected by laser excitation in the cathode fall region of dc discharges. The net increase in the steady-state current observed in this work was due to a combination of increase in ion mobility and increase in the rate of secondary electron emission at the cathode.

In the IR wavelength region, color center lasers operating near 2.6 μm were first used by Jackson et al. (1980) and by Begemann and Saykally (1982) to record optogalvanic spectra of atomic transitions. Early studies of the optogalvanic effect in a CO_2 laser plasma by Carswell and Wood (1967) have been put to use in the frequency stabilization of CO_2 lasers (Skolnick, 1970). The first IR LOG spectra of molecules were those recorded for NH_3 at 9.5 μm and NO_2 at 6.2 μm by Webster and Menzies (1983) using TDL sources. Figure 3.24 shows a portion of the LOG spectrum of NH_3, for example, recorded by irradiating the negative glow region of a pure ammonia discharge using a TDL with output power of $\simeq$0.8 mW. In the absence of information concerning the de-

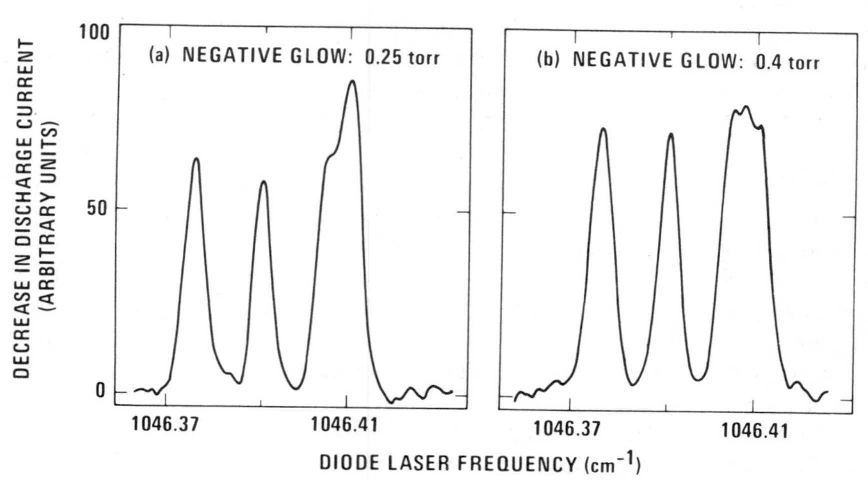

Figure 3.24. Portion of IR laser optogalvanic spectrum of NH_3. [From Webster and Menzies (1982).]

pendence of the various reactions among the neutral and charged species on the vibrational, rotational, and translational energies of molecular species, large changes in the discharge impedance resulting from excitation of molecular vibrational–rotational bands were not expected a priori, especially considering the low output powers ($\leqslant 1$ mW) available from TDLs in the 3–30-μm wavelength region.

By investigating the spatial dependence of the magnitude and sign of the optogalvanic signal resulting from transverse excitation of the NH_3 discharge, two contributions to this signal were identified. Mechanisms were proposed by which laser excitation of vibrational–rotational bands may cause changes in the impedance of the electrical discharge. In particular, Webster and Menzies (1983) found that even when exciting away from the discharge region, optogalvanic signals could be observed. They proposed that in this case the absorbed IR energy was transferred via collisions to translational energy, this local temperature increase propagating down the cell through the discharge as a pressure (optoacoustic) wave.

The range of applicability of IR optogalvanic spectroscopy has now been extended to include the use of CO_2 (9–11-μm), N_2O (10-μm), and color center (2.7-μm) laser sources. Muenchausen et al. (1984) observed optogalvanic signals using an isotopic CO_2 laser to probe rf discharges in D_2O, H_2CO, NH_3, SO_2, H_2S, and H_2O_2, this work allowing the identification of many new near ($\leqslant$200-MHz) coincidences in these molecules. Studies using off-axis laser excitation supported the V to T acoustic wave mechanism.

More recently, Hameau et al. (1984, 1985) have taken full advantage of the higher output powers ($\simeq$20 W) available from CO_2 and N_2O laser sources to record Doppler-free signals in NH_3 through observation of an inverted Lamb dip of a width of about 2 MHz HWHM at 75 mTorr gas pressure. This group also recorded the first IR microwave double-resonance spectrum (Hameau et al., 1984) using optogalvanic detection.

The ability to record optogalvanic spectra resulting from small absorptions is not limited to the mid-IR region. May (1985) used a color center laser at 2.7 μm to scan over features of the 2-0 overtone transition of NO in an rf cell. Although this laser source provided more single-mode output power (3–5 mW) than that used in the TDL work, the 2-0 overtone transitions are nearly two orders of magnitude weaker than those belonging to the 5.2-μm fundamental. A portion of the recorded spectrum is shown in Figure 3.25. The signals produced were associated with an increase in discharge current and identified with gas heating.

Much work remains to unambiguously identify the mechanisms responsible for the observed optogalvanic effect in molecules excited with IR radiation. However, once such an identification is made, these same mechanisms may themselves be studied quantitatively. That is, the technique may be used in the

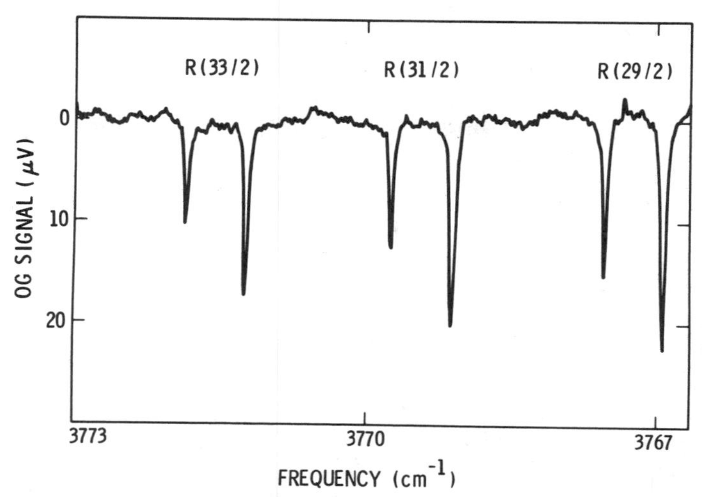

Figure 3.25. Laser optogalvanic spectrum of section of NO(2-0) band showing the *R*-branch doublets due to the $^2\Pi_{1/2}$ and $^2\Pi_{1/2}$ spin components of NO ground electronic state(s). [From May (1984).]

future to measure, for example, ionization, attachment, detachment, recombination, and relaxation rates.

Radio frequency discharges will find increasing use over dc discharges for optogalvanic studies for several reasons (Muenchausen et al., 1984): the uniformity of the discharge along the discharge axis, the reduced sensitivity of the background-noise level on gas pressure or type, and the reduced contribution from gas decomposition or surface deposition. For molecular ion detection, however, dc discharge cells with defined negative and cathode glow regions will be required.

3.4.6 Electric and Magnetic Field Modulation Spectroscopy

Stark and Zeeman effects in molecules are widely studied because they provide information on important molecular constants such as electric dipole moments, magnetic moments of paramagnetic molecules, *g* factors, and state quantum numbers. The effects also form the basis of very sensitive detection techniques. Laser Stark spectroscopy and laser magnetic resonance (LMR) spectroscopy, for example, have been used extensively in the IR wavelength region with fixed-frequency CO_2, N_2O, and CO gas lasers (see the review of Duxbury, 1983, and the book of Hirota, 1985, for examples). Here the electric or magnetic field is scanned linearly with time so that the molecular transitions are tuned into resonance with the stable, fixed-frequency laser. Small ac modulation of the field superposed on the scanning field and detection at the frequency of modulation allows Stark or Zeeman patterns to be recorded. For free radical detection and spectroscopic study, LMR has proved a very powerful technique, with additional advantages over Stark spectroscopy, which suffers from pressure and throughput restrictions and from radical loss through electrode surface reaction.

The arrival of tunable IR lasers such as color center, difference frequency, and lead–salt diode lasers has eliminated a problem encountered with fixed-frequency lasers, that of converting tuned field resonances to transition frequencies (wavelengths), and has also opened up new applications even closer to the techniques in microwave spectroscopy and electron paramagnetic resonance spectroscopy to which they owe their heritage.

Because of the dependence of the Stark and Zeeman effects on molecular constants, the applied field and the quantum numbers of the states coupled by laser frequency field modulation spectra often look quite different from direct absorption spectra, and indeed provide different information. For example, magnetic field modulation can pull out spectral features due to free radical absorption in the presence of a strong overlapping absorption by a nonparamagnetic molecule. This discriminating nature of field modulation techniques can also be invaluable for unraveling complex spectra by identifying rotational quantum number series, as is illustrated in the example that follows.

At JPL we have recently studied the high-resolution FTIR Bomem and TDL spectra of the HNO_3 ν_3 band near the 1326-cm^{-1} band origin in order to obtain the upper state rovibrational molecular constants (Webster et al., 1985). Heavier asymmetric top molecules like HNO_3 generally have small rotational constants and narrow Doppler widths (0.001 cm^{-1} for HNO_3 at room temperature). In addition, the large number of IR-active vibrations means that spectrally overlapping bands occur. For these molecules, the very high resolution of the TDL spectrometer is usually sufficient. However, unambiguous line assignment in the dense Q-branch regions is required for refinement of any preliminary rotational constants, to remove any remaining correlation among rotational constants, and in some cases to identify the band-type selection rules. Confirmation and simplification of the rotational analysis could be achieved if a line-labeling technique were applied to the congested spectral regions where lines overlap within their Doppler widths.

Figure 3.26 shows the high-resolution Bomem FTIR and TDL spectra of the HNO_3 7.5-μm band near the 1326-cm^{-1} band origin. By applying a small electric field modulation across the sample, the much simplified Stark modulation spectrum is recorded. The intensity patterns and phases of the field-modulated lines have allowed unambiguous assignment of the observed lines. The complementarity of the Bomem FTIR and TDL Stark spectrometer has here been fully exploited to determine a preliminary set of rotational constants of the excited vibrational state involved.

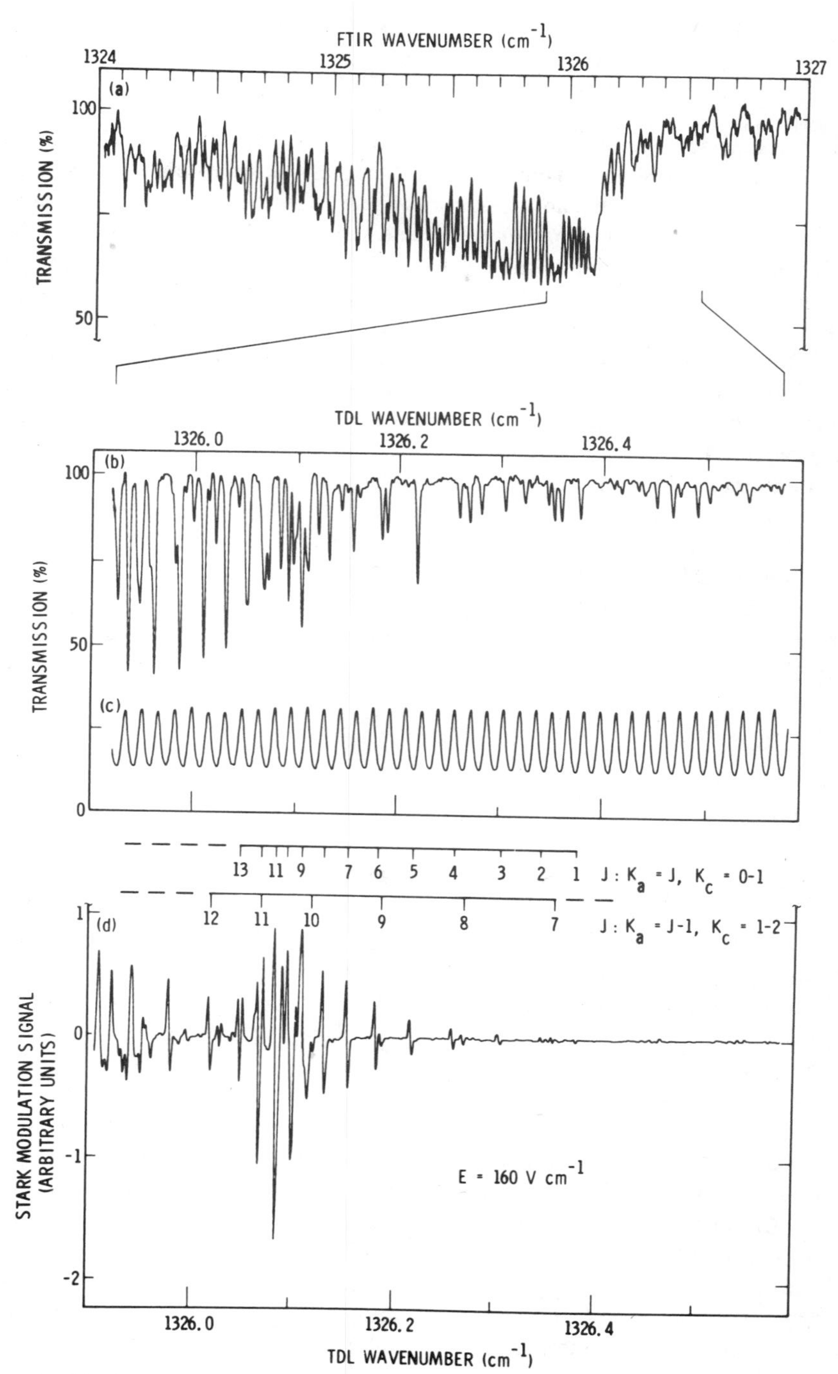

Figure 3.26. (*a*) Bomem, (*b*) TDL direct absorption, (*c*) etalon transmission, and (*d*) Stark modulation spectra of HNO_3 near 1326 cm^{-1}. Fringe FSR is 0.0160 cm^{-1}, and peak-to-peak applied field for lower trace is 160 V cm^{-1}. [From Webster et al. (1985).]

3.4.7 Ion Velocity Modulation Spectroscopy

One exciting new technique recently invented by Gudeman et al. (1983) has been applied to the sensitive detection and study of numerous positive and negative molecular ions. This absorption technique relies on the velocity modulation of ions within their Doppler width and is achieved by modulation of the high voltage across pressure gas discharges that produce the ionic species of interest (see the review by Gudeman and Saykally, 1984). This technique provides high discrimination against neutral absorbers. Phase-sensitive detection at the modulation frequency f clearly distinguishes absorptions due to negative ions from those due to positive ions by the symmetry of the first-derivative signals produced. Detection at $2f$ allows the concentration modulation spectra to be recorded for comparison.

For example, recently, at the University of California at Berkeley, the vibrational–rotational spectra of two of the most fundamental chemical species have been recorded, namely, the conjugate acid and base of water, the hydronium ion (H_3O^+) and the hydroxide ion (OH^-).

For both species, tunable color center lasers were used to record numerous transitions in the 2.8-μm wavelength region with a precision of 0.002 cm^{-1}, allowing molecular constants to be determined for comparison with ab initio calculations.

3.5 APPLICATIONS OF INFRARED LASER TECHNIQUES TO ANALYSIS

3.5.1 Stratospheric Measurements

It became widely recognized in the late 1970s that an improved understanding of stratospheric photochemistry and dynamics was necessary in order to assess anthropogenic influences on the ozone layer. Predictions of ozone destruction by oxides of nitrogen (Johnston, 1971) and by chlorine (Stolarski and Cicerone, 1974; Rowland and Molina, 1975), concentrations of which could be significantly increased by such activities as high-density supersonic transport flight patterns, heavy use of synthetic fertilizers, and continued use of long-lived chloro-fluoromethanes (CFMs) as propellants, stimulated research programs to improve photochemical models, measure relevant reaction rates in the laboratory with improved accuracy, and measure a host of reactive chemical species in the stratosphere with new and more powerful instrument techniques.

The complicated interactions among the trace species in the stratosphere can be organized into those that affect three families: NO_x, HO_x, and ClO_x (World Meteorological Organization, 1981). A catalytic ozone destruction cycle exists through which small concentrations (ppb) of certain species in each family, using analogous mechanisms, significantly affect the much larger (ppm) ozone concentrations. The large number of species belonging to these three families and their high reactivity and low concentrations are all factors contributing to the difficulty of stratospheric measurements. From high-resolution absorption spectroscopy and a knowledge of the IR spectral characteristics of the species studied, however, quantitative information about the molecular concentrations in the stratosphere may be obtained.

In order to enhance sensitivity and minimize interference effects when attempting to measure these trace species, it is desirable to look for sharp spectral features with a high-resolution instrument. Thus, IR laser techniques are particularly applicable. At stratospheric pressures, IR spectral linewidths are narrow, approaching Doppler linewidths as the altitude increases to the 40-km region. Instruments that have spectral resolution comparable to or less than Doppler linewidths in the mid-IR (0.001–$0.005\ cm^{-1}$) will attain maximum sensitivity provided enough radiation is available to identify a spectral feature with high signal-to-noise ratios. In this section two balloon-borne IR laser instruments are discussed that were developed with these properties in mind: a laser heterodyne radiometer and a tunable laser absorption spectrometer utilizing harmonic absorption spectroscopy.

Since the sun is such a strong source at mid-IR frequencies, when compared with the thermal emission of the stratospheric species themselves, a common operational model involves viewing the sun at large zenith angles, from stratospheric altitudes, so that long stratospheric pathlengths are available without interference from tropospheric absorbers. A balloon-borne laser heterodyne radiometer (LHR) was constructed at JPL to measure altitude profiles of selected stratospheric constituents using this mode of operation. The instrument was flown four times during the 1978–80 period, from Palestine, Texas, to measure chlorine monoxide (ClO) (Menzies, et al., 1981). ClO is the key free radical in the catalytic ozone destruction cycle by chlorine:

$$O_3 + Cl \rightarrow O_2 + ClO$$
$$\underline{ClO + O \rightarrow O_2 + Cl}$$
$$\text{Net: } O_3 + O \rightarrow 2O_2$$

The free Cl atom can be regenerated more than a hundred times in this cycle, before it finally reacts with another species and is removed by the formation of a stable "reservoir" molecule. The ClO radical's key role in the photochemical model predictions, combined with the fact that it had never been detected in the stratosphere when the ozone depletion predictions were first announced, motivated the construction of the LHR.

A heterodyne spectroradiometer is an attractive instrument to apply to the remote detection of species which absorb radiation at wavelengths $\lambda \geq 10\ \mu m$ with linewidths near the Doppler limit (Menzies, 1976). As explained in subsection 3.4.3, the efficiency of this technique increases rapidly with increasing wavelength in the infrared because its etendue is proportional to λ^2. When the desired spectral resolution is, e.g., $10^{-3}\ cm^{-1}$ (30 MHz), the noise-equivalent power (NEP) of approximately 2×10^{-16} watts (for a 1-s integration time) is much less than that of an infrared spectrometer instrument, which uses direct detection. The limited etendue is a drawback, but for species-specific detection purposes, when a small spectral interval is adequate for unique identification the advantages outweigh the drawbacks. As stated earlier, using the sun as a strong source of IR radiation (the effective brightness temperature of the solar disk being approximately 5500 K) provides high enough signal-to-noise ratio to detect absorptances as small as 10^{-3} with short integration times (~ 10 s).

The fact that the ClO fundamental vibration–absorption band is near 12 μm (850 cm^{-1}) makes it possible to use a line-tunable CO_2 laser as the local oscillator for the heterodyne radiometer, a propitious occurrence since the CW CO_2 laser properties make it ideal as a local oscillator. An isolated $^{35}Cl\ ^{16}O$ line, the $R(13/2)$ line of the $\Omega = \frac{3}{2}$ subband, was selected for observation by the balloon-borne LHR instrument, with the $P(14)$ line of the $^{14}C\ ^{16}O_2$ isotope, at 853.181 cm^{-1}, used as the local oscillator line.

A considerable geometric amplification of optical depth occurs when an instrument views the sun from a sufficiently high altitude so that the rays pass through the altitude region of interest when the solar zenith angle (the angular deviation from the vertical) is 90° or greater (Figure 3.27). The air mass tra-

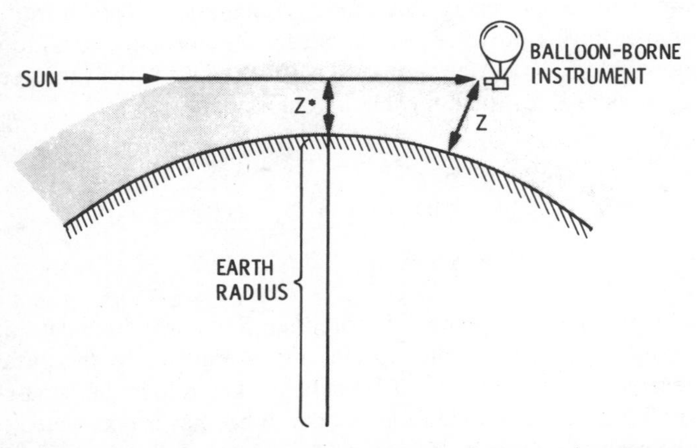

Figure 3.27. Solar occultation geometry for measurement of trace species concentrations using absorption spectroscopy.

versed by a ray from the center of the solar disk that reaches the instrument can be calculated using the Chapman function (Chapman, 1931; Menzies et al., 1981). For example, the amplification of pathlength (air mass) when the zenith angle $\theta_p = 90°$, compared with that obtained when $\theta_p = 0$ (direct overhead view) is $[(\pi/2)X_p]^{1/2}$ [where $X_p = (R + Z)/H_p$, R = Earth radius, Z = instrument altitude, H_p = scale height at observer altitude], which is approximately equal to 40. For $\theta_p > 90°$, the air mass traversed by the ray increases rapidly from this value.

Altitude profile information can be obtained using the solar occultation technique from a high-altitude platform, as shown in Figure 3.27, by properly analyzing the absorption data obtained at various zenith angles $>90°$. (Majumdar et al., 1981; Abbas et al., 1981). When the instrument field of view is less than a few millirads, the total integrated absorption along the path from sun to instrument, for $\theta_p > 90°$, is dominated by the absorbers that are at altitudes within 1–2 km of the tangent altitudes, Z^*. The sharpness of the geometric weighting function near its peak value at Z^* allows accurate profile determinations to be made with 2 km vertical resolution by the use of a layer-by-layer retrieval algorithm (Majumdar et al., 1981; Murcray et al., 1978), except in altitude ranges for which the specie mixing ratio is rapidly decreasing with decreasing altitude.

A stratospheric ClO profile obtained with the LHR instrument in November 1979 is shown in Figure 3.28, along with photochemical model predictions by Froidevaux et al. (1985) for various solar zenith angles, two data points from a balloon-borne Microwave Limb Sounder (Waters et al., 1981), and an envelope of several in situ ClO measurements made by Anderson and colleagues (1980) (World Meteorological Organization, 1981) using the resonance fluorescence technique. The agreement is quite good near the ClO number density maximum, although the rapid fall-off observed with decreasing altitude cannot be well explained with present photochemical models. It has become increasingly apparent, though, that natural variability of the concentrations of key trace species in the stratosphere hampers the use of single-species measurements made over a rather short period of time in photochemical model validation exercises. Simultaneous measurements of several species in the same air parcels are more useful for input to the theoretical models.

Recognizing the value of a multispecies instrument in evaluating the stratospheric photochemistry, the balloon-borne laser in situ sensor (BLISS) instrument, a high-resolution absorption spectrometer employing several tunable lead–salt semiconductor diode lasers, was developed to provide measurement of trace species concentrations and their diurnal variations (Menzies et al., 1983). This instrument probes a region of the stratosphere below the gondola by transmitting TDL radiation to a retroreflector suspended 500 m below. It is technically correct to state that the BLISS instrument measures species in situ (i.e., in their

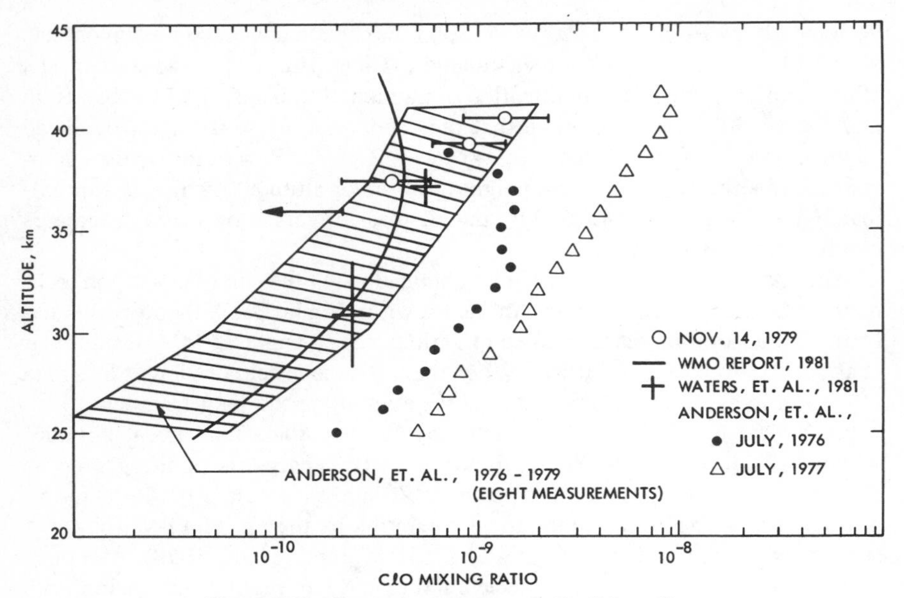

Figure 3.28. Comparison of stratospheric ClO profiles.

original undisturbed positions) since the major portion of the region probed is quite remote from the instrument. For this reason BLISS is well suited to the measurement of unstable (radical) species, and because it does not rely on solar radiation, it can make measurements over a full 24-h period. The high sensitivity to selected trace species is obtained by using second-harmonic detection of the absorption of TDL radiation (3–15 μm) in a 1-km pathlength defined by a retroreflector lowered 0.5 km below the BLISS gondola. A HeNe laser and coaligned TV camera with CID imaging are used for retroreflector tracking under microprocessor control. The use of a retroreflector, a 1-km pathlength, and a 2-kHz modulation frequency for harmonic detection represents the experimental arrangement used in the BLISS instrument. The TDL current is modulated at 2 kHz and ramped in time. A cell filled with a fraction of a Torr of the gas to be monitored allows a reference spectrum for line identification and wavelength calibration to be recorded. Simultaneously, a lock-in (phase-synchronous) amplifier detects the second-harmonic signal at 4 kHz, and a second lock-in amplifier detects the total returned signal.

The retroreflector technique in TDL monitoring of atmospheric constituents allows sensitive detection with lasers that have output powers of only 10–100 μW. When operating single mode, this power is concentrated into a very narrow spectral width of ~ 1 MHz for Dewar-cooled lasers with carefully regulated temperature and current such as those used in the BLISS instrument. To extract accurate concentrations of trace species from measurements using second-harmonic high-resolution TDL spectroscopy, the laser must operate on a single longitudinal mode, or the mode structure energy spectrum must be precisely known. Since the TDL output wavelength depends on both temperature and current, a region of single-mode operation can usually be found by judicious choice of these two parameters. Mode structure analysis as a function of TDL temperature and current is done in the laboratory using a scanning spectrometer.

A major consideration of this instrument technique is the movement of the suspended retroreflector and its influence on the stability of the reflected TDL signal (Johnson et al., 1985). The motion of a typical 15-cm aperture 4 kg retro suspended 500 m below the instrument gondola is dominated by the pendulum effect, which is driven by transient forces during the reeldown process. The damping of this pendulum motion is extremely small at middle stratospheric altitudes due to the low air density. The effects of wind shear and associated turbulent eddies on the retroreflector assembly are also correspondingly small because of the small amounts of kinetic energy in the low-density moving air panels. The period of a simple pendulum of 500 m length is ~45 s, and the maximum expected angular velocity of the retro assembly at the nadir point was calculated to be 4 mrad/s based on a model including maximum transient forces expected during the reeldown process. The concept of using a retroreflector lowered from the main instrument gondola is illustrated in Figure 3.29. A negative lens diverges the launched HeNe beam to cover an area of about 2 m diameter at 500 m from the gondola. The HeNe beam is coaligned with the IR laser beam and bore sighted to the center of the camera FOV. Tracking is accomplished using the visible HeNe laser source at tracking coordinates adjusted to maximize the IR return to the telescope. It is, therefore, assumed that the visible and IR beams do not move with respect to each other. Useful tracking requires a priori that the IR beam be aligned so that it falls within the HeNe cone, that is, so that IR light falls somewhere on the 2-m-dimaeter area at 500 m.

The camera CID information is not only sensed by the microprocessor control system but is also fed to a video monitor. While not strictly necessary, visual observation in this way is a definite advantage in understanding, testing, and confirming the tracking operation. When the tracking parameters are being set up, a narrow-bandpass optical filter (bandwidth 1 nm) centered at 632.8 nm is inserted over the camera lens so that the HeNe light returned by the retroreflector produces illumination levels at the CID much higher than those produced by, for example, sunlit clouds. Without this filter the contrast may not be good enough, depending on the power of the HeNe laser.

The software for tracking comprises a spiral search algorithm, a tracking algorithm, and a fine search/offset algorithm. Until 532.8-nm light returned by the 14-cm diameter retro is seen, the search algorithm is used to scan the track-

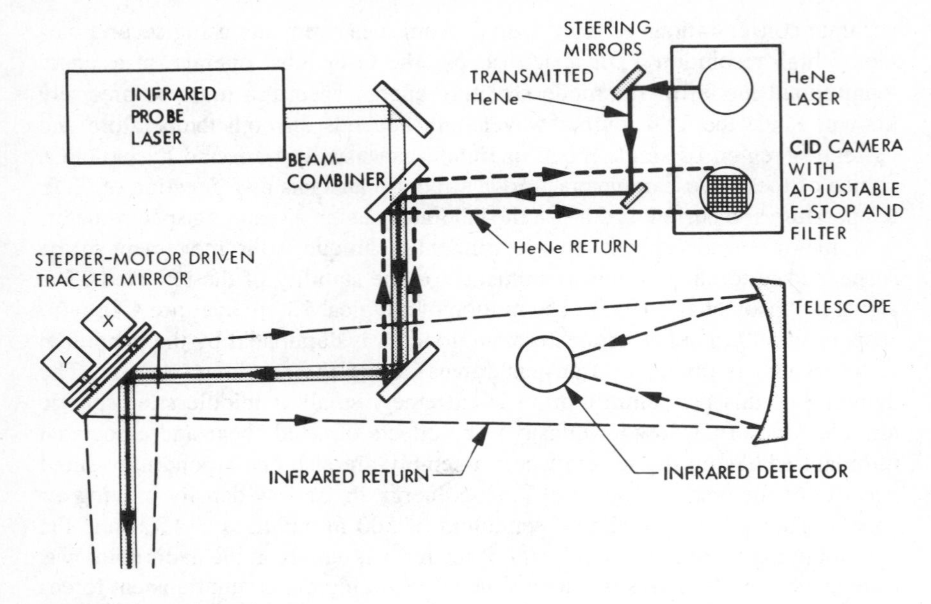

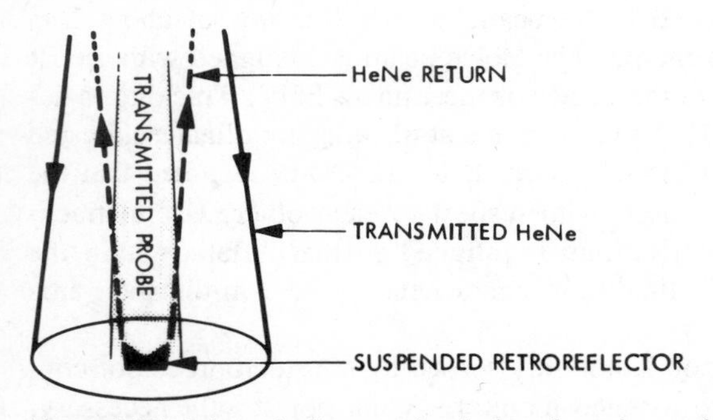

Figure 3.29. Schematic diagram of optical configuration used for tracking system. [From Johnson et al. (1985).]

ing mirror to cover a spiral pattern at 500 m. The tracker mirror movement is accomplished using stepper motors that are geared to lead screws. The red glint from the retro is acquired and then centered in a 32 × 32-pixel central section of the camera's 64,000-pixel array. Every 0.05 s the algorithm updates the position of the centroid of the red glint with respect to the center of the 32 × 32-pixel array and computes a new scan velocity to move the tracker mirror an appropriate amount to hold the retro image at the set coordinates. The resolution of the tracker mirror motion is 10^{-5} rad/step. With tracking fully maintained, a fine search/offset algorithm is then used so that the retro effectively samples (returns) a different part of the 2-m-diameter HeNe laser beam. This time the IR returned signal at the telescope receiver is monitored. When this signal is at a maximum, the final locking coordinates are defined.

This procedure is illustrated in Figure 3.30. In part (*a*), the tracker mirror would be in the spiral search mode. Part (*b*) of this figure shows that the retro has been "acquired" and the tracker mirror moved so that the retro image is at the center of the central 32 × 32-pixel area of the CID array. As the retro "pendulums," the optical relationship shown in continuously maintained by the tracker. In part (*c*), the fine search/offset algorithm has been used and stopped at the offset position where the IR return signal to the telescope is maximized. In the example shown, the IR beam is found by a HeNe beam offset from the central position of about +10 pixels in the x axis and 0 pixels in the y axis. Each TDL being used would have a slightly different image location (typically).

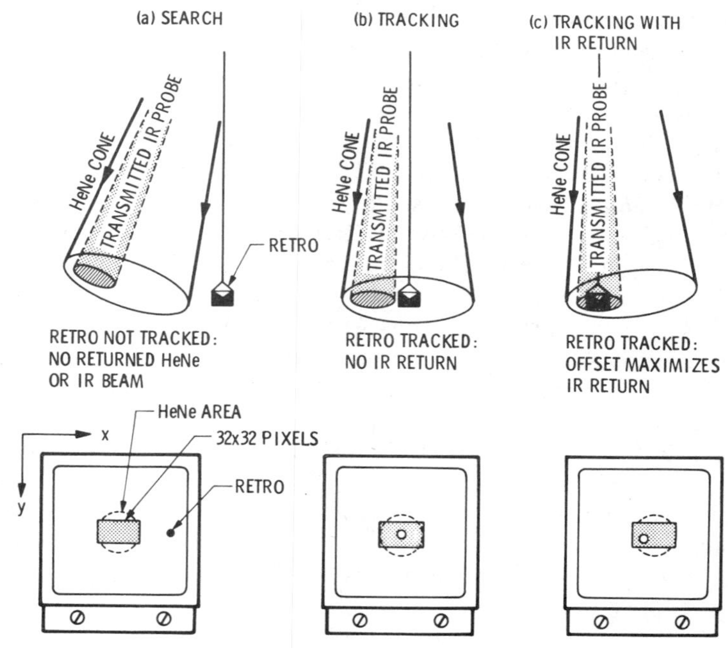

Figure 3.30. Schematic diagram of three basic steps in tracking procedure. Lower illustrations represent 2.2 × 1.7-in. FOV of camera as seen on video monitor. Although retroimage is marked on first such illustration, it would not be seen with narrow-bandpass optical filter in place. The dotted circle is drawn to represent (unobservable) spread and location of He–Ne beam. [From Johnson et al. (1985).]

The offset ability, either through the fine search/offset algorithm or a single two-coordinate command, is implemented without breaking the tracking control.

On October 16, 1983, the BLISS instrument was launched to an altitude of 36 km for a daytime flight over Palestine, Texas. With the retro at 300 m below the gondola, the average X and Y errors were typically 0.1–0.2 pixels. Retroreflector velocities were well below the instrument design limit; being typically 0.05–0.4 ms^{-1} at 300 m. The tracking ability was tested for the retroreflector lowered up to 460 m below the gondola. During the period of float at 36 km altitude, stratospheric nitric oxide was detected, with the measured mixing ratio being 17 ± 5 ppb (Webster and Menzies, 1984). Leaving the retroreflector at 300 m below the gondola, and the tracking in operation, the balloon and instrument were valved down from 36 to 30 km altitude at a mean descent rate of about 1 ms^{-1}. During this descent the tracking was maintained, although on two occasions large oscillations (errors greater than 2 pixels) were observed lasting several seconds, possibly resulting from ballast disposal. This descent test was particularly valuable in establishing the capability of the BLISS instrument to make vertical profile measurements.

The BLISS instrument was more recently flown (November 1985) from Palestine, Texas, and succeeded in making vertical profile measurements of NO_2, NO, and H_2O in the 35–28-km altitude range. Figure 3.31 shows flight scans of NO_2 lines taken at two altitudes, compared with a direct absorption reference cell scan for line identification. In a more recent flight, simultaneous in-situ measurements of NO, NO_2, O_3, jNO_2, CH_4, H_2O, CO_2, pressure, and temperature in the 40-26-km region of the stratosphere were made [Webster and May, 1987] from Palestine, Texas (32°N) on Oct. 16, 1986 using the the JPL Balloon-borne Laser In-Situ Sensor (BLISS) instrument, carrying piggy-back a U. of Denver jNO_2 instrument. Using tunable diode laser absorption spectroscopy over a long pathlength, measurements were made during a twenty one hour flight of: daytime NO, NO_2, O_3, and jNO_2 near 40 km; NO, NO_2 and jNO_2 through the sunset transition near 40 km; post-sunset measurements of the rate of decay of NO_2 into the night over the 35-29-km range; night-time measurements of NO_2, O_3, HNO_3, CH_4, H_2O, and CO_2 at 27 km; and of NO after sunrise. Measurements of NO, NO_2, O_3, CH_4, H_2O, and CO_2 showed good agreement with previous observations. Figure 3.32 shows a flight scan over NO_2 features near 1598 cm^{-1} compared to a synthetic spectrum, illustrating the excellent signal-to-noise ratios obtained for an NO_2 volume mixing ratio of only several parts per billion.

3.5.2 Industrial Applications of Tunable Diode Lasers

Early industrial applications of lasers were based on their high brightness rather than their tunability. For example, laser welding, machining, and annealing

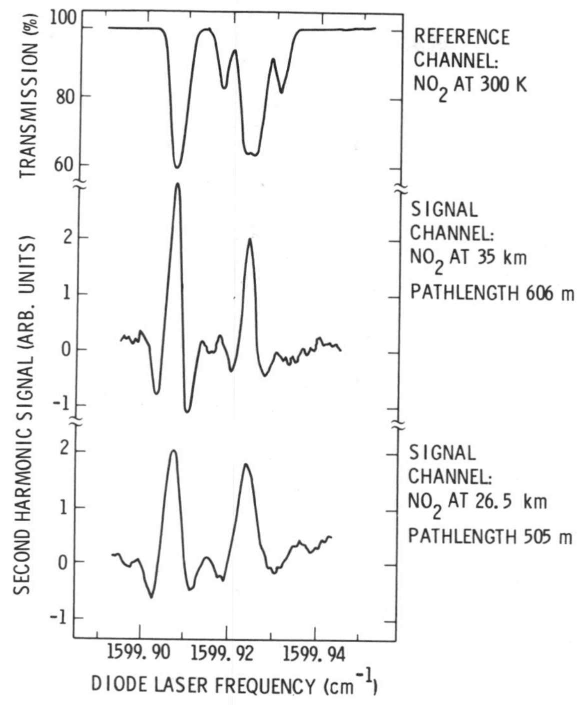

Figure 3.31. November 1985 flight data from BLISS experiment.

demand high power per unit area rather than high spectral purity or wavelength tunability. Other applications, such as laser printing, graphics, and recording are based primarily on the ability to focus a laser beam down to a tiny spot. More recently, III–V compound semiconductor diode lasers have been used to pump other lasers and for optical communications (see Byer, 1984) and tunable lead–salt diode lasers used for the detection of gases and vapors (Forrest and Wall, 1983). These applications require precise control of the laser wavelength.

The first applications of lead–salt tunable lasers involved laboratory measurements of IR spectral absorption lines of gaseous species at relatively low pressure (Hinkley et al., 1976). By reducing pressure below 1 atm, the full potential of the very narrow linewidth of a tunable laser could be utilized to resolve complex spectra of heavy molecular species. These spectroscopic experiments led to several unique industrial applications such as the identification of isotopic species during separation processes and the on-line measurement of the halogen gas partial pressure inside quartz bulbs.

Industrial applications of lead–salt diode lasers have been slow to materialize because of the need for cryogenic refrigeration. Recently, however, using ad-

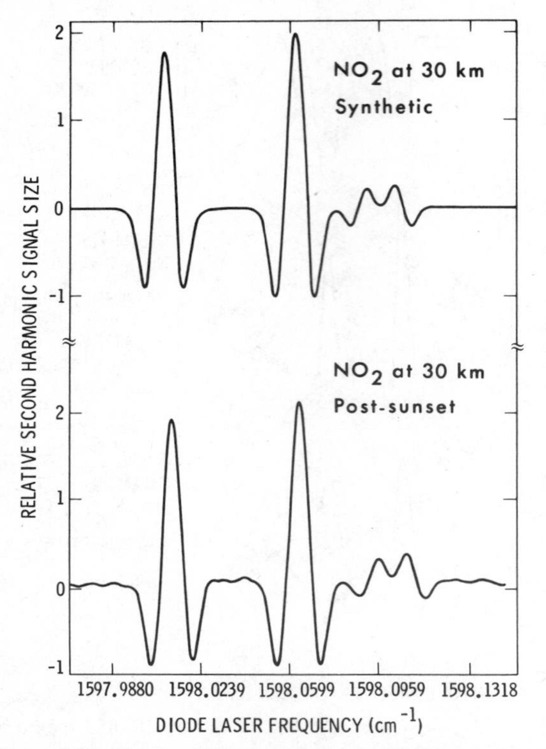

Figure 3.32. October 1986 flight data from BLISS experiment.

vanced quantum-well fabrication technology, it has been possible to produce lasers that operate at significantly higher temperatures. Table 3.8 lists several industrial process control, monitoring, and diagnostics applications of tunable lead–salt diode lasers that are either operational or have undergone prototype demonstration. (See also the reviews of Grant, 1983; Hinkley, 1976; Forrest and Wall, 1983.)

Most of the industrial process control applications utilize one or more lasers, a detector, a sample and reference optical path, and appropriate electronics for laser modulation, signal processing, and display. Most applications benefit by the use of second-derivative (or second-harmonic) spectroscopy, described in Section 3.3. This approach normally provides greater sensitivity by reducing the background signal amplitude as the laser is tuned.

3.5.2.1 Quartz Halogen Lamp Testing

For quality control during the manufacture of quartz halogen lamps, it is desirable to measure the halogen (HBr) gas concentration and background noble-gas

Table 3.8. Industrial Applications of Lead–Salt Diode Lasers

Application	γ (cm^{-1})	Reference
Process Control		
Halogen lamp fabrication	2635	Mantz, 1977
RTV manufacture	1781	Scweid and Hardman, 1983
Moisture in IC packages	1653	Mucha, 1983
Oxygen in silicon crystals	1106	Lo and Majkowski, 1984
Formaldehyde in ethylene gas		Wall and Jones, 1982
Acetylene in liquefaction plant		Maddox and Turner, 1983
Monitoring		
Sulfuric acid vapor		Pearson and Mantz, 1978
By-products in arced SF_6	Various	Mantz et al., 1980
Methane gas detection		Sano and co-workers, 1983
Automobile emissions	Various	Hill and co-workers, 1980
Catalytic converter status		Herz and co-workers, 1978
Explosive vapors		Claspy and co-workers, 1976
Diagnostics		
Nuclear reactor rod analysis	2211	Wahlen and co-workers, 1977
Combustion processes	Various	Schoenung and Hanson, 1982
Automobile engine operation	Various	Chang and Sell, 1978
Cigarette smoke analysis	Various	Forrest, 1980

(typically argon) pressure of each lamp or representative sample. In the past it has been necessary to select lamps at random, break them, and measure the concentration of HBr by a sampling technique. This is time consuming and clearly destructive. Tunable diode lasers permit the HBr to be measured by a nondustructive, automatic technique based on IR spectroscopy, resulting in a substantial cost reduction over current techniques and higher throughput. A diode laser is tuned to 2635 cm^{-1} to match the $R(4)$ line of the 1-0 fundamental band of $H\,^{79}Br$. In addition to measuring the HBr concentration, the total gas pressure within the cell can be determined simultaneously. The relative accuracy of this technique is about 0.5% for HBr concentrations of 10%, and 10% for HBr concentrations of around 1.0% (see Wall and co-workers, 1980).

3.5.2.2 Nuclear Reactor Rod Status

A fully automated tunable diode system has been developed for the monitoring of $^{14}CO_2$ from nuclear reactor rods for the purpose of determining the time for their replacement (Wahlen and co-workers, 1977). The test consists of measuring 1 part of $^{14}CO_2$ in 10^7 parts of $^{12}CO_2$ over a period of several months.

The apparatus consists of a modified version of a commerically available tunable diode laser spectrometer. One diode laser is selected for detection of $^{14}CO_2$ using an absorption line at 2210.885 cm^{-1}, and the other is tuned to N_2O as a reference gas at 2210.794 cm^{-1}. As with most of the industrial monitoring techniques, second-derivative modulation/detection is used, and Figure 3.33 shows a representative scan for a 23.8-ppb sample of $^{14}CO_2$. This system can operate unattended for three months.

3.5.2.3 RTV Process Monitor

In the manufacture of room temperature silicone rubber (RTV), the concentration of the cross-linking agent, methyltriacetoxysilane, is critical to produce performance. It can be monitored by measuring the absorption of laser radiation by an acetoxycarbonyl bond characteristic of the cross-linker, at 1781 cm^{-1}.

A diode laser system has been operating in an industrial plant for over 7 years (Forrest, 1985). In order to monitor the methyltreacetoxysilane, the laser beam must be passed through the viscous RTV material whose transmission at the requisite concentration is only about 10% for a 0.5-mm-thick sample. Consequently, laser brightness, as well as its wavelength, is important to measure the methyltriacetoxysilane concentration with reasonable accuracy.

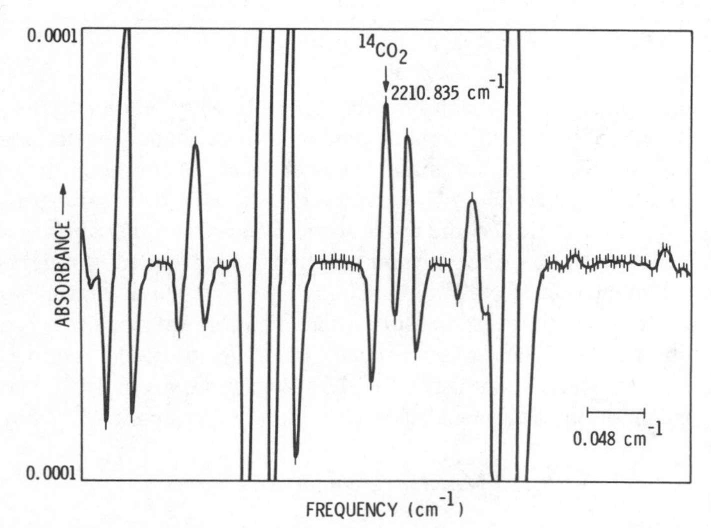

Figure 3.33. TDL scan of 24-ppbv sample of $^{14}CO_2$. [From Wahlen et al. (1977).]

3.5.2.4 Moisture in Integrated-Circuit Packages

Moisture is known to cause degradation in the performance of integrated circuits, and researchers at the AT&T Bell Laboratories have demonstrated that a commercial diode laser spectrometer can be modified to permit the detection of water vapor in small (10–100-μL) samples of gas to a concentration of 50 ppm (Mucha, 1983). The diode laser approach was found superior to the industry-wide standard of mass spectroscopy because, in addition to higher sensitivity and precision, the time-resolved method allowed by the diode laser is insensitive to the moisture exposure history of the system walls. It does not, therefore, require cumbersome procedures for evacuating and conditioning that are inherent in mass spectrometer systems. This has resulted in a marked improvement in the number of devices that can be analyzed per day. With respect to accuracy, the diode laser measurements were found to be within 5% of results based on kinetic analyses.

3.5.2.5 Oxygen in Silicon Wafers

In the Czochralski technique for growing silicon crystals, oxygen is an unintentional dopant whose concentration is typically around 10^{18} cm^{-3} but varies with certain growth parameters. If the oxygen content can be rapidly determined for each wafer, they can be appropriately selected; moreover, the growth process may be modified as required by the ultimate applications. The use of a diode laser spectrometer for fast (10-ms) measurements of oxygen in production line silicon wafers as reported in 1984 by Wayne Lo and Richard Majkowski. Figure 3.34*a* illustrates the difference in spectral transmittance in the 600–1400-cm^{-1} region taken with a Fourier transform spectrometer for silicon wafers with and without the oxygen contamination. The diode laser wavelengths used for production line monitoring are indicated. Figure 3.34*b* shows the resulting measurement of oxygen concentration as a function of distance across the sample; the spatial variation of the oxygen concentration is apparent.

3.5.2.6 Combustion Diagnostics

Tunable diode lasers have been used for dynamic monitoring of combustion processes since the mid-1970s. This type of application takes advantage of the nonintrusive, rapid-scan capability of these diode lasers to provide information on the strength and shape of spectral absorption lines of the species to be monitored. Such measurements can yield the gas temperature (Hanson and Falcone, 1978) and flow dynamics (Hanson, 1977), can be made under adverse background conditions (Hanson, 1980), and can correlate temporal and spatial data during combustion (Schoenung and Hanson, 1982).

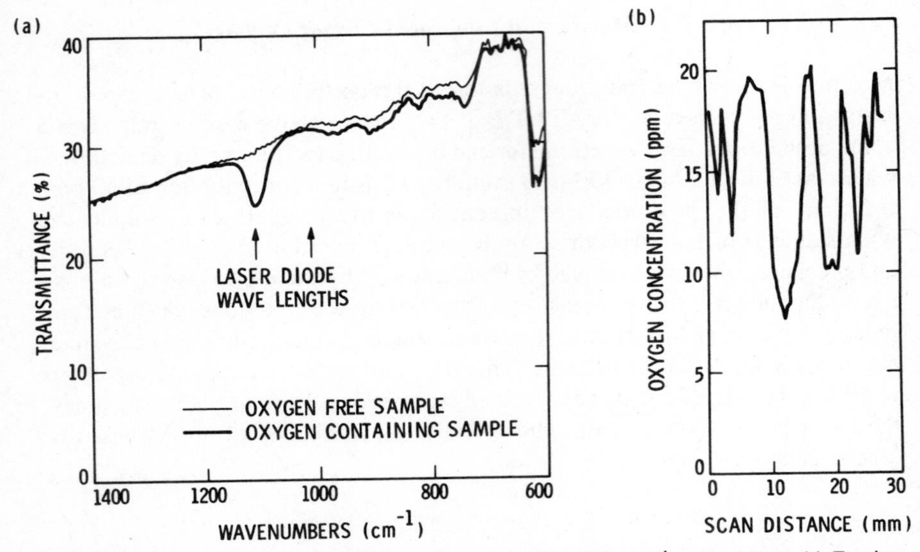

Figure 3.34. (*a*) Difference in spectral transmittance in 600–1400-cm^{-1} region taken with Fourier transform spectrometer for silicon wafers with and without oxygen contamination. Diode laser wavelengths used for production line monitoring are indicated. (*b*) Resulting measurement of oxygen concentration as function of distance across sample; spatial variation of oxygen concentration is apparent. [From Lo and Majkowski (1984).]

3.5.2.7 Automobile Emissions Monitoring

Tunable diode lasers have been used to detect specific gases in the exhaust of automobiles. Some of the species detected are carbon monoxide, nitric oxide, ethylene, and benzene. The General Motors Corporation has pioneered the use of this approach to provide real-time analyses of engine operation and catalytic converters in addition to exhaust emissions monitoring (Hill and co-workers, 1980).

Although in situ laboratory measurements are important for research purposes, laser techniques also permit real-time measurements of exhaust emissions to be made remotely, such as across highways to monitor the background concentrations of specific gases for health reasons. In fact, automatic measurements can be made of the exhaust products of individual automobiles or trucks as they pass by. An application of this type may be useful in providing a nonintrusive indication of whether or not a vehicle is in compliance with the emission ranges allowed by air pollution regulations.

3.5.3 Surface Geology Studies

Remote sensing from airborne or Earth-orbiting vehicles offers the ability to discriminate among rock and mineral types and to identify them on the basis of their IR spectral characteristics (Goetz and Rowan, 1981). The IR region between 8 and 13 μm is called the thermal IR because the maximum thermal emission from Earth's surface at terrestrial temperatures in the 250–320-K range occurs at wavelengths in this region. It is also a region for which the lower atmosphere is relatively transparent, with the stratospheric ozone band centered at 9.5 μm being the only strong absorption band observed when viewing the upwelling thermal radiation from a space platform. This spectral region contains the important restrahleun bands for silicates (Si–O stretch vibration).

Recently an airborne multispectral scanner was developed and flown over the Death Valley, California, area (Kahle and Goetz, 1983; Gillespie et al., 1984). This area had been the subject of a previous geologic mapping study (Hunt and Malvey, 1966), and various samples had undergone laboratory spectroscopic study. The airborne thermal infrared multispectral scanner (TIMS) instrument provided images that distinguished the mapped quartzites, carbonates, volcanic rocks, and saline deposits in Death Valley.

The passive multispectral scanner operating in the thermal IR, while providing an important advance in geologic remote sensing, does have certain fundamental limitations. Because of the limited upwelling thermal radiance, broad spectral bandwidths are necessary to achieve an acceptable signal-to-noise ratio. Also, the emission level depends strongly on temperatures. These limitations may be avoided by using an active IR multiwavelength laser reflectance spectrometer. The reflectance spectrum is insensitive to the surface temperature, and the dynamic range in spectral reflectance observations is much greater than in thermal emission observations. (For example, the emissivities may be from 0.8 to 0.99, corresponding to reflectances ranging from 1 to 20%.) Conceptually, one might conclude, however, that reflectance measured in the case of specific incidence and reflections angles, such as with the active laser instrument, may be much more dependent on surface morphology than are the emission measurements, rendering the technique questionable for use in composition studies.

An active airborne IR laser instrument was flown over Death Valley in 1983 in order to compare its 2-λ reflectance spectra (see Figure 3.35) with the TIMS instrument data, and the result was a very high degree of negative correlation between the reflectance measurements (at 9.23 and 10.27 μm) and the emissivity data, which is expected from Kirchhoff's law (Kahle et al., 1984). Both the laser instrument and the TIMS revealed the same compositional information about the surface materials. The laser instrument, originally developed at JPL for atmospheric trace-gas detection and mapping, had been flown on a number of prior flights in order to study tropospheric ozone transport from areas of high photochemical production (i.e., certain urban areas) to areas with relatively clean air (Shumate et al., 1981). It contains two waveguide Cl_2 lasers, each line tunable over the 9–11-μm region, and two optical heterodyne receivers, with the appropriate optics to direct the coaligned laser beams at the ground and detect the backscattered laser radiation. The use of heterodyne detection

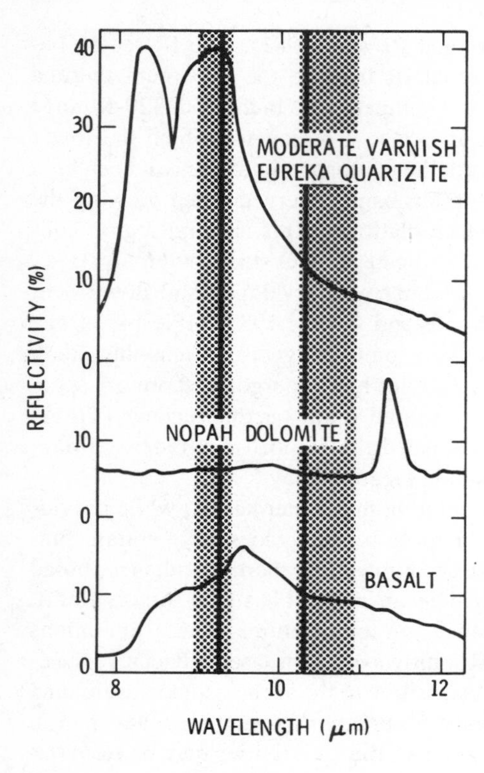

Figure 3.35. Laboratory reflectance spectra of various rock samples from Death Valley, CA, with two spectra bands of NASA/JPL thermal infrared multispectral scanner instrument (stippled) and selected NASA/JPL laser absorption spectrometer wavelengths (solid lines).

permits a very high degree of spectral filtering (with a spectral resolution in this case of 10^{-4} cm^{-1}), making the measurement possible with low-power (~200-mW) CW lasers.

Thus, the use of a CO_2 laser spectrometer operating from an Earth-orbiting platform, previously shown to be capable of measuring profiles of atmospheric constituents such as ozone and water vapor (Menzies and Chahine, 1974), now appears to have a large potential for geologic remote sensing. In addition to the classification of rock types, scientists working for the DFVLR in the Federal Republic of Germany have studied its usefulness for soil moisture measurements (Weisemann et al., 1985). Based on flights of an airborne CO_2 laser spectrometer developed at Batelle Institute in Frankfurt, FRG, and on laboratory spectral studies, there is a strong possibility for the DFVLR and the European Space Agency (ESA) to support the development of more sophisticated multiwavelength active CO_2 laser spectrometers for global measurements from airborne and space platforms (Lehmann and Weisemann, 1985).

ACKNOWLEDGMENTS

The authors would like to thank Jack S. Margolis for helpful suggestions and critical reading of the manuscript, and Darlene Padgett and Betty Citron for its careful preparation. The research described in this chapter was performed by the Jet Propulsion Laboratory, California Institute of Technology, under contract with the National Aeronautics and Space Administration.

REFERENCES

Abbas, M. M., G. L. Shapiro, and J. M. Alvarez (1981). *Appl. Opt.* **20,** 3755.

Abramowitz, M., and I. A. Stegun (1970). *Handbook of Mathematical Functions*, Dover Publications, New York.

Allario, F., S. J. Katzberg, J. M. Hoell, and J. C. Larsen (1980). *Appl. Phys.* **23,** 47.

Allen, H. C. Jr., and P. C. Cross (1963). *Molecular Vib-Rotors*, Wiley, New York.

Anderson, J. G., H. J. Grassl, R. E. Shetter, and J. J. Margitan (1980). *J. Geophy. Res.* **85,** 2869.

Arams, F. R., E. W. Sard, B. J. Peyton, and F. P. Pace (1967). *IEEE J. Quantum Electronics* **QE-3,** 484.

Arndt, R. (1965). *J. Appl. Phys.* **36,** 2522.

Atkins, P. W. (1978). *Physical Chemistry*, Freeman, San Francisco.

Begemann, M. A. and R. J. Saykally (1982). *Opt. Commun.* **40,** 277.

Bell, A. G. (1881). *Phil. Mag.* **11,** 510.

Bjorklund, G. C. (1980). *Opt. Lett.* **5,** 15.

Bjorklund, G. C., and M. D. Levenson (1982). *Phys. Rev. A* **24,** 166.

Bjorklund, G. C., W. Lenth, M. D. Levenson, and C. Oritz (1981). Frequency Modulated (FM) Spectroscopy, in SPIE Vol. 286, *Laser Spectroscopy for Sensitive Detection*, p. 153.

Blaney, T. G. (1975). *Space Sci. Rev.* **17,** 691.

Bragg, S. L. (1981). Ph.D. Thesis, Washington University, St. Louis, Missouri.

Brown, L. R., J. S. Margolis, R. H. Norton, and B. D. Stedry (1983). *Appl. Spectry.* **37,** 287.

Bruce, C. W., and R. G. Pinnick (1977). *Appl. Opt.* **16,** 1762.

Burkhardt, E. G., C. A. Lambert, and C. K. N. Patel (1975). *Science 188*, 1111.

Byer, R. L. (1984). Proceedings of the NASA Workshop on Tunable Solid State Lasers for Remote Sensing, Stanford University, October; see also E. D. Hinkley, J. R. Lesh, and R. T. Menzies (1985). Lasers and Sensors in Space, *Laser Focus*, February, p. 78.

Carswell, A. I., and J. J. Wood (1967). *J. Appl. Phys.* **38,** 3028.

Cassidy, D. T., and J. Reid (1982). *Appl. Phys.* **B29,** 279.

Chang, Man-Feng, and J. A. Sell (1978). A Linearized Model of Engine Torque and Carbon Monoxide Emissions. General Motors Research Laboratory Report GMR-4042.

Chapman, S. (1931). *Proc. Phys. Soc. London* **43,** 483.

Claspy, P. C., Y. Pao, S. Kwong, and E. Nodov (1976). Laser Optoacoustic Detection of Explosive Vapors. *Appl. Optics* **15,** 1506.

Corney, A. (1977). *Atomic and Laser Spectroscopy*, Clarendon Press, Oxford.

Deaton, T. F., D. A. Depatie, and T. W. Walker (1975). *Appl. Phys. Lett.* **26,** 300.

Demtroder, W. (1981). *Laser Spectroscopy*, Springer-Verlag, Berlin.

Dewey, C. F., Jr., R. D. Kamm, and C. E. Hackett (1973). *Appl. Phys. Lett.* **23,** 633.

Dicke, R. H. (1946). *Rev. Sci. Instrum.* **17,** 268.

Dicke, R. H. (1953). *Phys. Rev.* **89,** 472.

Duxbury, G. (1983). *Chem. Soc. Rev.* **12,** 453.

Elbaum, M., and M. C. Teich (1978). *Opt. Commun.* **27,** 257.

Eng, R. S., A. R. Calawa, T. C. Harman, and P. L. Kelley (1972). *Appl. Phys. Lett.* **21,** 303.

Fink, U., and M. J. S. Belton (1969). *J. Atm. Sci.* **26,** 952.

Foreman, J. W., Jr., W. W. George, J. L. Jetton, R. D. Lewis, J. R. Thorton, and H. J. Watson (1966). *IEEE J. Quantum Electron.* **QE-2,** 260.

Forrest, G. (1980). Tunable Diode Laser Measurement of Methane, Ethane, and Water Vapor in Cigarette Smoke. *Appl. Optics* **19,** 2094.

Forrest, G. (1985). Private communication.

Forrest, G. T., and D. L. Wall (1983). Automated Tunable Diode Laser Monitoring Systems, in *Tunable Diode Laser Development and Spectroscopy Applications*, Wayne Lo, ed., Proc. SPIE 438.

Freed, C., J. W. Bielinski, W. Lo, and D. L. Partin (1984). Output Characteristics of Lead Telluride Quantum-Well Diode Lasers. Paper ThEE3 at the 13th International Quantum Electronics Conference, Anaheim, CA, 18–21 June.

Frerking, M. A., and D. J. Muehlner (1977). *Appl. Opt.* **16,** 526.

Froidevaux, L., M. Allen, and Y. Yung (1985). *J. Geophys. Res.* **90,** 1299–13029.

Gerlach, R., and N. M. Amer (1980). *Appl. Phys.* **23,** 319.

Gerry, E. T., and D. A. Leonard (1966). *Appl. Phys. Lett.* **8,** 227.

Gillespie, A. R., A. B. Kahle, and F. D. Pallucone (1984). *Geophys. Res. Lett.* **11,** 1153.

Goetz, A. F. H., and L. C. Rowan (1981). *Science* **211,** 781.

Goodwin, F. E., and M. E. Pedinoff (1966). *Appl. Phys. Lett.* **8,** 60.

Grant, W. B. (1983). Laser Spectroscopy Techniques Make Industrial Appearance. *Indust. Res. Dev.* **25,** 154.

Green, R. B., R. A. Keller, G. G. Luther, P. K. Schenck, and J. C. Travis (1976). *Appl. Phys. Lett.* **29,** 747.

Gudeman, C. S., and R. J. Saykally (1984). *Ann. Rev. Phys. Chem.* **35,** 387.

Gudeman, C. S., M. H. Regemann, J. Pfaff, and R. J. Saykally (1983). *Phys. Rev. Lett.* **50,** 727.

Hall, J. L., L. Hollberg, T. Baer, and H. G. Robinson (1981). *Appl. Phys. Lett.* **39,** 680.

Hameau, C., J. Wascat, D. Dangoisse, and P. Glorieux (1984). *Opt. Commun.* **49,** 423.

Hameau, C., E. Arimondo, J. Wascat, and P. Glorieux (1985). *Opt. Commun.* **53,** 375.

Haner, D. A., C. R. Webster, P. H. Flamant, and I. S. McDermid (1983). *Chem. Phys. Lett.* **96,** 302.

Hanson, R. K. (1977). Shock Tube Spectroscopy: Advanced Instrumentation with a Tunable Diode Laser. *Appl. Optics* **16,** 1479.

Hanson, R. K. (1980). Absorption Spectroscopy in Sooting Flames Using a Tunable Diode Laser. *Appl. Optics.* **19,** 1264.

Hanson, R. K., and P. K. Falcone (1978). Temperature Measurement Technique for High-Temperature Gases using a Tunable Diode Laser. *Appl. Optics* **17,** 2477.

Herman, R., and R. F. Wallis (1955). *J. Chem. Phys.* **23,** 637.

Herman, R., R. W. Rothery, and R. J. Rubin (1958). *J. Mol. Spectry.* **2,** 369.

Herz, R. K., J. B. Kiela, and J. A. Sell (1980). Dynamic Behavior of Automotive Catalysts. II. Carbon Monoxide Conversion under Transient Air/Fuel Ratio Conditions. General Motors Research Laboratory Report GMR-4194, PCP-193.

Herzberg, G. (1945). *Molecular Spectra and Molecular Structure, II. Infrared and Raman Spectra of Polyatomic Molecules*, Van Nostrand Reinhold, New York.

Hill, J. C., W. Lo, and J. A. Sell (1980). Lead-Salt Diodes Analyze Automotive Exhaust. *Laser Focus*, November, p. 86; see also J. A. Sell (1983). Tunable Diode Laser Measurements of Carbon Monoxide in Engine Exhaust. *Proc. SPIE* **438;** and E. D. Hinkley and P. L. Kelley (1971). *Science* **171,** 635; see also E. D. Hinkley, K. W. Nill, and F. A. Blum (1976). Infrared Spectroscopy with Tunable Lasers, in *Laser Spectroscopy of Atoms and Molecules*, H. Walther, ed., Springer-Verlag, Berlin.

Hinkley, E. D. (1970). *Appl. Phys. Lett.* **16,** 351.

Hinkley, E. D., and P. L. Kelley (1971). *Science* **171,** 635.

Hinkley, E. D., ed. (1976). *Laser Monitoring of the Atmosphere*, Springer-Verlag, Berlin.

Hinkley, E. D., R. T. Ku, and P. L. Kelley (1976). in *Laser Monitoring of the Atmosphere*, by E. D. Hinkley, ed., Springer-Verlag, Berlin.

Hirota, E. (1985). *High-Resolution Spectroscopy of Transient-Molecules*, Springer Series in Chemical Physics **40.**

Hollas, J. M. (1982). *High Resolution Spectroscopy*, Butterworths, London.

Horowitz, P., and W. Hill (1980). *The Art of Electronics*, Cambridge University Press, Cambridge, Chapter 14.

Huffaker, R. M., A. V. Jelalian, and J. A. L. Thomson (1970). *Proc. IEEE* **58,** 322.

Humlicek, J. (1979). *J. Quant. Spectrosc. Radiat. Transfer*, **21,** 309.

Hunt, C. B., and D. R. Malvey (1966). Stratigraphy and Structure, Death Valley, California. U. S. Geol. Survey Prof. Paper 494-A.

Hurst, W. S., G. J. Rosasco, and W. Lempert (1984). *SPIE* **482,** 23.

Jackson, D. J., E. Arimondo, J. E. Lawler, and T. W. Hansch (1980). *Opt. Commun.* **33,** 41.

Jackson, W. B., N. M. Amer, A. C. Boccara, and D. Fournier (1981). *Appl. Opt.* **20,** 1333.

Jansson, P. A., and C. L. Korb (1968). *J. Quant. Spectrosc. Radiat. Transfer*, **8,** 1399.

Johnson, M. A., A. L. Betz, R. A. McLaren, E. C. Sutton, and C. H. Townes (1976). *Astrophys. J.* **208,** L145.

Johnson, R. A., C. R. Webster, and R. T. Menzies (1985). *Rev. Sci. Instrum.* **56,** 547.

Johnston, H. S. (1971). *Science* **173,** 517.

Kahle, A. B., and A. F. H. Goetz (1983). *Science*, **222,** 24.

Kahle, A. B., M. S. Shumate, and D. B. Nash (1984). *Geophys. Res. Lett.* **11,** 1149.

Kavaya, M. J. (1982). Ph.D. Thesis, California Institute of Technology, Pasadena, CA.

Kavaya, M. J., J. S. Margolis, and M. S. Shumate (1979). *Appl. Opt.* **18,** 2602.

Kim, K. C., E. Griggs, and W. B. Person (1978). *Appl. Opt.* **17,** 2511.

Kingston, R. H. (1978). *Detection of Optical and Infrared Radiation*, Springer-Verlag, New York.

Kizirnis, S. W., R. J. Brecha, B. N. Ganguly, L. P. Goss, and R. Gupta (1984). *Appl. Opt.* **23,** 3872.

Koga, R., M. Kosaka, and H. Sano (1981). Memoirs of the School of Eng., Okayama University, **16,** 21.

Korb, C. L., R. H. Hunt, and E. K. Plyer (1968). *J. Chem. Phys.* **48,** 4252.

Kostick, T., M. J. Mumma (1983). *Appl. Opt.* **22,** 2644.

Kreuzer, L. B. (1971). *J. Appl. Phys.* **42,** 2934.

Kroto, H. W. (1975). *Molecular Rotation Spectra*, Wiley, New York.

Laguna, G. A. (1984). *Appl. Opt.* **23,** 2155.

Lehmann, F., and W. Wiesemann (1985). LIMES: An Optical Multisensor for Earth Observations Using CO_2 Laser Techniques. Proceedings, Third Topical Mtg. on Coherent Laser Radar: Technology and Applications, Great Malven, V. K., July.

Lenth, W. (1983). *Opt. Lett.* **8,** 575.

Lo, W., and R. F. Majkowski (1984). Detection of Oxygen Concentration in Silicon Wafers using a Tunable Diode Laser. Research Publication GMR-4633, General Motors Research Laboratories, 2 March 1984; see also A. Ohsawa, K. Honda, S. Ohkawa, and R. Ueda (1980). Determination of Oxygen Concentration Profiles in Silicon Crystals Observed by Scanning IR Absorption using Semiconductor Laser. *Appl. Phys. Lett.* **36,** 147.

Lorentz, H. A. (1906). *Proc. Amst. Akad. Sci.* **8,** 591.

Lundqvist, S., J. Margolis, and J. Reid (1982). *Appl. Opt.* **21,** 3109.

Maddox, W., and C. M. Turner (1983). Diode Laser Spectrometry for Monitoring Acetylene On-Line in a Liquefaction Plant. *Proc. SPIE* 438.

Majumdar, A., R. T. Menzies, and S. L. Jain (1981). *Appl. Opt.* **20,** 505.

Maki, A. G., A. S. Pine, J. S. Wells, D. A. Jennings, A. Fayt, and A. Goldman (1985). Optical Remote Sensing of the Atmosphere, Technical Digest, Incline Village, Nevada, Jan. 15–18.

Malmstadt, H. V., C. G. Enke, and S. R. Crouch (1981). *Electronics and Instrumentation for Scientists*, Benjamin/Cummings, Menlo Park, CA.

Mantz, A. W. (1977). Laboratory HBr and Total Pressure Monitor for Halogenated Quartz Lamps. Report LAI-4 of Laser Analytics Division of Spectra Physics Corporation, August; see also, A. W. Mantz and R. S. Eng (1978). Tunable Laser Measurements of Line Intensities and Pressure-Broadened Linewidths of HBr: Ar in the HBr Fundamental Region. *Appl. Spectroscopy* **32,** 239.

Mantz, A. W. (1980). Use of Tunable Diode Lasers for the Determination of Low Level Gas By-Products in Arced Sulfur Hexafluoride. Laser Analytics Division/Spectra Physics Internal Report.

Mantz, A. W., and D. L. Wall (1980). Tunable Diode Laser Systems for Quality Control Specialized Analytical and Process Monitoring Applications.

May, R. D. (1984). Ph.D. Thesis, University of North Carolina, Chapel Hill.

May, R. D., (1987). Response Function of a Tunable Diode Laser Spectrometer from an Iterative Deconvolution Procedure, *J. Quant. Spectrosc. Radiat. Transfer*, (in press).

McDermid, I. S. (1983). Modern Photochemical Techniques: The Study of Fluorescence Decay, in *Comprehensive Chemical Kinetics, Vol. 24, Modern Methods in Kinetics*, C. H. Bamford and C. F. H. Tipper, eds., Elsinor Scientific Publishing, Amsterdam.

Measures, R. M. (1984). *Laser Remote Sensing*, Wiley, New York.

Menzies, R. T. (1971). *Appl. Opt.* **10,** 1532.

Menzies, R. T. (1976). Laser Heterodyne Detection Techniques, in *Laser Monitoring of the Atmosphere*, E. D. Hinkley, ed., Springer, New York.

Menzies, R. T., and M. T. Chahine (1974). *Appl. Opt.* **13,** 2840.

Menzies, R. T., and R. K. Seals, Jr. (1977). *Science* **197,** 1275.

Menzies, R. T., C. W. Rutledge, R. A. Zanteson, and D. L. Spears (1981). *Appl. Opt.* **20,** 536.

Menzies, R. T., C. R. Webster, and E. D. Hinkley (1983). *Appl. Opt.* **22,** 2655.

Molina, L. T., and W. B. Grant (1984). *Appl. Opt.* **23,** 3893.

Mucha, J. A. (1983). Trace Analysis of Moisture in Integrated Circuit Packages using Derivative Diode Laser Spectroscopy, *Proc. SPIE* 438; see also P. R. Bossard and J. A. Mucha (1981). Dynamic Measurement of the Water Vapor Content of Integrated Circuit Packages using Derivative Infrared Diode Laser Spectroscopy. *IEEE Proc. IRPS*, CH1619-6/81.

Muenchausen, R. E., R. D. May, and G. W. Hills (1984). *Opt. Commun.* **48,** 317.

Murcray, D. G., A. Goldman, G. R. Cook, D. K. Rolens, and L. R. Megill (1978). On the Interpretation of Infrared Solar Spectra for Altitude Distribution of Atmospheric Trace Constituents. Report FAA-EE-78-30, U.S. Dept. of Transportation, Washington, D.C.

Nieuwenhuijzen, H. (1970). *Opt. Tech.*, May, 68.

Olivero, J. J., and R. L. Longbothum (1977). *J. Quant. Spectrosc. Radiat. Transfer* **17,** 233.

Pao, Y.-H., ed. (1977). *Optoacoustic Spectroscopy and Detection*, Academic Press, New York.

Partin, D. L., R. F. Majkowski, and D. E. Swets (1984). G.M. Research Publication No. GMR-4804.

Patel, C. K. N., and R. J. Kerl (1977). *Appl. Phys. Lett.* **30,** 578.

Pearson, E. F., and A. W. Mantz (1978). Tunable Diode Laser Sulfuric Acid Stack Monitoring System. Proceedings Laser 79 Opto-Electronics Conference, Munich, July 1979; see also G. P. Montgomery, Jr., R. F. Majkowski (1978). Tunable Laser Spectroscopy of Foreign-Gas-Broadened Sulfuric Acid (H_2SO_4) Vapor. *Appl. Optics* **17,** 173.

Penner, S. S. (1959). *Quantitative Molecular Spectroscopy and Gas Emissitivities*, Addison-Wesley, Reading, MA.

Penning, F. M. (1928). *Physica* **8,** 137.

Pierluissi, J. H., P. C. Vanderwood, and R. B. Gomez (1977). *J. Quant. Spectrosc. Radiat. Transfer* **18,** 555.

Pine, A. S. (1980). *J. Mol. Spectry.* **82,** 435.

Podolske, J. R., M. Loewenstein, and P. Varanasi (1984). *J. Mol. Spectry.* **107,** 241.

Pugh, L. A., and K. N. Rao (1976). Intensities from Infrared Spectra, in *Molecular Spectroscopy: Modern Research*, Rao and Matthews, eds., Academic Press, New York, p. 165.

Ramirez, R., and L. DeWitt (1985). Handshake **10,** Spring, p. 10, Tektronix Inc., Beaverton, Oregon.

Reid, J., and D. Labrie (1981). *Appl. Phys.* **B26,** 203.

Reid, J., D. T. Cassidy, and R. T. Menzies (1982). *Appl. Opt.* **21,** 3961.

Reid, J., R. L. Sinclair, W. B. Grant, and R. T. Menzies (1985). *Opt. Quant. Electron.* **17,** 31.

Rosencwaig, A., and A. Gersho (1976). *J. Appl. Phys.* **47,** 64.

Rosengren, L.-G. (1975). *Appl. Opt.* **14,** 1960.

Rothman, L. S., A. Goldman, J. R. Gillis, R. R. Gamache, H. M. Pickett, R. L. Poynter, N. Husson, and A. Chedin (1983). *Appl. Opt.* **22,** 1616.

Rowland, F. S., and M. J. Molina (1975). *Rev. Geophys. Space Phys.* **13,** 1.

Sano, H., R. Koga, and M. Kosaka (1983). Portable Lead-Salt Diode Laser System and Temporal Fluctuation of Local Atmospheric Methane in the Field. OSA/IEEE Conference on Lasers and Electro-Optics, 17–20 May.

Schoenung, S. M., and R. K. Hanson (1982). Laser Absorption Sampling Probe for Temporally and Spatially Resolved Combustion Measurements. *Appl. Optics* **21,** 1767.

Schweid, N., and B. Hardman (1983). Unique Tunable Diode Laser Application: Room Temperature Vulcanizing (RTV) Sealant Production Control, *Proc. SPIE*, **438,** 61.

Shimoda, K., ed. (1976). High-Resolution Laser Spectroscopy, Topics in *Applied Physics*, Vol. 13, Springer-Verlag, Berlin.

Shumate, M. S., R. T. Menzies, J. S. Margolis, and L.-G. Rosengren (1976). *Appl. Opt.* **15,** 2480.

Shumate, M. S., R. T. Menzies, W. B. Grant, and D. S. McDougal (1981). *Appl. Opt.* **20,** 545.

Siegman, A. E. (1966). *Proc. IEEE* **54,** 1350.

Siegman, A. E. (1971). *An Introduction to Lasers and Masers*, McGraw-Hill, New York.

Skolnick, M. L. (1970). *IEEE J. Quantum Electron.* **6,** 139.

Steinfeld, J. I. (1974). *Molecules and Radiation: An Introduction to Modern Molecular Spectroscopy*, MIT Press, Cambridge, MA.

Stolarski, R. S., and R. J. Cicerone (1974). *Can. J. Chem.* **52,** 1610.

Sutton, E. C., S. Subramanian, and C. H. Townes (1982). *Astron. Astrophys.* **110,** 324.

Swofford, R. L., and J. A. Morrell (1978). *J. Appl. Phys.* **49,** 3667.

Tam, A. C., J. Hussla, and H. Sontag (1985). Photothermal Detection of the Onset of Photochemical Nucleation (Laser Snow) in Vapors. Paper TUJ3, OSA/IEEE Conf. on Lasers and Electro-Optics Digest of Technical Papers, May, 1985.

Teich, M. C. (1968). *Proc. IEEE* **56,** 37.

Tiuri, M. E. (1966). Radio-Telescope Receivers, in *Radio Astronomy*, J. D. Kraus, ed., McGraw-Hill, New York.

Toth, R. A., R. H. Hunt, and E. K. Plyler (1969). *J. Mol. Spectry.* **32,** 85.

Townes, C. H. (1970). Unpublished report.

Townes, C. H., and A. L. Schawlow (1975). *Microwave Spectroscopy*, Dover Publications, New York.

Valero, V. P. J., C. B. Suarez, and R. W. Boese (1979). *J. Quant. Spectrosc. Radiat. Transfer* **22,** 93.

Van Buren, H. H. (1975). *Space Sci. Rev.* **17,** 621.

Van Vleck, J. H., and V. F. Weisskopf (1945). *Rev. Mod. Phys.* **17,** 227.

Wahlen, M., R. S. Eng, and K. W. Nill (1977). *Appl. Opt.* **16,** 2350.

Walkup, R., R. W. Dreyfus, and Ph. Avouris (1983). *Phys. Rev. Lett.* **50,** 1846.

Wall, D. L., and Huw R. Jones (1982). Use of Tunable Diode Lasers for Trace Detection of Formaldehyde Impurity in Ethylene. Sponsor Report 7308, Laser Analytics Division, Spectra Physics, Inc., 3 May 1982.

Wall, D. L., E. F. Pearson, and A. W. Mantz (1980). Tunable Diode Lasers for Industrial Analysis and Monitoring Applications. IFAC PRP 4 Automation Conference, Ghent, Belgium, 1980.

Waters, J. W., J. C. Hardy, R. F. Jarnot, and H. M. Pickett (1981). *Science* **214,** 61.

Webster, C. R. (1985). *J. Opt. Soc. Amer. B* **2,** 1464.

Webster, C. R., and R. T. Menzies (1983). *J. Chem. Phys.* **78,** 2121.

Webster, C. R., and R. T. Menzies (1984). *Appl. Opt.* **23,** 1140.

Webster, C. R., and C. T. Rettner (1983). *Laser Focus* **19**(2), 41.

Webster, C. R., I. S. McDermid, and C. T. Rettner (1983). *J. Chem. Phys.* **78,** 646.

Webster, C. R., R. D. May, and M. R. Gunson (1985). *Chem. Phys. Lett.* **121,** 429.

Wiesemann, W., F. Lehmann, and Ch. Werner (1985). *Infrared Phys.* **25,** 467.

Wilson, E. B., Jr., and A. J. Wells (1946). *J. Chem. Phys.* **14,** 578.

Wilson, G. V. H. (1963). *J. Appl. Phys.* **34,** 3276.

Wolfe, W. L., and G. L. Zissis (1978). *The Infrared Handbook*, Office of Naval Research, Department of the Navy, Government Printing Office, Washington, D.C.

World Meteorological Organization (1981). *The Stratosphere 1981*, WMO Global Ozone Research and Monitoring Project Report No. 11 (WMO, NASA, FAA, NOAA cosponsors).

Yariv, A. (1976). *Introduction to Optical Electronics*, Holt, Rinehart and Winston, New York.

Yeh, Y., and H. Z. Cummins (1964). *Appl. Phys. Lett.* **4,** 176.

Zahniser, M. S., and A. C. Stanton (1983). Aerodyne Co., Final Report No. RR-348, Billerica, MA.

CHAPTER

4

DIFFERENTIAL ABSORPTION LIDAR FOR POLLUTION MAPPING

KENT A. FREDRIKSSON

National Department of Physics
Swedish Environment Protection Board
Box 118
S-22100 Lund, Sweden

4.1 INTRODUCTION

The problems associated with atmospheric pollution have stressed the need for efficient measurement methods for air pollutants. Remote sensing techniques utilizing lasers have proved to be powerful instruments in many cases. In this chapter the most-employed laser technique for measurements on air pollutants will be reviewed and discussed. The basic equations of importance are studied, and typical laser systems for remote sensing are described. The practical aspects of the technique and measurement activities are emphasized, and examples of measurements are shown. In particular, we will pick examples from laser remote sensing in the UV, the visible, and the near-IR regions, as there are several laser systems operating in these wavelength regions in use today.

Remote sensing of atmospheric properties, utilizing the absorption and scattering processes of particles and molecular species, was carried out before the invention of the laser with artificial or natural light sources. New and improved methods for atmospheric measurements resulted when laser sources became available in the early sixties. Lasers have some distinct advantages for atmospheric probing compared with other light sources. The possibility of generating short pulses of high intensity in narrow beams is the basis for several powerful optical remote sensing techniques. In addition, lasers have, in general, a very narrow spectral linewidth, and some are also tunable within a limited spectral region. In the last 10 years such lasers have become much more powerful and have made new measurement techniques possible in many different areas.

Several lidar techniques for measuring molecular and atomic species in the atmosphere have been developed. These involve the use of fluorescence, absorption, resonance scattering, or Raman scattering. The effective cross sections in the lower atmosphere for fluorescence and resonance scattering are low due to collision quenching and other effects, and therefore, the usefulness of these processes for pollution studies is limited. The Raman scattering cross sections are very low and the Raman lidar technique is applicable only in studies involving high molecular concentrations.

The most sensitive method is based on the absorption of the transmitted laser light by the atmospheric constituents. Absorption cross sections at molecular absorption wavelengths are often large enough for even minor constituents to cause a measurable attenuation. The method is usually referred to as DIAL, an acronym for differential absorption lidar, and utilizes the differential absorption properties of molecular species. The technique was introduced some 20 years ago by Schotland (1966), who made measurements on water vapor using a temperature-tuned ruby laser. However, at that time there were no tunable sources to match absorption lines of other atmospheric gases. The technique has developed rapidly since these first DIAL measurements (e.g., Kildal and Byer, 1971; Measures and Pilon, 1972; Byer and Garbuny, 1973; Hinkley et al., 1976; Collis and Russel, 1976; Measures, 1984). The development of the DIAL technique has been coupled to the progress in laser technology, and with the tunable lasers of today it has become a sensitive and powerful method of measuring specific atmospheric gases.

It is questionable whether measurements where topographic targets or retroreflecting mirrors are employed as the scattering medium should be considered as lidar measurements. However, since such long-path absorption measurements and the DIAL technique are closely connected regarding instrumentation and applications we include them in this context. Sometimes the differential absorption lidar technique is called DAS, which stands for differential absorption and scattering. DASE is another synonymous acronym that has been used to describe the technique.

4.2 DIFFERENTIAL ABSORPTION LIDAR TECHNIQUE

4.2.1 Lidar Equation for Elastic Scattering

A lidar system is a laser remote sensor made up of a transmitting laser source and an optical receiver for collecting scattered light. The general configuration of a typical lidar system is schematically illustrated in Figure 4.1. The figure shows a monostatic arrangement, in which the laser source and the receiver are at the same location. This configuration predominates greatly and is the only type that will be discussed in this chapter. In general, the transmitter is a pulsed laser source, and range information is provided by the time delay between the transmitted beam and the received backscattered radiation. A lidar system can be either coaxial or biaxial. In a coaxial system the laser beam coincides with the axis of the receiver telescope, whereas in a biaxial system the laser beam is transmitted beside this axis. Both of these types are supplied in atmospheric lidar systems.

Lidar techniques are limited to atmospheric windows for electromagnetic radiation. These windows are determined by scattering and molecular absorp-

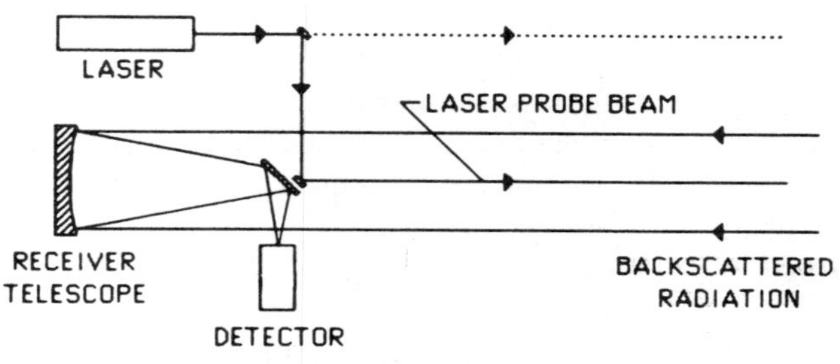

Figure 4.1. Configuration of coaxial lidar system. Dashed line: laser beam direction in biaxial system.

tion of the main atmospheric components. Below 200 nm the atmosphere is totally opaque due to absorption by oxygen. The near-UV and visible regions are free from strong attenuation by molecules, and the propagation here is determined essentially by Rayleigh and Mie scattering. There are several windows in the near- and mid-IR regions, which are interrupted by strong carbon dioxide and water vapor absorption bands. The region from around 5.3 μm to around 7.8 μm is totally opaque due to water vapor absorption. In the far-IR region there is a broad atmospheric window from 7.8 μm to around 13 μm. The interaction of radiation with atmospheric species and the propagation of radiation through the atmosphere has been discussed in detail in the recent book by Measures (1984).

The atmospheric scattering employed in lidar measurements has a strong wavelength dependence. The scattering coefficients at UV wavelengths are orders of magnitude higher than the corresponding coefficients at wavelengths in the far-IR region. A theoretical discussion on scattering and extinction properties with reference to relevant experimental work has been presented by Collis and Russell (1976).

Lidar scattering can be represented by an equation that in the most general form is rather complex. A detailed discussion of the lidar equation is provided by Measures (1984), who also relates it to basic electromagnetic theory. However, in many applications the lidar equation can be simplified to a less complex, but still appropriate version. In DIAL applications only elastic scattering has to be considered. The received power of the elastic backscattered radiation from a range R can be written as

$$\begin{aligned} P(\lambda, R) = {} & C(\lambda, R) \frac{A_r}{R^2} P_L(\lambda) \, \Delta R \sum_i \sigma_{B_i}(\lambda) N_i(R) \\ & \times \exp\left\{ -2 \int_0^R \left[\sum_m \sigma_{A_m}(\lambda) N_m(r) \right.\right. \\ & \left.\left. + \sum_p \sigma_{A_p}(\lambda) N_p(r) + \sum_i \sigma_{S_i}(\lambda) \, N_i(r) \right] dr \right\} \end{aligned} \tag{4.1}$$

where

$C(\lambda, R)$ is a system function determined by the geometric considerations of the receiver optics, the quantum efficiency of the detection system at wavelength λ, and the overlap of the transmitted laser beam with the field of view of the receiver.

A_r/R^2 is the acceptance solid angle of the receiver optics with a collecting area A_r.

$P_L(\lambda)$ is the average power in the transmitted laser pulse at wavelength λ.

ΔR is the effective range resolution of the lidar signal. This is limited to $c(\tau_L + \tau_d)/2$, where τ_L and τ_d are the laser pulse duration and the detector's integration period, respectively, and c is the speed of light.

$\Sigma_i \, \sigma_{B_i}(\lambda) \, N_i(R)$ is the backscattering coefficient, in which $\sigma_{B_i}(\lambda)$ is the backscattering cross section at wavelength λ and $N_i(R)$ represents the number density of scatterer species i.

The exponential factor accounts for the attenuation of the laser beam and the backscattered radiation. It is here divided into three summation terms, where the first two describe the extinction due to absorption by molecules and particles, respectively. Here $N_m(r)$ and $N_p(r)$ are the number density of absorbing species at range r, and $\sigma_{A_m}(\lambda)$ and $\sigma_{A_p}(\lambda)$ are the absorption cross sections. The third term represents the extinction due to scattering by molecular and particle species of number density $N_i(r)$ and scattering cross section $\sigma_{S_i}(\lambda)$.

This version of the lidar equation describes only single-scattering events. However, the contribution to the lidar signal from multiple scattering in high-turbidity atmospheres may sometimes be considerable. An adequate modification of the lidar equation is then not feasible, and quantitative measurements are much more difficult.

The temporal power profile of the transmitted laser pulse is not considered in Equation (4.1), and the effective range resolution, ΔR, is assumed to be small compared with the range R.

In very special cases there is one dominating absorbing molecule in a localized volume, while other extinction terms are negligible, and the scattering properties are approximately constant in the surrounding atmosphere. Then Equation (4.1) can be applied for a simple diagnosis of the molecular content in question. In other cases the increased content of a pollutant in a localized volume is to be monitored, and then a comparison can be made with the lidar signal for the surrounding reference atmosphere in the same or another direction. Strictly, this is a differential absorption technique based on lidar measurements at one wavelength only. The equations below can easily be adapted to cover this situation. However, in the general case a true DIAL measurement involving more wavelengths is required.

4.2.2 Principle of DIAL Technique

In a differential absorption measurement the attenuation by the atmosphere at an absorption wavelength of a certain molecule is compared with the attenuation at neighboring wavelengths, where the cross section for this specific absorption is considerably lower. The absorption caused by the molecule of interest can then often be separated from the attenuation associated with other atmospheric constituents.

Lidar signals, backscattered from the atmosphere or a target, are measured at two or more wavelengths in a DIAL measurement. In the case of two wavelengths, which so far is the most common approach, a DIAL equation can be formed as the ratio of the two lidar signals. One of the two wavelengths, λ_{abs}, is then normally chosen to coincide with the center wavelength of an absorption line of the molecule of interest. The reference wavelength, λ_{ref}, is detuned from λ_{abs}, so that the absorption cross section is considerably smaller.

The principle of the DIAL technique, when applied to measurements of a pollutant in an industrial plume, is illustrated in Figure 4.2. Two spatially identical laser pulses at λ_{abs} and λ_{ref} are transmitted into the atmosphere, and the corresponding backscattered signals are collected. The differential attenuation caused by the plume is reflected in the difference between the backscattered lidar signals for ranges beyond the plume. The two laser beams can be generated by the same laser or two different lasers.

We can often assume that the scattering factor and the extinction factor from the sum of particles are equivalent at neighboring wavelengths. From Equation (4.1) we then get

$$\frac{P(\lambda_{abs}, R)}{P(\lambda_{ref}, R)} = \frac{C(\lambda_{abs}, R)P_L(\lambda_{abs})}{C(\lambda_{ref}, R)P_L(\lambda_{ref})} \exp\left[-2\int_0^R \left(\sum_m [\sigma_{A_m}(\lambda_{abs}) - \sigma_{A_m}(\lambda_{ref})]N_m(r)\right) dr\right] \tag{4.2}$$

With the assumption that one molecular species dominates the absorption at these wavelengths, we can rewrite this equation in differential form and express the number density of absorbing molecules as

$$N_m(R) = \frac{1}{2\,[\sigma_{A_m}(\lambda_{abs}) - \sigma_{A_m}(\lambda_{ref})]} \frac{d}{dR}\left[\ln\frac{P(\lambda_{ref}, R)}{P(\lambda_{abs}, R)} + \ln\frac{C(\lambda_{abs}, R)P_L(\lambda_{abs})}{C(\lambda_{ref}, R)P_L(\lambda_{ref})}\right] \tag{4.3}$$

The integrated content of the absorbing species for a range increment $\Delta R = R_2 - R_1$ becomes

$$\int_{R_1}^{R_2} N_m(r)\,dr = \frac{1}{2\,[\sigma_{A_m}(\lambda_{abs}) - \sigma_{A_m}(\lambda_{ref})]}\left[\ln\frac{P(\lambda_{ref}, R_2)\,P(\lambda_{abs}, R_1)}{P(\lambda_{abs}, R_2)\,P(\lambda_{ref}, R_1)} + \ln\frac{C(\lambda_{abs}, R_2)C(\lambda_{ref}, R_1)}{C(\lambda_{ref}, R_2)C(\lambda_{abs}, R_1)}\right] \tag{4.4}$$

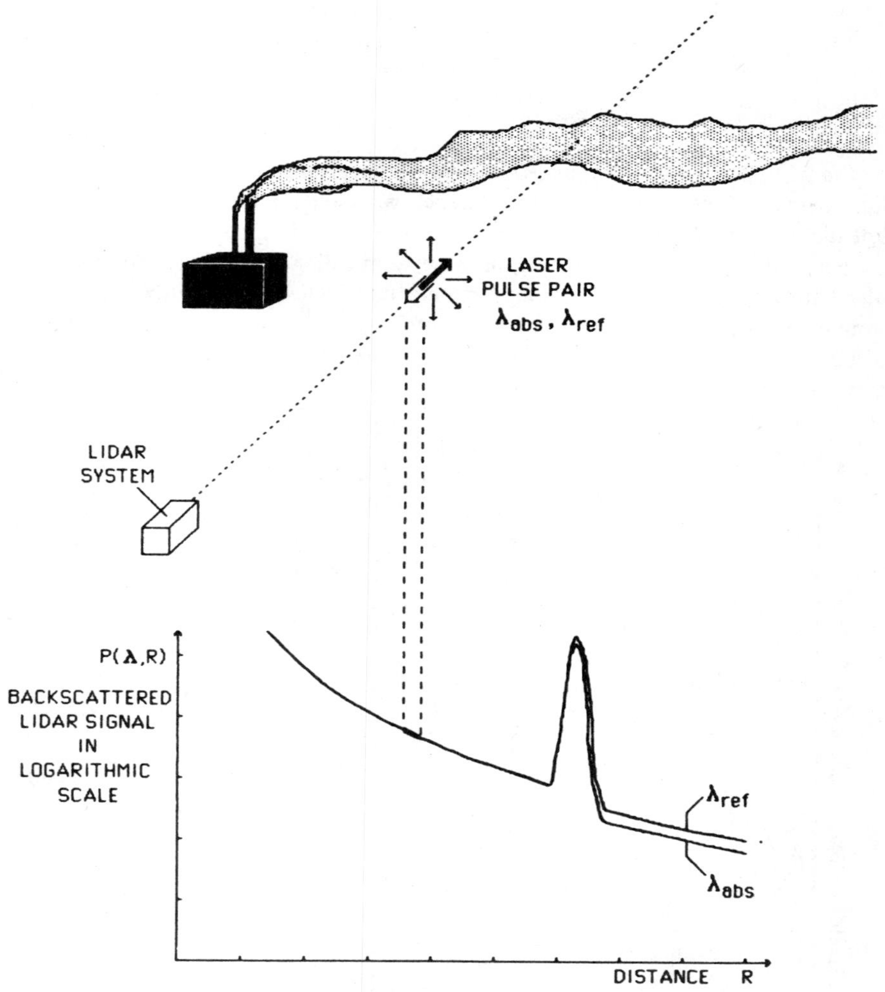

Figure 4.2. Principle of DIAL technique.

The second logarithmic term in this expression can in general be omitted. Only if the system constant has a wavelength dependence, which in turn is range dependent, does the term differ from zero. Such a dependence can be avoided through careful system design and appropriate measurement routines. The molecular number density, averaged over the distance $\Delta R = R_2 - R_1$ can then be written as a simple and useful expression:

$$\overline{N}_m = \frac{1}{2(R_2 - R_1)[\sigma_{A_m}(\lambda_{\text{abs}}) - \sigma_{A_m}(\lambda_{\text{ref}})]}\left[\ln \frac{P(\lambda_{\text{ref}}, R_2)P(\lambda_{\text{abs}}, R_1)}{P(\lambda_{\text{abs}}, R_2)P(\lambda_{\text{ref}}, R_1)}\right] \tag{4.5}$$

Figure 4.3 shows an example of a DIAL measurement of SO_2. Averaged lidar signals at the absorption and reference wavelengths are shown in the two left diagrams in the figure. The evaluation corresponding to Equation (4.2) is shown in the upper-right diagram, and the fourth diagram displays the molecular number density along the measurement path. The concentration was here evaluated for a range interval of 50 m, which was continuously swept with the distance, that is, the concentration was evaluated for the path $R \pm 25$ m for each range R.

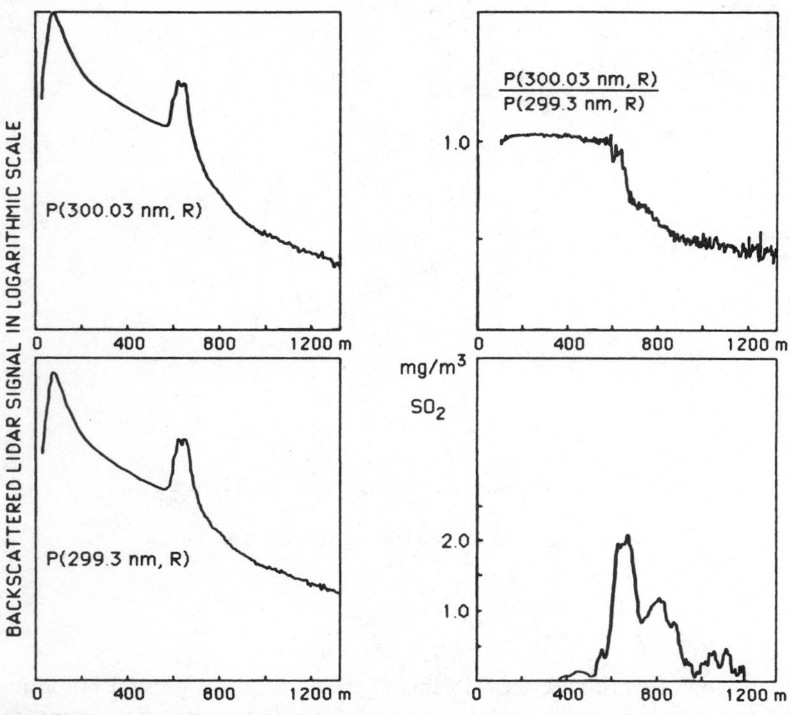

Figure 4.3. Example of DIAL measurement on SO_2 in plume. Lidar signals at absorption and reference wavelengths are shown in left diagrams. Upper-right diagram shows evaluated DIAL curve. SO_2 concentration evaluated with 50-m pathlength as shown in fourth diagram (bottom right).

The description given also applies, in the main, to the case where a topographic target or a retroreflecting mirror is employed as the scattering medium. Under these conditions there is a gain in sensitivity and range, but the range resolution is lost. Integrated concentrations along the path of the laser beam are then measured. The appropriate lidar equation corresponding to Equation (4.1) is

$$
\begin{aligned}
E(\lambda, R_T) = {} & C(\lambda, R_T) \frac{A_r}{R_T^2} E_L(\lambda) \frac{\rho_S(\lambda)}{\pi} \\
& \times \exp\left\{ -2 \int_0^{R_T} \left[\sum_m \sigma_{A_m}(\lambda) N_m(r) \right. \right. \\
& \left. \left. + \sum_p \sigma_{A_p}(\lambda) N_p(r) + \sum_i \sigma_{S_i}(\lambda) N_i(r) \right] dr \right\}
\end{aligned}
\tag{4.6}
$$

where $E(\lambda, R_T)$ = radiative energy received from topographic target at distance R_T, provided that integration period of detector, τ_d, is kept longer than laser pulse duration, τ_L

$E_L(\lambda)$ = transmitted energy of laser pulse

$\rho_s(\lambda)$ = backscattering efficiency of topographic target.

With the same assumptions as before, we can express the integrated concentrations along the path of the laser beam as

$$
\overline{N}_m = \frac{1}{2R_T[\sigma_{A_m}(\lambda_{\text{abs}}) - \sigma_{A_m}(\lambda_{\text{ref}})]} \left[\ln \frac{E(\lambda_{\text{ref}}, R_T) E_L(\lambda_{\text{abs}}) \rho_S(\lambda_{\text{abs}})}{E(\lambda_{\text{abs}}, R_T) E_L(\lambda_{\text{ref}}) \rho_S(\lambda_{\text{ref}})} \right]
\tag{4.7}
$$

In this expression we have included the wavelength dependence of the scatterer as this must be considered in many applications. When DIAL studies in the IR region are made with CO_2 lasers, the scattering efficiency of topographic targets in general differs for the different laser frequencies. However, it is not as critical for most DIAL measurements in the visible or the UV regions. The techniques employing atmospheric and topographic target scattering can then often be combined. The backscattered atmospheric return is studied for the shorter and medium distances, and for a longer distance, where this signal is weak, a target is used to improve the scattering efficiency.

In the reasoning leading to DIAL Equation (4.2), we made the assumption that the scattering factors are equal for the two lidar signals. This requires that the two laser beams are simultaneous and spatially identical. In practice, these requirements are difficult to fulfill exactly for two single-lidar measurements, but with proper signal averaging it is possible to carry out evaluations according to the equations above.

4.2.3 Factors Concerning Accuracy in DIAL Measurements

Accurate DIAL measurements require careful system design and data handling. Factors determining the accuracy and the sensitivity must be analyzed and optimized. A comprehensive error analysis for lidar has been presented by Measures (1984), and the accuracy optimization for DIAL systems has been discussed by Mégie and Menzies (1980) among others. Error sources in IR DIAL measurements have also been analyzed by Menyuk and Killinger (1983) and Menyuk et al., (1985). In this context we will point out and discuss the potential sources of error in the DIAL technique and place emphasis on the details of special relevance in pollution mapping. In Table 4.1 factors affecting the accuracy of DIAL measurements are listed. The table is somewhat of a check-list for realistic DIAL measurements, and though many potential sources of error are pointed out here, it is often possible to reduce their influence with careful system design and proper measurement routines. First we will summarize the limiting factors for the signal-to-noise ratio.

Table 4.1. Factors Determining Accuracy in DIAL Measurements

Signal-to-noise limits:
Signal shot noise
Background noise
Dark-current noise
Factors related to absorption properties:
Uncertainties in absorption cross sections
Overlapping absorption profiles
Uncertainties in laser wavelengths
Effects due to finite laser linewidth
Differential spatial-, temporal-, and wavelength-dependent factors:
Differential overlap functions and misalignment effects
Differential temporal power profiles
Fabry–Perot effects
Wavelength-dependent scattering properties
Factors determined by atmospheric conditions and detection system:
Multiple-scattering effects
Temporal noncorrelation
Temporal and spatial averaging
Characteristics of photodetectors and amplifiers
Effects of data acquisition

4.2.3.1 Signal-to-Noise Limits

The magnitude of the signal-to-noise ratio in DIAL measurements is determined by three factors: the signal shot noise, the background noise, and the dark-current noise. The signal shot noise is the statistical fluctuation of the scattered radiation and the detected signal. This may be comparatively large in a single-lidar measurement, and it is often not meaningful to evaluate single laser pulse pairs. In the extreme case the lidar signal is even captured as discrete single-photon scattering events in a photon-counting manner. Thus, several signals are normally averaged to increase the signal-to-noise ratio of the calculated DIAL signal. This must then be performed with great care to avoid the systematic errors discussed below. The spatial and temporal stability of the laser transmitter and the atmospheric conditions are important parameters related to the shot-noise limit. Signal amplifiers, employed in the detection system, may also influence the signal-to-noise ratio in some cases.

The background radiation from the sky or a topographic target introducing the background noise can be negligible under favorable conditions. In nighttime measurements and in spectral regions with low background illumination this does not usually affect the signal-to-noise ratio. However, the background radiation in the visible, the near-UV, and the near-IR regions may very well be greater than the backscattered laser light under daylight conditions. This must then be handled with delicate spectral rejection and a narrow field of view of the receiver. Electronic rejection of the collected signal is also often needed to suppress the average level of the background illumination.

The dark current of the detector limits the signal-to-noise ratio only in IR DIAL measurements, where solid-state detectors are employed. Photomultipliers can be used in the UV and the visible regions, and these generally have a very low dark current.

The detection limits determined by the signal-to-noise ratio have been analyzed in realistic measurements by Fredriksson and Hertz (1984), who used a one-laser DIAL system. In the study 280 pairs of lidar signals generated by a laser with low power (1–2 mJ) were averaged. Their results for NO_2 measurements in the visible region with two different background illuminations under daylight conditions are shown in Figure 4.4. In the diagram the product of the gas concentration and the absorption path is given as a function of the range. It can be seen that the detection limit due to noise is about 15 $\mu g/m^3$ for a 300-m absorption path below 1 km, but it increases quite drastically for greater distances when the background is a clear sky.

A similar study was made by Staehr et al. (1985). They used a two-laser DIAL system with high laser powers (100–150 mJ). Their results, from a study where 120 pulse pairs were averaged, are shown in Figure 4.5. With no special regard to the background noise, they showed that under favorable conditions

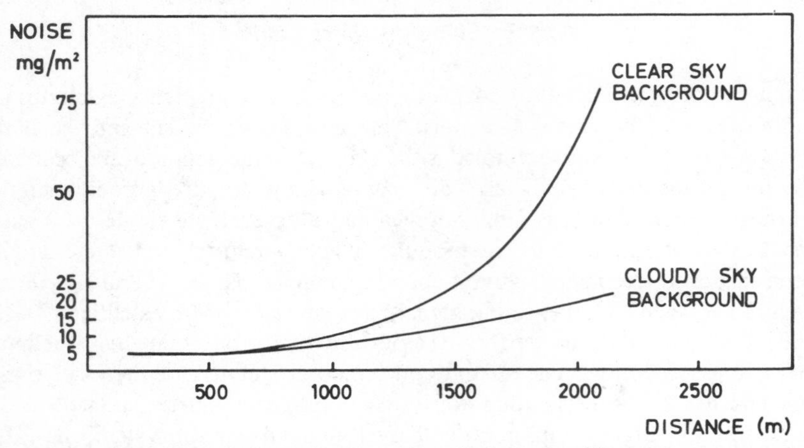

Figure 4.4. Absolute statistical error introduced by noise in daytime NO_2 DIAL measurements for two typical cases. Error given as product of concentration and absorption pathlength used for evaluation (Fredriksson and Hertz, 1984).

the detection limit for NO_2 over a 300-m path can be as low as 20 $\mu g/m^3$ up to ranges of 6 km.

4.2.3.2 Factors Related to Absorption Properties

From the equations above it is obvious that the accuracy in a DIAL measurement is directly related to the accuracy of the absorption cross-sectional values for λ_{abs} and λ_{ref}. Those molecules having strong electronic absorption lines in the visible or the near-UV regions are well suited in this respect. Absorption cross sections can then be determined with high accuracy in laboratory measurements, and with proper measurement routines in field work these can be employed without introducing any appreciable errors. However, many molecules do not possess such suitable absorption features in this wavelength region but have their strong electronic absorption lines below $\approx$220 nm, where absorption and scattering in the atmosphere completely prevent lidar measurements.

Essentially all molecules have vibrational–rotational absorption lines in the IR region, which offers other possibilities for DIAL studies. Unfortunately, these lines are subject to collision broadening, which sometimes destroys the sharp absorption profile, and also makes the absorption cross-sectional determination more difficult. Overlapping by absorption profiles of other molecules is an even more serious problem in the IR DIAL technique. Nevertheless, there are many instances where suitable differential absorption cross sections are available.

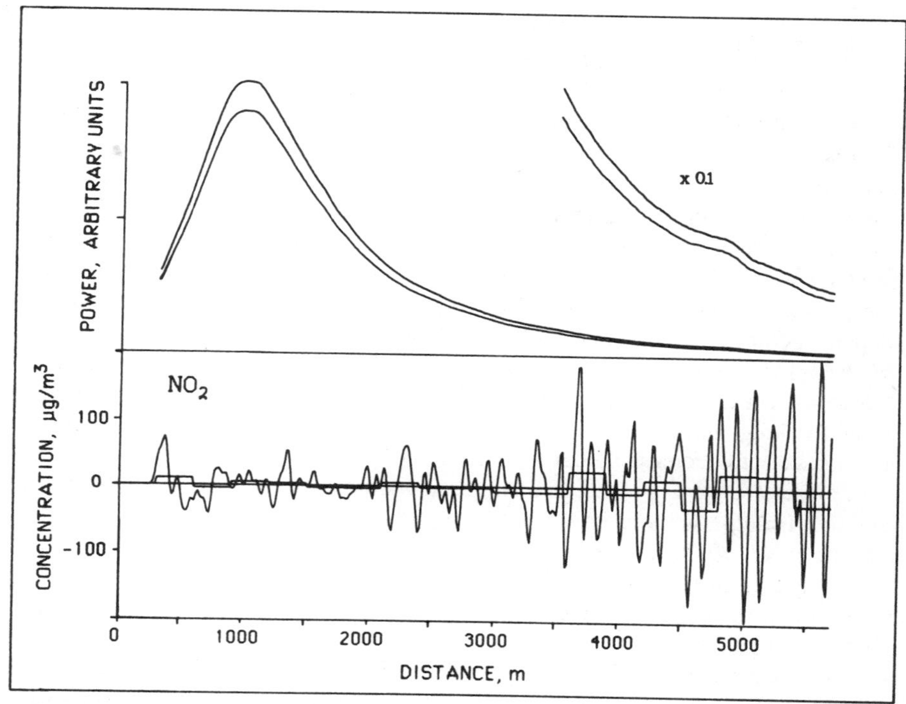

Figure 4.5. Statistical error in NO_2 measurements for two-laser DIAL system measured with both lasers tuned to same wavelength. Upper diagram shows averaged lidar signals for 120 pulse pairs; in lower diagram deviations from zero concentration are shown. Evaluations corresponding to Gaussian averaging function for 50-m pathlength and discrete 300-m absorption paths are shown (Staehr et al., 1985).

The requirement for distinct absortion cross-sectional values demands also a careful calibration of the laser wavelengths. In IR DIAL measurements and when gases with narrow absorption lines in the UV region are measured, the exact determination of the wavelengths can be of crucial importance for the accuracy. It is also required that the linewidth of the laser is narrower than the probed absorption features. This is often not too difficult to fulfill when electronic transitions are studied, even if there are exceptions. However, the vibrational–rotational absorption bands in the IR spectrum are normally extremely narrow, and then the absorption cross section may vary within the laser linewidth. The error introduced by finite laser linewidths has been calculated by Cahen and Mégie (1981) for several realistic cases. They conclude that this effect must be taken into account in IR DIAL measurements.

4.2.3.3 Differential Spatial-, Temporal-, and Wavelength-Dependent Factors

In the ideal case the DIAL expression does not depend on the range except for the distribution of the studied gas along the measurement path. This assumption was made for the derivation of Equations (4.5) and (4.7). In the realistic case there may be another differential range dependence in the lidar signals. As is indicated in lidar Equations (4.1) and (4.6), the system function C may be range dependent, and for short ranges this certainly is the case as the collecting efficiency here must be restricted due to the geometric optics of the receiver. This dependence can be expressed with an overlap function, describing the overlap of the transmitted laser beam with the effective field of view of the receiver telescope. Most DIAL systems are designed so that the overlap function is constant for long ranges, but for short ranges the overlap is sometimes limited intentionally to decrease the large dynamic range of the lidar signal. This effect is often referred to as geometric compression and has been discussed by several authors (Halldórsson and Langerholc, 1978; Harms et al., 1978; Harms, 1979; Sasano et al., 1979; Sassen and Dodd, 1982).

A misalignment of the laser beam and telescope axis also induces a range dependence in the system function and the overlap function. In principle, a slight misalignment gives no systematic error in DIAL measurements but merely reduces the signal-to-noise ratio because of the lost efficiency in the detection. However, if one of the wavelengths has a different overlap function compared with the other, there will be a range dependence in the DIAL expression. It can be difficult to align the two laser beams if different laser sources are used to generate the two wavelengths, and even a diminutive deviation may then cause a systematic error.

Differences in the beam direction may occur even if one laser is used to generate the two beams. Prisms and other wavelength-dispersive elements in the laser beam path are potential sources of differential misalignment, which may change the overlap functions. These effects are normally negligible for neighboring wavelengths.

In measurements at short distances it is also essential that the spatial intensity distribution of the laser beam be the same for the two wavelengths since the telescope field of view here is limited. A potential systematic error due to this geometric compression is more likely in a two-laser DIAL system, in which the two beam profiles are individually determined. However, the same problem exists in a single-laser system, where crystals are employed for frequency conversion of the transmitted beam (e.g., frequency doubling of visible to UV radiation or frequency mixing to UV or IR radiation). The individual tuning of these crystals for the laser beams then determines the beam profile. In principle, the same difficulty could also arise in the wavelength tuning of the laser itself.

From this discussion, on the effects due to misalignment and differences in the spatial distribution of the laser beams, it follows that the geometric compression of the lidar signal must be dealt with very carefully in a DIAL system to avoid systematic errors. A further restriction associated with geometric compression arises if multiple scattering contributes to the lidar signal. This contribution will constitute a larger portion of the signal for the ranges where the single backscattering from the beam is obscured.

When topographic targets are utilized as the scattering medium, a similar problem may arise. The backscattering properties of the topographic target must be identical for the two laser beams, and this requires identical spatial distributions of the beams if the target is nonuniform. When small targets or retroreflecting mirrors are employed, substantial errors can be introduced in this way.

It is also essential that the two transmitted laser pulses have an identical temporal power profile. If, as an example, one of the pulses has a longer pulse duration or a longer-lasting tail, the lidar signals attributed to a certain range R will represent an extended path, ΔR, compared with the other. This results in systematic errors because of the strong range dependence of the lidar signal. The effect must be observed especially when two laser sources are used.

When a light beam is transmitted through an optical window, there are, in general, wavelength-dependent intensity modulations due to interference or Fabry–Perot effects. In DIAL measurements where the backscattered intensities from different ranges are compared, the signals for the two ranges are transmitted through the same optical components and are subject to the same intensity modulations. However, if the laser power is measured before it is transmitted and used for the calculation of the concentration from zero distance, the signals may be influenced to a different degree by intensity modulating optical components. These effects can be large compared with the studied gas absorption and must be carefully examined in each such case. A similar situation exists when calibrations of DIAL systems are made with gas cells, where cell windows can introduce the same kind of error.

The backscattering and extinction factors in the lidar equation were considered to be the same for the two laser wavelengths in the derivation of DIAL Equation (4.2). This presumption is true on the average for neighboring wavelengths if there are no interfering absorbing gases. Backscattering and extinction from particles generally have a smooth wavelength dependence as they are well distributed over different sizes and shapes. However, Petheram (1981) and Yue et al. (1983) have examined the backscattering coefficient for certain pairs of CO_2 laser lines and showed that this is a sensitive function of line frequency and relative humidity. The dependence is due to the IR absorption by scattering aerosol particles, which is partly determined by their chemical composition. This implies the need for a continuously tunable IR laser source or that several carefully chosen CO_2 laser lines be employed for a DIAL measurement. The same need of tunability and several laser lines is often required by the overlapping absorption features of different molecules in the IR region. These circumstances are more favorable in the visible and in parts of the UV region, where there also exist continuously tunable sources today.

The wavelength dependence of the scatterer in DIAL measurements employing a topographic target was included in Equation (4.7) and must be considered in IR DIAL measurements in particular. In airborne nadir-pointing DIAL systems there are also differential albedo effects, which are often of great importance. These dependencies have been examined by Wiesemann et al. (1978), Boscher et al. (1980), Shumate et al. (1982), Grant (1982a), and Asai and Igarashi (1984).

4.2.3.4 Factors Determined by Atmospheric Conditions and Detection System

In the preceding discussion, we made the assumption that the lidar signals are made up of single-scattering events. However, multiple scattering may contribute considerably to the backscattered light in high-turbidity atmospheres. When measuring in a dense plume, the lidar signals attributed to the ranges behind the plume may, to some extent, be due to multiple-scattering events, and then a too low concentration will be inferred for the gaseous content. Appropriate measurement routines can usually be employed when concentrations in such plumes are to be studied. The same source of error appears in studies on distributed gases in very hazy and foggy atmospheres. It is then extremely difficult to make corrections for these effects, and thus realistic measurements are prevented under these atmospheric conditions. The problem is more serious for the shorter wavelengths in UV DIAL measurements, where the scattering cross sections are much larger.

Since each single lidar signal is unique and reflects the time-varying atmospheric conditions, two comparable recordings should be made at the same time or at least within tens of microseconds. Menyuk and Killinger (1981) and Menyuk et al. (1985) have examined this requirement for temporal correlation. When two laser sources are employed, this temporal correlation can be achieved, but then an exact correlation is in general prevented by the presence of a spatial difference in the density distribution for the two beams. In other measurements, when a single laser is used for both beams, the temporal requirement is often not fulfilled. It has been shown by several research teams that in most cases these limitations can be overcome by signal averaging. Usually, some period of averaging is needed anyway to gather enough statistics, and then a time separation of 0.1 s between sequential lidar signals does not appreciably influence the accuracy.

The averaging procedure itself may be a systematic source of error. Generally, the lidar signals in the form of Equation (4.1) or (4.6) are averaged, but the molecular concentration is proportional to minus the logarithm of this signal. In a stable atmosphere no error is introduced, but if the concentration varies greatly, a lower value than the actual mean concentration will be measured. If, on the other hand, the logarithm of the signal is averaged, there can be a systematic error due to the noise component of the signal. The extreme of this source of error is when there is a complete cutoff in only some of the lidar signals due to varying attenuation in a plume. The averaged lidar signal will then still have a nonzero value behind the plume, but it will of course give no information on the mean concentration. These errors are avoided by applying appropriate measurement routines; for example, measurements of a plume should be made at some distance from the source where the plume has spread somewhat.

An error can also arise due to the averaging procedure if the relative power of the two laser beams changes during the measurement. This effect is especially important when studying an inhomogeneous atmosphere, where the scattering and absorption properties change, for example, in an industrial area with several diffuse particle plumes. Then a change in the laser power of one beam relative to the other might coincide with a large change in these properties, and the averaged lidar signals will not be comparable.

When we wrote the lidar Equation (4.1), we made the asumption that the range resolution, ΔR, was short compared with the range R. If this is not the case because of a long-lasting laser pulse or a comparatively long integration period in the detection system, the equation should be expressed as an integral over ΔR. The proper evaluation routines then become more complicated as a lidar signal cannot be attributed to one range but will be scattered from an extended path. In physical situations, where there are several localized sources, only long-path concentrations will then be measurable.

The photodetector and the data acquisition electronics employed in a DIAL system are delicate instruments. Photodetectors in general do not give a signal strictly proportional to the number of photons collected. The nonlinearity is worse for strong signals and must then be compensated for. A signal amplifier might also have a nonlinear response as well as the transient recorder used. Zero drifting, limited effective bandwidths, and limited effective dynamic ranges of fast amplifiers and fast recorders are other potential sources of error. The triggering of transient recorders must be precise, as well as the modulation electronics for the detection efficiency, if such a device is employed. These and other similar effects must be carefully examined when designing a DIAL system. The effects are often quite easy to check in connection with field measurements, and then appropriate measurement routines can be chosen and adjustments can be made.

4.3 DIAL SYSTEMS FOR POLLUTION MONITORING

A DIAL system consists basically of a laser transmitter, a receiver, and electronics for data acquisition and system steering. In this section we will discuss the main components of a DIAL system and give examples of the equipment in use.

4.3.1 Laser Systems

The laser transmitter is the most critical component of a DIAL system. The ideal pulsed laser source produces short pulses of high energy and operates at a high repetition rate. It possesses a narrow spectral bandwidth, and the laser frequency is continuously tunable. Further, the transmitted beam has a low divergence. The shot-to-shot fluctuations in the laser energy should be small, and the stability of the laser frequency is critical. Fieldwork also requires reliable lasers that are easy to handle. The development of laser techniques for measurements on pollutants has, to a great extent, been dependent on laser performance in these respects. Currently there are several types of laser that have been employed for DIAL studies.

The laser source used in the UV and visible DIAL is a dye laser. In such a laser the active medium is a solution of organic molecules, which can be made to lase over a limited wavelength range, typically 30 nm. With different dyes it is possible to cover the range from the near UV to the near IR, (400–800 nm). The dye laser is pumped by either another laser or fast flashlamps. Laser pumping with Nd–YAG lasers is the common method in DIAL systems, but there are also excimer- and flashlamp-pumped dye lasers. Dye lasers pumped by nitrogen lasers give too low energies for most DIAL applications but have been employed for the development of the technique. The useful Nd–YAG laser radiation for pumping is obtained by harmonic generation. This yields either frequency-doubled radiation at 532 nm or a tripled output at 355 nm. The former is suitable for pumping yellow and red dyes and the latter for blue dyes.

The wavelength range of the dye laser can be extended by employing frequency-doubling techniques. Ultraviolet radiation is effectively generated with standard doubling crystals that also maintain the beam quality. The most frequent DIAL application has been the measurement of SO_2, employing such Nd–YAG pumped dye lasers. A schematic diagram of such a laser is shown in Figure 4.6. This setup is a standard dye laser with an extra tuning element to make the laser run at two wavelengths simultaneously. The organic dye laser cavity is pumped by a fraction of the 532-nm output from a Nd–YAG laser, while the main part of this energy pumps an amplifier dye cell. In this way the linewidth and beam quality are favored by the low energy in the cavity, and the high-energy output is obtained in the amplifying stage. The wavelength selection is accomplished with an echelle grating, which is rotated for tuning.

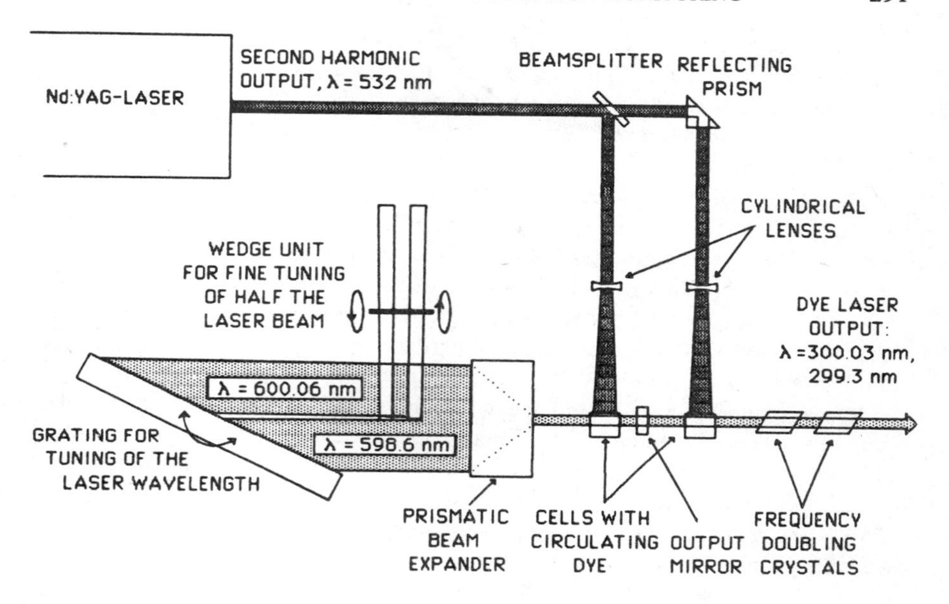

Figure 4.6. Condensed schematic diagram of Nd–YAG pumped dye laser operating at two wavelengths simultaneously. Wavelengths correspond to absorption and reference wavelength in SO_2 DIAL measurement (Fredriksson, 1985).

It would be a great advantage if the two lidar signals could be measured simultaneously with only one laser. Several methods for operating a dye laser at two wavelengths simultaneously have been reported (Schmidt, 1975; Lotem and Lynch, 1975; Lotem et al., 1976; Marx et al., 1976; Inomata and Carswell, 1977; Chandra and Compaan, 1979; Nair and Dasgupta, 1980; Sage and Aubry, 1982; Fredriksson, 1985). In all these configurations the beam in the dye laser cavity is split into two beams, which can be tuned separately. A wedge plate, two separate gratings, or a configuration with a beamsplitter and mirrors can be utilized to split the beam and create a cavity with two optical resonances. The extra tuning element in Figure 4.6 consists of two rotatable wedge plates, which change the wavelength of half the beam by changing its angle of incidence and reflection at the grating (Fredriksson, 1985). The wavelength is changed with high precision by rotating the two wedge plates in opposite directions. In this way the fundamental visible laser radiation for the absorption and reference beams can be generated simultaneously. Two frequency-doubling crystals, angle phase matched to the absorption and reference beams, respectively, are used for conversion of the visible output to the UV wavelengths in the SO_2 spectrum in this case.

The wedge arrangement in Figure 4.6 is also an example of a fast tuning device for a dye laser. The wedges can also cover the whole beam, and the laser wavelength can be changed between each shot, as is normally done in many DIAL measurements. The rotation of the wedges is accomplished with fast stepping motors.

The mixing of frequencies in nonlinear crystals is another technique used to increase the range of a dye laser. Frequency mixing with the radiation for the pumping Nd–YAG laser extends the feasible tuning range down to 220 nm and up to about 4 μm. However, the linewidth of the resulting radiation is generally too large, and thus spectral narrowing etalons are required in the laser cavities. This and other practical aspects make the mixing technique too complicated for the field work at the moment.

It is also possible to extend the wavelength range by applying stimulated Raman scattering techniques. A Raman shifting device in combination with a dye laser has a very broad spectral range, and a narrow linewidth can be maintained, but the available pulse energies at suitable wavelengths for DIAL monitoring are rather low. Combinations of Raman shifters and excimer lasers, which have a limited tuning range, have also been proposed, but the usefulness for pollution studies is yet to be determined.

An optical parametric oscillator (OPO) is a continuously tunable device for the near- and mid-IR region, which uses a crystal to down-convert the radiation from a Nd–YAG laser to the 1.4–4-μm region. The technique has to date only been used in basic research and is not easy to use in field measurements.

Hydrogen fluoride and deuterium fluoride lasers are gas lasers with tunability to distinct lines in the mid-IR region. In the cases where the absorption profile of a molecule coincides with the laser lines in the region 2.6–4 μm, they may be well suited for DIAL applications.

Semiconductor lasers have a limited applicability to DIAL, but they are being developed, and one can expect them to become more useful in measurements on gases in the future. A semiconductor laser can have a tuning capacity, and with a selection of lasers it is in principle possible to cover the wavelength region from 600 nm to 34 μm. The laser has a high efficiency, but the pulse energy is very low. Another disadvantage is the large divergence of the laser beam. This indicates that they will not be useful for DIAL remote sensing if *remote* is interpreted as a distance of kilometers, but in compact systems over short ranges they may prove to be very useful. The attractive possibility of mass-produced low-price lasers points to this, but the requirement on cryogenic cooling of the lasers is a disadvantage in a compact field system.

Much of the early work on DIAL was carried out with CO_2 lasers. These lasers can be either continuous wave (CW) or pulsed and are generally tunable over a large number of lines in the 9.1–11-μm region. They are very efficient lasers, producing high-power beams of adequate quality. Carbon monoxide lasers, which are gas lasers similar to the CO_2 lasers, have lines in the 5–7-μm region. The line-tunable lasers rely on the coincidences between the laser lines and the absorption lines of the pollutants. This is a serious restriction in DIAL due to overlapping absorption profiles and wavelength-dependent scattering properties discussed in the preceding section, and fieldwork is therefore limited to a few special gases and measurement situations. The continuously tunable CO_2 lasers, operating at high pressures, have so far only seen limited application.

Although frequency mixing techniques for CO_2 lasers are at a preliminary stage of development, the very large range in wavelengths that can be generated in the mid-IR with nonlinear crystals make this approach very attractive. It will also be possible to tune the sum frequency almost continuously with this technique.

There are many interesting new laser sources and laser techniques being developed. Intracavity spectral narrowing techniques to match requirements on narrow linewidths will certainly be improved, and injection frequency-locking techniques will be developed further. Repetition rate and pulse energy will be increased, and the efficiency of lasers will be improved. The solid-state lasers are now optically pumped with flashlamps, but there are more efficient pumping techniques under development. Very interesting and promising developments of Nd–YAG and other solid-state lasers are in progress, which will certainly extend the measurement capabilities with dye lasers. The alexandrite laser is a recently developed tunable solid-state laser for the 700–800-nm region, and development is in progress of such tunable lasers for other wavelength regions. It is likely that mixing and Raman shifting techniques will be applied in combination with different laser sources in the near future.

The calibration of the laser wavelengths in DIAL measurement is critical. The calibration can be made with a monochromator or a gas cell containing the gas studied or another gas with a suitable spectrum. Recently, glow discharge lamps have been applied as calibration units employing the optogalvanic effect (e.g., Nestor, 1982; Egebäck et al., 1984). These are very precise, easy to operate, inexpensive accessories to a DIAL system. A dedicated wavelength control system with active stabilization of a dye laser was developed by Cahen et al. (1981) and was incorporated in an airborne DIAL system by Browell et al. (1983).

The optical components for transmitting and steering the laser beam are normally right-angled prisms, dielectric mirrors, or metallic mirrors. It is worth noting that aluminum mirrors cannot be used for concentrated short pulses of high power (e.g., from a Nd–YAG pumped dye laser) since the aluminum layer will evaporate. In IR lidar systems gold mirrors are commonly used. A beam-expanding telescope is sometimes used to increase the diameter and decrease the divergence of the transmitted beam, which will make it possible to decrease the power on system optics and also to use a smaller receiver field of view. Beam expanders for high-power systems must be of the Galilean or similar type with no focused real images to avoid breakdown effects.

In fieldwork one also has to consider eye safety when transmitting laser beams. This is of particular importance in pollution monitoring as the beams are often transmitted close to the ground in urban areas. In many countries there are regulations for laser applications, and in any case there is a practical limitation on DIAL measurements. The most severe restriction on the laser output power is in the 400-nm to 1.4-μm region, where the cornea, the lens, and the vitreous body of the eye transmit. One of the most important pollutants, NO_2, is preferably measured at 448 nm, where the eye is very sensitive. However, in this case measurements have also been done with low-power lasers avoiding danger to the public. Besides limiting the output power it is possible to expand the laser beam before it is transmitted and thus keep the exposure level to a target of eye size low.

4.3.2 Directing and Receiver Systems

Several coaxial DIAL systems for pollution mapping are equipped with a large mirror to steer the laser beam and to collect the backscattered light. The laser and telescope are then fixed in the system, and the telescope is generally vertically mounted, as is shown in Figure 4.7*a*. The systems can be easily equipped with a quartz window for weather protection. The steering mirror, in general aluminum coated, must be protected against the high-power pulses as discussed above. Often the limited laser power and the beam divergence suffice to limit the spot intensity. Otherwise, this can be achieved with a beam expander. It is also recommended that an extra small mirror be mounted on the large mirror, which can be easily changed in the case of damage. Similar systems without the steering mirror are applied in measurements on vertical profiles.

Other systems have a fixed laser and a movable telescope, which is directed toward the target. The laser beam is then transmitted via a set of mirrors or prisms along the rotation axes of the telescope and made parallel to the telescope. The main disadvantage of this system design is that the laser power will be reduced due to multiple reflection losses. In practice, it is also difficult to maintain the overlap of the laser beam and telescope when the direction is changed.

The third type of system has a movable platform for the laser and the telescope, and the entire platform is directed toward the target. This requires a large and sturdy system if dye lasers or other large lasers are employed. The design will certainly be convenient for future compact laser systems. In an airborne DIAL system the movable platform is the entire aircraft.

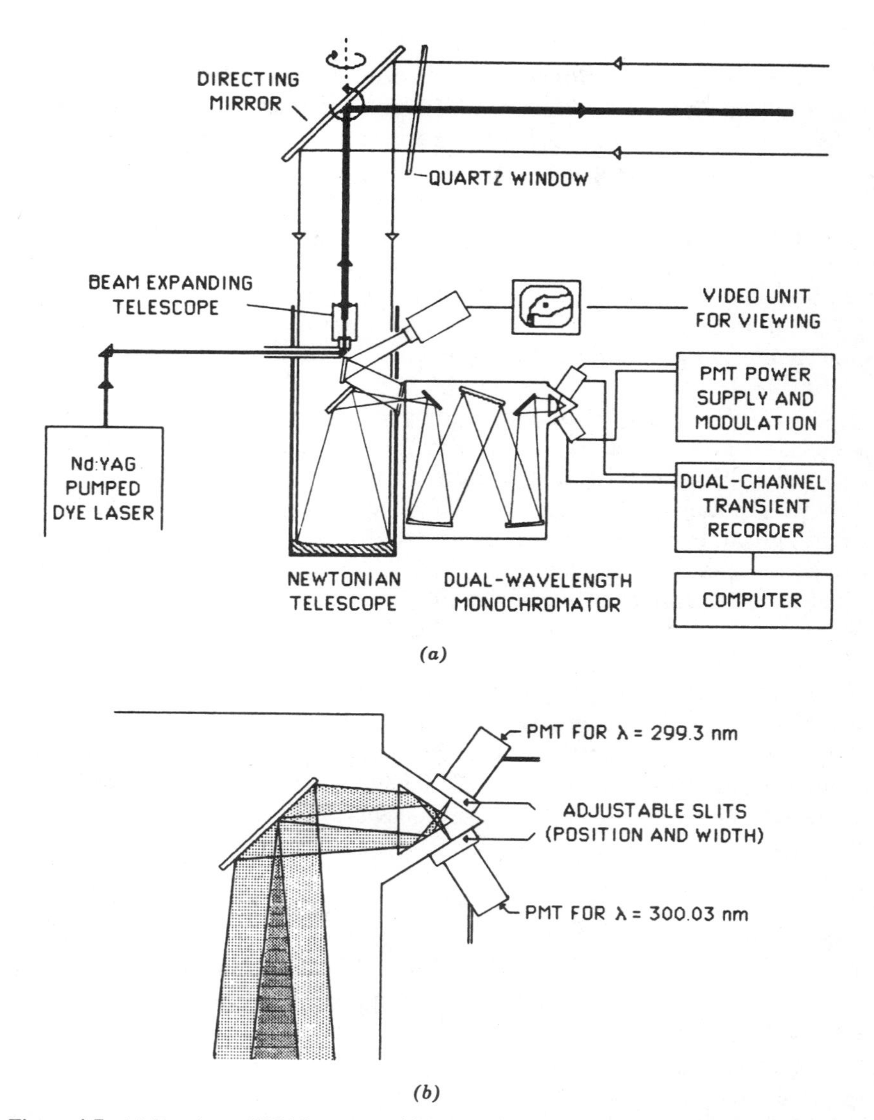

Figure 4.7. (*a*) Receiver of DIAL system with Newtonian telescope. Monochromator is designed for measurements at two wavelengths simultaneously. (*b*) Detail of monochromator with prism separator for two wavelength regions. Dispersion is exaggerated for sake of clarity. Wavelengths noted correspond to SO_2 DIAL measurement (Fredriksson, 1985).

The receiver telescope is of the Newtonian, Cassegrainian, or Gregorian type with a large parabolic mirror. A medium-sized telescope can have a spherical mirror as well if the field of view is small. As an alternative to the large mirror, a configuration with a Fresnel lens is sometimes used, which has the advantages of being lighter and cheaper.

The Newtonian configuration is exemplified in Figure 4.7*a*. The mirrors in the telescope, as well as any steerable mirrors, are coated with aluminum to cover the entire range up to 11 μm, but in the IR a gold coating may be superior. The aluminum itself is coated with a layer of magnesium fluoride or silicon dioxide to enhance the reflectivity in the UV region. The latter is also a good protective coating that will improve the durability and simplify cleaning operations.

A video system can be used for guiding when directing the lidar system, as is indicated in Figure 4.7*a*. A mirror in the focal plane reflects the telescope view, except for the lidar field of view, to a video camera. The lidar field of view will appear as a black spot on the video screen. A rotating device (e.g., a Dove prism) is needed to rotate the picture when the large directing mirror is rotated. Aligning the laser and the receiver telescope can present problems, which have been solved practically in different ways for existing DIAL systems. A precision method of aligning IR DIAL systems has been proposed by Oppenheim and Menzies (1982).

The telescope can be designed to reduce the collected signal from short distances and thus decrease the dynamic range of the lidar signal. There is always some geometric compression, which prevents the near-field scattering from overloading the photodetector, but because of the potential systematic errors discussed, the effect also makes measurements at short ranges more difficult. Thus, a larger geometric compression is hardly an improvement in the general case.

The multiple-scattering contribution to the lidar signal can be partly removed with a polarizer in the detection system, taking advantage of the fact that the plane polarization of the laser beam is retained only in the single scattering from spherical particles (e.g., Pal and Carswell, 1976, 1978). However, the applicability for a DIAL system in field conditions has yet to be examined in detail.

In general, the receiver must be equipped with a spectral discriminator or a spectrum analyzer to suppress the background light. In the UV, visible, and near-IR regions narrow-bandpass filters are universally used as discriminators. Interference filters are the common type, but Fabry–Perot interferometers are sometimes preferable, giving a higher resolution. Absorption filters can be added to improve spectral rejection. Combinations of filters are often employed, and it is in principle possible to separate the wavelengths of the two lidar signals with these techniques. The interferometric elements are designed for parallel beams, and a suitable lens is thus required in the light path for optimal enhancement of the desired wavelength.

A grating monochromator is an alternative to filters for selection of a wavelength interval for detection. It is more universal in the sense that one device in principle covers the whole spectral range, but it demands careful design of the receiver system to be effective enough. There are gratings that can be employed in simple and compact units in a DIAL receiver. Measurements on spectral profiles will be easier with a well-designed grating configuration, and the range of application of such a system will be improved.

There is also the need to separate the two lidar signals in a DIAL system with simultaneously transmitted beams at the two wavelengths. This can be done with a dispersive system of the type described by Fredriksson (1985), which is included in the system portrayed in Figure 4.7*a*. Here the wavelength separation after a diffraction grating is accomplished with a 60° prism as is schematically shown in Figure 4.7*b*. Two slits with adjustable positions and widths cover the wavelengths shorter and longer, respectively, than the center wavelength, which is determined by the grating and the prism edge position. This is an effective way of separating neighboring wavelengths, and even with large slits the wavelength ranges will be totally separated.

The background noise in IR lidar is mainly due to thermal radiation and increases with the detector area and the receiver field of view. According to Brassington (1983), there is little use in employing narrow-band filters for wavelengths greater than 3 μm. Liquid-nitrogen-cooled shields can be used to reduce the field of view of the detector.

There are two different modes of operation applied in IR laser technology. The detection technique can be either direct detection or heterodyne detection. Direct detection is the simple measurement of the number of photons detected by the receiver as a photocurrent, similar to the UV and visible techniques. In the heterodyne detection, or coherent lidar, a local oscillator is mixed with the lidar signal to produce an intermediate frequency signal. This improves the detectability of the scattered light but puts additional requirements on the lidar system concerning the frequency characteristics, beam divergence, and overall complexity.

A heterodyne system demands a more careful configuration of the receiver since the phase of the detected light must be matched to the local oscillator. The difficulties increase with the increasing size of the receiver and the field of view. The spectral discrimination is provided by the local oscillator, and very narrow spectral widths are then obtainable.

A photomultiplier tube is the best lidar detector for the UV and visible regions. It is possible to select a photomultiplier that has high gain, high quantum efficiency, and low dark current. The rise-time and frequency response of a fast photomultiplier (focused type) can be comparable to the digitizing rates of the very fast transient digitizers (100–500 MHz). The photomultiplier can be designed to be very close to linear when operated at low peak currents, but it can be very nonlinear in other cases. The specifications for the photodetector are very important and must be examined carefully in every DIAL system.

Often, no preamplifier is required, and the anode current can be fed directly to the transient recorder. This is advantageous since an extra preamplifier adds noise to the signal and may also decrease the bandwidth and be nonlinear.

The strong variation of the lidar signal with the range (an approximately $1/R^2$ dependence) can be partly compensated for by the photomultiplier tube. The high-voltage supply of the multiplier dynode chain or the focusing electrode is then gated or modulated so as to change the gain during the backscattering of the lidar signal. Thus, strong signals from short ranges are detected with low gain, while the weak signals from long ranges can be measured with a higher gain. Mechanical or electro-optical shutters to block the light path to the photodetector are alternatives to gating in some cases. For lidar over medium ranges a modulation of the high-voltage supply with a shaped voltage pulse to even-numbered dynodes, according to the proposal by Allen and Evans (1972), is effective. Besides decreasing the dynamic range of the signal, the change in gain is sometimes essential to keep the mean current due to background light on a reasonable level.

Semiconductor detectors, designed for optimal performance in a certain wavelength region, are employed for IR DIAL systems. The common types are InSb for atmospheric transmittance windows from 1–5.3 μm and HgCdTe for the range 8–13 μm. The detectors have to be cryogenic cooled to liquid nitrogen temperature (77 K), and preamplifiers, which invariably introduce noise, are required. The performance of a photodiode is, in general, a trade-off between bandwidth, detector size, and detector noise. The development of detectors for the IR region is progressing rapidly, and improvements on system designs can be foreseen.

A transient digitizer is normally employed to capture the lidar signals. There are digitizers available with a 500-MHz sampling rate, but the effective resolution of these is rather poor. The common instruments in DIAL systems are 50–100-MHz digitizers with an 8-bit digital resolution and 5–10-MHz digitizers with 10- or 12-bit digital resolution. It is worth noting that the sampling rate and the digital resolution, in a sense, are connected. To demonstrate the importance of this, we can look at an example: The lidar signal from a range interval of 15 m will be captured in a one-time channel in a 10-MHz, 10-bit digitizer, which will give a resolution of 1 part in 1024. In a 100-MHz, 8-bit digitizer the signal from the same range interval will be spread over 10 time channels. To be comparable, we must add theses channels, which will give a resolution of 1 part in 2560. That is, the poorer digital resolution and the faster sampling rate can sometimes be advantageous even if the maximum range resolution is not required. It is also of importance to note that the effective digital resolution of a commercial digitizer in a lidar application is lower than that indicated by the number of bits. Further, it is important that the low signals, which may correspond to single backscattered photons, are larger than the bit size of the digitizer. For a high-energy system this means that the signal from short ranges will saturate the digitizer if it is not suppressed. From the example given it follows that this problem is no less for a 10-bit digitizer than for an 8-bit digitizer with the above sampling rates.

The transient digitizer must be carefully adjusted to give a linear response, and the triggering of the unit and its internal time scale have to be precise. Zero-drifting and digitizing errors have to be adjusted or compensated for in the measurement routines. The digitized steps may differ for sequential time channels, which need not result in errors but must be considered when devising the routines. Every digitizer has some pecularities of its own, which must be examined closely. Recently, improved transient digitizers have been developed, and further progress is expected. There are now fast dual-channel devices, which make simultaneous dual-wavelength lidar measurements possible. Alternatively, these can be used to increase the effective dynamic range by employing one channel for the stronger signals and the other for the weaker signals from longer ranges.

4.3.3 Data Acquisition and System Steering

A number of lidar signals are generally averaged before the DIAL evaluation. Since the concentration is proportional to minus the logarithm of the signal, the averaging of the lidar signal might lead to a systematic error, as was discussed above. If, on the other hand, the logarithm of the signal is averaged, the signal-to-noise ratio of each individual signal must be rather good or else the logarithm will not be meaningful. This is generally not the case, and thus the best choice is to average the signal before the calculation. An exception to this may be the case where strong signals from topographic targets are measured. An alternative would then be to employ a logarithmic amplifier for the detection. The best universal technique will perhaps be the combination of the two averaging principles, which will be practically applicable with new powerful microcomputers. An example of this is discussed next.

The background noise signal must be subtracted from the lidar signal before the calculations are carried out. It is important that this background signal be recorded in exactly the same way as the lidar signal. The best technique is to record a few signals when the laser beam is physically blocked but with all the electronics, including photomultiplier modulation, running as normal. This procedure will also remove any synchronous electrical interference from the averaged signal. If only one detector is employed and if the background signal is the same for all wavelengths measured, the same averaged background signal should be subtracted from the averaged lidar signals. One will then introduce random noise into the lidar signals, but this extra noise will largely be eliminated in the DIAL evaluation since the same "noise component" is added to corresponding time channels of the lidar signals.

The new generation of 16-bit microcomputers will simplify the construction of data handling and system steering. Until now it has been necessary to use large computers or multichannel (hardware) memories for the fast averaging of lidar signals at repetition rates over 10 Hz, and in some DIAL systems continuous operation has either not been possible or the rate of data collection has had to be reduced. When working with microcomputers, the data are conveniently stored on floppy disks, but in measurements where large amounts of data have to be stored very fast (e.g., when charting a pollutant in a large area), the size of the internal memory in an 8-bit computer is a limitation. With current internal computer memories and compact hard disks (Winchester disks), available with 16-bit microcomputers, this restriction is removed.

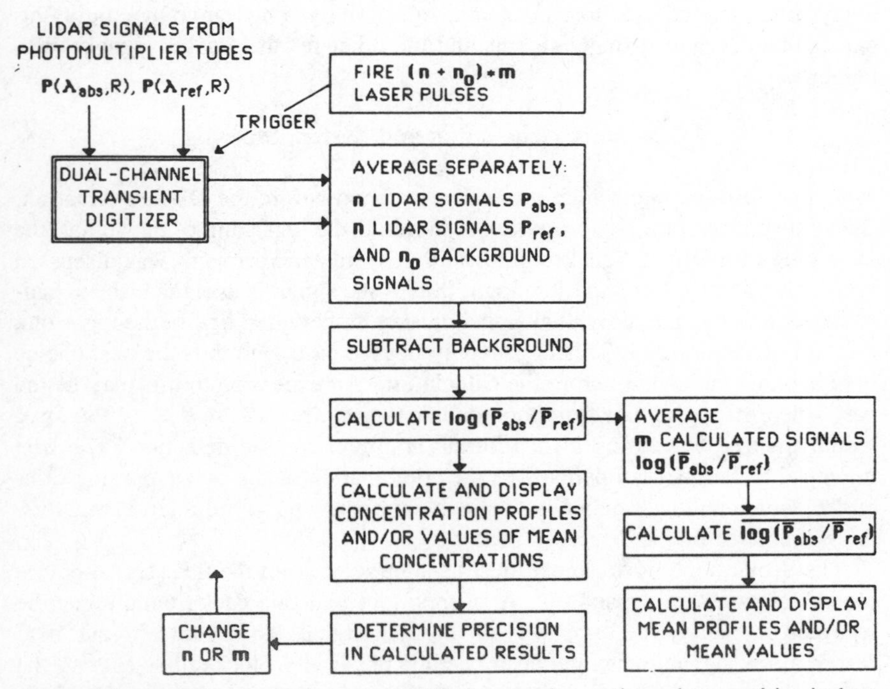

Figure 4.8. Program structure for proposed DIAL measuring routine, where precision is determined on-line and used to steer ongoing measurement.

It will be possible to apply new measurement and evaluation routines with the new microcomputers. Since these devices are quite inexpensive compared with other costs associated with lidar measurements, it should also, in many cases, be possible to incorporate them into existing systems. One possible improvement would be to carry out evaluation during the measurement cycle and let the evaluated data steer the continuing measurement routine. It would then also be possible to directly measure the precision in a measurement. Figure 4.8 shows a proposed program structure for DIAL measurements with presently available hardware components.

Most current DIAL systems are, to a large extent, controlled by minicomputers. There are systems that can be run automatically by the computer during

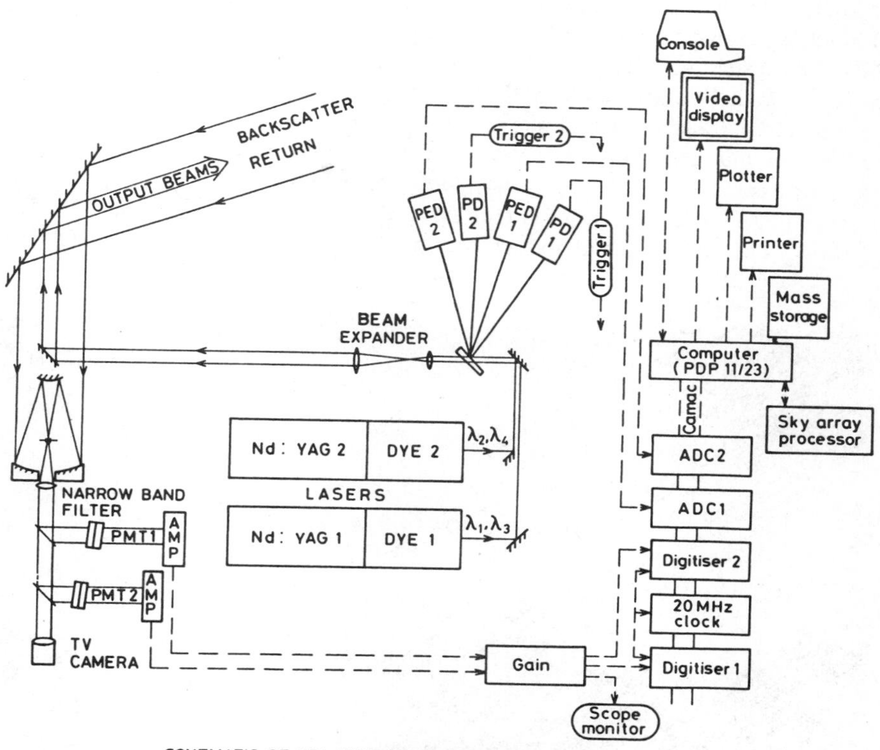

Figure 4.9. Schematic diagram of 4-wavelength DIAL system developed and operated by National Physical Laboratory, U.K. for simultaneous monitoring of two pollutants. Pyroelectric detectors (PED) used for normalization of backscattered lidar signals, and photodiodes (PD) used to trigger computer. (Reproduced by permission of the NPL, U.K.)

long measurement cycles and require manual assistance only to start the run. These systems control laser firing and wavelength selection in the dye laser, lidar direction changes during the measurement, blocking of the laser beam for measurement of background levels, and triggering of electronics and data acquisition and are also capable of presenting results on-line. In general, the main program is in a high-level language such as FORTRAN or Pascal while some subroutines for steering operations are in assembler language. Computer programs that are comprehensible and easy to use will facilitate the introduction of the technique into regular environmental control.

4.3.4 System Integration

Systems intended for pollution monitoring are often assembled in compact mobile units. An example of a mobile DIAL system setup is portrayed in Figure 4.9. This system was developed by (and is in operation at) the National Physical Laboratory in England (Jolliffe et al., 1984). It is equipped with a directing mirror for the transmitted and received light. The system can be operated when driving around, and it can measure two gases simultaneously, employing two identical laser systems. A compact system constructed by the National Swedish Environment Protection Board (Fredriksson et al., 1981; Egebäck et al., 1984) is shown during a series of measurements in Figure 4.10. The laser beam is

Figure 4.10. National Swedish Environment Protection Board's mobile DIAL system in typical measurement situation at paper mill (Egebäck et al., 1984).

Table 4.2. Typical Data for Lidar Systems Applied in Pollution Mapping

Measurement Example	SO_2 [a]	NO_2 [b]	O_3 [c]	HCl [d]
Laser	Nd–YAG-pumped (2 lasers)	Nd–YAG-pumped (1 laser)	Nd–YAG-pumped (1 laser)[e]	DF (1 laser)
Pulse energy, mJ	10	4	3	10
Pulse length, ns	7	7	20	300
Pulse repetition rate, Hz	10	10	10	1
Beam divergence, mrad	1	0.7	—	—
Wavelengths	300.05 nm, 299.38 nm	448.1 nm, 446.8 nm	283 nm, 300 nm	3.636 24 μm, 3.698 23 μm
Pointing arrangement	Movable telescope	Directing mirror	Vertically pointing[f]	Movable telescope
Telescope	Newtonian	Newtonian	Cassegrainian	Cassegrainian
Diameter, cm	51	30	50	50
Focal length, cm	127	100	—	—
Optical filter				
Width, nm	10	5	—	—
Transmission, %	15	54	—	—
Detector	Photomultiplier	Photomultiplier	Photomultiplier	InSb
Quantum efficiency, %	17	17	—	—
Transient recorder				
Digitization	10 bits	8 bits	8 bits	
Sampling rate, MHz	10	100	50	0.5–2[g]

[a] From Hawley et al. (1983).
[b] From Fredriksson and Hertz (1984).
[c] From Sutton (1986).
[d] From Michaelis and Weitkamp (1984).
[e] One Nd–YAG laser and two alternating dye lasers.
[f] Mobile system can be operated when driving around.
[g] Bandwidth of amplifier.

transmitted via a directing mirror in the dome construction of this system. Other systems have no directing mirror, but it is possible to move the telescope or both the laser and the telescope when directing the lidar. A system with a steerable telescope has been developed at SRI International, California (Hawley et al., 1983), and a system with a steerable laser and telescope unit has been built by CISE laboratories and is being operated by the research institute of ENEL in Milano, Italy (Marzorati et al., 1984).

Mobile DIAL systems for pollution mapping have also been constructed by the Central Electricity Research Laboratories in England (Brassington, 1983; Sutton, 1985) and by the National Institute for Environmental Studies in Japan. The GKSS-Forschungszentrum in Germany (Michaelis and Weitkamp, 1984) has a system housed in a container, and there are mobile DIAL systems being built in France, the Netherlands, Sweden, the United States, Switzerland, and other countries. An airborne DIAL system has been constructed by the NASA Langley Research Center in Virginia (Browell et al., 1983). All these systems are intended for measurements mainly in the UV and visible regions. Then there are also several stationary UV and visible DIAL systems and IR DIAL systems, which so far have been used mainly for basic research.

Projects with airborne CO_2 laser systems for IR DIAL measurements have been reported by the Jet Propulsion Laborabory in California (Shumate et al., 1981), the Goddard Space Flight Center in Maryland (Bufton et al., 1983b), and Battelle Institut in Germany (Englisch et al., 1983).

Technical specifications for a few DIAL systems when applied in specific pollution measurements are summarized in Table 4.2. The data are good examples of what can be achieved in a field measurement today.

4.4 MEASUREMENTS ON POLLUTANT GASES

4.4.1 Absorption by Molecular Pollutants

Absorption wavelengths for a number of molecules of interest in air pollution studies are listed in Table 4.3. Most of the wavelengths listed for the IR region correspond to wavelengths of line-tunable CO_2-, CO-, or DF lasers, which are currently the most commonly used lasers in this region. It may well be possible to employ larger absorption cross sections for some molecules when other laser sources become available. A comprehensive list of absorption properties of atmospheric and trace gases is provided in the U.S. Air Force Geophysical Laboratory (AFGL) compilations (Rothman, 1983a, b). A table of references to IR absorption spectra measured with tunable narrow-linewidth lasers has been put together by McDowell (1981).

The differential absorption cross section applicable in DIAL measurements is dependent on the molecular absorption features and the accessible laser wave-

Table 4.3. Absorption Cross Sections for Pollutant Gases at Wavelengths Relevant for Laser Monitoring

Molecule	Wavelength	Laser (Example)	Absorption Cross Section (10^{-22} m^2)	Reference
Nitric oxide, NO	226.8 nm	Dye laser[a]	4.6	Edner et al., 1983
Ozone, O_3	253.6 nm	Dye laser[a]	11.3	Griggs, 1968
	289.4 nm	Dye laser[a]	1.5	Griggs, 1968
Benzene, C_6H_6	250 nm	Dye laser[a]	1.3	Byer and Garbuny, 1973
Mercury, Hg	253.7 nm	Dye laser[a]	5.6×10^4	Byer and Garbuny, 1973
Sulfur dioxide, SO_2	300.0 nm	Dye laser[a]	1.3	Brassington, 1981
Chlorine, Cl_2	330 nm	Dye laser[a]	0.26	Jacobs and Giedt, 1965
Nitrogen dioxide, NO_2	448.1 nm	Dye laser	0.69	Woods and Joliffe, 1978
Methane, CH_4	3.270 μm		2.0	Hinkley et al., 1976
	3.391 μm	HeNe laser	0.6	Hinkley et al., 1976
Propane, C_3H_8	3.391 μm	HeNe laser	0.8	Hinkley et al., 1976
Hydrogen chloride, HCl	3.636 μm	DF laser	0.20	Krüger, 1982
Methane, CH_4	3.715 μm	DF laser	0.002	Murray et al., 1976
Sulfur dioxide, SO_2	3.984 μm	DF laser	0.42	Altmann and Pokrowsky, 1980
Carbon monoxide, CO	4.709 μm		2.8	Hinkley et al., 1976
	4.776 μm	CO_2 laser[b]	0.8	Hinkley et al., 1976
Nitric oxide, NO	5.215 μm	CO laser	0.67	Hinkley et al., 1976
	5.263 μm	CO_2 laser[b]	0.6	Hinkley et al., 1976
Propylene, C_3H_6	6.069 μm	CO laser	0.09	Hinkley et al., 1976
1,3-Butadiene, C_4H_6	6.215 μm	CO laser	0.27	Hinkley et al., 1976
Nitrogen dioxide, NO_2	6.229 μm	CO laser	2.68	Hinkley et al., 1976
Sulfur dioxide, SO_2	8.881 μm		0.2	Hinkley et al., 1976
	9.024 μm	CO_2 laser	0.25	Hinkley et al., 1976
Ammonia, NH_3	9.220 μm		3.6	Hinkley et al., 1976
Fluorocarbon-11, CCl_3F (Freon-11)	9.261 μm	CO_2 laser	1.09	Mayer et al., 1978

Table 4.3. (*Continued*)

Molecule	Wavelength	Laser (Example)	Absorption Cross Section (10^{-22} m^2)	Reference
Ozone, O_3	9.505 μm	CO_2 laser	0.45	Mayer et al., 1978
	9.508 μm		0.9	Hinkley et al., 1976
Fluorocarbon-113, $C_2Cl_3F_3$	9.604 μm	CO_2	0.77	Hinkley et al., 1976
Benzene, C_6H_6	9.621 μm	CO_2 laser	0.07	Mayer et al., 1978
MMH, $CH_3N_2H_3$	10.182 μm	CO_2 laser	0.06	Menyuk et al., 1982a
Ethyl-mercaptan, C_2H_5SH	10.208 μm	CO_2 laser	0.02	Mayer et al., 1978
Chloroprene, C_4H_5Cl	10.261 μm	CO_2 laser	0.34	Mayer et al., 1978
Monochlorethane, C_2H_5Cl	10.275 μm	CO_2 laser	0.12	Mayer et al., 1978
Ammonia, NH_3	10.333 μm	CO_2 laser	1.0	Mayer et al., 1978
Ethylene, C_2H_4	10.533 μm	CO_2 laser	1.19	Mayer et al., 1978
Sulfur hexafluoride, SF_6	10.551 μm	CO_2 laser	30.3	Englisch and Wiesemann, 1978
Trichlorethylene, C_2HCl_3	10.591 μm	CO_2 laser	0.49	Mayer et al., 1978
1,2-Dichlorethane, $C_2H_4Cl_2$	10.591 μm	CO_2 laser	0.02	Mayer et al., 1978
Hydrazine, N_2H_4	10.612 μm	CO_2 laser	0.18	Menyuk et al., 1982a
Vinyl chloride, C_2H_3Cl	10.612 μm	CO_2 laser	0.33	Mayer et al., 1978
UDMH, $(CH_3)_2N_2H_2$	10.696 μm	CO_2 laser	0.08	Menyuk et al., 1982a
Fluorocarbon-12, CCl_2F_2 (Freon-12)	10.719 μm	CO_2 laser	1.33	Mayer et al., 1978
Perchlorethylene, C_2Cl_4	10.742 μm	CO_2 laser	0.18	Mayer et al., 1978
1-Butene, C_4H_8	10.787 μm	CO_2 laser	0.13	Hinkley et al., 1976
Perchlorethylene, C_2Cl_4	10.834 μm	CO_2 laser	1.14	Hinkley et al., 1976
Fluorocarbon-11, CCl_3F (Freon-11)	11.806 μm		4.4	Hinkley et al., 1976
Acethylene, C_2H_2	13.890 μm		9.2	Hinkley et al., 1976

[a]Extended to UV region.

[b]Frequency doubled.

Absorption lines of several toxic compounds have also been measured by Loper et al. (1982).

lengths. The absorption features can be quite different, as will be demonstrated in two examples. The SO_2 absorption spectrum at wavelengths of about 300 nm is widely employed in DIAL measurements. Several neighboring absorption and reference wavelengths conceivable for DIAL measurements can be found in the region. Measurements of this spectrum have been performed (e.g., Woods et al., 1980; Brassington, 1981). The spectrum from 290–317 nm, measured with a linewidth of 0.05 nm (5 cm^{-1}) is shown in Figure 4.11. The measurements by Woods et al. (1980) were made with a resolution of 0.002 nm and revealed some narrow structures of the absorption lines, which are not resolved in Figure 4.11. Brassington et al. (1984) have discussed the errors introduced into SO_2 measurements in the case where the laser linewidth is larger than this spectral feature. This case can be contrasted with the absorption features of O_3 and Cl_2 in the UV region, which are very broad structures with no narrow spectral lines. The absorption spectrum of Cl_2 is shown in Figure 4.12. This spectrum was measured with a linewidth of 0.1 nm, but no narrow structure of importance was observed even with a narrow-band laser. The absolute scale for the absorption cross section in the diagram was fitted to the measurement by Jacobs and Giedt (1965). On the other hand, the absorption features in the IR region are in general extremely narrow. As an example, the absorption line of CH_4 at 3392.02 μm has a width (FWHM) of about 0.1 cm^{-1} at atmospheric

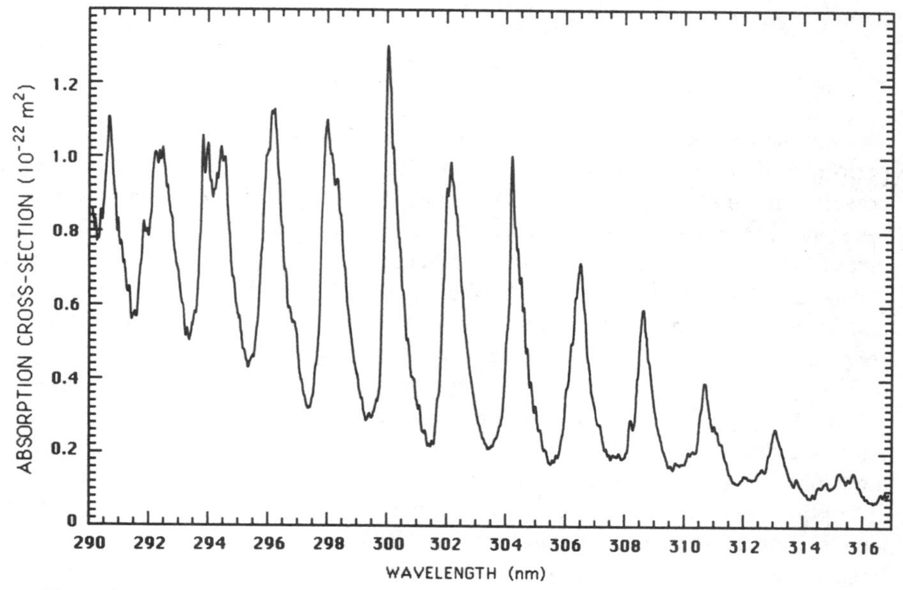

Figure 4.11. Absorption cross section of SO_2 as function of wavelength (Brassington, 1981).

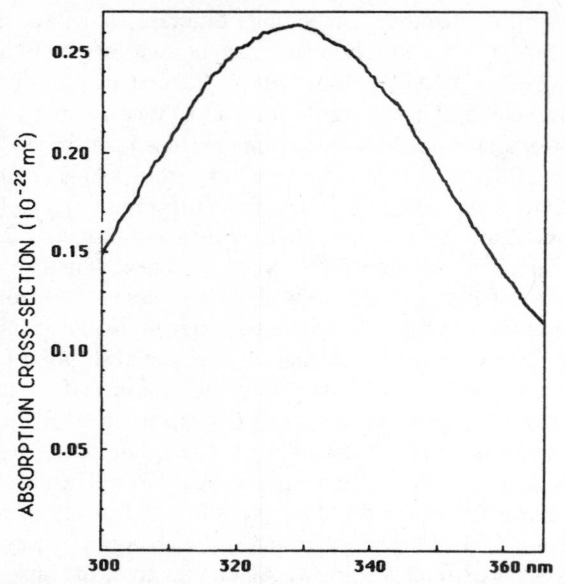

Figure 4.12. Absorption cross section of Cl_2 as function of wavelength.

pressure (McDowell, 1981). Some species (e.g., NO and Hg) also have extremely narrow structures in the UV.

4.4.2 Pollution Mapping

The pollutant most frequently measured with DIAL is SO_2. Many field tests performed by different research teams have revealed that the technique is very powerful for measurements on SO_2 plumes and ambient SO_2 concentrations. References to relevant works are given in Table 4.4(a) together with other DIAL measurement projects in the UV and visible regions.

There are several reasons why applications to SO_2 are favorable. The laser and detection techniques are well established for the 300-nm region, where frequency-doubled dye lasers are employed, and there are no interfering absorbing gases in the atmosphere with the exception of O_3, which has a comparatively low differential absorption cross section. Further, the background light is reduced due to absorption by the atmospheric ozone layers, and the atmospheric extinction is normally not too great. The conceivable differential absorption cross section of the gas is comparatively large, and thus a low detection limit is attainable. Egebäck et al. (1984) has estimated the detection limit to be 5 mg/m^3 m in a typical case. This equals 10 μg/m^3 for a 500-m absorption path. The accuracy in measurements on a plume from a power sta-

Table 4.4(a). Measurements on Pollutants with UV and Visible DIAL

Gas Measured	Reference
Nitric oxide, NO	Edner et al., 1983[a,b]
Mercury, Hg	Aldén et al., 1982[b]
Ozone, O_3	Grant and Hake, 1975[c]; Pelon and Mégie, 1982; Fredriksson, 1982; Browell et al., 1983; Ancellet et al., 1984; Jolliffe et al., 1984; Sutton, 1985
Chlorine, Cl_2	Edner et al., 1985
Sulfur dioxide, SO_2	Grant and Hake, 1975[c]; Hoell et al., 1975[c]; Fredriksson et al., 1979[c], 1981; Adrian et al., 1979[c]; Browell, 1982; Hawley et al., 1983; Egebäck et al., 1984; Jolliffe et al., 1984; Marzorati et al., 1984; Michaelis and Weitkamp, 1984[c]; Ancellet et al., 1984
Nitrogen dioxide, NO_2	Rothe et al., 1974a[c], 1974b[c]; Grant et al., 1974[c]; Tsuji et al., 1976[d]; Fredriksson et al., 1979[d]; Baumgartner et al., 1979[c]; Konefal et al., 1981[c]; Fredriksson et al., 1981; Fredriksson and Hertz, 1984; Sugimoto et al., 1981[c]; Engoyan et al., 1983[c]; Jolliffe et al., 1984; Staehr et al., 1984[c]

[a]Long-path absorption measurement.

[b]With Raman shifter.

[c]Flashlamp-pumped dye lasers.

[d]Nitrogen-laser-pumped dye laser.

All other measurements were made using Nd–YAG-laser-pumped dye lasers.

tion was found to be better than 20% by the National Physical Laboratory in England upon comparison with in-stack monitors. It must be taken into consideration that exact agreement cannot be expected in this case since the two measurement situations are not physically identical. A horizontal charting of the SO_2 concentration over an oil refinery performed by this laboratory (Jolliffe et al., 1984) is shown in Figure 4.13. Another example in another form of presentation is shown in Figure 4.14. This diagram represents an area scan of SO_2 concentrations over an urban area and was performed by the GKSS-Forschungszentrum in Germany (Michaelis and Weitkamp, 1984). This measurement was made with a flashlamp-pumped dye laser, while the former example was made with a Nd–YAG-pumped dye laser.

Since the SO_2 concentration in a plume can be very high, it is not unlikely that a laser beam is totally absorbed at the absorption wavelength, and if the beam is transmitted only part of the time, large averaging errors will be introduced that can be difficult to detect in the final result. The remedy for this is to employ wavelengths with lower absorption cross sections.

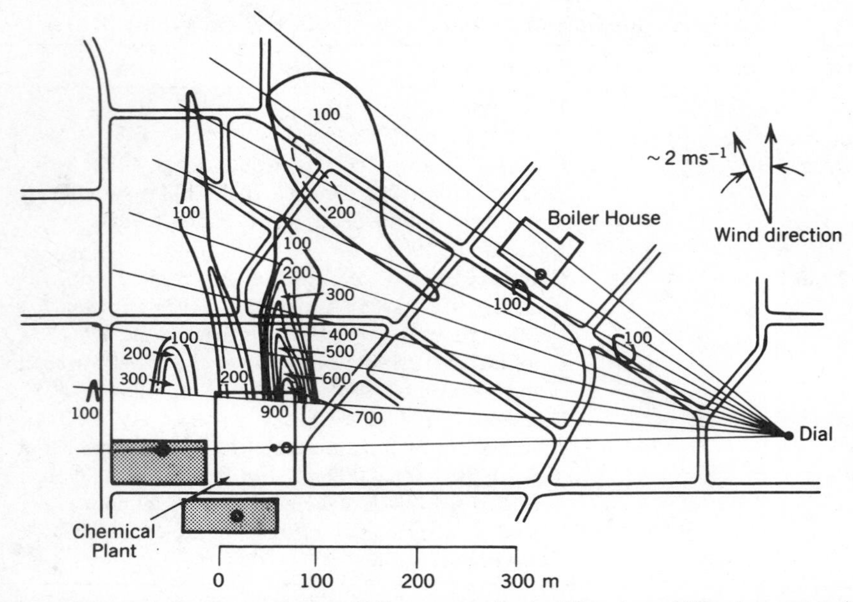

Figure 4.13. Area scan measurement of SO_2 concentration in 100 parts per billion increments (1 ppb = 1 part in 10^9 by volume) over an oil refinery. Total measurement time 10 min. (Reproduced by permission of National Physical Laboratory, U.K.)

Sulfur dioxide is one of the major air pollutants, and there is a great need for reliable instruments to monitor SO_2 emissions. Conventional point-monitoring instruments have limitations and are not well suited for measurements on flows and plumes. This is an important practical aspect, which has promoted the development of the DIAL technique. The large-scale problem of the acidification of soil and water due to acid rain certainly demands increased efforts.

The atmospheric molecule NO_2 is a rare exception in the sense that it has an absorption spectrum in the visible region. As for SO_2, the DIAL technique is also favored here because of the availability of tunable lasers and efficient detection systems. Acceptable differential absorption cross sections are to be found around 450 nm (Woods and Jolliffe, 1978; Takeuchi et al., 1978), and a maximum absolute error of 10 mg/m^3 m for a realistic measurement can be obtained (Fredriksson and Hertz, 1984). However, this limit may be impaired in measurements where a clear sky is the background light source, as was discussed in a previous section. The background rejection must be carefully designed to make daytime measurements possible.

Several NO_2 DIAL measurements have been reported [Table 4.4(a)], and since the dye laser and the receiver system, in principle, are universal for the

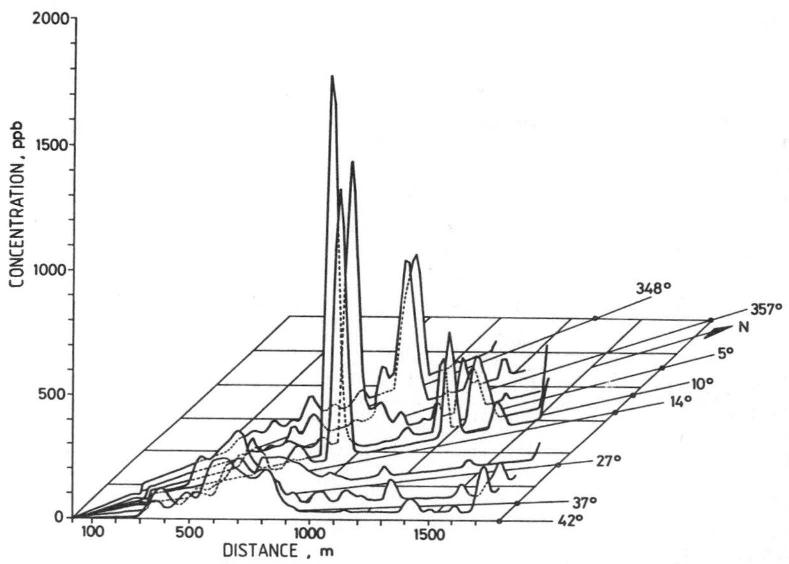

Figure 4.14. SO_2 concentration vs. range for few measurement directions in southern part of Hamburg. Elevation angle 5.3°. Measurement time for each direction 100 s (Michaelis and Weitkamp, 1984).

UV and visible regions, SO_2 DIAL systems could be applied to measurements on NO_2. Figure 4.15 shows a DIAL measurement on a widespread plume of NO_2 from a metropolitan area. Mean concentrations for several paths were measured, which revealed a uniform concentration in the vertical section.

Nitrogen oxides are emitted primarily from motor vehicle traffic, the burning of fossil fuels, and industrial processes. There are not that many concentrated plumes of NO_2, but the gas is normally distributed over larger areas. The measurement capabilities of the DIAL technique when applied to NO_2 match the measurement needs in many pollution situations.

The third pollutant measurable with dye lasers is O_3. As was previously mentioned, the spectrum of O_3 has no narrow absorption lines but a broad absorption band, which extends from 200–300 nm. The maximum absorption is at 255 nm. The maximum differential absorption cross section for neighboring wavelengths is to be found in the region from 265–285 nm. To get a reasonable differential absorption cross section, the two measurement wavelengths must be separated by several nanometers. A typical choice for the absorption wavelength is in the region 283–290 nm, and the reference wavelength is typically 300 nm. The choice is then not only determined by the absorption cross sections but also by the available efficient laser dyes. Since the separation of the two

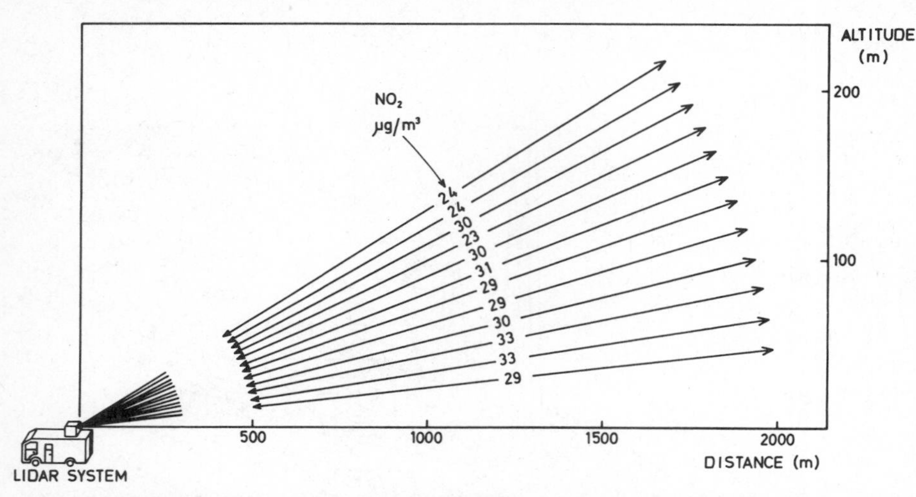

Figure 4.15. Measurements of NO_2 concentrations in widespread plume from metropolitan area. Average concentrations given for each direction, and evaluated paths indicated. Measurement time for each direction 1 min (Fredriksson and Hertz, 1984).

wavelengths is considerable, care must be taken to allow for differences in the extinction coefficient. The wavelength dependence of the extinction due to Rayleigh scattering can be calculated, but the contribution to the extinction from Mie scattering by particles is more difficult to estimate. In the practical case reasonable estimates can often be made, and it should, in difficult cases, be possible to develop measuring routines where three or more wavelengths are employed. An analysis of the O_3 DIAL technique and related uncertainties is presented by Pelon and Mégie (1982).

Ozone does not normally occur in localized concentrations from specific outlets but is connected with large-scale atmospheric processes. Figure 4.16 shows an example of a measurement on O_3 made by Ancellet et al. (1984). This measurement was made during a European campaign in a large industrial center in southern France with several oil refineries and industries (6th CEC Campaign on Remote Sensing at Fos-Berre, June 1983). The O_3 concentration, which was comparatively high on this occasion, was also measured with a point monitor. The results shown in the figure agree quite well, but an exact agreement between the lidar data and the ground measurement cannot be expected. Another measurement on O_3 is shown in Figure 4.17. This measurement, which was made by Sutton (1986), is a measurement in the vertical direction and shows the O_3 distribution in the atmospheric layer closest to the ground. A comparison was made with radiosonde measurements, and the agreement was very good, as can be seen in the figure.

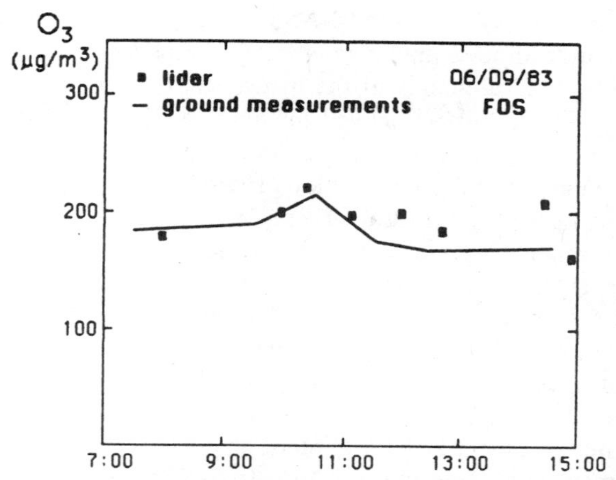

Figure 4.16. Comparison between DIAL and ground measurements of O_3 mass mixing ratio performed at Fos–Berre industrial area in southern France (Ancellet et al., 1984).

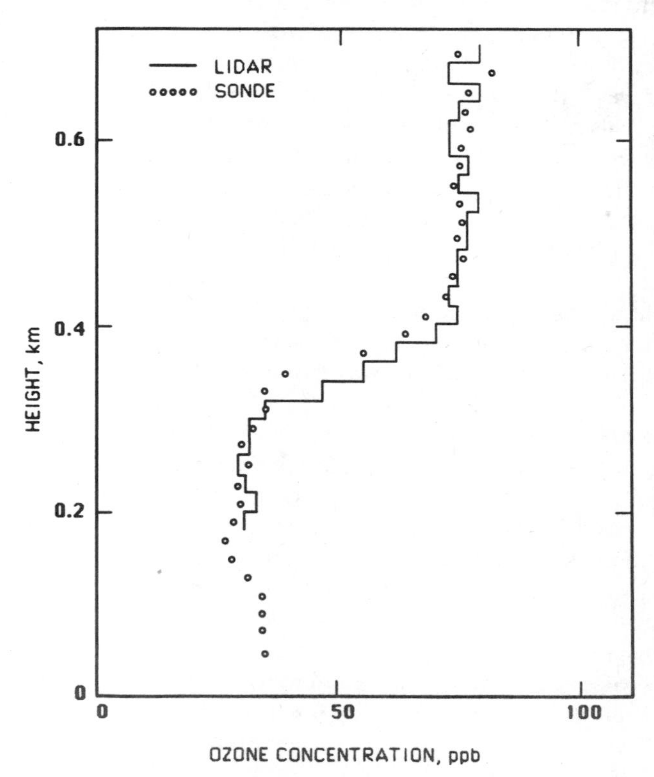

Figure 4.17. Measurement of vertical profile of O_3 in ground layer. Comparisons made with radiosonde measurements (Sutton, 1986).

The airborne DIAL system at the NASA Langley Research Center has been employed primarily for measurements of O_3 and aerosol vertical profiles (Browell et al., 1983). Measurements of the natural O_3 concentrations in the upper troposphere and the stratosphere have been made by Uchino et al. (1978, 1979, 1983a, b) and Werner et al. (1983), who employed excimer lasers and Raman shifting devices, and by Mégie and Pelon (1983) using Nd–YAG and dye laser technology (see also Mégie et al., 1977). Their objectives were not pollution mapping in the sense discussed here, but the technique was similar.

Measurements on atomic mercury have been carried out by Aldén et al. (1982). Since the absorption feature of an atom consists of extremely narrow lines with large absorption cross sections, the detection limit for atomic mercury can be several orders of magnitude lower than for a molecule. On the other hand, the demands placed on the laser linewidth and the wavelength calibration are much greater. Aldén et al. (1982) claim that the detection limit for a long-path measurement should be close to the atmospheric background concentration.

Nitrogen oxide has an absorption band at 226.8 nm that can be employed for DIAL studies. Edner et al. (1983) carried out measurements where retroreflecting mirrors were used to measure mean concentrations along a street. The transmitter was a Raman-shifted frequency-doubled dye laser, which produced low-energy pulses. Another problem besides that of generating laser power with a narrow linewidth in this wavelength region is the very strong Rayleigh scattering, which results in severe attenuation of the laser beam and considerable multiple-scattering effects in range-resolved measurements.

A few NO measurements of the mean concentration along a 700-m path are shown in Figure 4.18. In this case, the laser was repeatedly tuned to six different wavelengths during the measurement. The backscattered return signals from the retroreflector were fitted to the absorption profile, and the concentration was calculated. The absorption profile was measured with the DIAL system using a gas cell. It is obvious that a fitting of several wavelengths in this way will increase the measurement accuracy in any DIAL measurement. However, to be efficient enough for the field work, the fitting procedure requires high repetition rates or multiwavelength lasers and corresponding detection systems.

Another gas that has an absorption spectrum in the UV region is Cl_2. However, the differential absorption is too low to make realistic DIAL measurements on ordinary outlets and ambient concentrations (Edner et al., 1985). However, there are cases when accidental releases occur, and then the technique may be useful in mapping the pollutant, which is a heavy gas and thus does not diffuse rapidly. A somewhat similar case regarding IR DIAL measurements on CH_4, primarily from leak sources, has been discussed by Grant (1982b), who also constructed a system based on HeNe lasers emitting CW laser light at two wave-

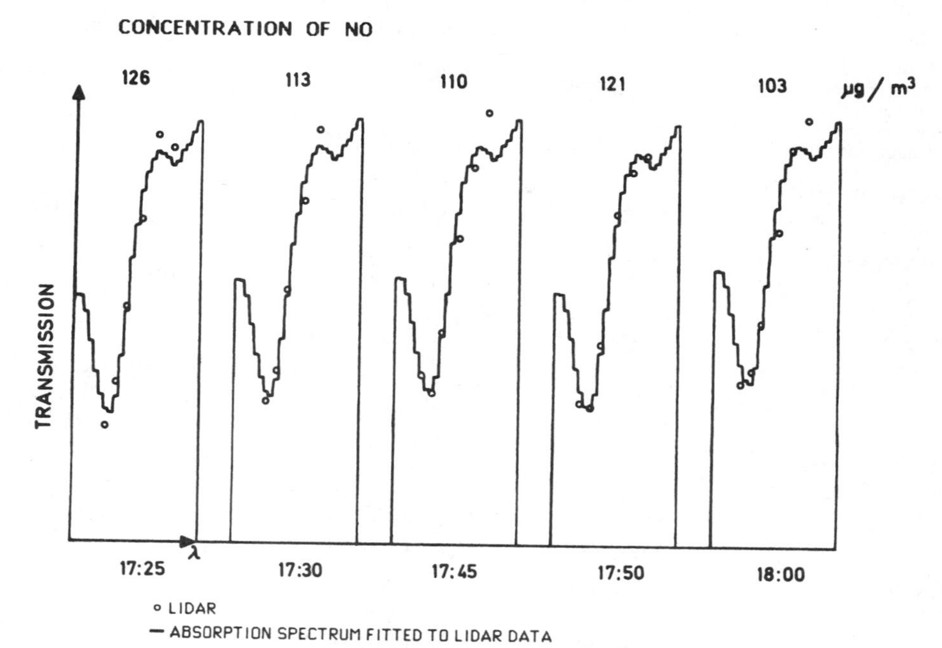

Figure 4.18. Five consecutive DIAL measurements of NO concentration along a street. Retroreflecting mirror at distance of 700 m employed for backscattering. Six different wavelengths fitted to NO spectrum in each 5-min measurement (Edner et al., 1983).

lengths near 3.39 μm. The system is to be used for measurements using close topographic targets or retroreflectors.

Several promising DIAL projects [listed in Table 4.4(b)] employing different kinds of lasers have been performed in the near- and mid-IR region. The most dedicated pollution studies are certainly the HCl measurements made by Weitkamp (1981) and Weitkamp et al. (1984). Using a DF laser, they measured gaseous HCl originating from the incineration of chlorinated hydrocarbons at sea. Figure 4.19 shows one example of such a mapping.

Several long-path absorption measurements employing tunable diode lasers in the mid-IR region have been carried out [see the references in Table 4.4(b)]. Cassidy and Reid (1982) have reported measurements on sensitivities and discussed ways of increasing the measurement capacities with diode lasers. They were able to detect absorptions as small as 0.01% over pathlengths of up to 250 m. Pollution measurements in this wavelength region have been performed by Byer and co-workers employing tunable optical parametric oscillators [see the references in Table 4.4(b)], but there are no such commercial laser systems

TABLE 4.4(b). Measurements on Pollutants with DIAL and Long-Path Absorption Methods in 1.5–5.3-μm Range[a]

Gas Measured	Laser Source	Reference
Hydrogen fluoride, HF	HF/DF laser	Rothe, 1980
Hydrogen chloride, HCl	DF laser	Murray et al., 1976
	DF laser	Weitkamp, 1981, Weitkamp et al., 1984
Methane, CH_4	DF laser	Murray et al., 1976
	OPO	Baumgartner and Byer, 1978
	Erbium–YAG laser	Watkins and White, 1981
	HeNe laser (CW)	Grant, 1982b
Sulfur dioxide, SO_2	OPO	Baumgartner and Byer, 1978
Carbon dioxide, CO_2	CO_2 laser[b]	Bufton et al., 1983a
Carbon oxide, CO	OPO	Henningsen et al., 1974
	Diode laser	Ku et al., 1975; Hinkley, 1976
	Diode laser	McClenny and Russwurm, (1978)
	CO_2 laser[b]	Killinger et al., 1980
Nitric oxide, NO	Diode laser	Hinkley, 1976
	CO laser	Menzies and Shumate, 1976
	CO_2 laser[b]	Menyuk et al., 1980

[a]Several measurements on H_2O and N_2O have also been performed with differential absorption methods in the mid-IR region.

[b]Frequency-doubled CO_2 laser.

available, and the technique is certainly too complicated for fieldwork. Watkins and White (1981) applied a different tuning technique for solid-state lasers and carried out measurements on CH_4 by employing the emission spike train in a long-pulse output from an erbium–YAG laser. The DIAL applications in the mid-IR region, in general, suffer from the lack of tunable, high-energy laser sources, and the detection technique is so far not as highly dedicated as in the UV and visible DIAL technique. This may very well change within the next few years.

Several DIAL projects have been carried out with CO_2 lasers. Since the aerosol backscattering is 2–3 orders of magnitude lower for the CO_2 laser wavelengths than for the UV wavelengths (Mégie and Menzies, 1980), range-resolved measurements with these lasers are much more difficult. Most measurements have been carried out on O_3 and C_2H_4, employing line-tunable lasers and topographic targets or retroreflectors for measurements of integrated concentrations along the path of the laser beam. The most extensive field measurement program on O_3 with CO_2 lasers has been performed by Menzies and

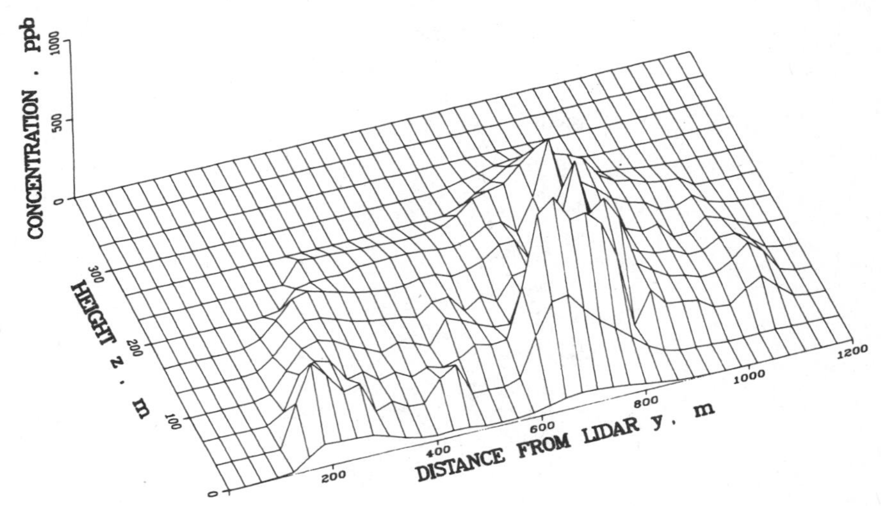

Figure 4.19. Isometric plot of HCl concentration in cross section of plume from incineration ship at distance of 2.25 km. Plume influenced by inversion layer at altitude of 200 m. Typical total measurement time per plot 1 h (Weitkamp et al., 1984).

Shumate (1978) and Shumate et al. (1981) with an airborne nadir-viewing laser system employing the ground as a topographic target. Their system had two CW lasers, and a heterodyne detection technique was employed. In some tests they made comparisons with point monitors in another aircraft (Shumate et al., 1981). However, there are several restrictions on the applications of line-tunable CO_2 lasers. The effects of differential backscattering from atmospheric aerosol and effects of differential reflectance of topographic targets can be severe limitations, as was discussed in a previous section. Continuously tunable CO_2 lasers have so far been operated only in research projects.

In Table 4.4(c) several relevant measurement projects employing CO_2 lasers are listed. Killinger, Menyuk, and co-workers have discussed errors and limitations of the technique in several papers (Menyuk et al., 1982b, 1985; Menyuk and Killinger, 1983). Fukuda et al. (1984) have recently discussed the sensitivity of range-resolved DIAL with CO_2 lasers. An experimental comparison of heterodyne and direct detection was made by Killinger et al. (1983). Korb and Weng (1982, 1984) have proposed a technique to reduce the error associated with the finite laser bandwidth, a problem that must be considered especially in IR DIAL.

A great deal of effort has been put into the IR DIAL technique by many research groups, but it is somewhat discouraging that so few field measurements have been made. However, everything points to a breakthrough when well-suited lasers and dedicated detection techniques become available.

Table 4.4(c). DIAL and Long-Path Absorption Measurements on Pollutants with CO_2 Lasers (9.1–11 μm)

Gas Measured	Laser Source	Reference
Ozone, O_3	CW laser	Guagliardo and Bundy, 1975
	CW laser	Asai and Igarashi, 1975
	CW laser	Menzies and Shumate, 1976, 1978
	CW laser	McClenny and Russwurm, 1978
	pulsed TEA laser	Asai et al., 1979
	pulsed TEA laser	Bufton and Stewart, 1980
	CW laser	Shumate et al., 1981
Ethylene, C_2H_4	CW laser	Menzies and Shumate, 1976
	pulsed TEA laser	Murray and van der Laan, 1978
	pulsed TEA laser	Rothe, 1980
	pulsed (Q-switched) laser	Lundqvist et al., 1981
	pulsed TEA laser	Killinger and Menyuk, 1981
	CW laser	Persson et al., 1982
Sulfur hexafluoride, SF_6[a]	CW laser	Wiesemann et al., 1978
Hydrazine, UDMH, and MMH	Pulsed TEA laser	Menyuk et al., 1982a

[a]Tracer molecule in conventional measurements on plume spreading.

An interesting measurement principle for DIAL has been proposed and tested by Byer and Shepp (1979) and Wolfe and Byer (1982). With several long-path absorption measurements in an area, a mapping is produced by an advanced computer routine in a manner similar to X-ray tomography. With this technique it would be possible to apply low-energy lasers and still produce a two-dimensional concentration map.

4.4.3 Measurements on Pollutant Flows

One of the big advantages of the DIAL technique is the ability to measure pollutant flows, and we will discuss this application particularly.

The pollutant flow from an industrial area or across a frontier is extremely difficult to measure and calculate with point monitors. The range-resolved DIAL technique offers an attractive alternative in this situation. A charting is then made of the gas in a vertical section, orthogonal to the wind direction. An example of a charting of SO_2 in a vertical section downwind of a chemical factory is shown in Figure 4.20 (Egebäck et al., 1984). The dispersed plume is due to diffuse emissions from the plant, and the dense plume originates from a

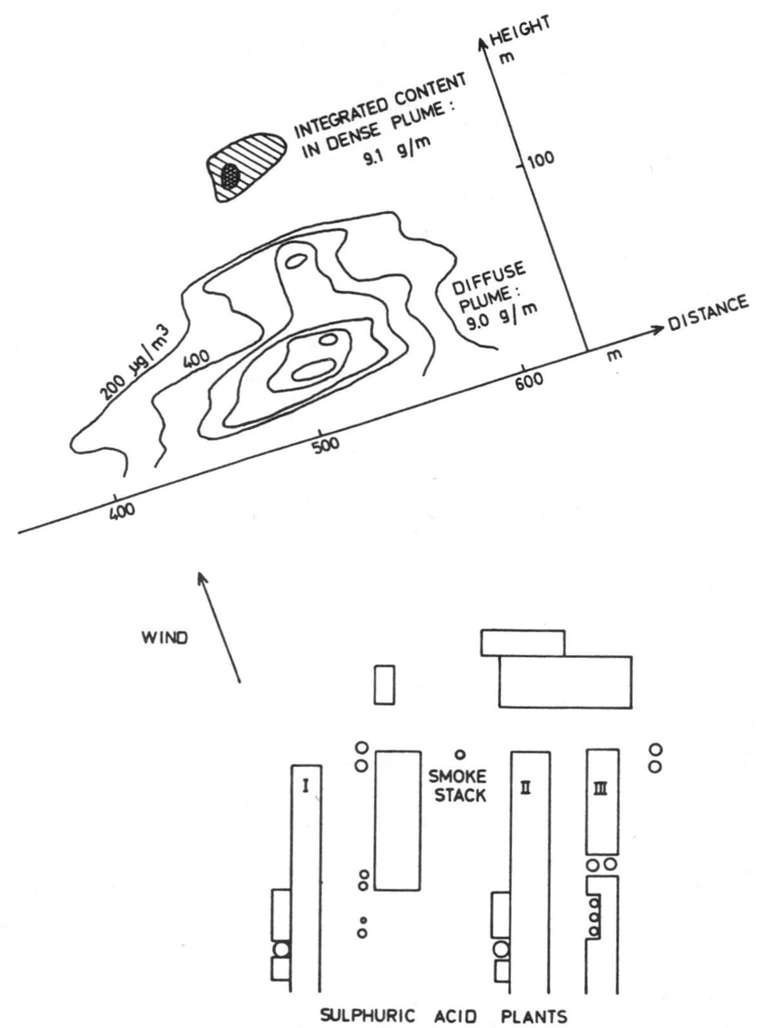

Figure 4.20. Charting of SO_2 in smokestack plume and diffuse emissions in vertical section downwind of chemical factory. Horizontal direction for charting is given by axis for distance in diagram. Integrated contents and wind velocity at different altitudes yield flow of gas (Egebäck et al., 1984).

tall stack. The readings in the diagram are the integrated SO_2 contents in the two plumes, which yield the flow of the gas when multiplied by the relevant wind velocities. Another example of a vertical charting of a plume, in a presentation form well suited for rapid evaluation and adapted to plume dispersion models, is shown in Figure 4.21. The vertical and horizontal projections of a

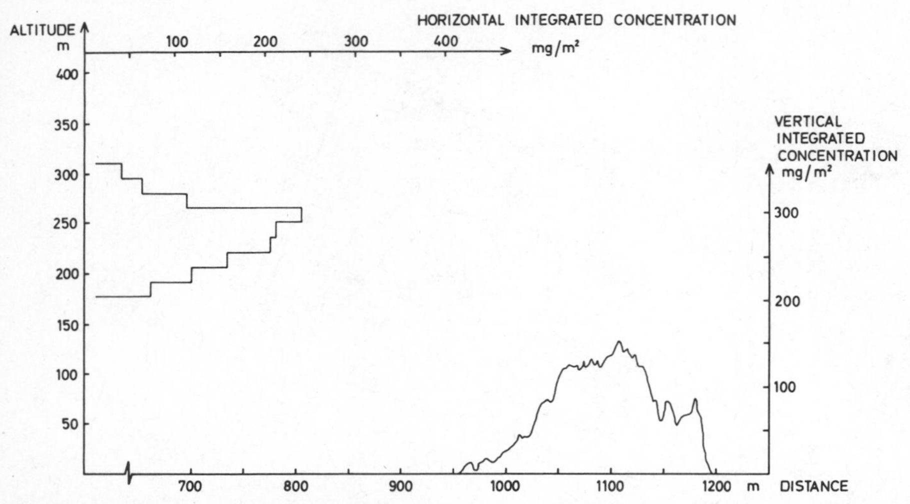

Figure 4.21. Vertical and horizontal projections of SO_2 content in plume from heating plant. Charting was made orthogonal to wind direction. Lidar range is indicated in diagram. Measurement time 10 min (Egebäck et al., 1984).

spreading plume from a heating plant are shown in the figure. Since the vertical direction was changed stepwise in the measurement, this projection is shown as a step function.

The accuracy in the wind velocity measurement will be as important as the accuracy in the DIAL measurement in a flow study. Point monitors for wind measurements are adequate only if they are mounted at the altitude of the plume studied, and this is in general not possible. Doppler sonar instrumentation for three-dimensional wind measurements provides an alternative as do Doppler lidar systems, but the complexity increases greatly with these extra systems. An attractive solution would be for the DIAL system to measure wind direction and wind velocity. Correlation techniques using lidar profiles of aerosol backscatter intensity have been proposed and studied with regard to wind measurements by several groups (Derr and Little, 1970; Eloranta et al., 1975; Armstrong et al., 1976; Sroga et al., 1980; Sasano et al., 1982). However, the correlation analysis, in general, requires large computers and a great deal of computing time. Fredriksson (1985) recently proposed a dual-beam lidar technique, which makes it possible to measure the wind parameters relevant for a plume with a simple cross correlation technique adapted for a microcomputer. The principal idea is shown in Figure 4.22. A wedge that can be rotated and has a center hole is used to split the laser beam into two beams with a small angular separation. One part of the beam is transmitted through the hole and is

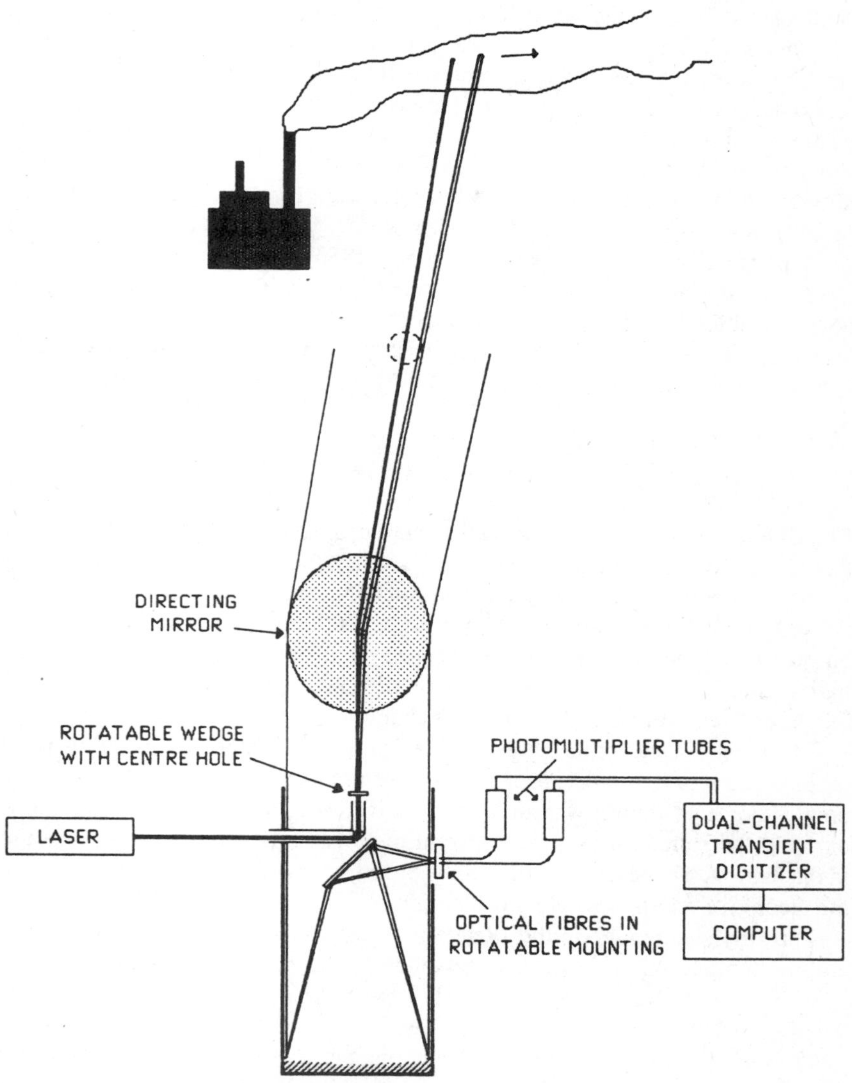

Figure 4.22. Schematic diagram of dual-beam lidar system for measurements of plume velocity by temporal correlation (Fredriksson, 1985).

unaffected by the wedge while the rest of the beam is refracted. The wide divergence angle is adjusted to be horizontal by rotating the wedge. Optical fibers are used to direct the two backscattered lidar signals to two photomultiplier tubes. The two fields of view of the detection are determined with high precision by the positions of the fibers. Two parallel sequences of backscattered lidar signals from the plume can be correlated using a simple program, and the wind direction and wind velocity are easily calculated. The technique is intended to be supplementary to an ordinary DIAL system.

Airborne DIAL systems (Browell et al., 1983) and mobile DIAL systems, which can measure when driving around (Jolliffe et al., 1984; Sutton, 1985), have the ability to carry out measurements over a large area in a short time. This can be very attractive in many situations (e.g., for chartings of plume spreading and for measurements of the flow of pollutants across a border).

4.5 CONCLUSIONS

The differential absorption and lidar techniques have become useful measuring techniques for air pollution mapping. DIAL systems operating in the UV and visible regions for measurements on SO_2, NO_2, and O_3 are useful instruments not only in dedicated campaigns but also in routine work for the monitoring of air quality and pollution emissions. These systems are also well suited for qualitative charting of aerosols (e.g., for plume spreading studies), which further increases the usefulness of the instrumentation.

There are several other gases with absorption wavelengths in the near-UV region, and it is likely that the UV DIAL technique will prove useful for mapping of some of these. Measurements on Hg, NO, and Cl_2 have been carried out, and there are, for example, several hydrocarbons with absorption features in this region. The laser development is progressing rapidly, which will improve the technique and make field work easier.

The DIAL technique in the IR region is limited today mainly because there are few tunable, high-power lasers available, and detection techniques are not as well developed as needed for DIAL applications. However, the potential of the IR DIAL technique is very large because of the availability of absorption features in atmospheric pollutants. Development is rapid in the field of IR technology, and we can hope for a breakthrough in IR DIAL applications in the next few years. Tunable diode lasers may provide a useful alternative for short-range applications if and when standard systems that are easy to handle become available.

Measuring lidar signals at several wavelengths of an absorption profile, and not only at two wavelengths as is generally employed today, would be a perfect way of improving the accuracy and detection limits in the DIAL technique. The future applicability of such a technique will be coupled to the advances made in laser and detector technology. There should be no severe optical restriction, but the difficulties may lie in the requirements of multichannel detectors and transient digitizers.

There are many other optical measurement techniques similar to DIAL in terms of both instrumentation and measurement demands. Differential optical absorption spectroscopy (Platt and Perner, 1983), in which other light sources are employed for measurements of trace gases, optical multichannel techniques (e.g., Onderdelinden and Strackee, 1978), IR spectroscopy techniques (e.g., Herget, 1982), and correlation techniques (e.g., Ward and Zwick, 1975; Davies et al., 1983), can very well be combined with DIAL systems to increase the measurement capabilities. There are also laser techniques based on Raman and fluorescence scattering, with several applications, that can be combined in a mobile laser laboratory together with the DIAL technique. A combined system will benefit from an enlarged range of applications, and this might help justify the high cost of a mobile lidar facility. Technical developments will also be favored by the wider range of techniques.

ACKNOWLEDGMENTS

The author has drawn on the experience of several years work at the National Swedish Environment Protection Board, the Chalmers University of Technology, and the Lund Institute of Technology and is grateful to all colleagues in the research teams during this period. He is also indebted to several researchers who have contributed material to this chapter. The help in proofreading the manuscript given by his colleague Helen Sheppard is also very much appreciated.

REFERENCES

Adrian, S., D. J. Brassington, S. Sutton, and R. H. Varey (1979). The Measurement of SO_2 in Power Station Plumes with Differential Lidar. *Opt. Quant. Electron.* **11,** 253–264.

Aldén, M., H. Edner, and S. Svanberg (1982). Remote Measurements of Atmospheric Mercury Using Differential Absorption Lidar. Opt. Lett. **7,** 221–223.

Allen, R. J., and W. E. Evans (1972). Laser Radar for Mapping Aerosol Structure. *Rev. Sci. Instrum.* **43,** 1422–1432.

Altmann, J., and P. Pokrowsky (1980). Sulphur Dioxide Absorption at DF Laser Wavelengths. *Appl. Optics* **19,** 3449–3452.

Ancellet, G., R. Capitini, D. Renaut, G. Mégie, and J. Pelon (1984). DIAL Lidar Measurements of Atmospheric Pollutants (SO_2, O_3) during the Fos-Berre 83 Experiment. Abstr. 12th International Laser Radar Conference, 13–17 August 1984, Aix en Provence, France, Service d'Aeronomie du CNRS, pp. 269–273.

Armstrong, R. L., J. B. Mason, and T. Barber (1976). Detection of Atmospheric Aerosol Flow Using a Transit Time Lidar Velocimeter. *Appl. Optics* **15,** 2891–2895.

Asai, K., and T. Igarashi (1975). Detecting of Ozone by Differential Absorption Using CO_2 Laser. *Opt. Quant. Electr.* **7,** 211–214.

Asai, K., and T. Igarashi (1984). Interference from Differential Reflectance of Moist Topographic Targets in CO_2 DIAL Ozone Measurement. *Appl. Optics* **23,** 734–739.

Asai, K., T. Itabe, and T. Igarashi (1979). Range-Resolved Measurements of Atmospheric Ozone Using a Differential-Absorption CO_2 Laser Radar. *Appl. Phys. Lett.* **35,** 60–62.

Baumgartner, R. A., and R. L. Byer (1978). Continuously Tunable IR Lidar with Applications to Remote Measurements of SO_2 and CH_4. *Appl. Optics* **17,** 3555–3561.

Baumgartner, R. A., L. D. Fletcher, and J. G. Hawley, (1979). A Comparison of Lidar and Air Quality Station Measurement. *ARCO J.* **29,** 1162–1165.

Boscher, J., W. Englisch, and W. Wiesemann (1980). Foreign Gas Interference and Differential Albedo Effects in CO_2 Laser Long Path Absorption Monitoring, Paper ThC2 in Technical Digest, Coherent Laser Radar for Atmospheric Sensing, Optical Society of America.

Brassington, D. J. (1981). Sulfur Dioxide Absorption Cross Section Measurements from 290 nm to 317 nm. *Appl. Optics* **20,** 3774–3779.

Brassington, D. J. (1983). Differential Lidar and Its Applications, in *Optical Remote Sensing of Air Pollution*, P. Camagni and S. Sandroni, eds., Elsevier Science, Amsterdam.

Brassington, D. J., R. C. Felton, B. W. Jolliffe, B. R. Marx, J. T. M. Moncrieff, W. R. C. Rowley, and P. T. Woods (1984). Errors in Spectroscopic Measurements of SO_2 Due to Nonexponential Absorption of Laser Radiation, with Applications to the Remote Monitoring of Atmospheric Pollutants. *Appl. Optics* **23,** 469–475.

Browell, E. V. (1982). Lidar Measurements of Tropospheric Gases. *Opt. Eng.* **21,** 128–132.

Browell, E. V., A. F. Carter, S. T. Shipley, R. J. Allen, C. F. Butler, M. N. Mayo, J. H. Siviter, Jr., and W. M. Hall (1983). NASA Multipurpose Airborne DIAL System and Measurements of Ozone and Aerosol Profiles. *Appl. Optics* **22,** 522–534.

Bufton, J. L., and R. W. Stewart (1980). Measurement of Ozone Vertical Profiles in the Boundary Layer with a CO_2 Differential Absorption Lidar. Abstr. 10th International Laser Radar Conference, 6–9 October 1980, University of Maryland, pp. 75–76.

Bufton, J. L., T. Itabe, L. L. Strow, C. L. Korb, B. M. Gentry, and C. Y. Weng (1983a). Frequency-Doubled CO_2 Lidar Measurement and Diode Laser Spectroscopy of Atmospheric CO_2. *Appl. Optics* **22,** 2592–2602.

Bufton, J. L., T. Itabe, and D. A. Grolemund (1983b). Airborne Remote Sensing Measurements with a Pulsed CO_2 DIAL System, in *Optical and Laser Remote Sensing*, D. K. Killinger and A. Mooradian, eds., Springer-Verlag, Berlin.

Byer, R. L., and M. Garbuny (1973). Pollutant Detection by Absorption Using Mie Scattering and Topographic Targets as Retroreflectors. *Appl. Optics* **12,** 1496–1505.

Byer, R. L., and L. A. Shepp, (1979). Two-Dimensional Remote Air-Pollution Monitoring via Tomography. *Opt. Lett.* **4,** 75–77.

Cahen, C., and G. Mégie (1981). A Spectral Limitation of the Range Resolved Differential Absorption Lidar Technique. *J. Quant. Spectrosc. Radiat. Transfer* **25,** 151–157.

Cahen, C., J. P. Jegou, J. Pelon, P. Gildwarg, and J. Porteneuve (1981). Wavelength Stabilization and Control of the Emission of Pulsed Dye Lasers by Means of a Multibeam Fizeau Interferometer. *Rev. Phys. Appl.* **16,** 353–358.

Cassidy, D. T., and J. Reid (1982). Atmospheric Monitoring of Trace Gases Using Tunable Diode Lasers. *Appl. Optics* **21,** 1185–1190.

Chandra, S., and A. Compaan (1979). Double-Frequency Dye Lasers with a Continuously Variable Power Ratio. *Opt. Comm.* **31,** 73–75.

Collis, R. T. H., and P. B. Russell (1976). Lidar Measurement of Particle and Gases by Elastic Backscattering and Differential Absorption, in *Laser Monitoring of the Atmosphere*, E. D. Hinkley, ed., Springer-Verlag, Berlin.

Davies, J. H., A. R. Barringer, and R. Dick (1983). Gaseous Correlation Spectrometric Measurements, in *Optical and Laser Remote Sensing*, D. K. Killinger and A. Mooradian, eds., Springer-Verlag, Berlin.

Derr, V. E., and C. G. Little (1970). A Comparison of Remote Sensing of the Clear Atmosphere by Optical, Radio, and Acoustic Radar Techniques. *Appl. Optics* **9,** 1976–1992.

Edner, H., K. A. Fredriksson, H. Hertz, and S. Svanberg (1983). UV Lidar Techniques for Atmospheric NO Monitoring. Report No. LRAP-21, Lund Institute of Technology, Sweden.

Edner, H., K. A. Fredriksson, A. Sunesson, and W. E. Wendt (1985). Cl_2 Monitoring Using a DIAL System. Unpublished report, Lund Institute of Technology, Sweden.

Egebäck, A.-L., K. A. Fredriksson, and H. M. Hertz (1984). DIAL Techniques for the Control of Sulfur Dioxide Emission. *Appl. Optics* **23,** 722–729.

Eloranta, E. W., J. M. King, and J. A. Weinman (1975). The Determination of Wind Speeds in the Boundary Layer by Monostatic Lidar. *J. Appl. Meteorology* **14,** 1485–1489.

Englisch, W., and W. Wiesemann (1978). Remote Sensing of Atmospheric Trace Gases by Differential Absorption Spectroscopy. Proc. Int. Conf. on Earth Observation from Space and Management of Planetary Resources, 6–11 March 1978, Toulouse, France, European Space Agency, (ESA SP-134), pp. 465–473.

Englisch, W., W. Wiesemann, J. Boscher, M. Rother, and F. Lehman (1983). Laser Remote Sensing Measurements of Atmospheric Species and Natural Target Reflectivities, in *Optical and Laser Remote Sensing*, D. K. Killinger and A. Mooradian, eds., Springer-Verlag, Berlin.

Engoyan, T. M., V. I. Kozintsev, V. G. Nikiforov, and A. F. Silnitskii (1983). Determination of NO_2 Concentrations in the Atmosphere by Differential-Absorption Lidar. *Zhurnal Prikladnoi Spektroskopii* **39,** 87–93, (in Russian).

Fredriksson, K. A. (1982). Conclusions from the Evaluation and Testing of the Swedish Mobile Lidar System. Report No. SNV PM 1639, National Swedish Environment Protection Board.

Fredriksson, K. A. (1985). DIAL Technique for Pollution Monitoring: Improvements and Complementary Systems. *Appl. Optics* **24,** 3297–3304.

Fredriksson, K. A., and H. M. Hertz (1984). Evaluation of the DIAL Technique for Studies on NO_2 Using a Mobile Lidar System. *Appl. Optics* **23,** 1403–1411.

Fredriksson, K. A., B. Galle, K. Nyström, and S. Svanberg (1979). Lidar System Applied in Atmospheric Pollution Monitoring. *Appl. Optics* **18,** 2998–3003.

Fredriksson, K. A., B. Galle, K. Nyström, and S. Svanberg (1981). Mobile Lidar System for Environmental Probing. *Appl. Optics* **20,** 4181–4189.

Fukuda, T., Y. Matsuura, and T. Mori (1984). Sensitivity of Coherent Range-Resolved Differential Absorption Lidar. *Appl. Optics* **23,** 2026–2032.

Grant, W. B., R. D. Hake, Jr., E. M. Liston, R. C. Robbins, and E. K. Proctor (1974). Calibrated Remote Measurements of NO_2 Using Differential Absorption Backscatter Technique. *Appl. Phys. Lett.* **24,** 550–552.

Grant, W. B. (1982a). Effect of Differential Spectral Reflectance of DIAL Measurement Using Topographic Targets. *Appl. Optics* **21,** 2390–2394.

Grant, W. B. (1982b). Helium-Neon Laser Remote Measurement of Methane at Landfill Sites. *Proceedings Conference on Resource Recovery from Solid Wastes*, S. Sengupta and K.-F.Y. Wong, eds., Pergamon, New York.

Grant, W. B., and R. D. Hake, Jr. (1975). Calibrated Remote Measurements of SO_2 and O_3 Using Atmospheric Backscattering. *J. Appl. Phys.* **46,** 3019–3023.

Griggs, M. (1968). Absorption Coefficients of Ozone in the Ultraviolet Regions. *J. Chem. Phys.* **49,** 857–859.

Guagliardo, J. L., and D. H. Bundy (1975). Earth Reflected Differential Absorption Using TEA Lasers: A Remote Sensing Method for Ozone. 7th International Laser Radar Conference, Palo Alto, California.

Halldórsson, T., and J. Langerholc (1978). Geometrical Form Factors for the Lidar Function. *Appl. Optics* **17,** 240–244.

Harms, J. (1979). Lidar Return Signals for Coaxial and Noncoaxial Systems with Central Obstruction. *Appl. Optics* **18,** 1559–1566.

Harms, J., W. Lahmann, and C. Weitkamp (1978). Geometrical Compression of Lidar Return Signals. *Appl. Optics* **17,** 1131–1135.

Hawley, J. G., L. D. Fletcher, and G. F. Wallace (1983). Ground-Based Ultraviolet Differential Absorption Lidar (DIAL) System and Measurements, in *Optical and*

Laser Remote Sensing, D. K. Killinger and A. Mooradian, eds., Springer-Verlag, Berlin.

Henningsen, T., M. Garbuny, and R. L. Byer (1974). Remote Detection of CO by Parametric Tunable Lasers. *Appl. Phys. Lett.* **24,** 242–244.

Herget, W. A. (1982). Remote and Cross-Stack Measurement of Stack Gas Concentrations Using a Mobile FT-IR System. *Appl. Optics* **21,** 635–641.

Herget, W. A. (1982). Remote and Cross-Stack Measurement of Stack Gas Concentrations Using a Mobile FT-IR System. *Appl. Optics* **21,** 635–641.

Hinkley, E. D. (1976). Laser Spectroscopic Instrumentation and Techniques: Long Path Monitoring by Resonance Absorption. *Opt. Quant. Electr.* **8,** 155–167.

Hinkley, E. D., T. Ku, and P. L. Kelley (1976). Techniques for Detection of Molecular Absorption Pollutants by Absorption of Laser Radiation, in *Laser Monitoring of the Atmosphere*, E. D. Hinkley, ed., Springer-Verlag,

Hoell, J. M., Jr., W. R. Wade, and R. T. Thompson, Jr. (1975). Remote Sensing of Atmospheric SO_2 Using the Differential Absorption Lidar Technique. Int. Conference on Environmental Sensing and Assessment, 14–19 Sept. 1975, Las Vegas, Nevada, IEEE (Cat. n. 75-CH 1004-1 ICESA).

Inomata, H., and A. I. Carswell (1977). Simultaneous Tunable Two-Wavelength Ultraviolet Dye Laser. *Optics Commun.* **22,** 278–282.

Jacobs, T. A., and R. R. Giedt (1965). Absorption Coefficients of Cl_2 at High Temperatures. *J. Quant. Spectr. Radiative Transfer* **5,** 457–463.

Jolliffe, B. W., R. C. Felton, N. R. Swann, and P. T. Woods (1984). Field Measurement Studies Using a Differential Absorption Lidar System. Abstr. 12th International Laser Radar Conference, 13–17 August 1984, Aix en Provence, France, Service d'Aeronomie du CNRS, 267–268, and personal communication.

Kildal, H., and R. L. Byer (1971). Comparison of Laser Methods for the Remote Detection of Pollutants. *Proc. IEEE* **59,** 1644–1663.

Killinger, D. K., and N. Menyuk (1981). Remote Probing of the Atmosphere Using a CO_2 DIAL System. *IEEE J. Quantum Electronics* **QE-17,** 1917–1929.

Killinger, D. K., N. Menyuk, and W. E. DeFeo (1980). Remote Sensing of CO Using Frequency-Doubled CO_2 Laser Radiation. *Appl. Phys. Lett.* **36,** 402–405.

Killinger, D. K., N. Menyuk, and W. E. DeFeo (1983). Experimental Comparison of Heterodyne and Direct Detection for Pulsed Differential Absorption CO_2 Lidar. *Appl. Optics* **22,** 682–689.

Konefal, Z., J. Szczepanski, and J. Heldt (1981). NO_2 Detection in the Atmosphere Using Differential Absorption Lidar. *Acta Phys. Pol. A* **A60,** 273–278.

Korb, C. L., and C. Y. Weng (1982). A Theoretical Study of a Two-Wavelength Lidar Technique for the Measurement of Atmospheric Temperature Profiles. *J. Appl. Meteorol.* **21,** 1346–1355.

Korb, C. L., and C. Y. Weng (1984). An Effective Frequency Technique for Representing Finite Bandwidth Effects in DIAL Experiments. Abstr. 12th International Laser Radar Conference, 13–17 August 1984, Aix en Provence, France, Service d'Aeronomie du CNRS, 177–180.

Krüger, G. (1982). Optoacoustic Measurement of HCl Specific Absorption Coefficients at DF Laser Wavelengths. *Appl. Optics* **21,** 2841–2844.

Ku, R. T., E. D. Hinkley, and J. O. Sample (1975). Long-Path Monitoring of Atmospheric Carbon Monoxide with a Tunable Diode Laser System. *Appl. Optics* **14,** 854–861.

Loper, G. L., G. R. Sasaki, and M. A. Stamps (1982). Carbon Dioxide Laser Absorption Spectra of Toxic Industrial Compounds. *Appl. Optics* **21,** 1648–1653.

Lotem, H., and R. T. Lynch, Jr. (1975). Double-Wavelength Laser. *Appl. Phys. Lett.* **17,** 344–346.

Lotem, H., R. T. Lynch, and N. Bloembergen (1976). Interference between Raman Resonances in Four-Wave Difference Mixing. *Phys. Rev. A* **14,** 1748–1755.

Lundqvist, S., C. O. Fält, U. Persson, B. Marthinsson, and S. T. Eng (1981). Air Pollution Monitoring with a Q-Switched CO_2-Laser Lidar Using Heterodyne Detection. *Appl. Optics* **20,** 2534–2538.

Marx, B. R., G. Holloway, and L. Allen (1976). Simultaneous Two-Wavelength Narrow-Band Output from a Pulsed Dye Laser. *Opt. Commun.* **18,** 437–438.

Marzorati, A., W. Corio, and E. Zanzottera (1984). Remote Sensing of SO_2 During Fields Tests at Fos-Berre in June 1983. Abstr. 12th International Laser Radar Conference, 13–17 August 1984, Aix en Provence, France, Service d'Aeronomie du CNRS, 259–262.

Mayer, A., J. Comera, H. Charpentier, and C. Jaussaud (1978). Absorption Coefficients of Various Pollutant Gases at CO_2 Laser Wavelengths; Application to the Remote Sensing of those Pollutants. *Appl. Optics* **17,** 391–393.

McClenny, W. A., and G. M. Russwurm (1978). Laser-Based Long Path Monitoring of Ambient Gases—Analysis of Two Systems. *Atmos. Environ.* **12,** 1443–1454.

McDowell, R. S. (1981). Vibrational Spectroscopy Using Tunable Lasers, in *Vibrational Spectra and Structure*, J. R. Durig, ed., Elsevier Scientific, New York.

Measures, R. M. (1984). *Laser Remote Sensing. Fundamentals and Applications*, Wiley, New York.

Measures, R. M., and G. Pilon (1972). A Study of Tunable Laser Techniques for Remote Mapping of Specific Gaseous Constituents of the Atmosphere. *Opto-electronics* **4,** 141–153.

Mégie, G., and R. T. Menzies (1980). Complementary of UV and IR Differential Absorption Lidar for Global Measurements of Atmospheric Species. *Appl. Optics* **19,** 1173–1183.

Mégie, G., J. Y. Allain, M. L. Chanin, and E. Blamont (1977). Vertical Profile of Stratospheric Ozone by Lidar Sounding from the Ground. *Nature* **270,** 329–331.

Menyuk, N., and D. K. Killinger (1981). Temporal Correlation Measurements of Pulsed Dual CO_2 Lidar Returns. *Opt. Lett.* **6,** 301–303.

Menyuk, N., and D. K. Killinger (1983). Assessment of Relative Error Sources in IR DIAL Measurement Accuracy. *Appl. Optics* **22,** 2690–2698.

Menyuk, N., D. K. Killinger and W. E. De Feo (1980). Remote Sensing of NO Using a Differential Absorption Lidar. *Appl. Optics* **19,** 3282–3286.

Menyuk, N., D. K. Killinger, and W. E. DeFeo (1982a). Laser Remote Sensing of Hydrazine, MMH, and UDMH Using a Differential-Absorption CO_2 Lidar. *Appl. Optics* **21,** 2275–2286.

Menyuk, N., D. K. Killinger, and C. R. Menyuk (1982b). Limitations of Signal Averaging Due to Temporal Correlation in Laser Remote-Sensing Measurements. *Appl. Optics* **21,** 3377–3383.

Menyuk, N., D. K. Killinger, and C. R. Menyuk (1985). Error Reduction in Laser Remote Sensing: Combined Effects of Cross Correlation and Signal Averaging. *Appl. Optics* **24,** 118–131.

Menzies, R. T., and M. S. Shumate (1976). Remote Measurements of Ambient Air Pollutants with a Bistatic Laser System. *Appl. Optics* **15,** 2080–2084.

Menzies, R. T., and M. S. Shumate (1978). Tropospheric Ozone Distributions Measured with an Airborne Laser Absorption Spectrometer. *J. Geophys. Res.* **83,** 4039–4043.

Michaelis, W., and C. Weitkamp (1984). Sensitive Remote and in situ Detection of Air Pollutants by Laser Light Absorption Measurements. *Fresenius Z. Anal. Chem.* **317,** 286–292.

Murray, E. R., and J. E. van der Laan (1978). Remote Measurement of Ethylene Using CO_2 Differential-Absorption Lidar. *Appl. Optics* **17,** 814–817.

Murray, E. R., J. E. van der Laan, and J. G. Hawley (1976). Remote Measurement of HCl, CH_4 and NO_2 Using Single-Ended Chemical-Laser Lidar System. *Appl. Optics* **15,** 3140–3148.

Nair, L. G., and K. Dasgupta (1980). Double Wavelength Operation of a Grazing Incidence Tunable Dye Laser. *IEEE J. Quantum Electronics* **QE-16,** 111–112.

Nestor, J. R. (1982). Optogalvanic Spectra of Neon and Argon in Glow Discharge Lamps. *Appl. Optics* **21,** 4154–4157.

Onderdelinden, D., and L. Strackee (1978). Remote Sensing of NO_2 Air Pollution Using an Optical Multi-Channel Analyser. *J. Phys. D: Appl. Phys.* **12,** 979–985.

Oppenheim, U. P., and R. T. Menzies (1982). Aligning the Transmitter and Receiver Telescopes of an Infrared Lidar: a Novel Method. *Appl. Optics* **21,** 174–175.

Pal, S. R., and A. I. Carswell (1976). Multiple Scattering in Atmospheric Clouds: Lidar Observations. *Appl. Optics* **15,** 1990–1995.

Pal, S. R., and A. I. Carswell (1978). Polarization Properties of Lidar Scattering from Clouds at 347 nm and 694 nm. *Appl. Optics* **17,** 2321–2328.

Pelon, J., and G. Mégie (1982). Ozone Monitoring in the Troposphere and Lower Stratosphere: Evaluation and Operation of a Ground-Based Lidar Station. *J. Geophys. Res.* **87,** 4947–4955.

Persson, U., J. Johansson, B. Marthinsson, and S. T. Eng. (1982). Ethylene Mass Flow Measurements with an Automatic CO_2-Laser Long Path Absorption System. *Appl. Optics* **21,** 4417–4425.

Petheram, J. C. (1981). Differential Backscatter from the Atmospheric Aerosol: the Implications for IR Differential Absorption Lidar. *Appl. Optics* **20,** 3941–3946.

Platt, U., and D. Perner (1983). Measurements of Atmospheric Trace Gases by Long Path Differential UV/Visible Absorption Spectroscopy, in *Optical and Laser Remote Sensing*, D. K. Killinger and A. Mooradian, eds., Springer-Verlag, Berlin.

Rothe, K. W. (1980). Monitoring of Various Atmospheric Constituents Using a CW Chemical Hydrogen/Deuterium Laser and a Pulsed Carbon Dioxide Laser. *Radio Electron. Eng.* **50,** 367.

Rothe, K. W., U. Brinkman, and H. Walther (1974a). Applications of Tunable Dye Lasers to Air Pollution Detection; Measurements of Atmospheric NO_2 Concentration by Differential Absorption. *Appl. Phys.* **3,** 115–119.

Rothe, K. W., U. Brinkman, and H. Walther (1974b). Remote Sensing of NO_2 Emission from a Chemical Factory by the Differential Absorption Technique. *Appl. Phys.* **4,** 181–184.

Rothman, L. S., R. R. Gamache, A. Barbe, A. Goldman, J. R. Gillis, L. R. Brown, R. A. Toth, J.-M. Flaud, and C. Camy-Payret (1983a). AFGL Atmospheric Absorption Line Parameters Compilation: 1982 Edition. *Appl. Optics* **22,** 2247–2256.

Rothman, L. S., A. Goldman, J. R. Gillis, R. R. Gamache, H. M. Pickett, R. L. Poynter, N. Husson, and A. Chedin (1983b). AFGL Trace Gas Compilation: 1982 Version. *Appl. Optics* **22,** 1616–1627.

Sage, J.-P., and Y. Aubry (1982). High Power Tunable Dual Frequency Laser System. *Opt. Comm.* **42,** 428–430.

Sasano, Y., H. Shimizu, N. Takeuchi, and M. Okuda (1979). Geometrical Form Factor in the Laser Radar Equation: an Experimental Determination. *Appl. Optics* **18,** 3908–3910.

Sasano, Y., H. Hirohara, T. Yamasaki, H. Shimizu, N. Takeuchi, and T. Kawamura (1982). Horizontal Wind Vector Determination from the Displacement of Aerosol Distribution Patterns Observed by a Scanning Lidar. *J. Appl. Meteorology* **21,** 1516–1523.

Sassen, K., and G. C. Dodd (1982). Lidar Crossover Function and Misalignment Effects. *Appl. Optics* **21,** 3162–3165.

Schotland, R. M. (1966). Some Observations of the Vertical Profile of Water Vapor by a Laser Optical Radar. Proc. 4th Symposium on Remote Sensing of the Environment 12–14 April 1966, Univ. of Michigan, Ann Arbor, pp. 273–283.

Schmidt, A. J. (1975). Simultaneous Two-Wavelength Output of an N_2-Pumped Dye Laser. *Opt. Comm.* **14,** 294–295.

Shumate, M. S., R. T. Menzies, W. B. Grant, and D. S. McDougal (1981). Laser Absorption Spectrometer: Remote Measurement of Tropospheric Ozone. *Appl. Optics* **20,** 545–553.

Shumate, M. S., S. Lundqvist, U. Persson, and S. T. Eng (1982). Differential Reflectance of Natural and Man-Made Materials at CO_2 Laser Wavelengths. *Appl. Optics* **21,** 2386–2389.

Sroga, J. T., E. W. Eloranta, and T. Barber (1980). Lidar Measurement of Wind Velocity Profiles in the Boundary Layer. *J. Appl. Meteorology* **19,** 598–605.

Staehr, W., W. Lahman, C. Weitkamp, and W. Michaelis (1984). Differential Absorption Lidar System for NO_2 and SO_2 Monitoring. Abstr. 12th International Laser Radar Conference, 13–17 August 1984, Aix en Provence, France, Service d'Aeronomie du CNRS, 281–284.

Staehr, W., W. Lahman, and C. Weitkamp (1985). Range-resolved Differential Absorption Lidar: Optimization of Range and Sensitivity. *Appl. Optics* **24,** 1950–1956.

Sugimoto, N., N. Takeuchi, and M. Okuda (1981). Remote Measurement of NO_2 Profile in a Stack Plume by a Differential Absorption Lidar. *Oyo Buturi* **50,** 923–928.

Sutton, S. (1986). Differential Lidar Measurements of Ozone in the Troposphere. Report No TPRD/L/2967/N85, Central Electricity Generating Board, UK.

Takeuchi, N., H. Shimizu, and M. Okuda (1978). Detectivity Estimation of the DAS Lidar for NO_2. *Appl. Optics* **17,** 2734–2738.

Tsuji, T., H. Kimura, Y. Higuchi, and K. Goto (1976). NO_2 Concentration Measurement in the Atmosphere Using Differential Absorption Dye Laser Radar Technique. *Jap. J. Appl. Phys.* **15,** 1743–1752.

Uchino, O., M. Maeda, J. Khono, T. Shibata, C. Nagasawa, and M. Hirono (1978). Observation of Stratospheric Ozone Layer by a XeCl Laser Radar. *Appl. Phys. Lett.* **33,** 807–809.

Uchino, O., M. Maeda, and M. Hirono (1979). Applications of Excimer Lasers to Laser-Radar Observations of the Upper Atmosphere. *IEEE J. Quantum Electronics* **QE-15,** 1094–1107.

Uchino, O., M. Maeda, H. Yamamura, and M. Hirono (1983a). Observation of Stratospheric Vertical Ozone Distribution by a XeCl Lidar. *J. Geophysical Research* **88,** 5273–5280.

Uchino, O., M. Tokunaga, M. Maeda, and Y. Miyazoe (1983b). Differential-Absorption-Lidar Measurement of Tropospheric Ozone with Excimer-Raman Hybrid Laser. *Opt. Lett.* **8,** 347–349.

Ward, R. Y., and H. H. Zwick (1975). Gas Cell Correlation Spectrometer: GASPEC. *Appl. Optics* **14,** 2896–2904.

Watkins, W. R., and K. O. White (1981). Wedge Absorption Remote Sensor. *Rev. Sci. Instrum.* **52,** 1682–1684.

Weitkamp, C. (1981). The Distribution of Hydrogen Chloride in the Plume of Incineration Ships: Development of New Measurement Systems, in *Wastes in the Ocean*, Vol. 3, Wiley, New York.

Weitkamp, C., W. Michaelis, H.-J. Heinrich, R. Baumgart, H. Lohse, H.-T. Mengelkamp, D. Eppel, A. Müller, U. Lenhard, H. J. Eberhardt, and C. Muschner (1984). Hydrogen Chloride Distribution in Plumes of Incineration Ships: Preliminary Results of a Campaign with R/V Tabasis in Summer 1982. Report No. GKSS 83/E/10 E, GKSS-Forschungszentrum, Geesthact, Germany.

Werner, J., K. W. Rothe, and H. Walther (1983). Monitoring of the Stratospheric Ozone Layer by Laser Radar. *Appl. Phys. B* **32,** 113–118.

Wiesemann, W., R. Beck, W. Englisch, and K. Gürs (1978). In-Flight Test of a Continuous Laser Remote Sensing System. *Appl. Phys.* **15,** 257–260.

Wolfe, D. C., Jr., and R. L. Byer (1982). Model Studies of Laser Absorption Computed Tomography for Remote Air Pollution Measurement. *Appl. Optics* **21,** 1165–1178.

Woods, P. T., and B. W. Jolliffe (1978). Experimental and Theoretical Studies Related to a Dye Laser Differential Lidar System for the Determination of Atmospheric SO_2 and NO_2 Concentrations. *Optics Laser Tech.* **10,** 25–28.

Woods, P. T., B. W. Jolliffe, and B. R. Marx (1980). High Resolution Spectroscopy of SO_2 Using a Frequency-Doubled Pulsed Dye Laser with Application to Remote Sensing of Atmospheric Pollutants. *Opt. Commun.* **33,** 281–286.

Yue, G. K., G. S. Kent, U. O. Farrukh, and A. Deepak (1983). Modeling Atmospheric Aerosol Backscatter at CO_2 Laser Wavelengths. 3: Effects of Changes in Wavelength and Ambient Conditions. *Appl. Optics* **22,** 1671–1678.

CHAPTER

5

LASER MEASUREMENTS OF ATMOSPHERIC TRACE CONSTITUENTS

G. MEGIE

Service d'Aéronomie
Centre National de la Recherche Scientifique
91371, Verrières le Buisson
Paris, France

5.1 INTRODUCTION

The behavior of trace constituents in Earth's upper atmosphere, governed by chemical, dynamical, and radiative processes, is of particular importance for the overall balance of the stratosphere and mesosphere. In particular, ozone plays a dominant role by absorbing the short-wavelength UV radiation, which might damage living organisms, and by maintaining the radiative budget equilibrium. The possible modifications of the ozone vertical distribution resulting from anthropogenic activities has emphasized during the last decade the need for a better understanding of middle-atmosphere processes and for accurate monitoring systems to detect early trends. Although the existence of atomic trace metals in the upper mesosphere may not affect directly the global equilibrium of the same regions due to their very low abundance, it has raised several questions in atmospheric science since its early detection in the nightglow emission in the 1920s. The actual theories that predict the existence of these trace metals in various compounds at lower stratospheric levels have also drawn the attention to their potential influence on the chemical equilibrium of trace species such as chlorine or bromine that directly affect the behavior of the ozone layer through regenerating catalytical cycles.

Remote measurements of trace constituents in Earth's upper atmosphere using an active technique like lidar have been made possible by the rapid development of powerful tunable laser sources that have opened a new field in atmospheric spectroscopy by providing sources that can be tuned to characteristic spectral features of atmospheric constituents. The interaction of a laser beam with the atmosphere is in principle dominated at all wavelengths by elastic processes, that is, Rayleigh scattering from atmospheric gas molecules and Mie scattering due to particles of different natures and shapes. However, considering the simplified lidar equation

$$n_r = n_e \frac{A_0}{R^2} (\beta_R + \beta_M + \beta_i) \, \Delta R \exp[-2(\tau_R + \tau_M + \tau_i)] \tag{5.1}$$

(where n_e, n_r are the number of photons emitted and received per pulse from range R; A_0 is the receiver area; β_R, β_M, and β_i are the volume backscattering coefficients for Rayleigh scattering, Mie scattering, and resonant scattering by a given species i; τ_R, τ_M, and τ_i being the corresponding optical thicknesses for the pathlength R; ΔR is the range resolution), the determination of a given constituent number density N_i is then possible by two means:

either β_i is much larger than $\beta_R + \beta_M$, allowing the measurement of N_i based on a scattering process,

or τ_i is much larger or of the same order of magnitude than $\tau_r + \tau_m$, allowing the measurement of N_i based on an absorption process.

This chapter will illustrate these two methods as they have been used for the measurement of stratospheric and mesospheric trace constituents from the ground:

1. Resonance scattering has allowed the measurement of trace metals between 80 and 100 km with a very large sensitivity down to the 10^{-12} mixing ratio range (less than 1 atom cm^{-3}). Historically, these developments went back to 1967 and the first apparition of flashlamp-pumped dye lasers followed by the measurement of the sodium atoms vertical distribution in 1969. This kind of measurement has been extended by both large increases in spatial and temporal resolution and the detection of other metallic species including potassium, lithium, and neutral and ionized calcium. This extended set of data has led to the solution of the questions raised for 50 years on the origin and behavior of these constituents.

2. Differential absorption led to the measurement of the ozone vertical distribution in both the troposphere and stratosphere. Continuous monitoring of ozone and other minor constituents in Earth's stratosphere and troposphere is of particular interest in present-day atmospheric physics, as a potential modification of the ozone layer due to man's activities can greatly modify Earth's environment and climate. Measurement of the total ozone column content and vertical profile by means of the ground based UV spectrometer network or by satellite-borne systems remains the fundamental basis for global observations and trend analysis. However, the required accuracy of these measurements will necessitate in the future the operation of new active systems, such as lidar, because they can provide a high degree of reliability in terms of accuracy and

absolute precision. Furthermore, the very large variability of the ozone number density in the troposphere and lower stratosphere and at the boundary between these two regions and its interpretation in terms of horizontal and vertical transport requires high spatial and temporal resolution measurements that are presently beyond the possibilities of passive systems.

5.2 RESONANT SCATTERING LIDAR OBSERVATIONS OF TRACE METALS IN UPPER ATMOSPHERE

The first lidar observation of trace metals (i.e., sodium) in the upper atmosphere was made by Bowman et al. (1969) following rapidly the development of the first flashlamp-pumped dye laser (Sorokin et al., 1967; Furumoto et al., 1969). This measurement was then complemented by more detailed observations made by several groups (Gibson and Sandford, 1971, 1972; Blamont et al. 1972a; Hake et al., 1972; Kirchhoff and Clemesha, 1973; Aruga et al., 1974), who successively revealed the behavior of the sodium layer at various temporal and spatial scales, that is, seasonal and diurnal variations, short-term enhancements due to meteoritic input, and the influence of dynamical processes such as tides or gravity waves. In 1973, Felix et al. made a first tentative measurement of a second alkali-potassium and a larger set of data was obtained in 1978 by Mégie et al. This last group extended the resonant scattering technique to the observations of lithium (Jégou et al., 1980) and neutral and ionized calcium (Granier et al., 1984). The different behavior of these metals was then explained by a general theory involving both neutral and ionized compounds that accounts for the observed variations (Jégou et al., 1985a,b).

We will successively review in this section the general theory of resonant scattering applied to lidar observations by studying the influence of limited laser bandwidth, pulse duration, and polarization effects on the scattering process efficiency, describing the various systems used during the last decade and giving illustrations of the significant results obtained and of their interpretation. Two types of laser sources have been used for these measurements—flashlamp-pumped dye laser and Nd–YAG laser-excited dye laser—which differ in terms of their emission characteristics. These system will be analyzed in more details in Section 5.2.2. Table 5.1 gives an average value of their parameters, which might be helpful for the determination of the various uncertainty sources reviewed hereafter. The receiver characteristics are also given to illustrate the calculation of statistical errors.

5.2.1 Basic Principles of Resonance Scattering Applied to Lidar Sounding

In this section we will only consider the case of atomic resonance scattering as it applies, taking into account upper atmospheric conditions. Due to the low

Table 5.1. Average Parameters of Resonant Lidar Systems for Sodium Measurements

Emitter	Flashlamp-Pumped Dye Laser	Nd^{3+}–YAG-Pumped Dye Laser
Output energy	500 mJ	120 mJ
Repetition rate	1 Hz	10 Hz
Pulse duration	1–3 μs	15 ns
Linewidth	1–6 pm	1.5–10 pm
Beam divergence	5.10^{-4} rad	$3\text{–}5.10^{-4}$ rad
Receiver		
Telescope diameter	0.8–1 m	
Telescope field of view	10^{-3} rad (nighttime) to 3.10^{-3} rad (daytime)	
Receiver bandwidth	0.5 nm (nighttime) to 20 pm (daytime)	

pressure of air in the 80–100-km-altitude range (10^{-5}–10^{-6} atm), we will consider that the scattering processes are independent and that only single scattering occurs. The independence of the scattering centers is related to their adequate separation—mean free path of 10^{-5} cm large as compared to atomic distances—and their random motion due to collision processes. The total scattered intensity can thus be calculated by simply adding the intensity from each scattering particle. Single scattering implies that a photon is only scattered once. This assumption holds true throughout the atmosphere except in the case of dense scattering medium such as clouds or fogs (Hinkley, 1976; Measures, 1984).

5.2.1.1 Resonance Scattering Processes

Resonance scattering of light by an atom involves the absorption of a photon to bring the atom in an excited state and its subsequent spontaneous emission. In this two-photon process, the emitted photon has essentially the same frequency as that of the original frequency of the absorbed photon. Furthermore, the scattering process is isotropic so that the probability of the photon being emitted in a given solid angle $d\Omega$ is equal to $d\Omega/4\pi$. To further analyze the interaction of light with atoms in the case of resonance scattering, we will consider the case of sodium atoms. This will permit a detailed description of this process to be given and provide a basis for generalization to any particular constituent.

Sodium atoms strongly absorb in the yellow part of the visible spectrum centered at 589.3 nm. The energy diagram presented in Figure 5.1 corresponds to the electronic transition between the ground state 3^2S and the first excited

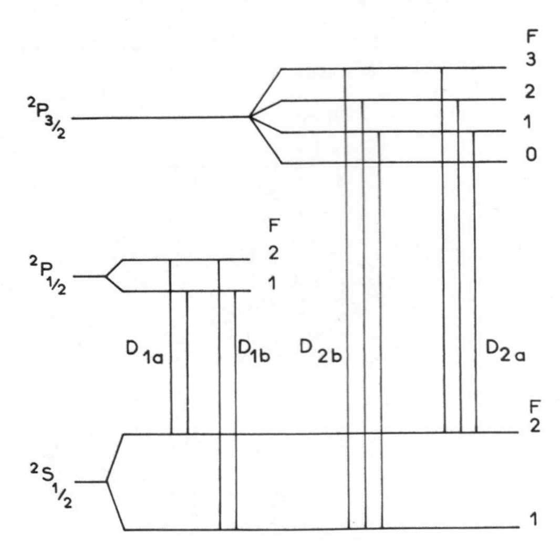

Figure 5.1. Energy level diagram of sodium D_1 and D_2 lines.

state 3^2P of sodium. Due to the spin orbit interaction, the excited state is split into two levels corresponding to two different values ($J = \frac{1}{2}$, $J = \frac{3}{2}$) of the angular momentum quantum number J. As the ground-state value of J is equal to $\frac{1}{2}$, two transitions are optically allowed corresponding to the well-known sodium doublet at wavelengths 588.997 nm (D_2) and 589.593 nm (D_1). Furthermore, the nucleus of the sodium atom has an odd number of protons (11) so that the nuclear spin I is different from zero: $I = \frac{3}{2}$. This results in a hyperfine coupling that, for example, in the case of the D_2 line, will split the lower level into two hyperfine states ($F = 1$; $F = 2$) separated by 1771.75 MHz (2.048 pm) and the upper level into four separate states ($F = 0$; $F = 1$; $F = 2$; $F = 3$). The frequency separation $\Delta\nu_H$ between the various hyperfine components is given by (White, 1934)

$$\Delta\nu_H = g_I \frac{R\alpha^2 Z_i^2 Z_0^2}{n^3(L + \frac{1}{2})J(J + 1)} \tag{5.2}$$

where g_I is the nuclear Landé factor, R the Rydberg constant, α the fine structure constant ($\alpha = 2\pi e^2/hc$), n the principal quantum number, Z_i and Z_0 the nuclear charges, respectively, inside and outside of the electronic envelope, L the orbital angular momentum quantum number ($L = 1$ for the $2P$ level), and

J the total angular momentum quantum number. Application of (5.2) for the two levels ${}^2S_{1/2}$ and ${}^2P_{1/2}$ leads to

$$\frac{\Delta\nu_H({}^2S_{1/2})}{\Delta\nu_H({}^2P_{3/2})} = 15 \tag{5.3}$$

Two series of lines called, respectively, D_{2a} and D_{2b} can thus be isolated. Within these series only transitions such as $\Delta F = 0, \pm 1$ are allowed, as represented in Table 5.2, where each of them is characterized by its frequency shift from the average center of the line and its relative intensity.

In the presence of Earth's magnetic field each hyperfine level is further split due to the Zeeman effect into $2F + 1$ sublevels. However, the maximum frequency shift between these sublevels is only 5 MHz for a 0.7-gauss (G) magnetic field and represents only 5% of the hyperfine splitting. It will not be considered hereafter except when we consider the application of resonant lidar technique to temperature measurements through the determination of the linewidth of the sodium absorption line (Section 5.2.3.3).

The natural linewidth of the excited state is directly related to its radiative lifetime of 16 ns and is equal to 20 MHz. At an average altitude of 100 km in the atmosphere the velocity distribution of the sodium atoms follows Maxwell's law, resulting in a frequency broadening of each line given by

$$\Gamma_D = 2\frac{\nu_0}{c}\left(\frac{2RT\ln 2}{M}\right)^{1/2} \tag{5.4}$$

where ν_0 is the central frequency of the transition, c the velocity of light, R the perfect gas constant, T the absolute temperature, and M the atomic mass. At 200 K in the case of sodium ($M = 23$) this leads to

$$\Gamma_D = 1075 \text{ GHz} \quad (1.246 \text{ pm}) \tag{5.5}$$

so that the lines are essentially Doppler broadened.

For each of the hyperfine components the frequency dependence of the atomic absorption coefficient per atom (or cross section) $\sigma_s(\nu)$ is then given by

$$\sigma_s(\nu) = \sigma_s(\nu_0)\exp\left[-\left(2\sqrt{\ln 2}\,\frac{\nu - \nu_0}{\Gamma_D}\right)^2\right] \tag{5.6}$$

Using Ladenburg's formula (Mitchell and Zemanski, 1934), the atomic absorp-

Table 5.2. Resonant Transitions of Metallic Species

	Transition	λ (nm)	Γ_D (200 K) (pm)	f	$\sigma_s(\nu_0)$ (cm^2)
Sodium, Na	$^2S_{1/2}$–$^2P_{3/2}$	588.9	1.245	0.655	1.52×10^{-11}
Potassium, K	$^2S_{1/2}$–$^2P_{1/2}$	766.4	1.245	0.34	1.34×10^{-11}
Lithium, Li	$^2S_{1/2}$–$^2P_{3/2}$	670.785	2.57	0.50	0.73×10^{-11}
Neutral Calcium, Ca(I)	$^1S_0$–$^1P_1$	422.673	0.68	1.75	3.85×10^{-11}
Ionized Calcium, Ca(II)	$^2S_{1/2}$–$^2P_{3/2}$	393.366	0.63	0.69	1.41×10^{-11}

tion coefficient $\sigma_s(\nu)$ is related, through its frequency integrated value, to the oscillator strength f of the transition

$$\int_0^\infty \sigma_s(\nu)\, d\nu = \frac{\pi e^2}{mc} f \tag{5.7}$$

Combination of (5.6) and (5.7) leads then to the value of the center frequency cross section

$$\sigma_s(\nu_0) = \frac{2}{\Gamma_D}\left(\frac{\ln 2}{\pi}\right)^{1/2} \frac{\pi e^2}{mc} f$$

As pointed out above, the hyperfine structure of the line has to be taken into account so that the atomic absorption coefficients $\sigma_s(\nu)$ is written as

$$\sigma_s(\nu) = \sum_i \sigma_{si}(\nu) = \sum_i \sigma_s(\nu_{0i})\exp\left[-\left(2\sqrt{\ln 2}\,\frac{\nu - \nu_{0i}}{\Gamma_{Di}}\right)^2\right] \tag{5.8}$$

where the summation is extended to the six hyperfine components. For the sodium D_2 line, the oscillator strength f equals 0.655 (Allen, 1973), and thus

$$\sigma_s(\nu_0) = \sum_i \sigma_s(\nu_{0i}) = 1.52.10^{-11}\ \text{cm}^2 \tag{5.9}$$

According to the relative values quoted in Table 5.2, one can then calculate each of the $\sigma_s(\nu_{0i})$ and thus the frequency dependence of $\sigma_s(\nu)$. This is presented in Figure 5.2 for different temperatures from 100 to 500 K and for both the D_1 and D_2 lines, showing that the hyperfine structure will have to be considered when calculating the efficiency of the scattering process.

The line parameters for resonant transitions for other atomic species of interest in atmospheric studies are given in the Table 5.3 and will allow similar calculations of the interaction of atoms with the laser light as presented in the next paragraphs.

5.2.1.2 Resonance Scattering Lidar Equation for Monochromatic Emission

The general lidar equation, which relates the received backscattered energy $E(\nu, R)$ at frequency ν (per unit frequency interval) from range R within a time interval τ_d to the emitted laser energy $E_L(\nu)$ per unit frequency interval, can be written in the case of scattering by a single species as (Measures, 1984)

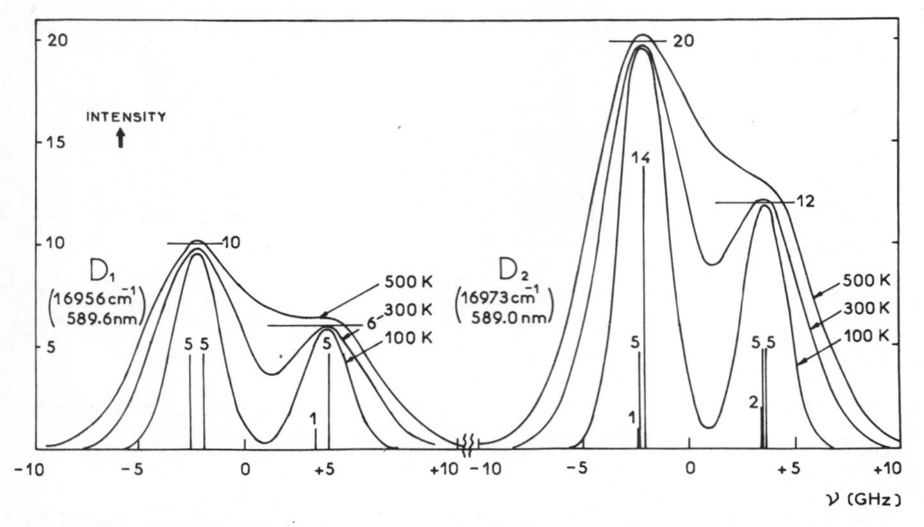

Figure 5.2. Spectral profile of D_1 and D_2 lines at various temperatures. Relative intensities of hyperfine components are also shown.

$$E(\nu, R) = E_L(\nu)\xi(\nu)\xi(R)T^2(\nu, R)\frac{A_0}{R^2}N_i(R)\frac{d\sigma_s(\nu)}{d\Omega}\frac{c\tau_d}{2} \tag{5.10}$$

where $\xi(\nu)$ = receivers spectral transmission factor

$\xi(R)$ = geometric factor, equal to 1 for total overlapping of emitter's and receiver's field of view

$N_i(R)$ = number density of species under study

Table 5.3. Characteristics of Sodium D_2 Resonance Line Hyperfine Structure

$^2S_{1/2}$	$^2P_{3/2}$	$\nu_i - \nu_0$ (GHz)	$\nu_i - \nu_0$ (pm)	f_i
		D_{2a}		
$F = 2$	$F = 1$	−0.7328	−0.8476	0.03125
$F = 2$	$F = 2$	−0.6962	−0.80513	0.15625
$F = 2$	$F = 3$	−0.6302	−0.72881	0.43750
		D_{2b}		
$F = 1$	$F = 0$	1.0333	1.19498	0.0625
$F = 1$	$F = 1$	1.0552	1.22031	0.15625
$F = 1$	$F = 2$	1.0919	1.26275	0.15625

$d\sigma_s(\nu)/d\Omega$ = differential scattering cross section under irradiation with laser radiation at frequency ν

c = velocity of light

A_0/R^2 = acceptance solid angle of receiver optics of area A_0

$T(\nu, R)$ = one-way atmospheric transmission factor at frequency ν and range R

given by

$$T(\nu, R) = \exp\left\{-2\left[\tau_R(\nu, R) + \tau_M(\nu, R) + \sum \tau_i(\nu, R)\right]\right\} \tag{5.11}$$

where $\tau_\alpha(\nu, R)$ is the one-way optical thickness due to Rayleigh extinction ($\alpha = R$), Mie extinction ($\alpha = M$), and absorption by atmospheric species ($\alpha = i$) at frequency ν.

In this equation the following assumptions have been made:

The detector response time τ_d and the laser pulse duration τ_L are small as compared to the time $2R/c$ required for the light to travel from the emitting point to the scattering volume and back (in the case of measurements between 80 and 100 km, the travel time is equal to ~ 1 ms). The effective range resolution for such a system is then limited to $c(\tau_d + \tau_L)/2$ but can be further decreased by integration over a larger time interval τ_S in the data acquisition process.

The temporal shape of the laser pulse has been approximated by a rectangle of duration τ_L.

The range-dependent parameters have been treated as constant over the time interval of range integration that we will consider to be large compared to τ_L.

The scattering medium is homogeneous over the zone of overlap between the laser beam and the receiver field of view.

The emitted and received radiations are considered as narrow band compared to the receiver spectral window.

The laser divergence is small enough so that the volume of space that comprises the range bin can be considered as a cylinder.

Furthermore, since we are considering only measurements made at higher altitude, above at least a few tens of kilometers, we will assume the following.

The geometric factor $\xi(R)$ is equal to 1 over the whole range of observation.

The scattering process is isotropic (cf. 5.2.1.1) so that

$$\frac{d\sigma_s(\nu)}{d\Omega} = \frac{\sigma_s(\nu)}{4\pi} \tag{5.12}$$

$\sigma_s(\nu)$ being considered as independent of the range R by neglecting the influence of temperature variations.

The metallic species under study is the only absorber at frequency ν, at least over the range of the measurement.

We will then rewrite (5.11) as

$$n_r(\nu, R) = n_e(\nu)\xi(\nu)T^2(\nu, R)\frac{A_0}{R^2}N_i(R)\frac{\sigma_s(\nu)}{4\pi}\,\Delta R \tag{5.13}$$

where ΔR is the range resolution equal to $\frac{1}{2}c\tau_S$, and the received signal has been expressed in terms of number of photons per unit frequency and per unit time, and $n_e(\nu)$ is the number of emitted photons in the same time and frequency interval. In this simple form the trace-metal number density at range R and within the altitude range ΔR can be thus directly determined assuming the knowledge of all experimental parameters including $\xi(\nu)$ and measuring the backscattering signal $n_r(\nu, R)$. However, this equation holds true only at a given frequency, and the effect of the finite laser and absorption linewidth should be considered in further detail. Furthermore, the exact value of $\xi(\nu)$ cannot usually be determined from simple experimental consideration, requiring instead a precise calibration procedure. All these effects will be studied in subsequent sections.

5.2.1.3 *Effect of Finite Linewidth*

The spectral shape of the absorption line has been described in detail in Section 5.2.1.1 in the case of sodium, which we will again consider as the reference case. The laser lineshape can be characterized by a distribution function $g(\nu)$ that, when evaluated at ν, represents the fraction of the total number of photons emitted in the frequency interval ν, $\nu + d\nu$. The distribution function $g(\nu)$ is normalized so that

$$\int_0^\infty g(\nu)\, d\nu = 1 \tag{5.14}$$

Equation (5.13) has then to be integrated over frequency to give the total number of photons backscattered at range R. In this derivation we will further assume that the frequency integration range is small enough to consider both $\xi(\nu)$

and $T(\nu, R)$ as constant over this interval [the contribution to the atmospheric transmission of the absorption $\tau_i(\nu, R)$ by the sodium itself will be considered in Section 5.2.1.6]. One then obtains

$$n_r(R) = n_e \xi T^2(R) \frac{A_0}{R^2} N_i(R)\, \Delta R \frac{1}{4\pi} \int_0^\infty g(\nu)\sigma_s(\nu)\, d\nu \tag{5.15}$$

where n_e is the total number of emitted photons per unit time interval. The integral term in (5.15) is generally referenced as the effective cross section $\sigma_{\text{eff}}(\nu)$, which relates the scatterer response to the frequency dependence of the atomic gas as a scattering combined with the frequency distribution of the excitation laser light. To quantitatively assess the effect of a finite laser bandwidth on the scattering process efficiency, we will approximate the laser spectral profile by a Gaussian lineshape of width Γ_L (FWHM) centered at a frequency ν_L:

$$g(\nu) = \frac{2\sqrt{\ln 2}}{\sqrt{\pi}\,\Gamma_L} \exp\left[-\left(2\sqrt{\ln 2}\,\frac{\nu - \nu_L}{\Gamma_L} \right)^2 \right] \tag{5.16}$$

According to the sodium absorption profile as given by (5.8), the effective cross section is then given for the D_2 line by

$$\sigma_{\text{eff}} = \frac{1}{\left(1 + \dfrac{\Gamma_L^2}{\Gamma_D^2}\right)^{1/2}} \sum_{i=1}^{6} k_{0i} \exp\left(- \frac{4 \ln 2 \left[(\nu_0 - \nu_i) - \Delta\nu_{0L} \right]^2}{\Gamma_L^2 + \Gamma_{Di}^2} \right) \tag{5.17}$$

where $\Delta\nu_{0L}$ is the frequency shift $\nu_0 - \nu_L$ between the center frequencies of the absorption line and the laser line. The summation is extended to the six hyperfine components of the D_2 line. Figure 5.3 gives the variation of σ_{eff} as a function of Γ_L for the case $\Delta\nu_{0L} = 0$. The effective cross section remains constant over a 1-GHz interval corresponding roughly to the value of Γ_D and then decreases rapidly with Γ_L.

The measurement of the laser lineshape and central frequency are thus required if one wants to calculate σ_{eff} for a given emitting system. However, these determinations will not be made exactly, and one has to consider the possible uncertainties on σ_{eff} related to experimental errors. The sensitivity of σ_{eff} to the laser line parameters is given in Figures 5.4 and 5.5, where we have presented the relative deviation of σ_{eff} for variations both in Γ_L and ν_{0L} for three different values of the laser linewidths (1, 4, and 10 GHz). As an example, a 10% determination of the laser linewidth for a 4-GHz line, the center frequency of which is determined within 1 GHz, will lead to an uncertainty of 20% in the σ_{eff} value.

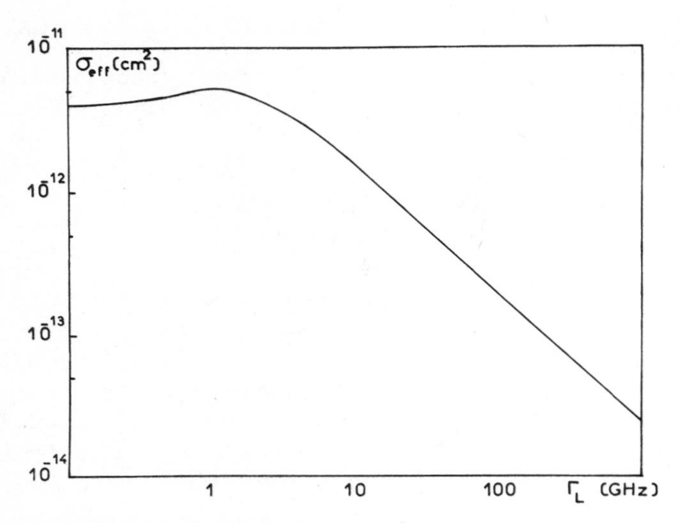

Figure 5.3. Variation of effective cross section σ_{eff} (cm^2) as function of laser linewidth Γ_L (GHz) for sodium D_2 line.

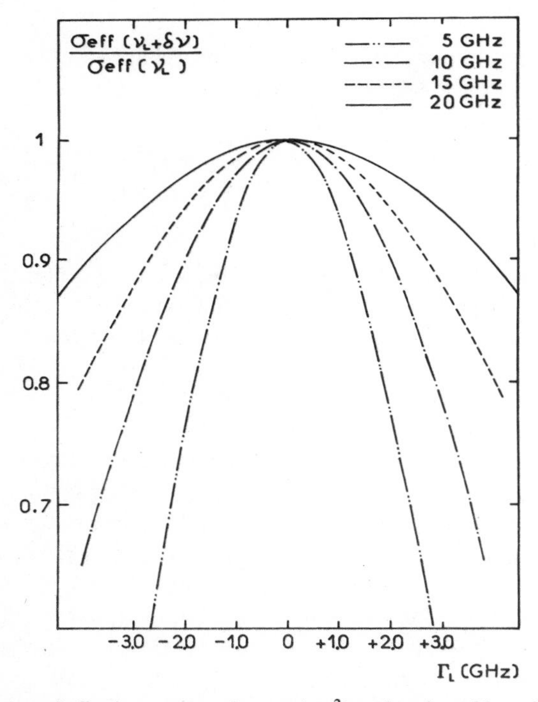

Figure 5.4. Variation of effective cross section σ_{eff} (cm^2) as function of laser line frequency with respect to central frequency of sodium D_2 line for various values of laser linewidth Γ_L (GHz). Vertical axis is ratio of effective cross-sectional values at frequency $\nu_L (= \nu_0) + \delta\nu$ and ν_0, where $\delta\nu$ is frequency shift (GHz).

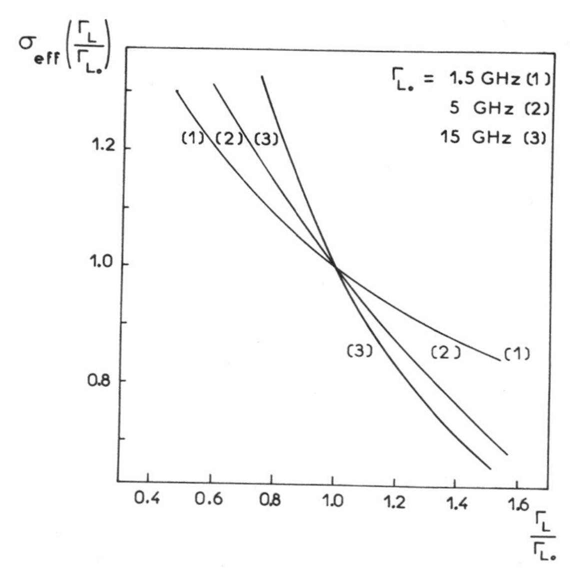

Figure 5.5. Sensitivity of effective cross section σ_{eff} (cm^2) to variations in laser linewidth Γ_L. The three curves relate ratio of cross-sectional values $\sigma_{\text{eff}}(\Gamma_L)/\sigma_{\text{eff}}(\Gamma_{L0})$ to value of Γ_L/Γ_{L0} for various values of Γ_{L0} (GHz).

5.2.1.4 Effects of Finite Lifetime of Excited State

Equation (5.12) has been derived assuming that the scattering process has an infinitely small duration. However, for a laser pulse duration in the nanosecond range, this does not hold true as the radiative lifetime τ_R of the atomic species in its excited state is generally on the order of tens of nanoseconds (16 ns for sodium, 27 ns for potassium). The excited-state population N_i^* is then not in steady state during the pulse duration τ_L, and its temporal variation is given by (Mégie et al., 1978a)

$$\frac{dN_i^*}{dt} = -\frac{N_i^*}{\tau_R} + (N_i - N_i^*)\frac{n_e(r)T}{R^2\Omega_L}\sigma_{\text{eff}} - N_i^*\frac{n_e(r)T}{R^2\Omega_L}\sigma_{\text{eff}} \tag{5.18}$$

The first term corresponds to radiative deexcitation by spontaneous emission, the second is due to absorption of the laser flux at range R equal to $n_e T/R^2\Omega$, and the third is related to stimulated emission from the excited level. Here Ω_L is the solid angle illuminated by the laser beam related to the full angle divergence θ_L by

$$\Omega_L = \frac{\pi}{4}\theta_L^2 \tag{5.19}$$

To evaluate the effects of the saturation of the excited-state population, an analytical solution of (5.18) can be derived assuming a rectangular temporal shape of the laser pulse:

$$n_e(t) = \frac{n_e}{\tau_L} \qquad 0 \leqslant t \leqslant \tau_L$$

$$n_e(t) = 0 \qquad t \geqslant \tau_L \tag{5.20}$$

The excited-state population $N_i^*(t)$ is then given by

$$N_i^*(t) = \frac{N_i \tau_T}{2\tau_e}(1 - e^{-t/\tau_T}) \qquad t \leqslant \tau_L \tag{5.21}$$

$$N_i^*(t) = \frac{N_i \tau_T}{2\tau_e}(1 - e^{-\tau_L/\tau_T})e^{-(t-\tau_L)/\tau_T} \qquad t \geqslant \tau_L \tag{5.22}$$

where the two time constants τ_e and τ_T are defined as follows:

$$\tau_e = \frac{\Omega_R R^2}{2\sigma_{\text{eff}} n_e T}\tau_L \tag{5.23}$$

$$\frac{1}{\tau_T} = \frac{1}{\tau_R} + \frac{1}{\tau_e} \tag{5.24}$$

The number of reemitted photons per unit time in the scattered volume is then $(R^2\Omega_R/\tau)N_i^*$, and the number of received photons on the collecting area is given by

$$n_r^s(R) = n_e T^2 \frac{A_0}{R^2}\int_0^{\tau_S} R^2\Omega_R \frac{N_i^*}{\tau_e}\, dt \tag{5.25}$$

where τ_S is the temporal resolution of the receiving system and is generally on the order of a few microseconds, which permits the integration to effectively be extended to infinity. One can then express $n_r^s(R)$ as a function of $n_r(R)$ as given by (5.13):

$$n_r^s(R) = n_r(R)\frac{\tau_T}{\tau_R}\left\{1 - \frac{\tau_R \tau_T}{\tau_L \tau_e}\left[\exp\left(-\frac{\tau_L}{\tau_T}\right) - 1\right]\right\} \tag{5.26}$$

The direct effect of the finite lifetime of τ_R is thus to decrease the number of received photons depending on the ratio τ_R/τ_e of the excited state lifetime τ_R to a "saturation" time τ_e as defined by (5.23). The term τ_e is directly related to the laser parameters and decreases with decreasing divergence ($\Omega_R \sim \theta_R^2$), increasing output power (n_0), and decreasing linewidth through the increase of σ_{eff}. Considering the D_2 absorption line parameters of sodium, the effective cross section varies from 5.10^{-12} cm^2 for a 1-GHz laser line to 10^{-12} cm^2 for a 10-GHz line. Assuming a laser output power of 10 MW (0.1 J, 10 ns) with an output divergence of 7.10^{-4} rad, τ_e is then given by (for a 90-km average altitude and a one-way transmission of 0.5)

$$\tau_e = \begin{cases} 104 \text{ ns} & (\Gamma_L = 10 \text{ GHz}) \\ 21 \text{ ns} & (\Gamma_L = 1 \text{ GHz}) \end{cases}$$

The ratio n_r^s/n_r then varies between 0.96 (10 GHz) and 0.83 (1 GHz) so that when one uses a narrow-linewidth short-duration laser emitter, the effect of saturation, which is independent of the atom number density N_i, cannot be neglected and should be accounted for in the inversion of the lidar equation.

5.2.1.5 *Effects Related to Polarization of Incident Beam*

The polarization of resonance radiation is a direct consequence of the hyperfine structure of the transitions and of the Zeeman effect. It has been treated theoretically by various authors (Van Vleck, 1925; Weisskopf, 1931; Mitchell and Zemansky, 1934) and summarized by Chamberlain (1961). In the case of hyperfine structure, the interchange among different levels cannot be adequately accounted for by treating each hyperfine line as due to a resonance transition and adding then the separate intensity components. The geometry of the scattering is illustrated in the Figure 5.6. The incident flux F has two components respectively polarized in the plane of scattering, $F_\parallel$, and perpendicular to this plane, $F_\perp$. If θ is the scattering angle, the polarization may be computed by means of a scattering matrix (Chandrasekhar, 1950; Chamberlain, 1961), where the intensities are given by

$$\begin{Vmatrix} I_\parallel \\ I_\perp \end{Vmatrix} = \sum_F \left\{ 4\beta(F) \begin{Vmatrix} \cos^2\theta & 0 \\ 0 & 1 \end{Vmatrix} + [\alpha(F) - 2\beta(F)] \begin{Vmatrix} 1 & 1 \\ 1 & 1 \end{Vmatrix} \right\} \begin{Vmatrix} F_\parallel \\ F_\perp \end{Vmatrix} \tag{5.27}$$

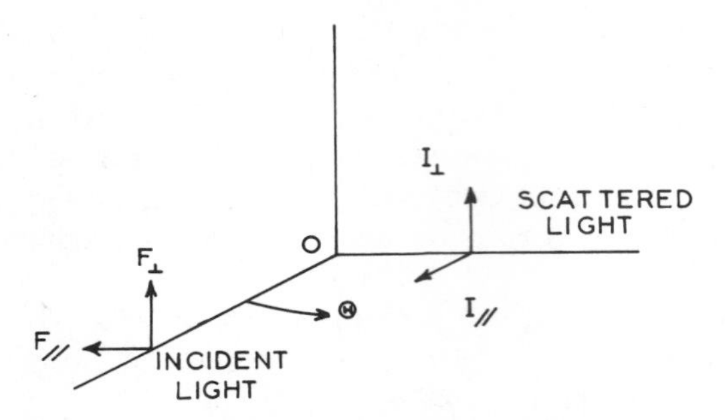

Figure 5.6. Geometry of scattering process.

where $\alpha(F)$ and $\beta(F)$ are dimensionless quantities that depend on the transition probabilities and statistical weights of the various sublevels involved in the scattering process. It takes into account the response of π (respectively, σ) components to linearly (respectively, circularly) polarized light, summation being extended to all hyperfine sublevels. The total intensity $I = I_{\parallel} + I_{\perp}$ can be calculated from (5.27) for different polarization states of the incident beam

$$F_{\perp} = 0 \qquad I = (2\alpha - 4\beta \sin^2\theta)F$$

$$F_{\parallel} = 0 \qquad I = 2\alpha F$$

$$F_{\parallel} = F_{\perp} \qquad I = (2\alpha - 2\beta \sin^2\theta)F$$

In the case of monostatic lidar observations, the scattering angle θ is equal to zero, and the scattered intensity is thus independent of the polarization of the incident beam. This does not preclude that the scattered beam is polarized depending on the incident beam polarization (see Section 5.2.3.3). Furthermore, in the presence of Earth's magnetic field and depending on the geometry of the lidar observations, the relative intensities of the hyperfine components can be modified due to the Hanle effect or zero-field level crossing (Hanle, 1924). This is of particular importance for high-spectral-resolution measurements of the backscattering lineshape, leading, for example, to atmospheric temperature determinations as developed in Section 5.2.3.3.

5.2.1.6 Effect of Laser Absorption within Metallic Layer

The most abundant metal in the 80–100-km range is atomic sodium, which has a maximum integrated content of $\sim 10^{10}$ cm^{-2}. The maximum optical thickness τ_i of the sodium layer for an effective cross section of 5.10^{-12} cm^2 is then 5.10^{-2}. The scattering medium can thus be considered as optically thin so that one need not take into account any variation of the laser lineshape throughout the layer. The number of photons absorbed over a range interval ΔR is then given by

$$n_a(R) = \int_0^\infty n_e(R) g(\nu) \left\{1 - \exp\left[-\sigma_s(\nu) N_i(R)\, \Delta R\right]\right\} d\nu \quad (5.28)$$

The transmission $T(R)$ of the layer ΔR at range R can then be calculated as

$$T(R) = 1 - \frac{n_a(R)}{n_e(R)} \quad (5.29)$$

Following Mitchell and Zemansky (1971), such calculations can be performed, assuming the two lines to be of Gaussian shapes (cf. 5.2.1.2) and using a series expansion of the integral, which, when the laser line is centered at the center of the absorption line, leads to

$$T(R) = 1 - \frac{\sigma_s(\nu_0) N_i(R)\, \Delta R}{\left(1 + \Gamma_L^2/\Gamma_D^2\right)^{1/2}} \quad (5.30)$$

(the second term in the expansion is five orders of magnitude smaller than the first). The transmission is then calculated, up to given range R_k, by combining the effect of each range cell from the bottom of the layer at range R_0 so that

$$T_k(R_k) = 1 - \sum_{j=0}^{k-1} \frac{\sigma_s(\nu_0) N_i(R_j)\, \Delta R}{\left(1 + \Gamma_L^2/\Gamma_D^2\right)^{1/2}} \quad (5.31)$$

where $R_j = R_0 + j\,\Delta R$.

This equation accounts for the transmission of the laser beam upward through the sodium layer. However, the light that travels downward has a different lineshape that corresponds to the absorption lineshape due to the scattering process. The downward transmission $T_k^d(R_k)$ can then be expressed as

$$T_k^d(R_k) = 1 - \sum_{j=0}^{k-1} \frac{\sigma_s(\nu_0) N_i(R_j)\, \Delta R}{\sqrt{2}} \quad (5.32)$$

As compared to the case when no absorption within the layer is considered $[N_i(R)]$, the sodium number density $N_i^c(R_k)$ at a given level R_k will then be given by

$$N_i^c(R_k) = \frac{N_i(R_k)}{T_k T_k^d} \tag{5.33}$$

Such calculations have been performed by Simonich et al. (1983) and are presented in Figure 5.7, which shows the full correction for both upward and downward propagation. The peak concentration is increased by 10%, mostly due, in their case ($\Gamma_L \leqslant \Gamma_D$), to the downward propagation correction. It thus seems worthwhile to include the transmission terms as calculated here in the expression for sodium number density.

5.2.1.7 Normalization

In Equation (5.13), from which the sodium vertical distribution can be determined as $N_i(R)$, two factors are difficult to either measure or calculate: atmospheric transmission $T^2(\nu, R)$ and number of received photons $n_r(\nu, R)$.The atmospheric transmission will vary considerably for different times of observations due to haze, aerosols, or clouds. The determination of the exact number of photons will require the exact knowledge of the optics transmission, the quantum efficiency of the photomultiplier tube, and the amplification factors of the various stages of the electronic chain. Taking these difficulties into account, one usually normalizes the sodium echo to the signal obtained by pure Rayleigh scattering due to atmospheric molecules at a given range R_0. Using an atmo-

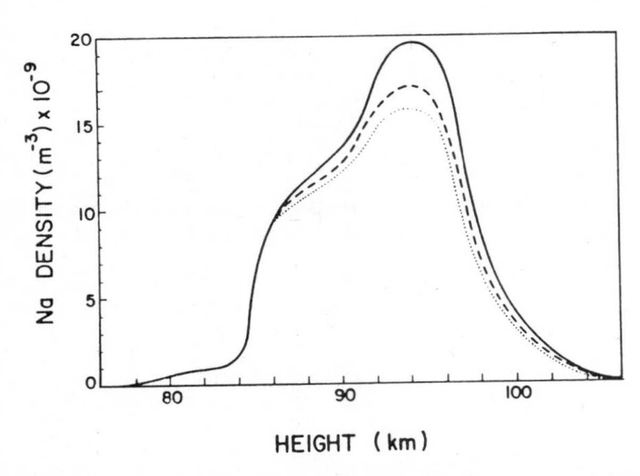

Figure 5.7. Average density profile measured at Sao José dos Campos on night of July 13, 1978 (dashed line) showing effect of correction for resonant extinction for wide-band laser (solid line) and profile that would have been measured by monochromatic laser (dotted line) (from Simonich and Clemesha, 1983).

spheric density model (CIRA, 1972; U.S. Standard Atmosphere, 1976) or an in situ determination of the same parameter by balloon-borne instruments, the absolute number density $N_i(R)$ can then be determined using the following procedure: The signal n_r' backscattered by atmospheric molecules at a range R_0 is given by

$$n_r'(\nu, R_0) = n_e(\nu)\xi(\nu)T^2(\nu, R_0)\frac{A_0}{R_0^2}N_T(R_0)\left(\frac{d\sigma_R(\nu)}{d\Omega}\right)_{180°}\Delta R \quad (5.34)$$

where $N_T(R_0)$ is the total density at range R_0 and $[d\sigma_R(\nu)/d\Omega]_{180°}$ is the Rayleigh differential cross section at frequency ν for a scattering angle of 180°.

By ratioing (5.13) and (5.34), one obtains

$$N_i(R) = N_T(R_0)\frac{n_r(\nu, R)}{n_r'(\nu, R_0)}\frac{T^2(\nu, R_0)}{T^2(\nu, R)}\frac{R^2}{R_0^2}\frac{d\sigma_R(\nu)/d\Omega}{\sigma^s(\nu)/4\pi} \quad (5.35)$$

The choice of the altitude range R_0 used for the normalization will be made based on the following considerations:

1. The atmosphere at R_0 should be free of aerosol particles to avoid any contribution due to Mie scattering; altitudes above 30 km will be used to be higher than the stratospheric aerosol layer.
2. The signals due to both Rayleigh and resonant scattering should be of the same order of magnitude to avoid any effect due to the signal dynamic so that the ratio n_r/n_r' could be directly related to the ratio of the return signals. The value of the Rayleigh differential cross section $[d\sigma_R(\nu)/d\Omega]_\theta$ can be calculated following Penndorf (1957) using the formula

$$\left(\frac{d\sigma_R}{d\Omega}\right)_\theta = \frac{2\pi^2(n^2-1)^2}{\lambda^4 N_T^2}\frac{2+\rho_n}{6-7\rho_n} \times 0.7629(1 + 0.932\cos^2\theta) \quad (5.36)$$

where n_s is the index of refraction of air, N_T the total atmospheric density, ρ_n a depolarization factor equal to 0.035 for air, and θ the scattering angle. For $\lambda = 589$ nm and for backscattering, (5.36) leads to

$$\left(\frac{d\sigma_R}{d\Omega}\right)_{180°} = 4.33 \times 10^{-28}\ \text{cm}^2 \quad (5.37)$$

This value is about 5×10^{-15} times smaller than the value of the resonant scattering effective differential cross section $\sigma_s/4\pi$, so that the two back-scattered signals will be equivalent for $N_T \sim 10^{15} N_i$ or $N_T \sim 10^{17}\ \text{cm}^{-3}$, which corresponds to the altitude range 35–40 km.

3. The atmospheric transmission ratio $T^2(\nu, R_0)/T^2(\nu, R)$ should be equal to 1, which means that the residual absorption or extinction between R_0 and R should be small. For $R_0 = 30$ km and $R = 90$ km, its value is due mainly to Rayleigh extinction at 589 nm is 3×10^{-3} so that (5.35) can be simplified as

$$N_i(R) = N_i(R_0)\frac{n_r(\nu, R)}{n_r'(\nu, R_0)}\frac{R^2}{R_0^2}\frac{d\sigma_R(\nu)/d\Omega}{\sigma_s(\nu)/4\pi} \tag{5.38}$$

The fact that $N_T(R_0)$ is taken from an atmospheric model implies that only an average value is considered; the true value at the time of the measurement can then be different, which taking into account the variability of the atmospheric density at the 40-km level will introduce an uncertainty in the determination of $N_i(R)$ of less than 10%.

The last effect to consider is related to the alignment of the emitted and received beams. To apply (5.38), the scattering volumes at 30 and 90 km should be such that a complete overlap exists between the two beams. Figure 5.8 shows the relative error introduced in the sodium number density when there is a misalignment $\phi_{\parallel}$ in the perpendicular plane. These calculations have been made for a laser divergence of 5×10^{-4} rad and a telescope field of view of 10^{-3} rad. Although such uncertainties do not affect the relative time evolution of the sodium vertical distribution provided the alignment is kept constant, they might affect the determination of the absolute number density.

5.2.1.8 Conclusion

We have considered here the various parameters that directly influence the lidar equation for resonant scattering and might introduce systematic errors in the retrieval of the species vertical distribution. A summary of such errors is given in the Table 5.4 for various laser sources relevant to the measurement of the sodium number density. These lidar systems (the characteristics of which were presented in Table 5.1) will be discussed in the next section. These errors are compared to the statistical error due to the signal noise calculated assuming Poisson statistics for the number of received photons within unit time. These lead to the usual expression for the signal-to-noise ratio for X independent laser shots (Measures, 1984):

$$\text{SNR} = \frac{S_r}{(S_r + S_B + S_N)^{1/2}}\frac{1}{\sqrt{X}} \tag{5.39}$$

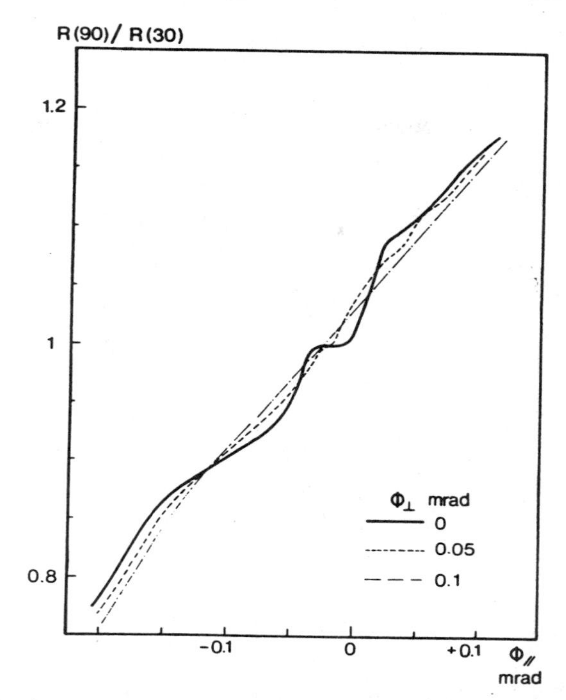

Figure 5.8. Ratio of geometric form factor R at 90 and 30 km as function of detuning angle ϕ between emitted and received beams. $\phi_{\parallel}$ is detuning angle in vertical plane P containing parallel emitter and receiver axis when aligned; $\phi_{\perp}$ is detuning angle in vertical plane perpendicular to P. Laser beam divergence 2×10^{-4} rad; telescope field of view 3×10^{-4} rad; distance between emitter and receiver 0.75 m (from Granier and Mégie, 1981).

where S_r is the received signal in the photon counting mode, S_B the signal due to the skylight background, and S_N the photomultiplier tube nosie.

5.2.2 Lidar Systems for Resonance Scattering Measurements

In this section we will briefly describe the lidar systems used for resonant scattering observations in the upper atmosphere. We will mainly emphasize the specific characteristics of the laser sources relevant to such observations, whereas the more common parts of lidar systems such as optical receivers, electronic time analyzers, or signal processing systems will not be discussed in detail.

Table 5.4. Uncertainties on Sodium Altitude Distribution and Total Content for Various Laser Emitters (See Table 5.1)[a]

Emitter	Integration Time (min)	ϵ_1 (%)	ϵ_1' (%)	ϵ_2 (%)		ϵ_3 (%)		ϵ_4 (%)		ϵ_5 (%)		ϵ_6 (%)		ϵ (%)		ϵ' (%)	
				A	*R*	*A*	*R*	*A*	*R*	*A*	*R*	*A*	*R*	*A*	*R*	*A*	*R*
Flashlamp-pumped dye laser	60	5	1	6	3	8	3	8	2	~0	~0	4	2	29	15	25	11
Nd^{3+}–YAG-pumped dye laser ($\Gamma_L \sim 15$ pm)	5	5	1	4	2	3	2	2	2	~0	~0	4	1	19	12	15	8
Nd^{3+}–YAG-pumped dye laser ($\Gamma_L \sim 1.5$ pm)	$1\frac{1}{2}$	5	1	4	2	3	2	2	2	2	1	4	~0	21	12	17	8

[a]Notation: ϵ_1, statistical error on sodium number density; ϵ_1', statistical error on sodium total content; ϵ_2, uncertainty on σ_{eff} related to frequency shift $\nu_1 - \nu_0$ of laser line; ϵ_3, uncertainty on σ_{eff} related to variations in laser linewidth Γ_L; ϵ_4, uncertainty related to misalignment; ϵ_5, uncertainty related to nonconsideration of temperature effects in sodium layer; ϵ_6, uncertainty on total density used for calibration, N_R; ϵ, total uncertainty on sodium number density; ϵ', total uncertainty on sodium total content; *A*, error on absolute value of number densities; *R*, error on relative variations.

5.2.2.1 *Flashlamp-Pumped Dye Lasers*

Although experiments to measure concentrations of individual species by resonance scattering using either a neodymium glass laser (Hirono, 1964; Kato et al., 1964), a pulsed gas laser (Young, 1964), or a Raman-shifted laser [Nugent, 1966] were proposed in the mid-1960s, the first detection of atmospheric sodium followed shortly after the discovery of the continuously tunable narrowband organic dye laser (Soffer and Mc Farland, 1967; Sorokin et al., 1967; Bradley et al., 1968).

Several groups subsequently built laser sources adapted to the measurement of the neutral sodium vertical distribution in the upper atmosphere (Bowman et al., 1969; Blamont et al., 1972a; Hake et al., 1971; Kirchhoff and Clemesha, 1973; Schuler et al., 1971) using flashlamp-pumped dye (Rhodamine 6G) lasers with high output energy and high mean output power. The flashlamps used were simple ablating wall lamps as designed by Ferrar (1969) suitable for dye laser pumping in the 1–10-μs pulse duration range. Several arrangements were proposed for improving the optical coupling between the excitation lamps and the circulating dye cell. These included close coupling in collinear arrangement with two to six flashlamps surrounding the cell (Gibson, 1972); the use of coaxial flashlamps to reduce the total inductance, thereby decreasing the risetime and increasing the efficiency of pumping (Furumoto and Ceccon, 1969); and the use of elliptical cavity arrangements (Schuler et al., 1971; Loth and Mégie, 1974; Allain et al., 1979).

To permit a high-repetition-rate operation of the laser, the flashlamps were usually surrounded by silica cooling jackets through which deionized water was circulated and cooled in an external refrigerator. The flashlamp firing circuits included a triggering system with air spacing spark gap and the appropriate charging capacitors. Further improvements in the system efficiency were made by using preionizing discharges or a simmer circuit (Jethwa et al., 1978). The dye cells were generally made of 5–10 mm bore diameter and 10–15 cm length silicon tubes through which a solution (10^{-4}–10^{-3} M/L) of Rhodamine 6G in various solvents (ethanol, methanol, water, isopropanol) was circulated. Such arrangements (Figure 5.9) resulted in broadband emission efficiencies between 1.5 and 2.5% (Gibson, 1972; Allain, 1979). However, the bandwidth of such emissions is very large ($\sim$8 nm), which resulted in a very low value of the effective cross section as defined in Section 5.2.1.3. Spectral narrowing and tuning were thus carried out using in most cases Fabry–Perot etalons inside the laser cavity. Techniques for constructing the etalons have been discussed by several authors (Gibson, 1969; Bradley et al., 1971; Mégie and De Witte, 1971). Their thicknesses and plate reflectivities are chosen so that the corresponding free spectral ranges (FSR) and spectral resolution lead to the emission of a

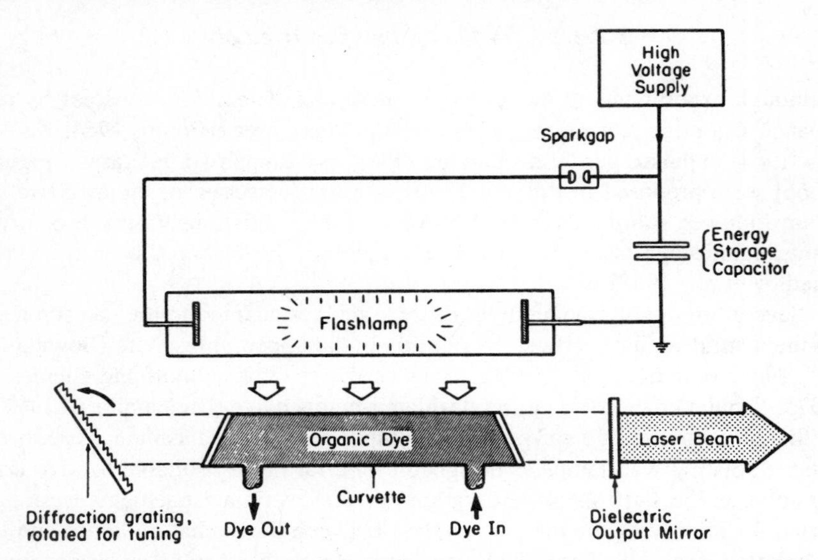

Figure 5.9. Schematic arrangement of flashlamp-pumped dye laser (from Measures, 1984).

single line within the dye emission bands. The general arrangement included a thin etalon ($e \sim$ 10–15 μm; FSR = 14 nm), either solid or air spaced, and a thicker one (e = 175 μm; FSR = 0.1 nm) and resulted in an emission linewidth of a few picometers. A third etalon has been added in some cases to further reduce the emitted linewidth (Gibson et al., 1979). Other wavelength tuning systems were also designed that included coarse-wavelength selection using a grating in the Littrow configuration (Schuler et al., 1971; Hake et al., 1972) or Lyot filters (Soep, 1970). Introduction of tuning elements in the laser cavity reduced the overall efficiency of the system below 1% for a single oscillator cavity mode (see Table 5.1 for the average emission parameters of flashlamp-pumped dye lasers).

The requirements for upper atmospheric resonant scattering measurements imply that one has to obtain simultaneously high output energies, narrow line-widths, and low divergences. These are obviously contradictory as the first implies a high pump energy that usually results in high divergence and thus low spectral resolution of the system. Therefore, amplification techniques were used by several groups (Hake et al., 1972; Loth and Mégie, 1974) to further increase the output energy by factors between 2 and 3 without degrading the spectral and spatial characteristics of the laser emission. More efficient techniques such as injection locking using very low energy oscillator sources were also investigated and developed (Flamant and Mégie, 1980).

Furthermore, the same systems were used to generate other wavelengths using different dyes in order to undertake the measurement of other minor constituents by resonant scattering. Gibson (1972) and Jégou and Chanin (1980) developed a flashlamp-pumped dye laser at 670.8 nm for observing the resonance line of lithium using a solution of rhodamine 640 in methanol. Attempts were also made to build a radiant source at 769.9 nm for potassium excitation using a cresylviolet dye solution (Gacoin and Flamant, 1972).

Although such flashlamp-pumped dye lasers are still in use in several groups for sodium observations (Shelton et al., 1980; Clemesha et al., 1982), they have tended to be replaced since the mid-1970s by more reliable solid-state laser-pumped sources.

5.2.2.2 Laser-Pumped Dye Lasers

Due to the difficulty of obtaining direct laser emission at 769.9 nm from flashlamp-pumped systems, two groups developed ruby laser-pumped dye laser sources for potassium excitation (Felix et al., 1973; Loth et al., 1976). The system design was then adapted to other excitation lasers such as Nd^{3+}–YAG and to others with an emission range from 355 nm to 800 nm (Bos, 1981; Granier et al., 1984, 1985).

The usual design includes an oscillator cavity and a three-amplifier chain (Figure 5.10) to increase the output energy. The excitation energy repartition between the various stages depends primarily on the efficiency of the dye used as an active medium; it has to be adapted so that the output energy of a given stage will be sufficient to completely lock the following amplifier to avoid spontaneous emission within this latter stage, which would destroy the spectral and spatial characteristics of the emitted beam (Lefrère et al., 1984). In the case of

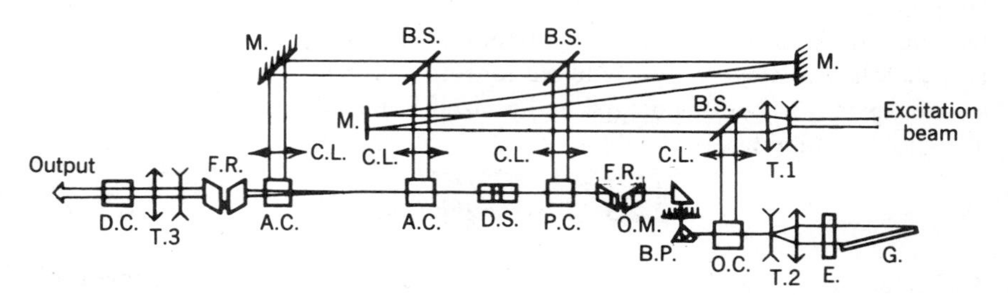

Figure 5.10. Schematic arrangement of Nd^{3+}–YAG-pumped dye oscillator: T_1, input telescope; BS, beam splitters; M, mirrors; Cl, cylindrical lenses; G, grating; E, etalon; T_2, intracavity telescope; OC, oscillator dye cell; BP, Broca prism; OM, output mirror; FR, Fresnel rotator; PC, preamplifier dye cell; DS, dispersive systems (Broca prisms); AC, amplifier dye cell; T_3, output telescope; DC, doubling crystal (from Bos, 1981).

Rhodamine 6G, which will provide laser emission at 590 nm, and for a total pump energy of 350 mJ at 0.53 μm (second-harmonic generation from a 0.8-J emission at 1.06 μm), only 20 mJ are injected transversally into the oscillator cell.

This cavity includes a grating at grazing incidence in the Littrow arrangement either directly or with a reflecting mirror, which is used to generate a narrow laser line with typical width of 1.5 pm for an 89.5° incidence. This linewidth is enhanced if the grating is used at a smaller incidence to values on the order of 5–6 pm, the operating configuration depending on the number of groves of the grating (2500–3200 groves mm^{-1} for operation in the visible wavelength range). Wavelength tuning is facilitated by a grating rotation that can be automatically driven by the wavelength monitoring system. The remaining excitation energy is then divided between the three amplifier cells (Dujardin and Flamant, 1978; Bos, 1981) and the overall efficiency of the dye laser energy conversion ranges to about 40% providing an output energy of 120 mJ in the case of Rhodamine 6G (see Table 5.1).

Such systems have also been extended with the same spectral characteristics to other wavelength ranges: recently Granier et al. (1984) have used a combination of a laser-pumped dye laser and frequency mixing techniques to generate laser emission at wavelengths of neutral calcium (423 nm) and ionized calcium (393 nm). The first is directly obtained by pumping a solution of Stilbene 420 in a methanol–ethanol mixture with the third harmonic of a Nd^{3+}–YAG laser at 355 nm. The overall efficiency is 13%, producing an output energy of 25 mJ from a pump energy of 180 mJ. Emission at 393 nm is obtained by first generating laser emission at 624 nm by direct pumping, using the Nd^{3+}–YAG second harmonic at 530 nm, of a Rhodamine 640 dye solution in water. Wavelength mixing is then performed in a KDP crystal using the Nd^{3+}–YAG fundamental emission at 1.06 μm and laser emission at 624 nm, which provides laser energy (~20 mJ) at 393 nm. Wavelength tuning in this case is provided by tuning the dye laser emission and rotating the mixing crystal. One can note here that the two wavelengths required for resonant scattering measurements of a same species in both its neutral and ionized forms can be obtained by using a single excitation laser.

5.2.2.3 Wavelength Control

The tunability of the laser source and the relatively narrow spectral width of the metallic species absorption lines require an exact positioning and knowledge of the laser emitted wavelength and spectral profile to compute the resonant scattering effective cross section. This can be done using high-resolution spectrometers and spectral lamps for calibration. In the case of sodium, however, a different technique has been developed using a sodium resonance cell to which part of the emitted beam is sent for reference purposes. A strong scattered signal is then observed in a direction perpendicular to the excitation light when the laser wavelength is properly tuned to the sodium D_2 line (Mégie and De Witte, 1971; Clemesha et al., 1975). The measurement of the intensity ratio before and after the cell could in principle lead to a direct determination of the effective cross section if one knows the temperature of the cell that determines the line-width of the scattered laser light. However, such resonance cells are generally operated at high temperatures (400 K) different from the atmospheric temperature at the mesopause level (200 K), and further calculations are thus required. The most advanced systems use newly developed wavemeters to directly provide either the absolute wavelength or the spectral profile of the laser light (Snyder, 1977; Cahen et al., 1981).

5.2.2.4 *Emitting and Receiving Optics*

Due to the low values of the backscattered signals, large optics are generally used for the upper atmospheric measurements. The average diameter of the telescope is in the 0.6–1-m range. Some authors have also used refractive optics such as large-diameter Fresnel lens (Rowlett and Gardner, 1978). The optical detection scheme is quite straightforward and includes focusing optics, an iris that determines the telescope field of view, a narrow-bandwidth interference filter, and a photomultiplier tube. The characteristics of a typical system are given in Table 5.1. For daytime measurements, these performances are increased to reduce the sky background light by reducing the receiver field of view to values below 3×10^{-4} rad and by a complementary spectral filtering using a Fabry–Perot interferometer with a spectral resolution of 10–20 pm (Granier et al., 1981).

Although most of the systems have been used in a fixed zenith-pointing configuration, steerable lidar systems have also been developed to study the horizontal variations of the metallic layers (Thomas et al., 1976). More recently an airborne system has been designed and operated from a low-altitude flying aircraft (Segal et al., 1984).

5.2.2.5 *Electronics and Computing Systems*

Resonant scattering measurements of the metallic layers in the upper atmosphere have taken place over more than 15 years. During such a period of time the rapid growth in electronic and computing technologies has obviously considerably modified the design of the electronic receiving part of the lidar systems. Whereas photon counters remain the basic elements of the time analysis system due to the small values of the detected signals, the use of microprocessors and newly developed computing systems allows presently a quasi-automatic handling of the data-processing systems. As in most present-day lidars, the data acquisition sequence is completely controlled by a minicomputer with an appropriate selection of the experimental parameters including the emitting source (see, e.g., Measures, 1984).

5.2.3 Results of Metallic Layers Lidar Observations

The existence of layers of metallic atoms in the upper atmosphere had been demonstrated well before the advent of lidar measurements (Slipher, 1929). Airglow observations either ground based or balloon and rocket borne had already given some information on the behavior of sodium, for example, at the mesopause level (Blamont and Donahue, 1964; Hunten, 1967; Gadsden, 1970). However, numerous questions including the origin of these metals and their diurnal and short-term variations were still to be solved due to the lack of measurement continuity and resolution and to the difficulty raised by radiative transfer inversion procedures in the retrieval of the species number density from the observed emitted or scattered signals (Chamberlain, 1961). As would be expected from the rapid development of both laser sources and electronic receiving subsystems, lidar measurement accuracy and time resolution have been greatly improved during the 16 years of resonant scattering lidar observations, which start back in 1969. After the first demonstration of the potential of the resonant lidar technique (Bowman et al., 1969; Sandford and Gibson, 1970; Blamont et al., 1972a), measurements were mainly directed towards studies of the long-term evaluation of the sodium layer as the seasonal variation of the total abundance and of the peak and altitude width of the distribution (Gibson and Sandford, 1971; Mégie and Blamont, 1977; Simonich et al., 1979). These measurements were then further complemented by observations of potassium (Felix et al., 1973, Mégie et al., 1978a) and lithium (Jegou et al., 1980), which gave new indications on the metallic species origin and equilibrium (Richter and Sechrist, 1979a,b; Liu and Reid, 1979; Jégou et al., 1985a,b). Those observations were confirmed by measurements of neutral and ionized calcium (Granier et al., 1984). Shorter term variations were also studied by several groups and included diurnal variations of the sodium layer (Gibson and Sandford, 1972; Granier et al., 1981; Granier and Mégie, 1981; Kirchhoff and Clemesha, 1983a), influence of tides and gravity waves on the sodium vertical distribution (Blamont et al., 1972a; Kirchhoff and Clemesha, 1973; Rowlett and Gardner, 1978; Shelton et al., 1980; Juramy et al., 1981), sporadic enhancements of sodium abundance due to meteoritic input (Hake et al., 1972; Mégie and Blamont, 1977; Clemesha et al., 1978b, 1980; Granier and Mégie, 1982). Finally, sodium atoms have also been used as trackers of the global atmospheric parameters such as temperature (Blamont et al., 1972b; Mégie et al., 1974; Gibson et al., 1979; Thomas and Batthacharyya, 1980; Fricke and

Von Zahn, 1985), eddy diffusion coefficients (Mégie and Blamont, 1977; Kirchhoff et al., 1983b,c; Kirchhoff and Takamashi, 1984) and winds (Rees and Sandford, 1974; Clemesha et al., 1980). We will review briefly in this section each of these aspects of resonant scattering lidar measurements showing their impact on our understanding of the behavior of trace species at the mesopause level.

5.2.3.1 Seasonal Variations of Metallic Species

The first measurements of the sodium height distribution over a full annual cycle were performed by Gibson and Sandford at Winkfield (0.7 W, 51°4 N) between July 1969 and July 1970 (Gibson and Sandford, 1971). They gave evidence for a maximum abundance during winter and a minimum in summer in agreement with observations of the twilight glow from the layer (Blamont and Donahue, 1964). The increase in content during winter is associated with a change in the layer shape corresponding to a lowering of the peak and a concentration increase on the lower side. These results were then confirmed by longer term observations covering three years between 1973 and 1976 at the Observatoire de Haute Provence (5° E, 44° N) (Mégie and Blamont, 1977): a fourfold increase occurs between June (total abundance of 1.5×10^9 cm^{-2}) and December (6×10^9 cm^{-2}) on the average (Figure 5.11). The peak of the height distribution then decreases from 93 km in summer to 88 km during wintertime associated with an average enhancement of the layer width from 7 to 14 km. Observations performed at lower latitude in the Southern Hemisphere (Sao José dos Campos, 23° S, 46° N) over 344 nights since 1972 showed an annual variation by a factor of 2 in the sodium abundance with maximum in local winter (Simonich et al., 1979; Clemesha, 1979; Kirchhoff et al., 1981a). All these authors underlined large day-to-day variations, whereas when averaging over a large number of measurements, the sodium equilibrium could be described by the classical photochemical theory, already elaborated from twilight and nightglow data, which predicted a photochemical control of the bottomside of the layer related to the sodium oxydo-reduction cycle involving mainly ozone and atomic oxygen,

$$Na + O_3 \rightarrow NaO + O_2 \qquad NaO + O \rightarrow Na + O_2 \tag{5.40}$$

and a diffusive control of the topside associated with an influence of the photoionization processes during daytime (Mégie and Blamont, 1977; Kirchhoff et al., 1981b). However, as atomic sodium is the most abundant form of sodium compound between 80 and 100 km, these models were unable to explain the observed seasonal variation without involving external causes such as dust par-

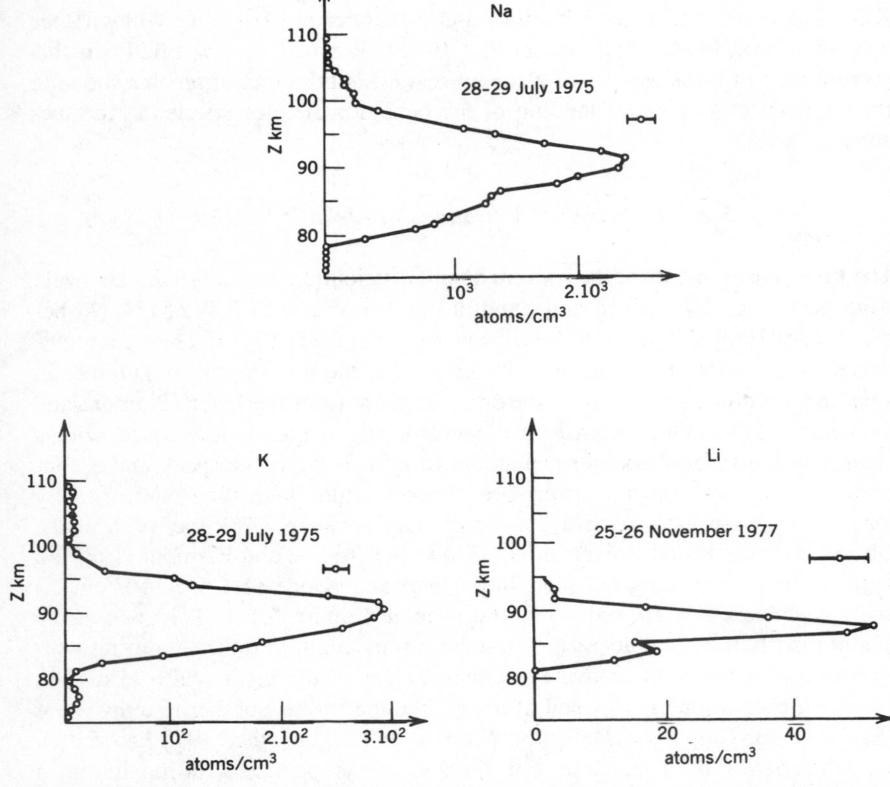

Figure 5.11. Seasonal variations of sodium, potassium, and lithium observed at Observatoire de Haute Provence by lidar soundings (dots) as compared to model simulations of Jégou et al. (1985a, b) (solid lines).

ticle equilibrium including changes in the rate of sublimation of sodium at the mesopause (Fiocco and Visconti, 1976).

In 1977, the seasonal variation of potassium was measured at the Observatoire de Haute Provence (Mégie et al., 1978a) to see if the same day-to-day variations were also observed for this constituent. Its long-term behavior in terms of total column abundance was very different as no increase was seen over the year corresponding to a constant value of the integrated content of 10^7 cm^{-2} (Figures 5.11 and 5.12). Taking into account the fact that the photochemical properties of these two alkalis are identical due to the similarity of their external electronic layers, their differentiation was then related to their sources. The abundance ratio was found to vary from 50 in winter, close to the terrestrial

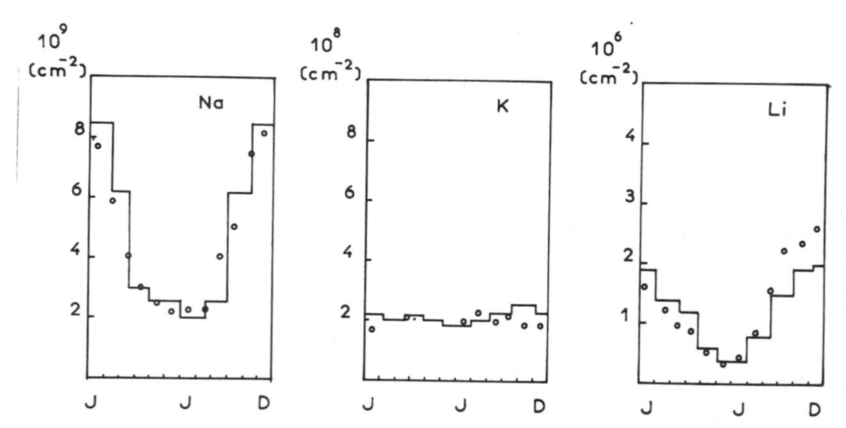

Figure 5.12. Altitude distribution of three alkalis (sodium, potassium, and lithium) as observed by resonant scattering lidar at Observatoire de Haute Provence.

one, down to 10 in summer, closer to the abundance ratio in the meteorites. This implied that these were two different origins for the alkalis in the upper atmosphere: a meteoritic one constant over the year and a terrestrial source related to the vertical transport of particles at high latitudes that only exists during wintertime (Mégie et al., 1978a). However, such a hypothesis failed again when the seasonal variation of lithium, the abundance of which is below 1 atom cm^{-3}, was shown to be similar to the sodium one (Jégou et al., 1980). These measurements were performed from November 1977 to December 1979 at the Observatoire de Haute Provence showing an average summer total abundance of 10^6 cm^{-2}, increasing to 4×10^6 cm^{-2} during winter (Figures 5.11 and 5.12). Meanwhile, many authors had observed sporadic enhancements of the sodium abundance that could be related to various meteor showers occurring throughout the year: quadrantids in January (Mégie and Blamont, 1977), perseids in summer (Mégie et al., 1978a; Clemesha et al., 1978b; Clemesha et al., 1980; Granier et al., 1982), orionids in October (Mégie and Blamont, 1977), and geminids in December (Hake et al., 1972; Mégie and Blamont, 1977). As an example, Figure 5.13 shows the isoline of sodium number densities as observed on July 24, 1982, after a sudden increase that occurred at 10:00 P.M. at an altitude of 102 km (Granier et al., 1982). Similar increases were observed for potassium (Mégie et al., 1978a) and lithium (Jégou and Chanin, 1980), giving evidence for an extraterrestrial origin of the metallic compounds.

Photochemical models were quite successful in explaining the sodium vertical distribution even in their early stage of development (Mégie and Blamont, 1977; Kirchhoff et al., 1979). In 1979, Liu and Reid, following Ferguson (1978), included in the sodium photochemistry the sodium hydroxide molecule

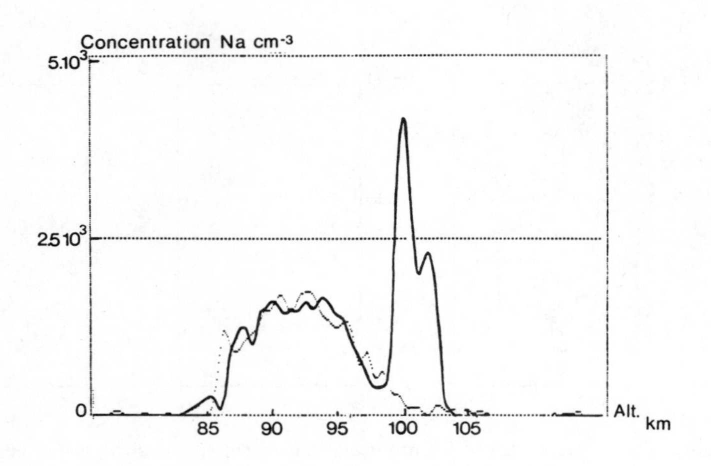

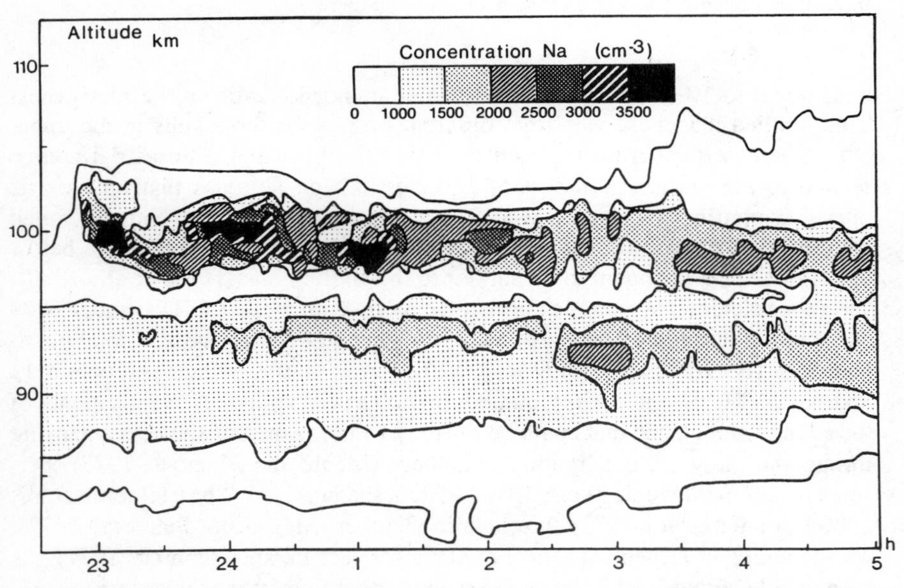

Figure 5.13. Instantaneous increase in sodium content as observed on July 24, 1981, at Observatoire de Haute Provence. Upper part: sodium altitude distribution as recorded at 21:55 (dotted line) and 22:01 (solid line) giving evidence for large extra terrestrial input at 101 km. Lower part: temporal evolution (local time from 10:00 P.M. to 5:00 A.M.) in terms of sodium density isocontours as function of altitude and time.

as a sink for sodium compounds at lower altitude (Liu and Reid, 1979). This model assumed that sodium is injected into the atmosphere through meteor ablation and is transformed in various oxidized compounds and that the equilibrium is maintained by a downward flux equal to the meteoritic influx (Mitra, 1973; Kirchhoff et al., 1981b; Hunten, 1981; Clemesha et al., 1981). However, either in the equilibrium form or in the transient flux form none of these models was able to explain or even address the different seasonal behavior of the alkalis as observed at the Observatoire de Haute Provence (Sze et al., 1982). In addition to neutral species, clustering of sodium ions as a potential sink was then introduced by Richter and Sechrist (1979a,b). This idea was further developed by Jégou et al. (1985a,b) in a general flux model that emphasizes the role of the meteoritic input source and of the long-term changes of the ionic clustering processes acting as a sink for the metallic species by producing ionic hydrates (Figure 5.14). Due to the different values of the clustering rates with the various alkalis and to their dependence on the temperature and winds through the Laplacian ion transport, the simulated seasonal and meridional variations of the alkalis total abundance and peak concentration altitudes as simulated by this model are in very close agreement with experimental data (Figure 5.11).

The validity of this model has been further confirmed by the observations of neutral and ionized calcium (Granier et al., 1984), which present as predicted an abundance ratio of about 1, whereas for the alkalis the neutral form is one to two orders of magnitude more abundant (Figure 5.15). Finally, the observation of the sudden increase in sodium content observed on July 24, 1982, also show the importance of the transport of ionic species in the equilibrium of the neutral layer. As it can be seen on the isodensity lines in Figure 5.13, each

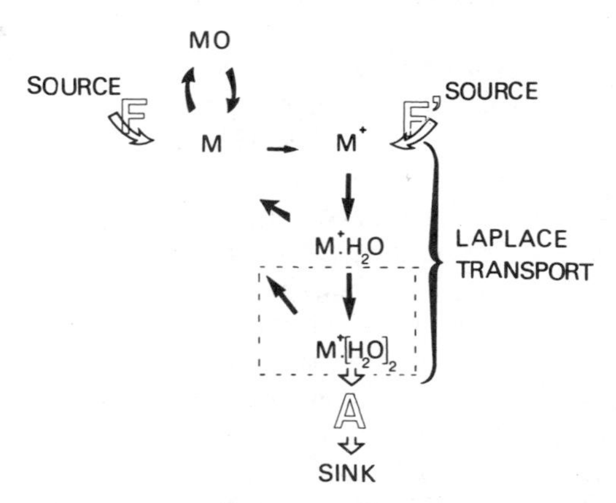

Figure 5.14. Schematic representation of various chemical and dynamical processes involved in flux model of metallic layers (M) as given by Jégou et al. (1985a, b).

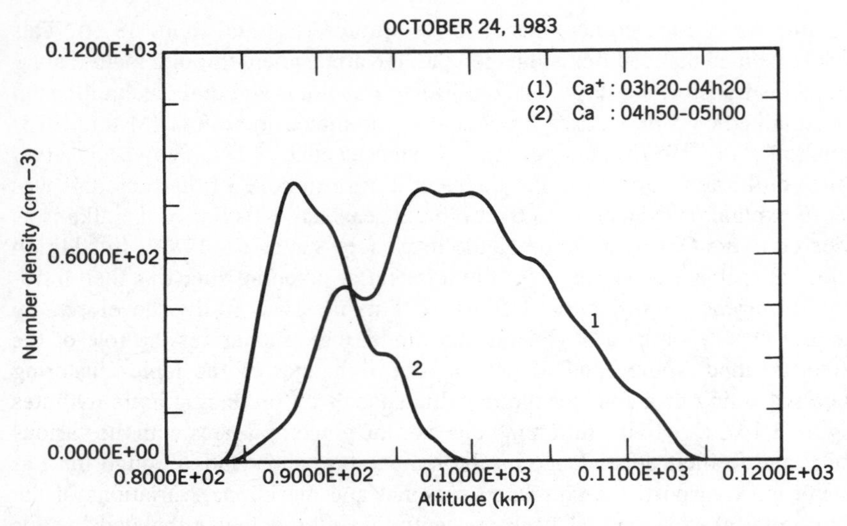

Figure 5.15. Vertical density distribution (altitude on horizontal axis) of ionized calcium, Ca^+ (curve 1), and neutral calcium, Ca (curve 2), during night of October 23–24, 1983, at Observatoire de Haute Provence (from Granier et al., 1984).

increase at 100 km is followed with a 2-h delay by an increase at lower altitude (92 km) with no enhancement in between. This transfer has been explained using the clustering chemical scheme in the presence of a sporadic E layer (Granier et al., 1982).

5.2.3.2 Diurnal and Short-Term Variations

Daytime measurements using the resonant lidar technique are further complicated as compared to nighttime measurements by the enlarged value of the sky background. Filtering techniques using wavelength-selecting interferometers have been developed leading to measurements by several groups (Gibson et al., 1972; Granier et al., 1981; Granier and Mégie, 1981; Clemesha et al., 1982). It was generally accepted from dayglow observations that the column density of atomic sodium was enhanced during the day by a factor of 2–6 relative to that at twilight (Blamont et al., 1964) whereas there had been some discussions on the possibility that the ring effect—partial filling of the Fraunhofer lines in scattered sunlight—may have affected the passive measurements (Grainger and Ring, 1962). Resonant lidar observations at midlatitudes (Granier and Mégie, 1981) showed, however, that the long-term behavior of the sodium layer during daytime is similar to the one already observed during nighttime. No regular variations of the characteristic parameters of the sodium layer—total content altitude of the concentration peak, topside and bottomside scale heights—during the diurnal cycle or during day–night transitions were observed, suggesting that photochemical processes are not dominant in governing the behavior of the short-term variations of the sodium layers except at lower altitudes below 82 km (Figure 5.16). Similar observations made at lower latitudes (Clemesha et al., 1982) in the Southern Hemisphere show the existence of a diurnal oscillation in both total abundance (15%) and height that might be related to the 2,2 mode of the semidiurnal tide. The small variation in abundance between day and night is reproduced by the model of Jégou et al. (1985a) adapted for diurnal variations (Granier et al., 1985). If the neutral chemistry alone, including the sodium hydroxide molecule, is used in a time-dependent model, the measured diurnal behavior can only be reproduced by introducing a height-dependent loss function for sodium resulting either from the reaction of NaOH with hydrogen (Kirchhoff and Clemesha, 1983a) or as more recently suggested from the three-body production of NaO_2 (Kirchhoff, 1983). However, because many of the sodium reaction rates are still uncertain, such calculations are still speculative. As previously shown, the main advantage of the ion-neutral model is its ability to explain all the observed behavior of the various metals already detected by resonant lidar scattering.

Lidar observations of resonance scattering from atmospheric sodium during the night have shown evidence of short-term changes in the height distribution (Blamont et al., 1972b), related to the propagation of gravity waves or tidal oscillations through the sodium layer. Two-dimensional signal-processing techniques have thus been developed that utilize spatial and temporal filtering to

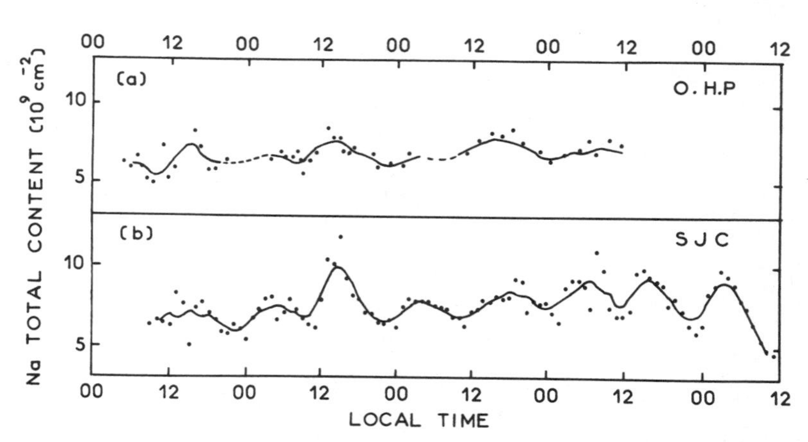

Figure 5.16. Variations of sodium total content during full diurnal cycles as observed at Observatoire de Haute Provence (December 16–19, 1980) and at Sao José dos Campos (May 11–15, 1981) (from Granier and Mégie, 1981, and Clemesha et al., 1982).

reduce shot noise and increase resolution (Gardner and Shelton, 1981) and have been applied to sodium lidar data. They give evidence for wavelike structures moving downward with time, with typical wavelengths of 3–15 km and phase velocities of less than 1 m s^{-1} (Rowlett et al., 1978; Shelton and Gardner, 1981) (Figure 5.17).

Because of the thickness of the sodium layer and the opposed density gradients occurring on the top and bottom sides, a nonlinear model is not adequate to describe gravity wave–sodium layer interactions, and nonlinear effects have to be included (Chiu and Ching, 1978; Shelton et al., 1980) in order to retrieve the wave characteristics from the lidar observations. To extend the spatial scale of these measurements in the horizontal direction, experiments have also been conducted using either steerable lidar (Thomas et al., 1976) or airborne lidar systems (Segal et al., 1984).

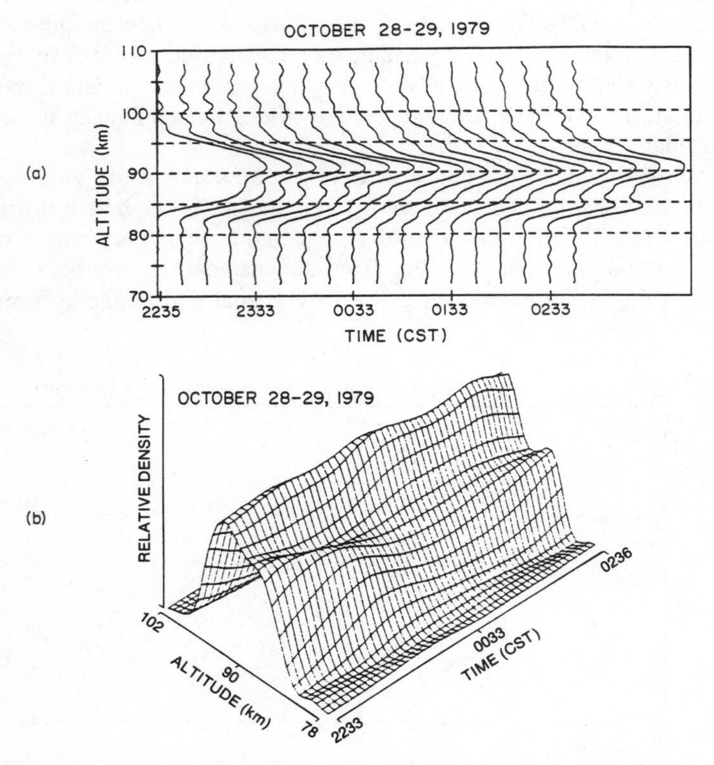

Figure 5.17. Time history of estimated altitude profiles of sodium density observed at Urbana on October 28–29, 1979 (upper part). Spatial and temporal filter cutoffs 0.44 km^{-1} and 0.33 min^{-1}. Profiles plotted at 15-min intervals. Three-dimensional view of estimated profiles of sodium density observed on same night (lower part) (from Shelton and Gardner, 1981).

5.2.3.3 Determination of Global Atmospheric Parameters from Resonant Lidar Observations of Sodium Layer

From resonant scattering lidar observations of the free sodium atoms, global properties of the atmosphere at the mesopause level can also be determined either directly, using a spectroscopic analysis of the return signal, or indirectly, using sodium layer theories. In the first case parameters such as temperature and wind motions, which directly influence the shape and position of the scattered line, will be accessible. In the second case parameters that directly govern the spatial and temporal variations of the layer such as eddy diffusion coefficients, winds, or ozone number density will be determined based on our present knowledge of the sodium layer behavior. The most important parameter to be determined using the resonant scattering lidar technique is atmospheric temperature as the knowledge of its height variation is a major requirement in different aspects of the aeronomy of the mesosphere, which include, for example, ion chemistry, airglow excitation processes, noctiluscent cloud formation, gravity wave activity, and thermal balance.

In Section 5.2.1.1, it was shown that the sodium D_2 resonance line consists of two groups (D_{2a} and D_{2b}) of three hyperfine lines each separated by about 2 pm with a Doppler-broadened width at 200 K of 1.2 pm. Measurement of these spectroscopic features will then lead to a direct determination of the translational temperature T. Two approaches have been taken in the past that correspond to two very different line parameters of the laser emission. Mégie et al. (1974) used a laser linewidth of 10 pm to excite all the hyperfine components of the D_2 line. They further transformed the lineshape measurement of the backscattered line into an intensity measurement by using an absorption technique that consists in partially absorbing the return signal in an absorption cell containing sodium vapor at a known temperature T_c. If the intensity entering the cell is I_0 and the intensity after traversing the cell is I, it can be shown that the reduction factor $R = I/I_0$ is only a function of the optical thickness of the cell τ_c and of the ratio T/T_c as given by Mégie et al. (1974),

$$R = \frac{\int_0^\infty \exp(-x^2 - \tau_c e^{-x^2 T/T_c})\, dx}{\int_0^\infty e^{-x^2}\, dx} \tag{5.41}$$

where $x = (\nu - \nu_0)/\Gamma_D$ (see Section 5.2.1.1 for notation).

The temperature T_c of the absorption cell is known, and its optical depth τ_c can be measured using a calibration sodium resonance cell of known temperature. Such measurements of atmospheric temperature have been made from 1972 at the Observatoire de Haute Provence (Blamont et al., 1972b; Mégie et al., 1974; Mégie and Blamont, 1977) and in the high-latitude region (Mégie et al., 1978b).

However, this approach requires a large sampling period, and new methods have more recently been developed that use a narrow-band ($\Gamma_L \leqslant 0.2$ GHz) laser emission tunable throughout the sodium absorption line (Gibson et al., 1979; Thomas et al., 1980; Fricke and Von Zahn, 1985).

The probing of the hyperfine structure can be performed at several wavelengths if one does not know accurately the laser absolute wavelength or by positioning the laser emission at the maximum of the D_{2a} component and at the intermediate minimum. As already pointed out (Section 5.2.1.5), the Zeeman splitting frequency is small when compared to the inverse of the lifetime of the excited state, and so the Hanle effect has to be accurately taken into account when calculating the spatial distribution of the intensities, which will depend on the polarization state of the laser emission (Hanle, 1924). Fricke and Von Zahn (1985) have calculated this effect for several assumptions of geographic location and polarization of the lidar that will directly affect the ratio of maximum to minimum intensity (Figure 5.18). In terms of temperature the difference

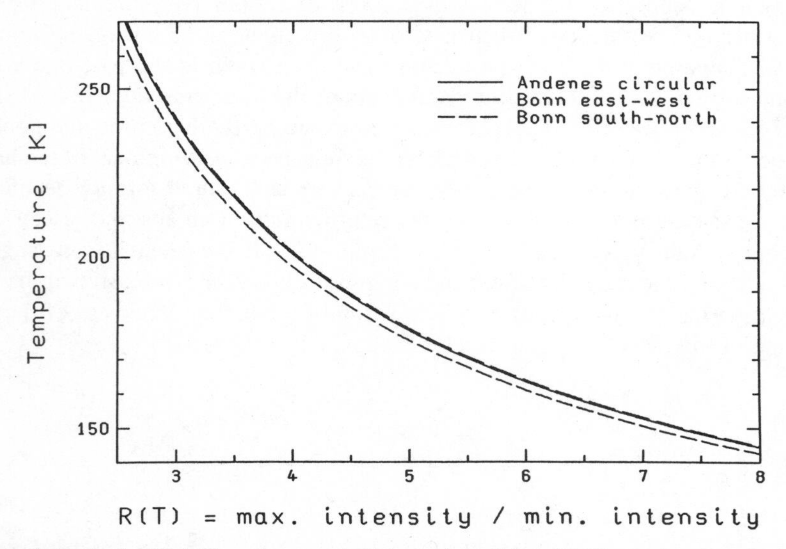

Figure 5.18. Ratio $R(T)$ of maximum and minimum intensities of D_2 line as function of temperature for three locations and different polarization of the emitted laser light showing the influence of the Hanle effect: Andenes (69° N, 16° E), circular polarization (full line); Bonn (51° N, 7° E), linear polarization aligned in north–south direction (interrupted line); Bonn (51° N, 7° E), linear polarization aligned in east–west direction (large interrupted line) (from Fricke and Von Zahn, 1985).

thus introduced varies from 6 K for an atmospheric temperature of 250 K down to 2.5 K near 150 K. For an accuracy of ±5 K at 200 K the intensity ratio needs to be measured with a relative accuracy better than 4.5%.

In their experiment Fricke and Von Zahn (1985) used a 0.13-pm FWHM laser line, the wavelength of which was measured for each pulse with a 0.02-pm precision using a Fabry–Perot etalon and a sodium atomic beam. Temperature profile obtained at Andoya in northern Norway (69° N, 16° E) during the winter of 1983–1984 is plotted in Figure 5.19. The statistical error in the derived temperature is 4–6 K except for the edge intervals, where it increases up to 10 K. Near the peak of the layer the maximum altitude resolution is 1 km, but as the signal decreases toward the boundaries of the layers, the integration interval had to be increased to 5 km. The corresponding integration time is about 90 mn. Such measurements are thus very promising as they allow the determination of the temperature in a region of the atmosphere that had previously been inaccessible.

The neutral sodium layer is very sensitive to the eddy diffusion coefficient

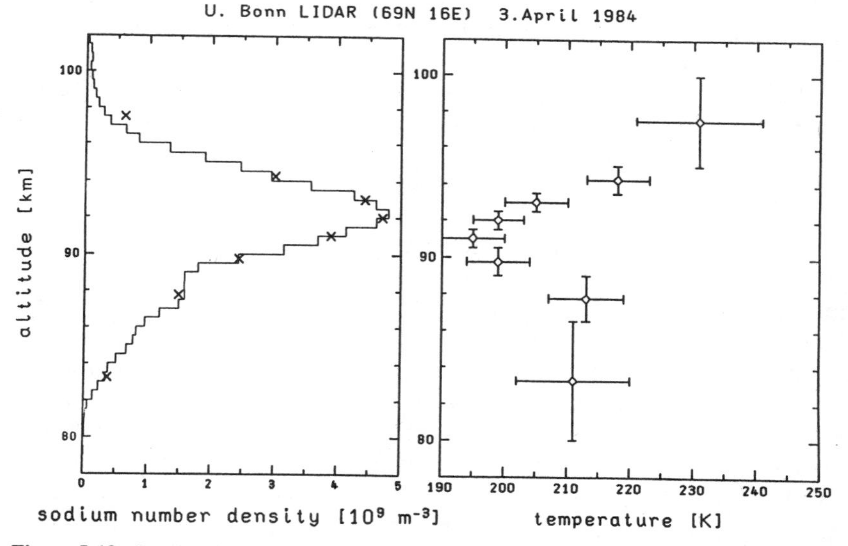

Figure 5.19. Results obtained for April 3, 1984, between 21 h 06 and 22 h 52 UT at Andoya (69° N, 16° E). Sodium layer subdivided into eight slices of about equal signal strengths to obtain comparable signal-to-noise ratios throughout layer. (*a*) Average sodium number densities in each of eight slices (crosses) based on comparison to Rayleigh scattered signal from 30 km altitude. Line is solid-angle corrected count rate normalized to sodium density in 91 ± 0.5 km slice. Differences to crosses are mainly due to different altitude resolution of two data sets. (*b*) Mesospheric temperature profile. Note that absolute value of temperature scale is uncertain by about ±10 K because of remaining uncertainty in lineshape of lidar transmitter. Error bars in figure show statistical errors from fit procedure only (from Fricke and Von Zahn, 1985).

K, which parametrizes the transport of minor constituents related to turbulent processes. The maximum sodium density and the shape of the layer are fixed, for a given influx at the source, by the absolute value of *K*, and Mégie and Blamont (1977) have thus derived a statistic of this coefficient in the 82–95-km range. Kirchhoff and Clemesha (1983b) further extended this method to determine the seasonal variation of *K* for a height of 94 km showing a maximum value during the summer of 23° S latitude. The same authors have also shown that from simultaneous ground-based observations of neutral sodium by lidar and of the sodium nightglow intensity, one can deduce using a photochemical equilibrium model the ozone density at the 88-km altitude level (Clemesha et al., 1978a,b; Kirchhoff et al., 1981c). A mean annual value of 8.10^7 cm^{-3} has been thus determined, whereas a strong seasonal variation with well-defined maxima at the equinoxes is measured.

Finally, laser radar observations of sodium atoms using either horizontally spaced measurements of the sodium layer structure (Clemesha et al., 1980) or steerable systems for sodium clouds tracking (Rees and Sandford, 1974) have led to the measurements of neutral winds in the mesosphere and thermosphere. Such measurements are, however, rather indirect, and a potential method for direct determination also exists based on the determination of the Doppler shift of the backscattered laser line, whereas no measurements have yet been performed due to the very high value of the requested spectral resolution.

5.2.4 Conclusion

Due to the very high efficiency of the atomic resonance scattering process, a large number of lidar observations not only of sodium, which is certainly among the most documented minor constituent in Earth's atmosphere, but also of other trace metals at the mesopause level have been conducted during the last decade. They have led to a better knowledge of the morphology and behavior of these constituents at various temporal and spatial scales and have brought a new insight in our understanding of their origin and interactions with neutral and ionized species in the 80–100-km altitude range. Whereas more measurements are still needed, particularly to increase the latitudinal coverage for potassium, calcium, and lithium observations, which have been restricted to a single location, and for a better determination of the reaction rate constants involved in the photochemical equilibrium, convincing explanations have yet been given for the different variations observed on a seasonal and diurnal scale. A major uncertainty that still remains is the identification of the chemical form of the sink of these trace metals at the upper stratospheric level and its potential interactions with stratospheric trace constituents involved in ozone-reducing catalytic cycles. However, such problems are presently beyond the potentialities of lidar observations and will have to be studied using more conventional in situ techniques.

The present accuracy and time–space revolution of resonant scattering lidar sodium measurement also open a new field of investigation concerning the dynamical and thermal structure of the mesosphere. The resonant scattering lidar has thus to be considered as a very powerful tool for continuous studies of an altitude region where up to now only snapshot measurements using rocket techniques or low-spatial-resolution satellite observations have been completed.

5.3 DIFFERENTIAL ABSORPTION LIDAR OBSERVATIONS OF TRACE CONSTITUENTS IN MIDDLE ATMOSPHERE

The resonant scattering lidar technique as developed in the previous section can only be used for the detection of atoms or ions that are the dominant constituents in the upper atmosphere. At lower altitudes the molecular or radical forms are generally the most abundant, and due to the low efficiency of the fluorescence processes, the measurements of trace species by optical techniques in the stratosphere is mainly based on the measurement of their absorption properties. This has been the case for a long time using passive instruments, whereas the recent development of the differential absorption lidar technique (Schotland, 1964) for stratospheric applications has allowed during the last seven years a very rapid growth in lidar observations of the ozone vertical distribution from the ground up to the 50-km altitude level. The purpose of this section is to give a detailed analysis of the method itself and to describe the various systems presently in operation. An analysis of their advantages and disadvantages with respect to one another will also be presented, and an overview of already obtained data will be given. Perspectives in terms of ozone and other constituent monitoring will be considered in the last section.

5.3.1 Methodology: The DIAL Technique

The basic principles of the differential absorption laser technique have been described by various authors (Schotland, 1964; Byer and Garbuny, 1973). Its application to ozone measurements for both UV and IR systems has been analyzed by Mégie and Menzies (1980). They concluded that as far as higher tropospheric and stratospheric measurements are concerned, the UV wavelength range is the best candidate for ground-based observations. Therefore, only such systems operating in the Hartley–Huggins bands of ozone between 280 and 320 nm will be considered here. The forthcoming analysis will follow the study made by Pelon and Mégie (1982a).

The usual lidar equation that relies the total number $n_{\lambda R}$ of backscattered photons at wavelength λ from the cell at the range R and of thickness ΔR to the number of emitted photons in the laser pulse $n_{e\lambda}$ at wavelength λ is written as

$$n_{\lambda R} = n_{e\lambda} \, \Delta R \frac{A_0}{R^2} \beta_{\lambda R} \xi \xi^* \exp[-2(\tau^0_{\lambda R} + \tau^e_{\lambda R})] \tag{5.42}$$

where $\beta_{\lambda R}$ = atmospheric backscattering coefficient at wavelength λ and range R

ΔR = thickness of range cell corresponding to time gate interval of 2 $\Delta R/c$ generally larger than the pulse duration τ_L

A_0 = receiver area

ξ = detector efficiency

ξ^* = optical efficiency of transmitter–receiver system

$\tau^0_{\lambda R}$ = integrated optical thickness due to absorption by constituent under study,

that is,

$$\tau^0_{\lambda R} = \int_0^R \sigma_{\lambda R} N_0(R) \, dR \tag{5.43}$$

where $\sigma_{\lambda R}$ = ozone absorption cross section and $N_0(R)$ the ozone number density

$\tau^e_{\lambda R}$ = integrated optical thickness excluding absorption by ozone

$\tau_{\lambda R}$ = total integrated optical thickness $\tau_{\lambda R} = \tau^0_{\lambda R} + \tau^e_{\lambda R}$

If one considers the ratio of the two backscattered signals corresponding to two successive cells R_1, and $R_2 = R_1 + \Delta R$ for the same laser pulse, one can write

$$\Delta\tau^0_{\lambda 1} = \frac{1}{2} \ln \frac{n_{\lambda 1}}{n_{\lambda 2}} + \ln \frac{\beta_{\lambda 2} R_1^2}{\beta_{\lambda 1} R_2^2} - \Delta\tau^e_{\lambda 1} \tag{5.44}$$

where $\Delta\tau_{\lambda R}$ is defined as the local optical thickness of the constituent within the range cell R, $R + \Delta R$. Assuming the knowledge of the ozone absorption cross section $\sigma_{\lambda R}$ at this range, $\Delta\tau^0_{\lambda R}$ is then directly proportional to the average ozone number density within the same range cell as

$$\Delta\tau^0_{\lambda R} = \sigma_{\lambda R} N_0(R) \, \Delta R \tag{5.45}$$

Thus, a one-wavelength lidar measurement can lead to the determination of the ozone vertical profile from the laser backscattered signals $n_{\lambda 1}$, $n_{\lambda 2}$ corresponding to ranges R_1 and R_2 if the last two terms of Equation (5.44) are known. If one assumes a pure molecular scattering atmosphere (no aerosol particles), $\beta_{\lambda R}$ and $\Delta\tau^e_{\lambda R}$ are then proportional to the atmospheric number density as the Rayleigh scattering process is the only one contributing to the laser light scattering. Using thus an atmospheric model or the atmospheric parameters as given by local radiosonde measurements, such a technique can be used for ozone profiling (Uchino et al., 1979). However, aerosol particles are always present in the troposphere and in the lower stratosphere around 20 km, and their abundance can be greatly increased by more than one order of magnitude during the one or two years following large volcanic eruptions (Mc Cormick and Swissler, 1983). Then single-wavelength measurements cannot be considered as fully reliable for any atmospheric situation and one has to use a second wavelength, λ_2, different from λ_1, to discriminate the ozone absorption from other potential interfering species. Writing Equation (5.44) for a second wavelength λ_2 and taking the difference between the two then leads to

$$\Delta\tau_{1R} - \Delta\tau_{2R} = \ln\frac{n_{22}n_{11}}{n_{12}n_{21}} + \ln\frac{\beta_{12}\beta_{21}}{\beta_{11}\beta_{22}} \tag{5.46}$$

where the first subscript refers to the wavelength and the second, when present, to the range.

The signal-to-noise ratio $S_{\lambda R}$ for a single-wavelength measurement is given, in the case of an incoherent detection scheme and assuming that the fluctuations of the various sources of the photodetector current are governed by Poisson statistics, by

$$S_{\lambda R} = \frac{n_{\lambda R}}{(n_{\lambda R} + n_B + n_D)^{1/2}} \tag{5.47}$$

where n_B is the number of background photons and n_D is the square of the photodetector dark-current fluctuations within the time gate interval $2\,\Delta R/c$.

In the usual derivation of the DIAL method, one considers that the wavelength variations of $\beta_{\lambda R}$ and $\tau^e_{\lambda R}$ between λ_1 and λ_2 contribute only to the systematic error term ϵ_2 (see below) (see, e.g., Schotland, 1974) so that (5.46) can be rewritten as

$$\Delta\tau^0_{1R} - \Delta\tau^0_{2R} = \frac{1}{2}\ln\frac{n_{22}n_{11}}{n_{12}n_{21}} \tag{5.48}$$

The average ozone number density $N_0(R)$ within the range cell R, $R + \Delta R$, is then given by

$$N_0(R) = \frac{\Delta\tau_{1R}^0 - \Delta\tau_{2R}^0}{(\sigma_1 - \sigma_2)\,\Delta R} \tag{5.49}$$

The relative uncertainty $\epsilon = \delta N_0/N_0$ depends on a large number of interrelated parameters and can be expressed as the sum of two terms, $\epsilon_1 + \epsilon_2$.

1. ϵ_1 is the statistical error due to the signal and background-noise fluctuations and is related to the signal uncertainty by

$$\epsilon_1 = \frac{1}{2\tau_R^0\sqrt{n_0}}\left(f_1^2 + f_2^2\right)^{1/2} \tag{5.50}$$

where

$$K = \frac{\tau_R^0}{\Delta\tau_R^0} = \frac{\int_0^R N_0(R)\, dR}{N_0(R)\,\Delta R} \tag{5.51}$$

is wavelength independent as far as one considers that the ozone absorption cross sections are altitude independent. One can also notice that the product $K\,\Delta R$ does not depend on the range resolution of the lidar system:

- τ_R^0 is the differential integrated optical thickness ($\tau_R^0 = \tau_{1R} - \tau_{2R}^0$).
- $n_0 = A_0\,\Delta R\,\xi\xi^*/R^2$ is the wavelength-independent part of $n_{\lambda R}$.
- f_λ is a wavelength-dependent function given by

$$f_\lambda^2 = \frac{e^{2\tau_{\lambda 1}}}{\beta_\lambda n_{e\lambda} P_l\, e^{-2\tau_{\lambda 1}}}\left[\left(1 + e^{2(\tau_\lambda^0/K)}\right) + X_\lambda^0\left(1 + 4e^{\mu_{\lambda 1}} e^{4(\tau_{\lambda 1} + \tau_\lambda^0/K)}\right)\right] \tag{5.52}$$

$$\mu_\lambda = \ln\left[\left(1 + \frac{\Delta R}{R}\right)\left(\frac{\beta_{\lambda 1}}{\beta_{\lambda 2}}\right)^{1/2}\right] + \Delta\tau_\lambda^e \tag{5.53}$$

$$X_\lambda^0 = \frac{n_B + n_D}{n_{\lambda 1}}\, e^{-2\tau_\lambda^e} \tag{5.54}$$

and is independent of the ozone absorption.

- P_λ is the number of laser pulses emitted at wavelength λ.

2. ϵ_2 is a systematic error due to the wavelength dependence of the scattering and absorbing (other than ozone) properties of the atmospheric medium, which has been neglected in the derivation of Equation (5.46). A general expression for ϵ_2 has been given by Mégie and Menzies (1980) as

$$\epsilon_2 = \frac{K\,\Delta R}{\tau_R^0}\, G(R, \lambda)\, \frac{\Delta\lambda}{\lambda_1} \tag{5.55}$$

where $G(R, \lambda)$ is a function of range and wavelength that depends on the scattering and extinction properties of the atmospheric gas and particles and is written as

$$G(R, \lambda) = 2\left(1 - \frac{m}{4}\right) \frac{1 - r_{\lambda R}}{(r_{\lambda R})^2} \left(\frac{1}{h} - \frac{1}{H}\right) + m\alpha_\rho + 4\alpha_M \tag{5.56}$$

Using these expressions of the various uncertainties, an error analysis of the DIAL measurement taking into account both experimental parameters and atmospheric characteristics can now be performed. A minimum value of ϵ can be obtained for a given range by an appropriate choice of the laser emission wavelengths λ_1 and λ_2. In the general case, this optimization of the DIAL measurement requires a numerical computation to include the experimental parameters of the lidar system and the atmospheric parameters related to molecular and aerosol scattering and ozone absorption. The description of such computations are beyond the scope of this section; however, a simple analysis can be made by considering separately the lower (0–2 km) and higher altitude ranges, as the relative contribution of the molecular and aerosol extinctions will be different in these two cases. Although this study is mainly devoted to the middle atmosphere, one has also to consider the lower tropospheric levels if a determination of the ozone total content is to be undertaken.

5.3.1.1 Boundary Layer and Lower Tropospheric Measurements

Below 2 km, ϵ_2 becomes the most important error term as the molecular and aerosol extinctions are large and as the backscattered signal is intense enough so that the statistical error ϵ_1 can be reduced below 10^{-2}. By using the extinction values given by Elterman (1968), the differential molecular extinction for a wavelength interval $\Delta\lambda = \lambda_1 - \lambda_2 = 5$ nm can be as large as 0.1 in the wavelength range considered (270–310 nm) and thus cannot be neglected. A correction has to be made by measuring the ground-level pressure and temperature to derive the atmospheric density in the first kilometers. An uncertainty of a few percentage points in this determination will reduce the error due to molecular extinction to less than 5×10^{-3}.

The remaining error in ϵ_2 is then related to the aerosol extinction, and a further distinction should be made between ozone measurements performed in rural (nonpolluted) or urban areas.

1. *Rural areas:* We have calculated the variation of ϵ_2 as a function of wavelength and range by using aerosol particles concentration as given by Elterman (1970) and typical of rural conditions. These variations are given in Figure 5.20 for different values of λ_1 and $\Delta\lambda = \lambda_2 - \lambda_1$ for a vertical range $R = 1$ km. The optimization of λ_1 and $\Delta\lambda$ in the wavelength range 265–285 nm leads to values of ϵ_2 less than 2%, which can be further reduced, if needed, by using the experimental procedure adopted for urban or polluted areas and described in the following section.

2. *Urban or polluted areas:* Owing to the large aerosol particles concentration in the boundary layer, ϵ_2 can reach values as large as 10^{-1}, and a complementary measurement will be needed to eliminate the aerosol differential extinction. This can be done, for example, by using three wavelengths in the 265–285-nm range with the same wavelength interval $\Delta\lambda = 5$ nm. The aerosol differential extinction can then be substracted if one assumes a linear variation for the aerosol scattering and extinction properties over 10 nm. This will avoid

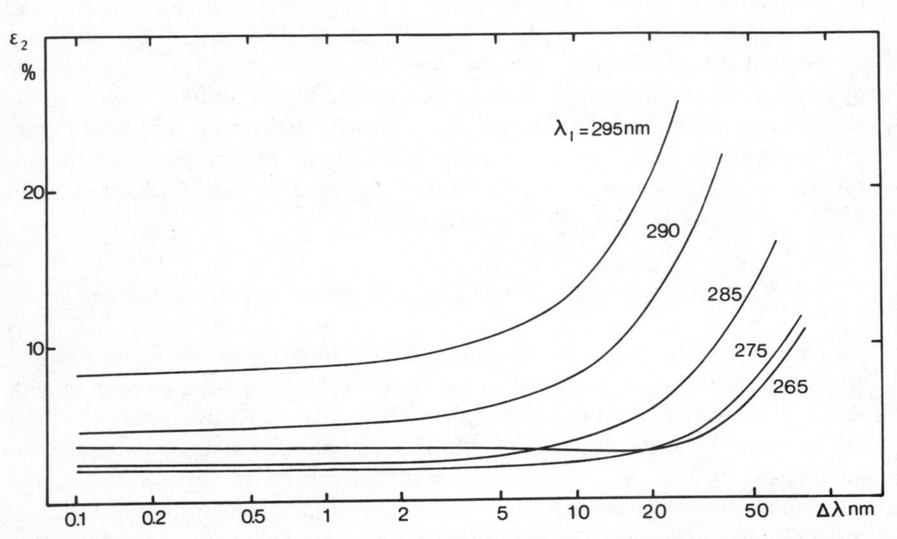

Figure 5.20. Variation of systematic error ϵ_2 as function of wavelength difference, $\Delta\lambda = \lambda_2 - \lambda_1$, for different values of absorbed wavelength λ_1 for range $R = 1$ km (rural aerosol model) (from Pelon and Mégie, 1982a).

the use of a theoretical model that depends on the nature and shape of the particles. Taking into account the experimental errors and the modeling uncertainties, the estimated upper limit of ϵ_2 will then be reduced to 2–3%.

5.3.1.2 *Upper Tropospheric and Stratospheric Measurements*

Above 2 km, ϵ_2 decreases rapidly with altitude, as represented in Figure 5.21, for average values of higher tropospheric and stratospheric aerosol content. The choice of the operating wavelength is then determined by calculating the minimum value of ϵ_1 as a function of the three interrelated and wavelength-dependent parameters τ_{1R}^0, τ_{2R}^0, $\Delta\lambda$. For a given value of the larger absorption wavelength λ_1, the minimum is obtained for $\tau_2^0 = 0$. This condition cannot be experimentally achieved in the case of the ozone UV absorption bands, which present continuous absorption features over a wide wavelength range. Thus, the optimization procedure will consist of first calculating the value of τ_{1R}^0 and thus λ_1, corresponding to the minimum value of ϵ_1 for $\Delta\lambda = \infty$ (i.e., $\tau_{2R}^0 = 0$), and then evaluating the decrease in accuracy that results in the choice of a finite value of $\Delta\lambda$ compatible with both the experimental constraints and a maximum value of ϵ_2 as calculated above.

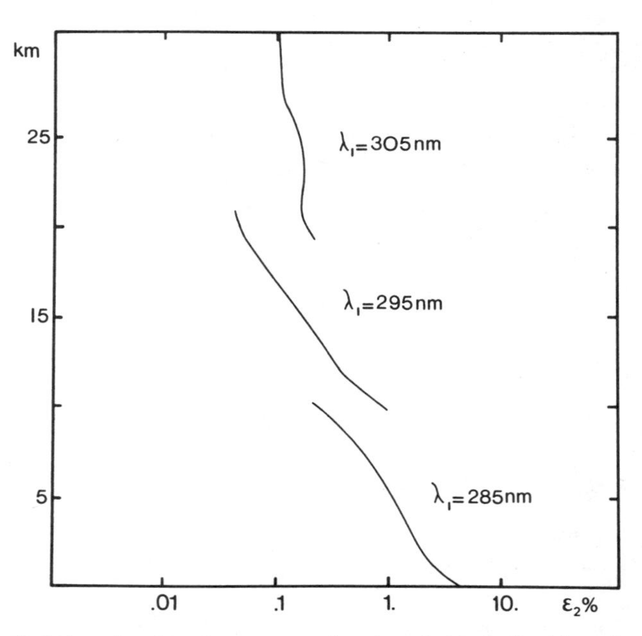

Figure 5.21. Variation of systematic error ϵ_2 as function of altitude for $\Delta\lambda = 5$ nm and three values of λ that optimize statistical error ϵ_1 (from Pelon and Mégie, 1982a).

Mégie and Menzies (1980) have shown that the optimum value of τ^0_{1R} is 1.28, in the case of shot-noise-limited signal when one neglects the off-line (λ_2) absorption (for ozone measurements this corresponds to large values of $\Delta\lambda$). We have represented in Figure 5.22 the variations of ϵ_1 relative to its minimum value as a function of $\Delta\lambda$. The values of atmospheric density as a function of altitude required for the calculations were taken from the *U.S. Standard Atmosphere* (1976). By comparing these variations with the ones obtained for ϵ_2 (Figure 5.20), one can see that the optimum wavelength interval has to be in the range 5–15 nm. The final choice of 5 nm has been adopted by considering the maximum tuning range of the laser systems and the possibility of using several wavelength pairs for the measurement within this range.

The optimum value of λ_1 as determined above is only valid for a given altitude range R. From an experimental point of view, it seems impossible to use as many wavelength pairs as the number of altitude levels of the measurement. Therefore, we have calculated the altitude variations of $\epsilon_1(\lambda_1, \Delta\lambda)$ by using the mean ozone profile of Krueger and Minzer (1976) typical of midlatitude regions. These relative variations represented on an arbitrary scale, correspond-

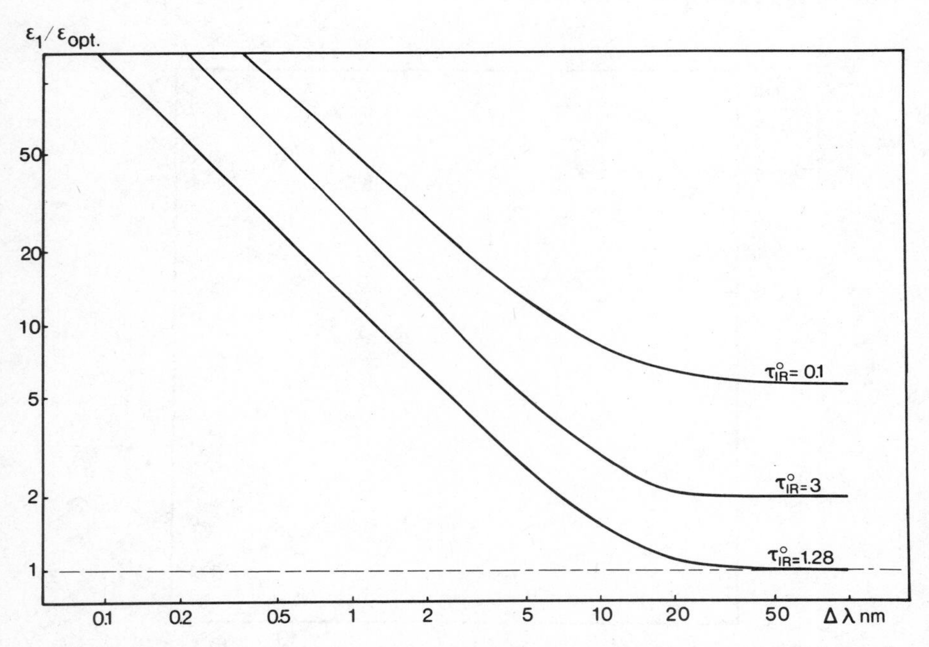

Figure 5.22. Variation of statistical error ϵ_1 as function of $\Delta\lambda$ for different optical thickness τ^0_{1R} corresponding to different wavelengths λ, or ranges. ϵ_1 is normalized to its minimum value (from Pelon and Mégie, 1982a).

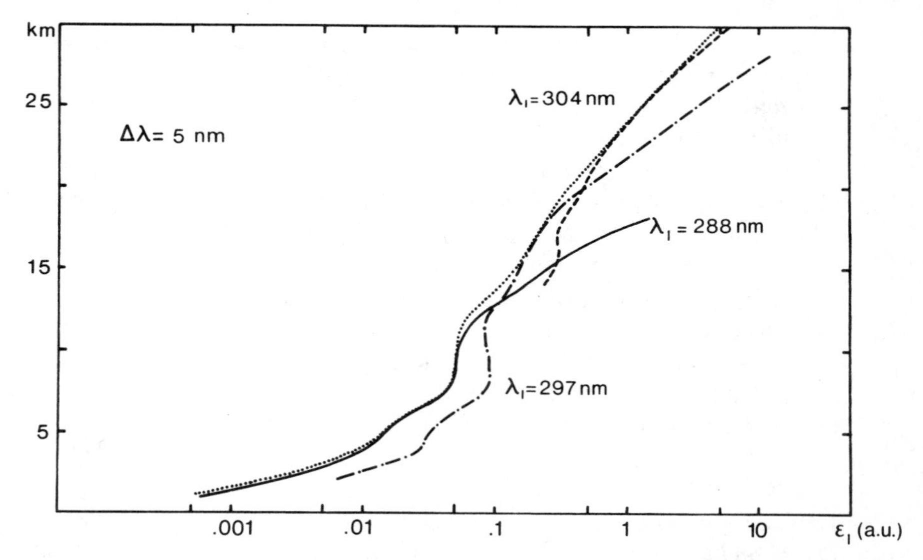

Figure 5.23. Variation of statistical error ϵ_1 as function of height for three different range-optimized wavelengths. Dotted curve represents optimum value of ϵ_1 for each altitude (from Pelon and Mégie, 1982a).

ing to experimental conditions, are given in Figure 5.23 for three wavelength pairs (λ_1, λ_2). The shorter wavelengths are used to probe the lower altitude levels, and the useful range of a given pair can be extended up to 7–8 km so that the measurement accuracy stays between its optimum value ϵ and 1.2ϵ.

5.3.1.3 Upper Stratospheric Measurements and Ozone Total Content

Attempts to measure ozone number density above its maximum using a ground-based UV system are complicated by the high extinction of the laser-emitted light due to the absorption in the lower altitude levels. Furthermore, the rapid decrease in the ozone number density above 28 km requires very rapidly increasing acquisition times for the measurement. The choice of higher wavelengths (wavelength pair *B* with λ_1 = 305.8 nm and λ_2 = 310.8 nm) allows these major problems to be overcome and to probe the upper levels with appropriate integration times for geophysical applications (Pelon and Mégie, 1982b). The ozone total content can also be derived from the measured vertical distribution using a combination of the two wavelengths pairs (pair *A* at lower altitudes, 288–294 nm; pair *B* at higher levels). Various sources of uncertainty have to be considered:

1. *The statistical fluctuations of the detected signals that constitutes the major error source, estimated to be less than 10 Dobson units in a 95% confidence interval.* This error can be further reduced by increasing the acquisition time.
2. *The Rayleigh extinction by the atmospheric gas molecules.* This uncertainty is reduced below 0.1 Dobson unit by correcting the lidar data using the pressure, temperature, and humidity distribution measured by radiosonde or lidar measurements.
3. *The Mie differential extinction by aerosol particles.* This uncertainty, which can be large at lower altitudes, is reduced when using a wavelength pair to measure the total content up to 15 km, which leads to high ozone optical thickness as compared with the aerosol contribution. It can be estimated to be less than 1 Dobson unit, whereas the use of the *B* pair in the same altitude range will have resulted in a one order of magnitude higher uncertainty.
4. *A systematic bias due to the discrepancies between the two wavelength pairs and the absolute cross-sectional determinations.* A comparison has been made at both wavelength pairs *A* and *B* can be used to measure the ozone number density between 13 and 18 km. Considering several profiles obtained during successive nights, a systematic bias can be detected showing ozone values 2% higher when measured using the *A* pair. This discrepancy can be related (1) to uncertainties in the absolute values of the cross sections and (2) to the very high spectral resolution of the laser measurements as compared with the spectrometric ones. Due to experimental uncertainties, the detected bias between the *A* and *B* pairs could only be considered as estimations in this first approach.
5. *A constant value of 10.4 Dobson units, which corresponds to the average ozone content above 40 km (Krueger and Minzer, 1976).* The 2σ standard deviation results in a 3.2-Dobson-unit uncertainty.

5.3.1.4 Interference with Other Absorbing Gases

Two minor constituents have to be considered as potential interference absorbers in the 300-nm wavelength range.

Sulfur Dioxide. SO_2 absorption spectrum presents a band structure with a spectral width for the individual features of 1 nm (Thompson et al., 1975). The differential cross section between an absorption maximum and a minimum ($\Delta\lambda \sim 1.5$ nm) can reach 10^{-18} cm^2. In the case of large amounts of SO_2 in the troposphere over urban areas, the corresponding differential absorption coefficient can reach values as large as 0.1 km^{-1} at the ground level (Stewart et al., 1978). To avoid any interference, the two laser wavelengths λ_1 and λ_2 should be chosen so that SO_2 differential absorption vanishes. This choice is always possible within 0.3 nm of a predetermined wavelength.

Nitrogen Dioxide. NO_2 presents in the same wavelength range absorption bands with cross sectional values on the order of 10^{-19} cm^2 (Bass et al., 1976). The differential cross section for a 5-nm wavelength interval is then 2.5 10^{-20} cm^2. Its results, if one considers the higher mixing ratios typical of urban or polluted areas (~0.3 ppm; cf. Seinfeld, 1975), is a differential absorption coefficient of 0.015 km^{-1} at the ground level. The corresponding error on the ozone measurement will be 2% and can be taken care of by the same procedure as for SO_2.

In any case, the high-altitude measurements will not be affected due to the very low atmospheric content in these interfering species above the boundary layer.

5.3.1.5 *Temporal Variations of Scattering Medium*

If the two wavelengths are not simultaneously emitted, variations in the optical properties of the scattering medium might occur between the two laser shots owing to atmospheric transport. This will be of particular importance at lower altitudes if the aerosol content is high or at the tropopause level in the presence of cirrus clouds. Thus, even if the integration time required for the measurement is large, the switching between the two wavelengths will have to be made with a time constant much smaller than the time constant characteristic of the dynamical transport. The experimental system will have to take this requirement into account (see Section 5.3.2), and the possible design of a dual-cavity laser emitting simultaneously at two wavelengths should also be investigated.

5.3.1.6 *Altitude Dependence of Ozone Absorption Cross Sections*

The value of the absorption cross sections that we used to derive the ozone number densities are taken from Inn and Tanaka (1953). The spectral resolution of their measurements is lower than the resolution of the lidar measurements as given by the laser linewidth (~50,000). The assumption has thus to be made that the mean value of the absorption cross section is the same over these different wavelength intervals. This is not experimentally confirmed, but the precise knowledge of the emitted wavelength (see the following section) will also allow to correct data from this systematic error using high-resolution absorption spectra of ozone, which have become available in this wavelength range very recently (Bass, 1984).

The ozone absorption cross sections are temperature dependent mainly in the Huggins bands above 310 nm where variations as large as a factor of 2 can be observed for a 100-K temperature difference. Low-resolution measurements (Vigroux, 1953) have shown that the absorption minima are more sensitive than the maxima to the temperature variations. In the Hartley bands, the relative temperature variation of the cross section is much lower, 1% for a 10-K variation (Vigroux, 1953; Bass, 1984). Therefore, the use of an atmospheric model to take into account this temperature dependence of the absorption cross section and an a posteriori control using the radiosonde measurements of the nearby meteorological stations will reduce this uncertainty on the ozone profile to less than 0.5%.

5.3.2 Experimental Systems

The system evaluation conducted in the previous section shows that the measurement of the ozone number density profile by the DIAL technique is achievable using presently available laser systems with a temporal resolution compatible with scientific geophysical objectives. Two altitude domains might be considered:

1. Below 30 km and due to the presence of aerosol layers in the troposphere and stratosphere, a dual-wavelength operation is requested with a wavelength interval between the two emitted laser lines not larger than 5 nm.
2. Above 30 km, where the atmosphere can be generally considered as purely molecular—although after large volcanic eruptions aerosol layers have been observed up to 40 km—a single wavelength lidar might give information on the ozone profile. However, such a determination will imply the use of an atmospheric density model that can have appreciable uncertainties in the ozone values, especially during perturbed conditions. This difficulty might be overcome using (again) a second laser wavelength with, however, the possibility of an increased wavelength interval.

5.3.2.1 Transmitter

The design of an experimental system must also take into account the available laser sources. Two types of lasers are presently potential candidates for such a system:

1. frequency-doubled dye lasers pumped by a Nd^{3+}–YAG laser, which have the advantage of being tunable over a broad wavelength range but is somewhat limited in terms of available output power, and
2. exciplex lasers, which are frequency fixed on a given transition of the active medium but can emit very large average and peak powers.

Both lasers have been used for the measurement of atmospheric ozone. The first, and presently the only, ground-based lidar system using dye lasers has been operational at the Observatoire de Haute Provence since 1977 (Mégie et al., 1977; Pelon and Mégie, 1982a,b). The active part of the transmitter is a laser-pumped frequency-doubled dye laser. The pump laser is a Nd^{3+}–YAG laser (Quantel model 480) emitting an energy pulse of 750 mJ at 1.06 μm with a repetition rate of 10 Hz. This IR emission is frequency doubled with an efficiency of 40% resulting in an available pump energy of 300 mJ at 0.53 μm. The transversely pumped dye laser (Jobin Yvon model HPHR) includes one oscillator cavity and three amplifier stages (Bos, 1981). To cover the wavelength range from 570 to 620 nm, two dye solutions are used as the active medium (1) between 570 and 600 nm, a 5×10^{-4} M/L solution of Rhodamine G in water + 5% ammonix, and (2) between 590 and 620 nm, a 5×10^{-4} M/e solution of Rhodamine 610 in water + 5% ammonix.

The energy conversion efficiency is 40% corresponding to an output energy of 120 mJ. The wavelength selection and spectral narrowing of the emitted laser line are made by using a 2750-groves mm^{-1} grating in a Littrow configuration. The emission characteristics of this laser are summarized in Table 5.5. To adapt these characteristics to the requirements brought out from Section 5.2, two experimental achievements remain to be made:

1. The output wavelength has to be converted to a value in the near-UV

Table 5.5. Characteristics of Lidar Systems in Operation at Observatoire de Haute Provence for Lidar Measurements of Ozone Vertical Distribution

	Nd^{3+}–YAG-Pumped	Exciplex Laser/Nd^{3+}–YAG Laser
Emitter		
Output energy	40 mJ	250 mJ (λ_{on}) to 200 mJ (λ_{off})
Repetition rate, Hz	10	20
Pulse duration, ns	15	15
Emitted wavelength	280–310 nm	308 nm (λ_{on}) to 355 nm (λ_{off})
Emission linewidth	5 pm	0.7 nm
Beam divergence, rad	5×10^{-4}	10^{-3}
Receiver		
Telescope diameter, cm	80	
Telescope field of view, rad	2×10^{-3}	
Receiver bandwidth, nm	3–70	

wavelength range by second-harmonic generation using a KDP crystal. Owing to the high peak power of the fundamental emission (20 MW), the energy conversion efficiency is close to 35%. The output energy between 285 and 310 nm is equal to 40 mJ.

2. The laser system has to emit sequentially two wavelengths. It is designed so that the full sequence of wavelength switching is automatic. The output wavelength of the dye laser fundamental emission is monitored by using a spectrometer and a Fizeau interferometer, giving patterns recorded on diode arrays. A computer-controlled servomechanism (Cahen et al., 1981) is used to both ensure the stability of the output wavelength from λ_1 to a second preprogrammed value λ_2. To allow this switching to be fast ($t \leqslant 1$ s) and to avoid the experimental difficulties connected to the accuracy of the doubling crystal positioning as a function of wavelength, two KDP crystals are used that are preset to the optimum value of the phase-matching angle for λ_1 and λ_2. The laser beam is then mechanically switched from one crystal to the other, depending on the incident wavelength. When a second wavelength pair has to be used, the values of λ_1 and λ_2 as programmed for the servocontrol loop are changed as are the crystal angles. By using this system, the stability of the laser emission wavelength is better than 1 pm, and the switching operation takes place in less than 0.5 s. As the laser divergence is smaller than 5×10^{-4} rad, no transmitting optics is used, and the beam is sent directly upward by using a total reflection prism.

Exciplex lasers for ozone measurements have been first used by Uchino et al. (1980). They used an XeCl laser emitting at 308 nm. The maximum output energy was 128 mJ/pulse for the first system developed with a rather low repetition rate of 0.1 pulse/s so that the average power was of the same order of magnitude as for the dye laser. However, improvement of these systems have now led to an XeCl laser with an output enegy of 130 mJ and a repetition rate up to 100 Hz (Rothe et al., 1983; Pelon and Mégie, 1984). Dual-wavelength systems are presently developed that use either:

- *Stimulated Raman scattering generation.* The 308 nm radiation is focused into a high-pressure methane cell to generate the reference line (off absorption). The wavelength interval is then equal to 2880 cm^{-1} leading to an emission at 338 nm that is absorbed less significantly by ozone. The conversion efficiency is 15% for a pressure of 35 atm and a focusing length of 125 cm. The two lines can be emitted simultaneously (Rothe et al., 1983).
- *The third harmonic of a ND^{3+}–YAG laser at 355 nm that then requires an additional laser source.* The large wavelength interval (47 nm) allows only measurement above 25 km if aerosol-free conditions can be assumed (Pelon and Mégie, 1984).

A summary of the characteristics of both operating lidar systems is given in the Table 5.5.

5.3.2.2 Optical Receiver, Electronic Processing, and Data Acquisition

The optical receiver, electronic processing, and data acquisition parts of the lidar system are similar for the various laser systems that have been used to date (see Table 5.5). The following description refers to the system presently in use at the Observatoire de Haute Provence. The backscattered signal is collected by either a 36- or an 80-cm-diameter Cassegrain configuration. The distance between the emitting point and the telescope axis can be varied depending on the altitude range of the observations to increase or decrease the range at which the fields of view of the transmitter and the receiver begin to overlap. This will avoid saturation of the photomultiplier tube from the low-altitude backscattered signals. The telescope field of view is adjusted by using a remotely controlled iris and can thus be reduced to its limit value compatible with the laser divergence. The spectral bandwidth of the receiver can be reduced by either using a wide-band interference filter (70 nm) or several narrow-bandwidth interference filters (3 nm) that can be automatically changed when the laser emission wavelengths are changed. The signal is detected by using a photomultiplier tube. The dynamical range of the backscattered signal between the ground and 30 km can be as large as 10^6 according to Equation (5.42). Consequently, two acquisition modes are thus used:

1. For the lower altitude range the electrical signal delivered by the PMT is analyzed by using a transient waveform recorder (Biomation 1010) with a sampling frequency of 10 MHz, which corresponds to a maximum-altitude resolution of 15 m. A 10-bit converter is used for the analog-to-digital conversion.
2. For the altitude levels above 10–12 km, a 256-channel photon counter is used in parallel with the transient recorder so that the two acquisition modes overlap with respect to the altitude range. The timegate of the photon counter can be varied from 1 to 8 μs, and the maximum altitude resolution is thus 150 m.

The data provided by the two acquisition systems are then fed into a PDP 11-34 computer and are stored on a floppy disk. A presummation of the single-laser shot signals is made to reduce the volume of stored data. The PDP 11-34 computer is used to control the full sequence of a DIAL measurement. The experimental parameters (acquisition time, altitude resolution, values of the various laser wavelengths, switching time, etc.) are typed in to automatically start the sequence of laser firings. During the experiment, the laser energy and emission wavelengths are continuously monitored, and the data acquisition takes place only if all these parameters are within the range of predetermined values.

5.3.3 Results of DIAL Ozone Measurements

The reliability of the various systems presented here for ozone monitoring has been tested on an operational basis for the last three to four years. As a result, the DIAL technique has proved its ability to currently provide the altitude profile of ozone number density from the ground up to 50 km.

The measurements performed at the Observatoire de Haute Provence using dye lasers as the transmitter are obtained in successive steps within integration times on the order of 1 hr. The profiles shown in Figures 5.24 and 5.25 correspond to the altitude ranges 0–17 and 15–40 km and are obtained sequentially. This experimental procedure results from the wavelength optimization as a function of height. The first profile is obtained within 15 minutes with a relative accuracy better than 5% for a vertical resolution of 450 m. The higher altitude profile requires an integration time of about 45 minutes. The vertical resolution decreases from 1 km at 25 km to 3 km at 35 km and above. Here again the 1σ standard deviation is better than 5% up to 25 km and decreases down to 20% at the uppermost level. Several comparisons with in situ measurements have been performed especially during the June 1981 Intercomparison campaign held at Gap (Chanin, 1983). The results show a good agreement between in situ and remote sensing instruments (Mégie and Pelon, 1983, Pelon and Mégie, 1983).

Similar comparisons with balloon-borne ozonosondes have also been performed by Uchino et al. (1983) covering the altitude range of XeCl measurements between 15 and 30 km. Here again the agreement is rather good (Figure 5.26). The XeCl laser developed by Rothe et al. (1983) has been implemented at a high altitude (3 km a.s.l.) station at the top of the Zugspitze in the German Alps. This system has been in operation since the winter of 1982 and has provided high-altitude measurements between 30 and 50 km. These have been compared to classical measuring techniques (Werner et al., 1985). A similar system has also been operated at the Observatoire de Haute Provence in September 1983, and comparison of the ozone vertical profile obtained by lidar Brewer Mast sonde and Umkehr measurements are shown in the Figure 5.27 in the altitude range 25–50 km (Pelon and Mégie, 1984).

One of the first applications of ground-based lidar systems for ozone measurements was in monitoring the ozone vertical distribution on a routine basis. The altitude extension of the lidar measurements up to the 50-km level will allow the study of long-term ozone variations in a region where the potential effects of man-made activities is maximum due to catalytic cycles involving

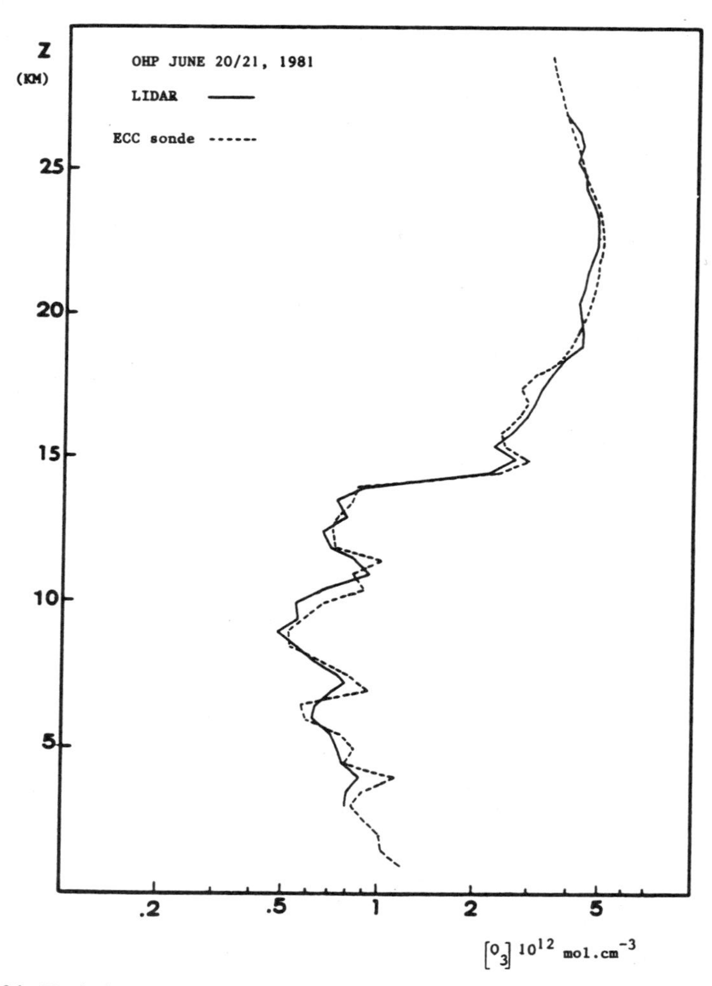

Figure 5.24. Vertical ozone distribution as measured by ECC sonde and lidar at Observatoire de Haute Provence (June 1981) (from Pelon and Mégie, 1982a).

more particularly chlorine species. With respect to passive systems presently in use, the lidar system presents, for such observation, several advantages (Mégie and Pelon, 1985):

1. the control of the emitting source, which results in a possible independent autocalibration of each system;

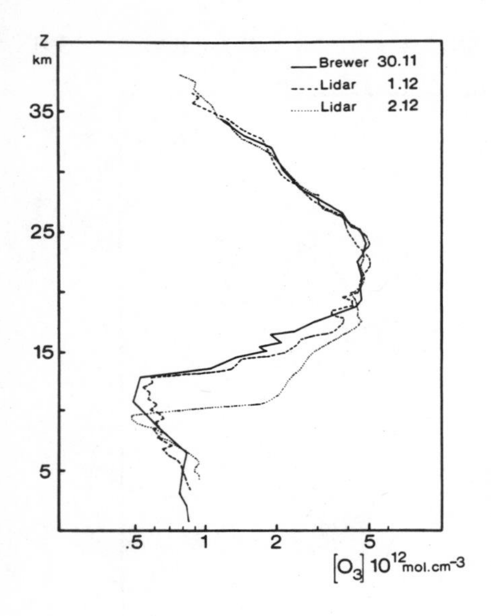

Figure 5.25. Ozone concentration profile measured by lidar at Observatoire de Haute Provence (44° N, 50° E) on December 1 (dashed line) and December 2 (dotted line), 1981, at 0 h UT compared with the profile obtained by Brewer Mast Sonde launched at Biscarosse (44° N, 0, 5° W) on November 30, 1981 (solid line) (from Pelon and Mégie, 1982b).

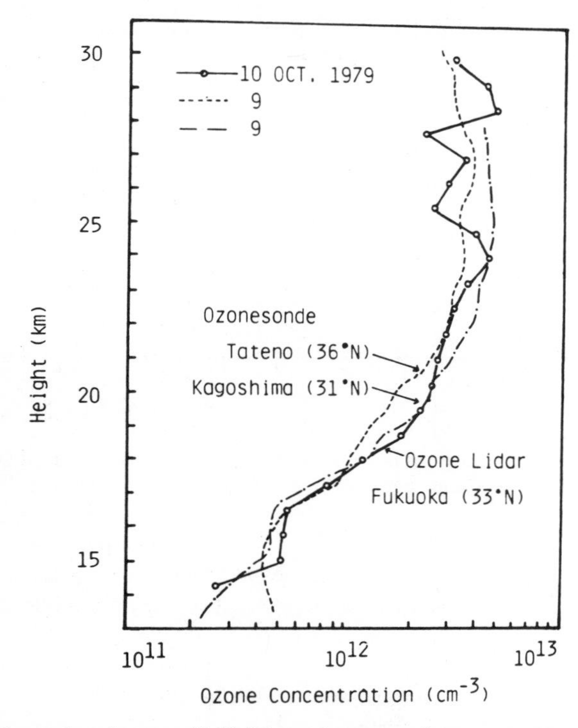

Figure 5.26. Comparison between XeCl lidar and ozonosonde measurements of ozone vertical distribution in lower stratosphere (from Uchino et al., 1980).

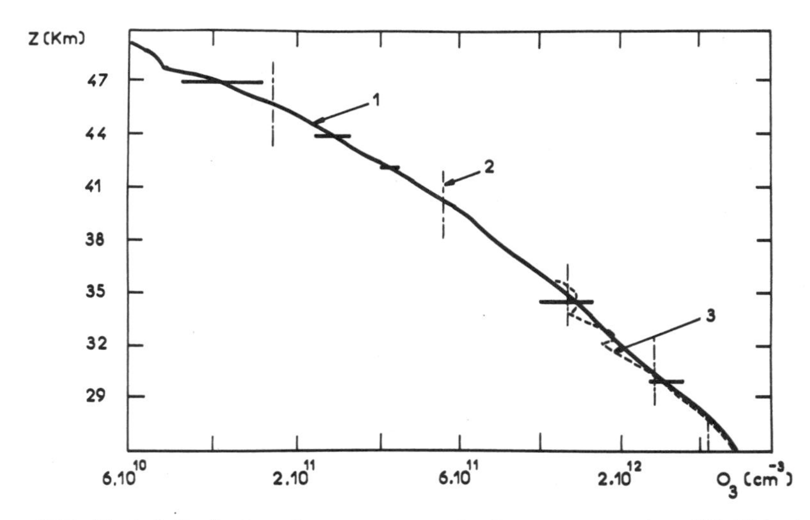

Figure 5.27. Vertical distribution of ozone concentration in upper stratosphere (25–50 km) as measured by XeCl lidar (full line), Brewer Mast Sonde (dotted line), and the Umkehr technique (solid vertical line) at Observatoire de Haute Provence. Statistical errors are shown for each system.

2. the possibility by a careful choice of the emitting wavelength to avoid interferences by aerosols, sulfur dioxide, or nitrogen dioxide; and
3. the direct altitude resolution of the system.

Table 5.6 summarizes the expected accuracies for ozone distribution and ozone total content measurements and emphasizes the requested corrections that can be provided by simultaneously obtained additional lidar determinations of the temperature and the aerosol contents.

Several studies have also been undertaken using lidar data that refer to various temporal and spatial scales. For example, short-term variability of the ozone number density below 16 km has been observed by Pelon et al. (1982a); successive profiles recorded during the night show a twofold increase in the ozone number density at 10 km between 9 P.M. and 2:30 A.M. on the following day (Figure 5.28). Such ozone variations, already observed by other techniques (Dütsch, 1979), are related to horizontal transport at the tropopause level. For such studies the temporal continuity in the observations that characterizes the lidar systems is of great advantage as it allows a determination of the horizontal extension of such ozone structures with a very high spatial accuracy.

The increase in the temporal resolution of the lidar measurements at the tropopause level presently allows integration times on the order of a few minutes that lead to the determination of time-height isocontours of the ozone number

Table 5.6. Errors Related to Lidar Measurements of Ozone Vertical Distribution for Various Altitude Ranges and Emitters

		Ozone Vertical Distribution (%)	Ozone Total Content (%)
Systematic Errors			
Uncertainty on ozone absorption cross sections absolute values		±3	±3
Temperature dependence on ozone cross sections with correction using lidar/radiosonde data		$\leqslant 0.3$	$\leqslant 0.3$
Influence of the linewidth fluctuations		$\leqslant 0.1$	$\leqslant 0.1$
Rayleigh extinction and scattering corrected from lidar/radiosonde data	0–25 km (290–295 nm)	$\leqslant 0.4$	
	25–50 km (305–310 nm)	$\leqslant 0.15$	$\leqslant 0.3$
Mie extinction and scattering corrected using lidar data and aerosol models	0–30 km	$\leqslant 1$	$\leqslant 0.4$
	30–50 km	$\leqslant 0.3$	
Absorption by minor constituents in nonpolluted areas		$\leqslant 0.1$	$\leqslant 0.1$
Residual ozone			$Z_{max} = 36$ km $\leqslant 0.8\%$ $Z_{max} = 50$ km $\leqslant 0.2\%$

Statistical Errors (Integration Time Requested for 3% Accuracy)

	5 km	10 km	15 km	20 km	25 km
Nd^{3+}–YAG-pumped dye laser ($E \sim 40$ mJ, $\Delta Z = 1$ km)	$\leqslant 1$ mn	2 mn	15 mn	80 mn	5 h
	25 km	30 km	35 km	40 km	45 km
Exciplex laser ($E \sim 250$ mJ, $\Delta Z = 3$ km)	$\leqslant 1$ nm	1 mn	15 mn	1 h	4 h

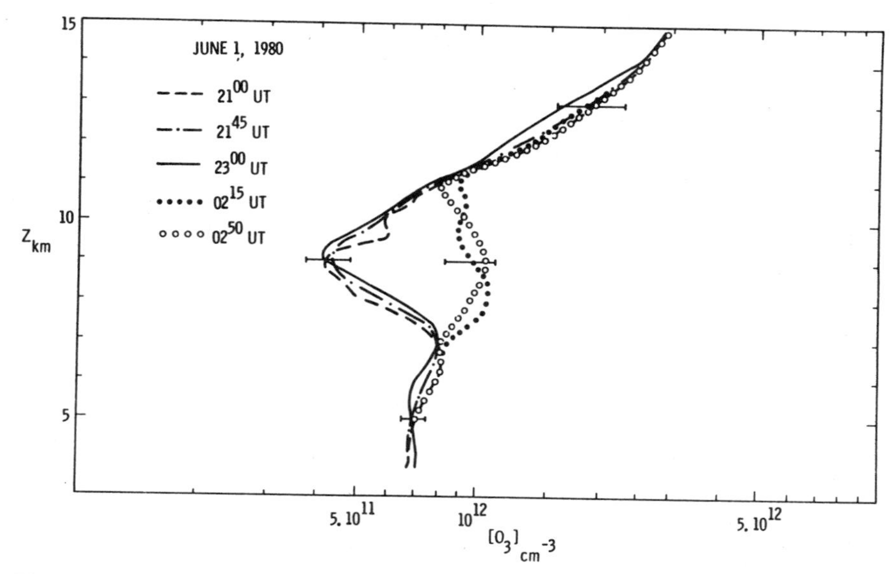

Figure 5.28. Short-term variations in ozone number density at tropopause level as observed on June 1, 1980, at Observatoire de Haute Provence (from Pelon and Mégie, 1982a).

density as a function of time, as shown in Figure 5.29. These measurements were obtained at the Observatoire de Haute Provence during the night of December 2, 1981, and give evidence for a large ozone increase between 6 and 8 km. The temporal and spatial scales associated with this event are related, as shown from a synoptic analysis of the meteorological charts, with a moving cold front and the associated intrusion of stratospheric air into the troposphere at the junction between the front and the tropopause boundary (Pelon and Mégie, 1984).

Large variations in ozone number density are also observed on a day-to-day basis. They are related to larger scale horizontal transport. As an example, the vertical ozone profile recorded during the night of July 9, 1980 (Figure 5.30), shows the presence of a very large ozone bulge with concentration up to 3×10^{12} cm^{-3} at the 10-km altitude level. For comparison, the average profiles observed during the beginning of July 1980 are plotted on the same figure. Such increases of the ozone concentration at the tropopause level have been observed on several occasions during field experiments. The altitude of the ozone bulge may vary from 10 to 12 km down to 6–8 km, as on March 11, 1981 (Figure 5.31); in this latter case the peak concentration decreases when the bulge is observed at lower altitudes. From the meteorological network data, one can

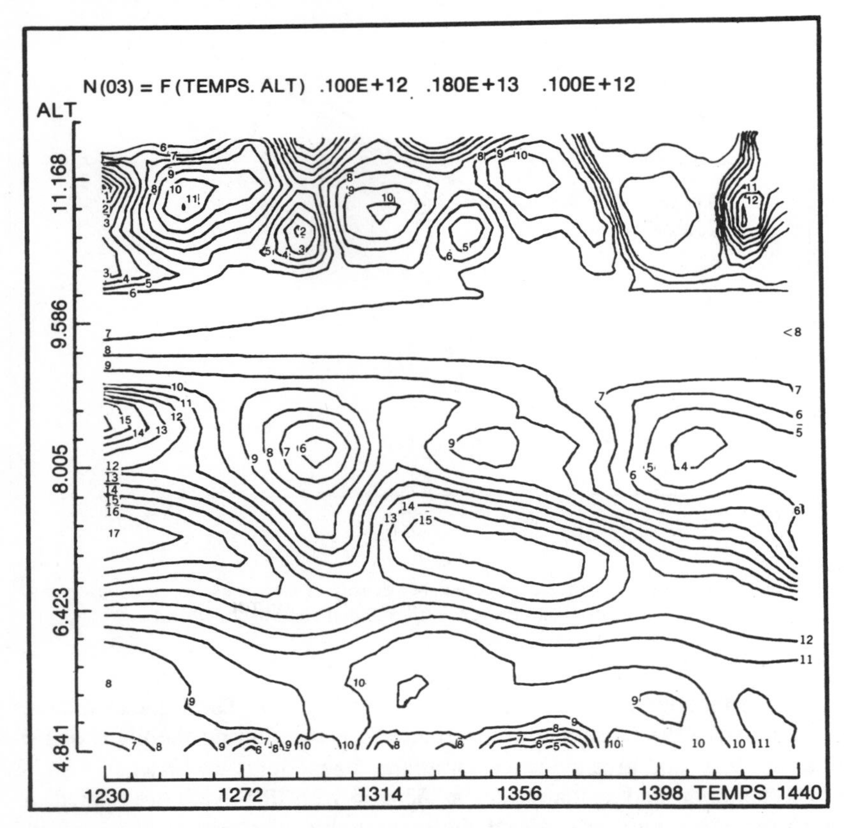

Figure 5.29. Isocontours of ozone number density (in 10^{11} cm^{-3}) as function of altitude (vertical axis) and time (min) for 3-h periods starting at 22:00 on December 2, 1981, at Observatoire de Haute Provence (from Pelon and Mégie, 1984).

show that these increases are systematically correlated with a 2–3-km decrease of the tropopause height, corresponding thus to the presence of warmer air at these levels over southern France.

A detailed analysis of one of these situations (July 1980) has already been performed by using the meteorological charts at various altitude levels in the troposphere and lower stratosphere (Pelon et al., 1981). For such events also the potentiality of a ground-based system can be fully utilized as shown by the study of Uchino et al. (1983), which gives evidence for a high positive correlation between the ozone column density in the altitude range of 15–25 km with the total ozone observed by a nearby Dobson spectrophotometer (Figure 5.32).

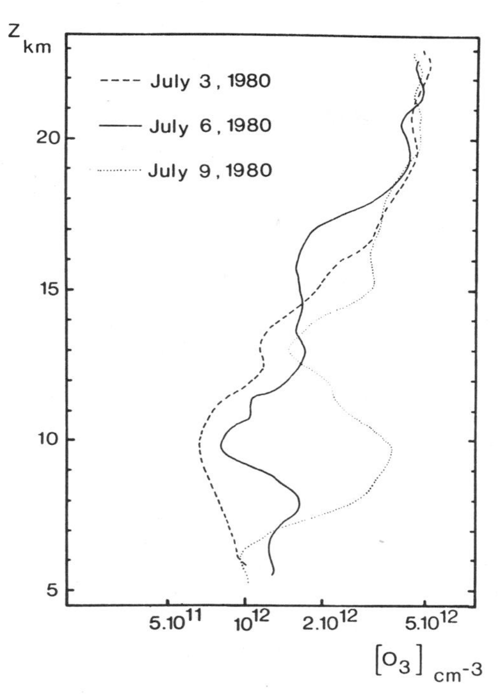

Figure 5.30. Ozone concentration profiles recorded during July 1980 in troposphere and lower stratosphere at Observatoire de Haute Provence giving evidence for large variation on day-to-day basis at tropopause level associated with synoptic dynamical processes (from Pelon and Mégie, 1982a).

Other correlations have also been studied by the same group such as a high positive correlation with temperature at 17 and 25 km, all these features being closely related to dynamical movements at the tropopause height (Pelon et al., 1983).

The emphasis is this presentation has been put on ozone measurements. When considering minor species in the stratosphere, the relative abundance ratios vary from the ppm range down to the ppb or even ppt ranges. As shown above, the ability of a ground-based active system measurement is strongly dependent on the local absorption to be measured. If one considers that the absorption cross sections are generally in the 10^{-18}-cm^2 range, an abundance ratio of 1 ppm at 30 km will lead to a 1-km optical thickness of 2.5%. A reasonable detection limit for lidar absorption measurement can presently be fixed around 0.25% for a two-way absorption, which thus limits the potential measurements to minor

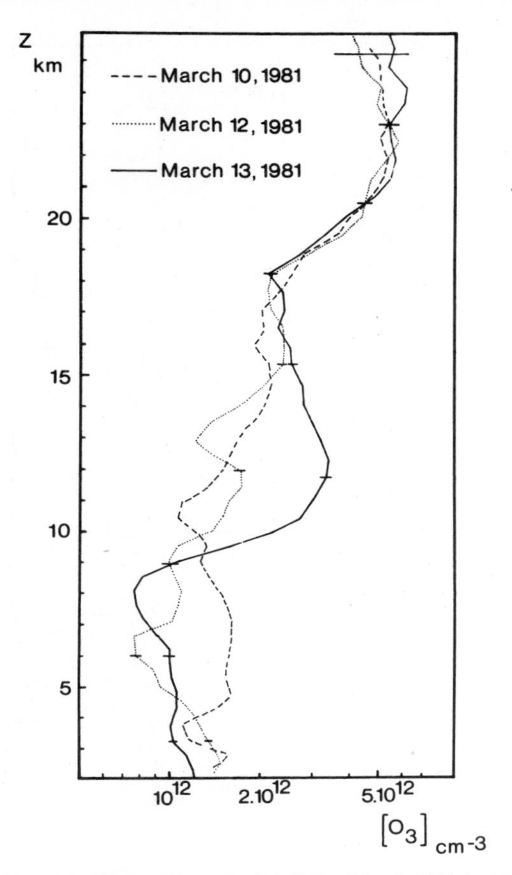

Figure 5.31. Ozone concentration profiles recorded during March 1981 in troposphere and lower stratosphere at Observatoire de Haute Provence (from Pelon and Mégie, 1982a).

constituents in the 50–100-ppb range. Furthermore, the vertical profile of the constituent under study is also to be taken into account as the detection at a given altitude level is strongly dependent on the absorption below. For example, water vapor measurements in the stratosphere from the ground that could be possible in the 940-nm spectral region are prohibited by the very large tropospheric content. Besides ozone, the only stratospheric constituent that has been measured to date is nitrogen dioxide using a laser source that emits at 440 nm (Bucchia and Mégie, 1983). However, only the NO_2 total content between 20 and 40 km has been directly measured, the altitude profile being derived using only inversion techniques. The potentiality of active systems for trace-

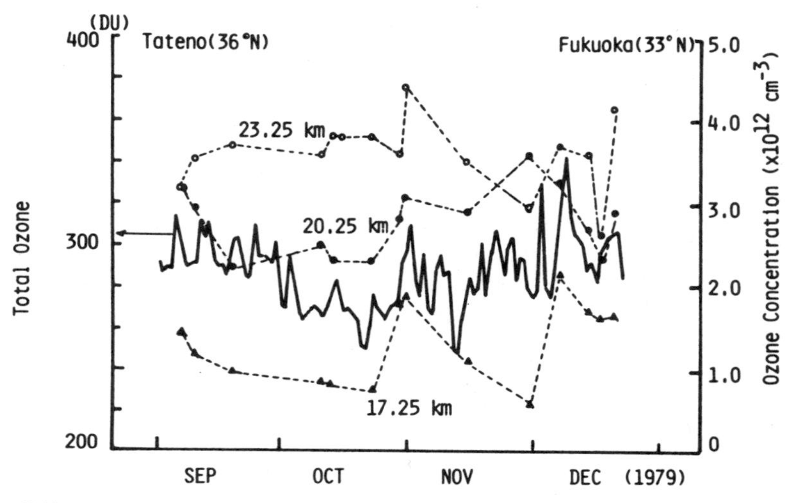

Figure 5.32. Comparisons between total ozone observed by Dobson spectrophotometer at Tateno (Japan) and ozone mean concentrations at various heights as measured by XeCl lidar for 3-month period in 1979 (from Uchino et al., 1980).

species measurement in the stratosphere is however much larger if one considers airborne, balloon-borne, or space-borne systems (Measures, 1984).

5.3.4 Conclusion

Due to the rapid development of powerful laser sources in the UV wavelength range, which includes Nd–YAG-pumped dye lasers and exciplex lasers, ground-based lidar systems are now operational for ozone monitoring in the troposphere and stratosphere. They can provide a direct measurement of the ozone vertical distribution from the ground up to the 50-km-altitude level. Such systems will certainly play an important role in the determination of long-term ozone trends in the photochemical region and thus in the evaluation of anthropogenic effects in Earth's environment. Furthermore, the unique capacities of active lidar systems in terms of high temporal and spatial resolution and measurements continuity allow the observation of the ozone variations at various time and space scales, which are of importance in such present-day areas of interest as troposphere–stratosphere exchanges, long-range transport, and global budget of ozone or correlations between ozone number densities and other atmospheric parameters. Considering the already operational character of the lidar systems, they will constitute in the very near future the basis for the development of a new-ground based network for midatmospheric observations of ozone and other trace gases.

ACKNOWLEDGMENTS

The author is greatly indebted to his colleagues who have contributed to the data collection, analysis, and interpretation as presented in this chapter as part of the work done at the service d'Aéronomie du C.N.R.S. and particularly to J. P. Jegou and C. Granier (study of metallic species) and J. Pelon (ozone measurements). The author also acknowledges their participation in fruitful discussions during the preparation of this manuscript, which also involved P. Flamant and J. Lefrere.

REFERENCES

Allain, J. Y. (1979). High Energy Pulsed Dye Lasers for Atmospheric Sounding. *Appl. Opt.* **18**(3), 287–290.

Allen, C. W. (1973). *Astrophysical Quantities*, Athlone Press, London.

Aruga, T., M. Kamiyama, M. Jyumonji, T. Kobayashi, and H. Inaba (1974). Laser Radar Observation of the Sodium Layer in the Upper Atmosphere. *Rep. Ionosph. Space Res. Japan* **28**(1–2), 65–68.

Bass, A. M. (1984). *Ultraviolet Absorption Cross Sections of Ozone: Measurement Results and Error Analysis*. Proc. Quadriennal Ozone Symp., Halkidiki, Grèce, p. 7.2.

Bass, A. M., A. E. Ledford, and A. H. Laufer (1976). Extinction Coefficients of NO_2 and N_2O_4. *J. Res. NBS* **80A,** 143.

Blamont, J. E., and T. M. Donahue (1964). The Sodium Airglow: Observation and Interpretation of a Large Diurnal Variation. *J. Geophys. Res.* **69,** 4093–4127.

Blamont, J. E., M. L. Chanin, and G. Mégie (1972a). Excitation par Laser des Atomes de Sodium Atmosphériques. *C. R. Acad. Sci. Paris* **274,** 93–96.

Blamont, J. E., M. L. Chanin, and G. Mégie (1972b). Vertical Distribution and Temperature Profile of the Nighttime Atmospheric Sodium Layer Obtained by Laser Backscatter. *Ann. Géophys.* **28**(4), 833–838.

Bos, F. (1981). Versatile High-Power Single-Longitudinal-Mode Pulsed Dye Laser. *Appl. Opt.* **20**(10), 1886–1890.

Bowman, M. R., A. J. Gibson, and M. C. W. Sandford (1969). Atmospheric Sodium Measured by a Tuned Laser Radar. *Nature* **221,** 456–457.

Bradley, D. J., A. J. F. Durrant, G. M. Gale, M. Moore, and P. D. Smith (1968). Characteristics of Organic Dye Lasers as a Tunable Frequency Source of Nanosecond Absorption Spectroscopy. *IEEE J. Quant. Elec.* **QE-4**(11), 707–715.

Bradley, D. J., W. G. I. Caughey, and J. I. Vukusic (1971). High Efficiency Interferometric Tuning of Flashlamp Pumped Dye Lasers. *Opt. Comm.* **4,** 150–153.

Bucchia, M., and G. Mégie (1983). Ground Based Active Remote Sensing of the Nighttime NO_2. *Annal. Geophys.* **1,** 4–5, 411–414.

Byer, R. L., and M. Garbuny (1973). Pollution Detection by Absorption Using Mie Scattering and Topographic Targets as Retroreflectors. *Appl. Opt.* **12,** 1496.

Cahen, C., J. P. Jégou, J. Pelon, P. Gildwarg, and J. Porteneuve (1981). Wavelength Stabilization and Control of the Emission of Pulsed Dye Lasers by Means of a Multibeam Fizeau Interferometer. *Revue Phys. Appl.* **16,** 353–358.

Chamberlain, J. C. W. (1961). *Physics of the Aurora and Airglow*, Academic Press, New York.

Chandrasekhar, S. (1950). *Radiative Transfer*, Clarendon Press, Oxford.

Chanin, M. L. (1983). The Intercomparison Campaign Held in France in June 1981: Description of the Campaign. *Planet. Space Sci.* **31**(7), 707.

Chiu, Y. T., and B. T. Ching (1978). The Response of Atmospheric and Lower Ionospheric Structures to Gravity Waves. *Geophys. Res. Lett.* **5,** 539–544.

CIRA (1972). *Cospar International Reference Atmosphere*, North Holland Publishing Company, Amsterdam.

Clemesha, B. R. (1979). Concerning the Seasonal Variation of the Mesospheric Sodium Layer at Low Latitudes. *Planet. Space Sci.* **27,** 909–910.

Clemesha, B. R., V. W. J. H. Kirchhoff, and D. M. Simonich (1975). Automatic Wavelength Control of a Flashlamp-Pumped Dye Laser. *Opt. Quant. Electr.* **7,** 193–196.

Clemesha, B. R., V. W. J. H. Kirchhoff, and D. M. Simonich (1978a). Simultaneous Observations of the Na 5893-Å Nightglow and the Distribution of Sodium Atoms in the Mesosphere. *J. Geophys. Res.* **83**(A6), 2499–2503.

Clemesha, B. R., V. W. J. H. Kirchhoff, D. M. Simonich, and H. Takahashi (1978b). Evidence of an Extra-terrestrial Source for the Mesospheric Sodium Layer. *Geophys. Res. Lett.* **5**(10), 873–876.

Clemesha, B. R., V. W. J. H. Kirchhoff, D. M. Simonich, H. Takahashi, and P. P. Batista (1979). Simultaneous Observations of Sodium Density and the NaO, OH (8.3) and OI 5577-Å Nightglow Emissions. *J. Geophys. Res.* **84**(A11), 6477–6782.

Clemesha, B. R., V. W. J. H. Kirchhoff, D. M. Simonich, H. Takahashi, and P. P. Batista (1980). Spaced Lidar and Nightglow Observations of an Atmospheric Sodium Enhancement. *J. Geophys. Res.* **85**(A7), 3480–3484.

Clemesha, B. R., V. W. J. H. Kirchhoff, and D. M. Simonich (1981). Comments on "A Meteor-Ablation Model of the Sodium and Potassium Layers" by D. M. Hunten. *Geophys. Res. Lett.* **8**(9), 1023–1025.

Clemesha, B. R., D. M. Simonich, P. P. Batista, and V. W. J. H. Kirchhoff (1982). The Diurnal Variation of Atmospheric Sodium. *J. Geophys. Res.* **87**(A1), 181–186.

Dujardin, G., and P. Flamant (1978). Amplified Spontaneous Emission and Spatial Dependence of Gain in Dye Amplifiers. *Opt. Commun.* **24**(3), 243–247.

Dütsch, H. U. (1979). Vertical Ozone Distribution and Tropospheric Ozone, in *Pro-*

ceedings of the NATO Advanced Study Institute on Atmospheric Ozone, M. Nicolet and A. C. Aikin, eds., U.S. Department of Transportation, Washington, D.C.

Elterman, L. (1968). UV, Visible and IR Attenuation to 50 km. Report AFCRL 68-1053, Cambridge Res. Lab., U.S. Air Force, Bedford, MA.

Elterman, L. (1970). Relationships Between Vertical Attenuation and Surface Meteorological Range. *Appl. Opt.* **9,** 1804.

Felix, F., W. Keenslide, G. Kent, and M. C. W. Sandford (1973). Laser Radar Observations of Atmospheric Potassium. *Nature* **246,** 345–347.

Ferguson, E. E. (1978). Sodium Hydroxide Ions in the Stratosphere. *Geophys. Res. Lett.* **5,** 1035–1038.

Ferrar, C. M. (1969). Simple High Intensity Short Pulse Flashlamps. *Rev. Sci. Inst.* **40**(11), 1436–1440.

Fiocco, G., and G. Visconti (1976). Origin of the Upper Atmospheric Na from Sublimating Dust: A Model. *J. Atmos. Terr. Phys.* **38,** 1279–1287.

Flamant, P., and G. Mégie (1980). Frequency Locking by Injection in Dye Lasers: Analysis of the CW and Delta Pulse Regimes. *IEEE J. Quantum Elec.* **QE-16**(6), 653–660.

Fricke, K. H., and U. von Zahn (1985). Mesopause Temperatures Derived from Probing the Hyperfine Structure of the D_2 Resonance Line of Sodium by Lidar. *J. Atm. Terr. Phys.* **47,** 499–512.

Furumoto, H. W., and H. L. Ceccon (1969). Optical Pumps for Organic Dye Lasers. *Appl. Opt.* **8,** 1613–1624.

Gacoin, P., and P. Flamant (1972). High Frequency Cresyl Violet Laser. *Opt. Commun.* **5**(5), 351–353.

Gadsden, M. (1970). Metallic Atoms and Ions in the Upper Atmosphere. *Ann. Geophys.* **26,** 141–152.

Gardner, C. S., and J. D. Shelton (1981). Spatial and Temporal Filtering Technique for Processing Lidar Photocount Data. *Opt. Lett.* **6**(4), 174–176.

Gibson, A. J. (1969). A Flashlamp Pumped Dye Laser for Resonance Scattering Studies of the Upper Atmosphere. *J. Phys. E: Sci. Instrum.* **2,** 802–806.

Gibson, A. J. (1972). Flashlamp-Pumped Dye Lasers for Investigations of the Upper Atmosphere. *J. Phys. E. Sci. Instrum.* **5,** 971–973.

Gibson, A. J., and M. C. W. Sandford (1971). The Seasonal Variation of the Nighttime Sodium Layer. *J. Atm. Terr. Phys.* **33,** 1675–1684.

Gibson, A. J., and M. C. W. Sandford (1972). Daytime Laser Radar Measurements of the Atmospheric Sodium Layer. *Nature* **239,** 509–511.

Gibson, A. J., L. Thomas, and S. K. Bhattachacharyya (1979). Laser Observations of the Ground-State Hyperfine Structure of the Sodium and of Temperatures in the Upper Atmosphere. *Nature* **281,** 131–133.

Grainger, J. R., and J. Ring (1962). Anomalous Fraunhofer Line Profiles. *Nature* **193,** 762–764.

Granier, C. (1982). Variation diurne et évolution à courte échelle de temps du sodium atomique à l'altitude de la mésopause: observation par sondage laser et modélisation. Thèse 3ème Cycle, Université Pierre et Marie Curie, Paris-France.

Granier, C., and G. Mégie (1981). Daytime Lidar Measurements of the Mesospheric Sodium Layer. *Planet. Space Sci.* **30**(2), 169–177.

Granier, C., J. Porteneuve, and G. Mégie (1981). Sondages laser de jour dans le domaine des longueurs d'onde visibles: application à l'étude du sodium atomique dans la mésosphère. *C. R. Acad. Sci. Paris*, **292,** serie II, 401–404.

Granier, C., J. P. Jegou, and G. Mégie (1984). Resonant Lidar Detections of Ca and Ca^+ in the Upper Atmosphere. Proc. 12th Intern. Laser Radar Conf., éd. G. Mégie, Aix en Provence, France, 229–232.

Granier, C., J. P. Jégou, M. L. Chanin, and G. Mégie (1985). A General Theory of the Alkali Metals Present in the Earth Upper Atmosphere. III Diurnal Variations. *Ann. Geophys.* **3,** 4, 445–450.

Hake, R. D., Jr., E. K. Proctor, and R. A. Long (1971). Tunable Dye Lidar Techniques for Measurements of Atmospheric Constituents. Proc. of the Soc. of Photo-Optical Instrumentation Engineers, Palo Alto, CA, vol. 27, Y. H. Katz, ed.

Hake, R. D., D. E. Arnold, D. W. Jackson, W. E. Evans, B. P. Ficklin, and R. A. Long (1972). Dye Laser Observations of the Nighttime Atomic Sodium Layer. *J. Geophys. Res.* **77**(34), 6839–6843.

Hanle, W. (1924). Uber magnetische beeinflüssung der polarisation der resonanzfluorebzenz. *Zeit. Phys.* **30,** 93–105.

Hinkley, E. D., ed. (1976). *Laser Monitoring of the Atmosphere*, Springer-Verlag, Berlin.

Hirono, M. (1964). On the Observation of the Upper Atmospheric Constituents by Laser Beams. *J. Radio Res. Lab.* **11**(56), 251–271.

Hunten, D. M. (1967). Spectroscopic Studies of the Twilight Airglow. *Space Sci. Rev.* **6,** 493–576.

Hunten, D. M. (1981). A Meteor-Ablation Model of the Sodium and Potassium Layers. *Geophys. Res. Lett.* **8**(4), 369–372.

Inn, E. C. Y., and Y. Tanaka (1953). Absorption Coefficient of Ozone in the Ultraviolet and Visible Regions. *J. Opt. Soc. Am.* **43,** 870.

Jégou, J. P., and M. L. Chanin (1980). Observation par sondage laser de lithium atmospherique créé artificiellement dans la haute atmosphère: mise en évidence des processus d'interaction ion-neutre. *C.R. Acad. Sc. Paris* **290,** Série B, 325–329.

Jégou, J. P., M. L. Chanin, G. Mégie, and J. E. Blamont (1980). Lidar Measurements of Atmospheric Lithium. *Geophys. Res. Lett.* **7**(11), 995–998.

Jégou, J. P., C. Granier, M. L. Chanin, and G. Mégie (1985a). A General Theory of the Alkali Metals Present in the Earth Upper Atmosphere. I. Flux Model: Chemical and Dynamical Processes. *Ann. Geophys.* **3,** 2, 163–176.

Jégou, J. P., C. Granier, M. L. Chanin, and G. Mégie (1985b). A General Theory of the Alkali Metals Present in the Earth Upper Atmosphere. II. Seasonal and Medirional Variations. *Ann. Geophys.* **3,** 3, 299–312.

Jethwa, J., F. P. Schafer, and J. Jasny (1978). A Reliable High Power Dye Laser. *IEEE J. Quant. Elec.* **QE-14**(2), 119–121.

Juramy, P., M. L. Chanin, G. Mégie, G. F. Toulinov, and Y. P. Doudoladov (1981). Lidar Sounding of the Mesospheric Sodium Layer at High Latitude. *J. Atm. Terr. Phys.* **43**(3), 209–215.

Kato, Y., and Y. Mori (1964). On the Possibility of the Upper Atmospheric Sounding by Optical Radar. *Rep. Ionos. Space Res. Japan* **18,** 103–108.

Kirchhoff, V. W. J. H. (1983). Atmospheric Sodium Chemistry and Diurnal Variations: An Up-date. *Geophys. Res. Lett.* **10**(8), 721–724.

Kirchhoff, V. W. J. H., and B. R. Clemesha (1973). Atmospheric Sodium Measurements at 23°S. *J. Atm. Terr. Phys.* **35,** 1493–1498.

Kirchhoff, V. W. J. H., and B. R. Clemesha (1983a). The Atmospheric Neutral Sodium Layer: 2. Diurnal Variations. *J. Geophys. Res.* **88**(A1), 442–450.

Kirchhoff, V. W. J. H., and B. R. Clemesha (1983b). Eddy Diffusion Coefficients in the Lower Thermosphere. *J. Geophys. Res.* **88**(A7), 5765–5768.

Kirchhoff, V. W. J. H., and B. R. Clemesha (1983c). The Dissipation of a Sodium Cloud. *Planet. Space Sci.* **31**(4), 369–372.

Kirchhoff, V. W. J. H., and H. Takamashi (1984). Sodium Clouds in the Lower Thermosphere. *Planet. Space Sci.* **32**(7), 831–836.

Kirchhoff, V. W. J. H., B. R. Clemesha, and D. M. Simonich (1979). Sodium Nightglow Measurements and Implications on the Sodium Photochemistry. *J. Geophys. Res.* **84**(A4), 1323–1327.

Kirchhoff, V. W. J. H., B. R. Clemesha, and D. M. Simonich (1981a). Average Nocturnal and Seasonal Variations of Sodium Nightglow at 23°S, 46°W. *Planet. Space Sci.* **29**(7), 765–766.

Kirchhoff, V. W. J. H., B. R. Clemesha, and D. M. Simonich (1981b). The Atmospheric Neutral Sodium Layer: 1. Recent Modelling Compared to Measurements. *J. Geophys. Res.* **86,** 6892–6898.

Kirchhoff, V. W. J. H., B. R. Clemesha, and D. M. Simonich (1981c). Seasonal Variation of Ozone in the Mesosphere. *J. Geophys. Res.* **86**(A3), 1463–1466.

Krueger, A. J., and R. A. Minzer (1976). A Mid-Latitude Ozone Model for the 1976 U.S. Standard Atmosphere. *J. Geophys. Res.* **81,** 4477.

Lefrère, J., G. Mégie, C. Cahen, and P. Flamant (1984). Evidence of Spectral Density Limits in Dial Measurements of the Humidity in the Boundary Layer. Proc. 12th Intern. Laser Radar Conf., éd. G. Mégie, Aix en Provence, France, 161–164.

Liu, S. C., and G. C. Reid (1979). Sodium and Other Minor Constituents of Meteoritic Origin in the Atmosphere. *Geophys. Res. Lett.* **6**(4), 283–28.

Loth, C., and G. Mégie (1974). A High Spectral Luminance Dye Amplifier. *J. Phys. E* **7,** 80–83.

Loth, C., Y. H. Meyer, and F. Bos (1976). High Power Dye Laser in the Near Infrared. *Opt. Comm.* **16,** 310–313.

McCormick, M. P., and T. J. Swissler (1983). Stratospheric Aerosol Mass and Latitudinal Distribution of the El Chichon Eruption Cloud for October 1982. *Geophys. Res. Lett.* **10**,(9), 877–880.

Measures, R. M. (1984). *Laser Remote Sensing—Fundamentals and Applications.* Wiley-Interscience, New York.

Mégie, G., and J. E. Blamont (1977). Laser Sounding of Mesospheric Sodium: Interpretation in Terms of Global Atmospheric Parameters. *Planet. Space Sci.* **25,** 1093–1109.

Mégie, G., and R. T. Menzies (1980). Complementarity of UV and IR Differential Absorption Lidar for Global Measurements of Atmospheric Species, *Appl. Opt.* **19,** 1173.

Mégie, G., and J. Pelon (1983). Measurements of the Ozone Vertical Distribution (0–25 km): Comparison of Various Instruments, GAP-Observatoire de Haute Provence, June 1981. *Planet. Space Sci.* **39**(7), 791.

Mégie, G., and J. Pelon (1985). Lidar Measurements of Ozone Profiles Optical Remote Sensing of the Atmosphere, *Tech. Digest*, OSA Publication, Monterey, CA, USA.

Mégie, G., and O. De Witte (1971). Excitation d'une résonance atomique par laser fluorescent à fréquence accordable. *Rev. Phys. Appl.* **6,** 341–343.

Mégie, G., J. E. Blamont, and M. L. Chanin (1974). Neutral Temperature Measurements between 80 and 100 km by Laser Sounding from the Ground, in *Methods of Measurements and Results of Lower Ionosphere Structure*, K. Rawer, ed., Adademie-Verlag, Berlin, pp. 27–32.

Mégie, G., J. Y. Allain, M. L. Chanin, and J. E. Blamont (1977). Vertical Profile of Statospheric Ozone by Lidar Sounding from the Ground. *Nature* **270,** 329.

Mégie, G., F. Bos, J. E. Blamont, and M. L. Chanin (1978a). Simultaneous Measurements of Atmospheric Sodium and Potassium. *Planet. Space Sci.* **26,** 27–35.

Mégie, G., M. L. Chanin, G. Y. Tulinov, and Y. P. Doudoladov (1978b). High Latitude Measurements of the Atomic Sodium Concentration and Neutral Temperature at the Mesopause Level by the Lidar Technique. *Planet. Space Sci.* **26,** 509–511.

Mitchell, A. C. G., and M. W. Zemansky (1934). *Resonance Radiation and Excited Atoms.* Cambridge University Press, London.

Mitra, V., (1973). Origin and Height Distribution of Sodium in the Earth's Atmosphere. *Ann. Geophys.* **29,** 341–351.

Nugent, L. J. (1966). Laser Radar Probe for the Measurement of Metastable Nitric Oxide in the Upper Atmosphere. *Nature* **211,** 1349–1351.

Pelon, J., and G. Mégie (1982a). Ozone Monitoring in the Troposphere and Lower Stratosphere: Evaluation and Operation of a Ground Based Lidar Station. *J. Geophys. Res.* **87**(C7), 4947.

Pelon, J., and G. Mégie (1982b). Ozone Vertical Distribution and Total Content as Monitored Using a Ground Based Active Remote Sensing System. *Nature* **299,** 137.

Pelon, J., and G. Mégie (1983). Lidar Measurements of the Vertical Ozone Distribution during the June 1981 Intercomparison Campaign GAP/OHP. *Planet. Sci.* **31**(7), 717.

Pelon, J., and G. Mégie (1984). Lidar Monitoring of the Ozone Vertical Distribution in the Troposphere and Stratosphere. Proc. 12th Intern. Laser Radar Conf., éd. G. Mégie, Aix en Provence, France, 247–250.

Pelon, J., P. Flamant, L. Chanin, and G. Mégie (1981). Intrusion d'ozone d'origine polaire aux latitudes moyennes: Mise en évidence par sondage laser. *C.R. Acad. Sci. Paris* **292,** 319.

Penndorf, R. (1957). Tables of Refractive Index for Standard Air and the Rayleigh Scattering Coefficient for the Spectral Region between 0.2 and 20.0 μ and Their Application to Atmospheric Studies. *J. Opt. Soc. Amer.* **47**(2), 176–182.

Rees, D., and M. C. W. Sandford (1974). Thermospheric Winds from Laser Tracking of Sodium Clouds. *Nature* **252,** 291–292.

Richter, E. S. and C. F. Sechrist, Jr. (1979a). A Meteor Ablation Cluster Ion Atmospheric Sodium Theory. *Geophys. Res. Lett.* **6**(3), 183–186.

Richter, E. S., and C. F. Sechrist, Jr. (1979b). A Cluster Ion Chemistry for the Mesospheric Sodium Layer. *J. Atm. Terr. Phys.* **41,** 579–586.

Richter, E. S., J. R. Rowlett, C. S. Gardner, and C. F. Sechrist, Jr. (1981). Lidar Observation of the Mesospheric Sodium Layer over Urbana, Illinois. *J. Atm. Terr. Phys.* **43**(4), 327–337.

Rothe, K. W., H. Walther, and J. Werner (1983). Differential Absorption Measurements with Fixed Frequency IR and UV Lasers, in *Optical and Laser Remote Sensing*, D. K. Killinger and A. Mooradian, eds., Springer-Verlag, New York.

Rowlett, J. R., and C. S. Gardner (1978). Lidar Observations of Wavelike Structure in the Atmospheric Sodium Layer. *Geophys. Res. Lett.* **5**(8), 683–686.

Sandford, M. C. W., and A. J. Gibson (1970). Laser Radar Measurements of the Atmospheric Sodium Layer. *J. Atm. Terr. Phys.* **32,** 1423–1430.

Schotland, R. M. (1964). The Determination of the Vertical Profile of Atmospheric Gases by Means of a Ground-based Optical Radar, in *Proceedings of the Third Symposium on Remote Sensing of Environment*, University of Michigan, Ann Arbor.

Schotland, R. M. (1974). Errors in the Lidar Measurement of Atmospheric Gases by Differential Absorption. *J. Appl. Meteorol.* **13,** 71.

Schuler, C. J., C. T. Pike, and H. A. Miranda (1971). Dye Laser Probing of the Atmosphere Using Resonant Scattering. *Appl. Opt.* **10**(7), 1689–16.

Segal, A., O. Voelz, C. S. Gardner, and C. F. Sechrist (1984). Airborne Lidar Observations of the Mesospheric Sodium Layer. Proc. 12th Intern. Laser Radar Conf., éd. G. Mégie, Aix en Provence, France, pp. 243–246.

Seinfeld, J. H. (1975). *Air Pollution-Physical and Chemical Fundamentals*. McGraw-Hill, New York.

Shelton, J. D., and C. S. Gardner (1981). Theoretical and Lidar Studies of the Density Response of the Mesospheric Sodium Layer to Gravity Waves Perturbations. Aeronomy Report 99, Library of Congress ISSN 0568-0581, University of Illinois, Urbana.

Shelton, J. D., C. S. Gardner, and C. F. Sechrist, Jr. (1980). Density Response of the

Mesospheric Sodium Layer to Gravity Wave Perturbations. *Geophys. Res. Lett.* **7**(12), 1069–1072.

Simonich, D. M., and B. R. Clemesha (1983). Resonant Extinction of Lidar Returns from the Alkali Metal Layers in the Upper Atmosphere. *Appl. Opt.* **22,** 1387–1390.

Simonich, D. M., B. R. Clemesha, and V. W. J. H. Kirchhoff (1979). The Mesospheric Sodium Layer at 23°S: Nocturnal and Seasonal Variations. *J. Geophys. Res.* **84**(A4), 1543–1550.

Slipher, V. M. (1929). Emissions in the Spectrum of the Light of the Night Sky. *Publ. Astr. Soc. Pacif.* **41,** 262–268.

Snyder, J. J. (1977). Fizeau Wavelength Meters, in *Laser Spectroscopy*, vol. III, J. L. Hall and J. L. Carlstend, eds., Springer, New York.

Soep, B. (1970). Sélection de la fréquence d'émission de lasers à colorants à l'aide d'une lame plusieurs fois demi-onde. *Opt. Commun.* **1**(9), 433–436.

Soffer, B. M., and B. B. McFarland (1967). Laser Second Harmonic Induced Stimulated Emission of Organic Dyes. *Appl. Phys. Lett.* **10,** 266–271.

Sorokin, P. P., J. R. Lankard, E. C. Hammond, and V. L. Morruzzi (1967). Laser Pumped Stimulated Emission from Organic Dyes: Experimental Studies and Analytical Comparisons. *I.B.M. J. Res. Develop.* **II,** 130–147.

Stewart, R. D., S. Hameed, and J. Pinto (1978). The Natural and Perturbed Troposphere. *IEEE Trans. Geosci. Electron.* **GE-16,** 30.

Sze, N. D., M. K. W. Ko, W. Swider, and E. Murad (1982). Atmospheric Sodium Chemistry. I. The Altitude Region 70–100 km. *Geophys. Res. Lett.* **9**(10), 1187–1190.

Thomas, L., and S. K. Bhattacharyya (1980). Mesospheric Temperatures Deduced from Laser Observations of the NaD_2 Line Profile. Proceedings of the Vth ESA-PAC Symposium on european rocket and balloon programmes and related research. Bournemouth (U.K.), 14–18 April 1980 (ESA SP-152-June 1980).

Thomas, L., A. J. Gibson, S. K. Bhattacharyya (1976). Spatial and Temporal Variations of the Mesospheric Sodium Layer Observed with a Steerable Radar. *Nature* **263,** 115–115.

Thomas, L., M. C. Isherwodd, and M. R. Bowman (1983). A Theoretical Study of the Height Distribution of Sodium in the Mesosphere. *J. Atm. Terr. Phys.* **45**(8/9), 587–594.

Thompson, R. T., J. M. Hoell, and W. R. Wade (1975). Measurements of SO_2 Absorption Coefficients Using a Tunable Dye Laser. *J. Appl. Phys.* **46,** 3040.

Uchino, O., M. Maeda, and M. Hirono (1979). Applications of Excimer Lasers to Laser-Radar Observations of the Upper Atmosphere. *IEEE, J. Quant. Elec.* **QE-15,** 1094–1104.

Uchino, O., M. Maeda, T. Shibata, M. Hirono, and M. Fujiwara (1980). Measurement of Stratospheric Vertical Ozone Distribution with a XeCl Lidar: Estimated Influence of Aerosols. *Appl. Opt.* **19,** 4175.

Uchino, O., M. Maeda, H. Yamamura, and M. Hirono (1983). Observation of Stratospheric Vertical Ozone Distribution by a XeCl Lidar. *Opt. Lett.* **8,** 347–351.

U.S. Standard Atmosphere (1976). NOAA, NASA, USAF, U.S. Government Printing Office, Washington, D.C.

Van Vleck, J. M. (1925). On the Quantum Theory of the Polarization of Resonance Radiation in Magnetic Fields. *Proc. Natl. Acad. Sci.* **11,** 612–618.

Vigroux, E. (1953). Contribution experimentale à l'absorption de l'ozone. *Ann. Phys. Paris* **72,** 709–717.

Weisskopf, V. (1931). Zur Theorie der Resonanzfluoreszenz. *Ann. Phys.* **9,** 23–66.

Werner, J., K. W. Rothe and H. Walther (1985). Measurements of the Ozone Profile up to 50 km Altitude by Differential Absorption Laser Radar. *Atmospheric Ozone*, Proc. Quadrennial Ozone Symposium, Halkidiki, Greece, 3–7 septembre 1984, D. Reidel Publishing, Dordrecht, Holland.

Young, R. A. (1964). Optical Radar Study of the Upper Atmosphere. *Disc. Faraday Soc.* **37,** 118–121.

CHAPTER

6

OCEANIC AND TERRESTRIAL LIDAR MEASUREMENTS

FRANK E. HOGE

NASA Goddard Space Flight Center
Wallops Flight Facility
Wallops Island, Virginia

The principal emphasis of this chapter will be on oceanic lidar measurements since most of the technical work described to date has been performed over marine targets. Terrestrial applications are, however, also included for the following reasons. First, some of the initial lidar Earth observations were performed as airborne metric profilometry measurements over land and sea ice. Second, recent airborne laser-induced fluorescence (LIF) experiments over living terrestrial plants indicate the strong potential of lidar techniques for the measurement of plant stress, senescence, pigment content ratios, and so on. When coupled with recently demonstrated metric plant height (including crown thickness and profile), this potential is further enhanced. Third, during the normal course of performing oceanic lidar investigations, an instrument can easily be configured for, and applied to, terrestrial targets such as trees, bushes, grasses, and/or foodcrops. Fourth, an oceanic fluorescence lidar contains all of the necessary subsystems needed to perform terrestrial metric and spectral measurements. Accordingly, terrestrial lidar measurements serve to complete the total lidar Earth observation picture.

While an ultimate goal of oceanic lidar instrument scientists is to obtain global measurements while operating from a space-borne platform, present technology will not allow this to occur in the very near term. Airborne lidars, however, can contribute to passive (or nonlidar) satellite ocean color sensor and algorithm development, validation, and calibration. An excellent example of the cooperative nature of passive and active (lidar) ocean color sensors is presented in the curvature algorithm efforts by Campbell and Esaias (1983). They used laser-induced fluorescence data acquired with the NASA Airborne Oceanographic Lidar (AOL) to help establish the physical basis for a spectral curvature algorithm for the quantification of chlorophyll from an airborne passive ocean color sensor. Also, Grew and Mayo (1983) used the same lidar data to show that the algorithm was valid in open ocean areas and that his passive ocean color instrument had continued to remain stable over a long period. These are the first reported uses of lidar and passive data taken simultaneously by separate active and passive sensors on the same airborne platform.

At present most remote sensing and analysis of marine phytoplankton is performed using passive sensors such as multispectral scanners. A noticeable example is the Coastal Zone Color Scanner (CZCS) operating from the NIMBUS-7 satellite. This is in fact the only sensor ever placed in orbit specifically to study living marine resources. However, airborne lidar sensors are presently under development to assist (a) in improved understanding of existing ocean color remote sensing methodology; (b) in the development of new, passive ocean color sensors; (c) in the validation and/or verification of satellite-derived phytoplankton concentration measurements; (d) in the development of new computational algorithms; and (e) in long-range space-borne lidar ocean color sensor development.

In their own right lidar color sensors can, as will be shown, obtain biooptical measurements that cannot be made by satellite and furthermore obtain these measurements with higher spatial and temporal resolution than is possible with satellite sensors. Additionally, lidar fluorosensors can operate in total darkness, allowing studies of diurnal effects. Moreover, the lidar-induced emissions are highly specific, unlike the broad, almost featureless solar-induced ocean color spectra. Finally, lidars offer the prospect of performing high speed (5–100-ns) time domain measurements indicative of molecular relaxation or transition rate processes in the target material.

Although the lidar sensor possesses very high potential, existing passive sensors still have the advantages of lower weight, volume, and power together with reduced complexity with its attendant high reliability. Supplying an additional observational degree of freedom, the passive sensor can provide complementary data for validating the lidar measurements. This active and passive sensor complementarity should ultimately be possible on all platforms including spacecraft.

While lidars are at present being used primarily to validate and improve passive sensors, they are also used to assist in the development of state-of-the-art microwave altimeters and scatterometers. The excellent range resolution provided by narrow pulse widths, the lidar can, for example, yield excellent ocean surface elevation and backscatter power measurement for truthing or comparison purposes. Further, the relatively small laser beam divergence (<1 mrad) and resulting receiver field of view afforded by a lidar allow the assessment of spatial averaging due to the much larger microwave sensor footprint.

Some of the first published accounts of airborne lidar geophysical measurements were given in the late 1960s time frame. Hickman and Hogg (1969) demonstrated that underwater topography measurements could be made by concurrently observing the water surface return and the pulse energy backscatter from the bottom. Still earlier, Ross et al. (1968) began making sea surface elevation measurements with a modulated CW helium–neon lidar system developed and sold by a commercial laser manufacturer. Even in the mid-1960s this latter lidar system was demonstrating the high potential of laser profiling in terrestrial applications (Jensen and Ruddock, 1965). A sustained but generally small level of research in oceanic and terrestrial lidar development and application has continued subsequent to these early experiments. While the total amount of marine and terrestrial lidar geophysical applications reported to date is rather sparse, fluorosensing experiments are even fewer in number. Furthermore, in some of these latter cases the only field work executed has been performed with the NASA AOL. Accordingly, in many areas of this chapter examples of field experiments utilizing the AOL will be presented. Whenever possible, the remote lidar measurement efforts by other research groups will be discussed herein. It is the sincere hope of this author that no important marine and terrestrial lidar research and applications have been overlooked. Any such omissions are purely inadvertent.

6.1 LIDAR INSTRUMENTATION FOR OCEANIC AND TERRESTRIAL APPLICATIONS

An oceanic lidar is herein defined as a laser transceiver operating from a remote platform situated in the atmosphere or in space at some finite distance from the water's surface. Thus, we will not consider lidars operating from submerged towed fish or submarine vessels but will limit our attention to those located on ground (near water's edge), pier, ship, and satellite platforms. Of these, airborne platforms will be emphasized since (1) the most advanced experiments performed to date have been conducted from them and (2) they are the natural forerunner to satellite lidar systems.

The lidar laser transmitter is assumed to be a pulsed type so that adequate excitation energy is available in a pulse width satisfactory for timing synchronism (ocean surface location and ranging). Furthermore, the laser wavelength is assumed to be within the so-called blue-green window of the ocean. Thus, the surface of the sea can be detected using the reflected Fresnel component while the remainder of the pulse probes the ocean volume. The lidar receiver is generally equipped with a telescope having sufficient aperture to provide enough gain for signal-to-noise ratio enhancement. The detector itself is usually a sensitive, wide-bandwidth, gain-multiplying device such as a photomultiplier or avalanche photodiode. Specific lidar instrumentation configurations fulfilling these general criteria will be referenced in later sections.

The atmosphere is primarily considered to be an intervening medium whose absorption and scattering properties interfere with the lidar oceanic and terrestrial measurements. Thus, unlike other lidars discussed in this book, oceanic and terrestrial lidar design and data analysis techniques are usually configured to avoid or reduce atmospheric effects. For airborne platforms the atmospheric effects for the frequently used 150-m altitude offers few problems. Precipitation and fog are the principal problems encountered. For specific lidar atmospheric measurement techniques, the reader is referred to other chapters in this book.

6.1.1 Oceanic and Terrestrial Lidar Measurements Concepts

Figure 6.1 conceptually illustrates the fundamental observational coordinates for a multichannel lidar. The time coordinate is considered to be the one that uniquely distinguishes an active sensor from a passive one. In a truly literal sense it provides the active or lidar sensor with a three-dimensional measurement capability. A passive sensor essentially possesses only two of these observational coordinates: the signal intensity and the emission wavelength. (In Section 6.1.2 the establishment of the temporal and spectral measurement coordinates within a lidar hardware configuration will be discussed in more de-

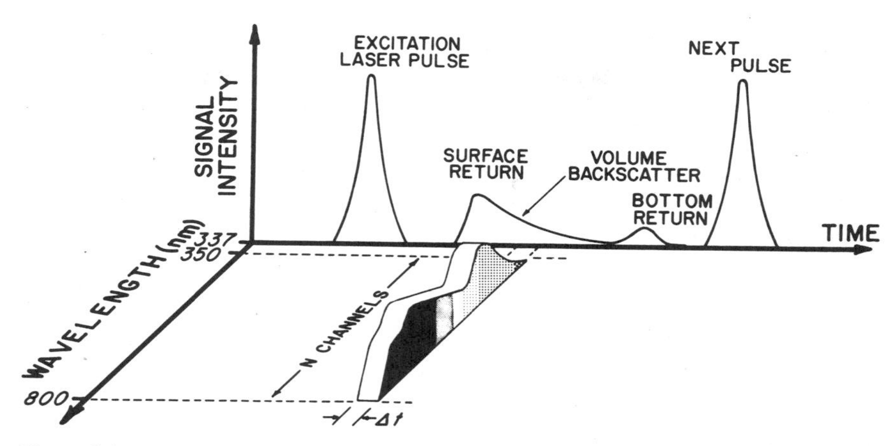

Figure 6.1. Fundamental observational coordinates of multichannel oceanographic lidar. For terrestrial applications the surface return comes from vegetation canopy while volume backscatter and bottom returns are obtained from intervening branches and forest floor, respectively. For both oceanic and terrestrial applications fluorescence spectrum sample gate time Δt can in principle be set to any temporal position from surface to bottom return or sufficiently widened to include entire surface-to-bottom time frame.

tail.) The time coordinate in Figure 6.1 shows a transmitted laser pulse that, in most lidars, is used to initiate temporal measurements and electronically arm the lidar receiver subsystem for impending signal reception, detection, and recording. The strength of this pulse is generally recorded to allow the return signals to be corrected for laser transmit power variations. (No atmospheric volume backscatter is depicted in the trailing edge of the exciting laser even though it may actually be present. Anyway, lidar system operating parameters cannot easily be optimized for simultaneous atmospheric and oceanic measurements.)

For an oceanic lidar the target is the water surface, volume, and, in the case of shallow water, the bottom. At the water's surface the transmitted pulse is partially reflected, but the major portion continues into the water volume to be scattered and absorbed by the water molecules as well as dissolved and entrained materials. The surface return pulse provides several important functions and/or measurements. First, when temporally compared or used with the transmitted pulse, metric measurement capability is realized since multiplication of one-half the elapsed time by the velocity of light yields the slant range to the target. When the vertical component of airborne platform motion is removed (by simple onboard accelerometer, ground-based radar, and/or ranging systems), the slant range measurement can yield sea surface wave, land topography, and vegetation canopy profiles. For a conically scanning oceanic lidar, the roll-and-pitch attitude of the remote platform can even be recovered from the slant range data.

Second, the strength of the initial portion of the surface return is indicative of the scattering properties (e.g., capillary waves) of the water surface. The subsequent portion of the return pulse yields volume backscatter data and, in the case of sufficiently clear and shallow water, a bottom return. (For deep-water bodies one may consider the illustrated bottom return as, e.g., backscatter from entrained particulates captured at the thermocline.) In Figure 6.1 a 337-nm (nitrogen) laser with a pulse width of ~ 8 ns is assumed to provide the exciting laser pulse. This wavelength was assumed for the purposes of this discussion so that the laser-induced spectrum would essentially be shown in its entirety. This wavelength is not the most appropriate one for all applications. In fact, while a bottom return is shown in Figure 6.1 for conceptual purposes, one would not utilize 337 nm for water depth measurements because the attenuation at this wavelength is larger than at blue-green wavelengths.

A third use of the surface return is to initiate the high-speed digitization of the entire surface return, volume backscatter, and bottom return. A high-speed waveform recorder can be activated to accomplish digitization of the backscattered laser signal representing the water's surface and water column to a depth of ~ 10–30 m. The reader will notice that the backscatter distances or attenuation lengths in the water column are short compared to the kilometer lengths observed in the atmosphere. High-speed electronic timing is therefore crucial. Of course, to accomplish the recording of itself, a facsimile of the surface return must first be sent through a delay line. The surface return is also used to initiate the temporal gating of the spectral fluorescence detection and recording system. In general, all N channels of a fluorosensor can be simultaneously activated so that they can be effectively positioned above, at, or below the water's surface. This is accomplished by varying the time delay portion of the fluorosensor detection system. The integration aperture time Δt (Figure 6.1) may also be adjusted so that the thickness of the horizontal ocean layer can be varied as well. Thus, as seen in Figure 6.1, both the spatial position and width of the water layer being observed can in principle be selected by a combination of delay lines and integration time selection. If the fluorosensor aperture gate is temporally adjusted to correspond to the bottom, the fluorescence from submerged vegetation can conceivably be acquired.

Should one desire temporal (or depth) resolution of the water Raman (or the constituent fluorescence), the on-wavelength waveform digitizer subsystem discussed above can be quickly and easily connected to any chosen spectral channel. Alternatively, the temporal waveform digitization of a chosen wavelength from the laser-induced fluorescence spectrum of, for example, an optically thick oil film would necessarily be interpreted as fluorescence time decay data, not depth-resolved information.

In terrain and vegetation mapping applications the ranging and temporal waveform-recording capabilities of the lidar can be used to make metric measurements of the surface elevation and tree/bush height. For example, in Figure 6.1, if the water's surface is, instead, a tree top, the lidar system can be used to obtain vertical terrestrial measurements in much the same manner as was done over water bodies. This is particularly true if the tree crown and/or leaf density is sufficiently sparse to allow downward penetration of portions of the laser beam. Of course, if some portion of the beam reaches the ground (as is frequently the case), the bathymetry or depth-resolved portion of the lidar system can be used to record the tree height and scatter from intervening leaves as well. Conceptually, one could at the same time record the fluorescence of a segment of the tree crown and understory at some chosen ''depth'' from the tree top.

6.1.2 Oceanic and Terrestrial Lidar Instrumentation

Depicted in Figure 6.2 is a generic form of a lidar system that can fulfill many, if not all, the measurement concepts outlined in Figure 6.1 and discussed in Section 6.1.1. The lidar in Figure 6.2 contains all the elements needed to spatially scan the surface, transmit and receive optical radiation, and record the resulting temporal and spectral data pretaining to an oceanic or terrestrial target.

At the heart of this oceanic and terrestrial lidar system is the computer and timing electronics that initiates measurements by commanding the laser to fire and completes the final recording of the data upon receipt of the backscattered radiation. Once the laser is fired, a small portion of the beam is diverted to detector 1 to (a) start the slant range measurement to the target and (b) measure the laser pulse energy. The lidar system is sometimes constructed in such a flexible manner that any one of several types of lasers may be interfaced depending on the experiment being performed. These lasers may have different (1) beam shapes (rectangular and/or oval, e.g., for gas discharge lasers) and (2) beam divergences. Thus, the lidar transmitter is provided with a beam shape and beam divergence control. To facilitate alignment, the transmit beam is made coaxial with the telescope. The scanner subsystem is common to both the transmitter and receiver and may consist of either reflective or refractive optical components depending on weight, volume, and spatial pattern required.

The receiver section is composed of a telescope having sufficient aperture to provide enough gain for signal-to-noise ratio enhancement. A Cassegrain-type telescope is a frequent choice for size, volume, and desirable co-linear axis configurations. The focal plane, adjustable field of view (FOV) allows the re-

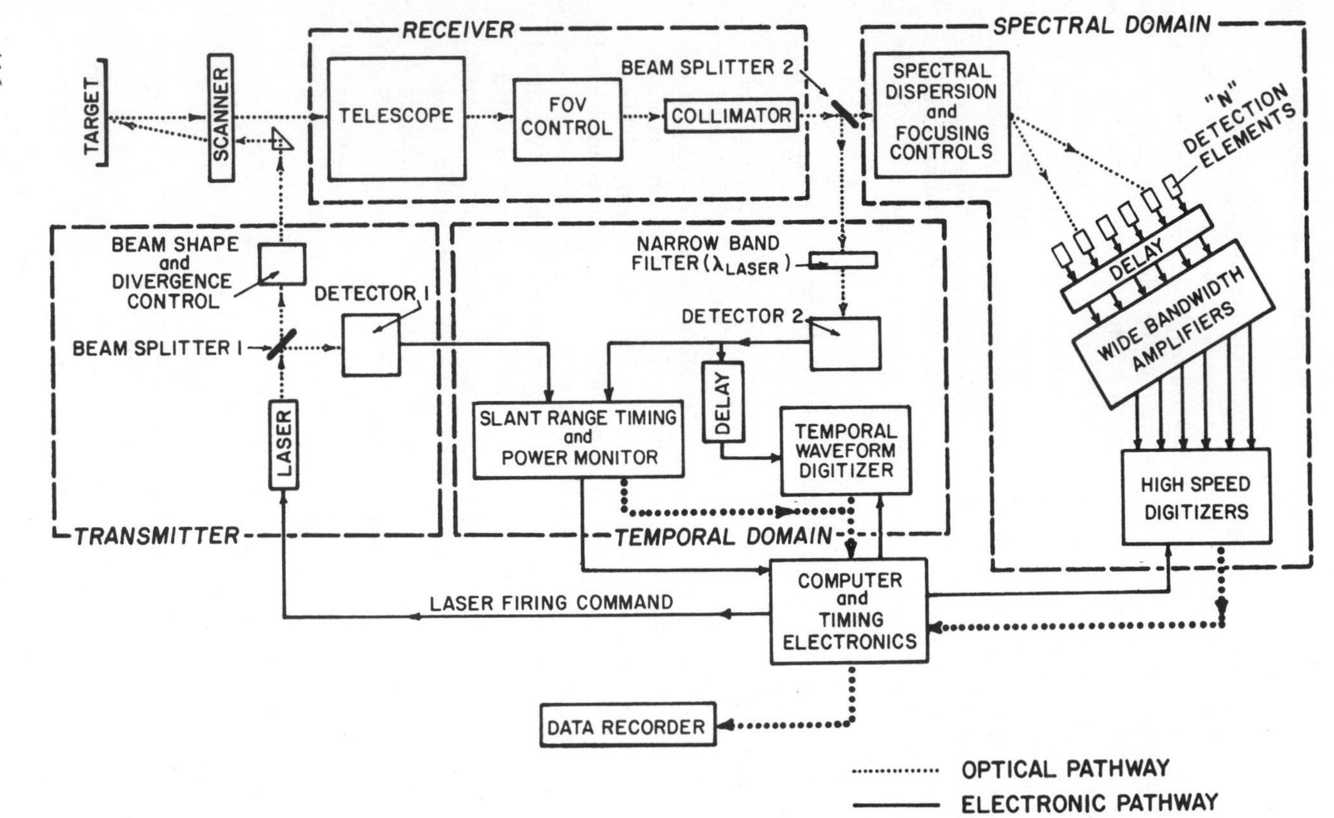

Figure 6.2. Block diagram of generic lidar system capable of making measurements described in Figure 6.1 and discussed in Section 6.1.1. See text for references to specific lidar hardware.

ceiver to view only the laser-irradiated spot on the target surface. Thus, the FOV can be made to closely match the transmit beam divergence. To accommodate symmetrical (e.g., rectangular) beam shapes, the FOV should be adjustable in both the horizontal and vertical dimensions. The receiver FOV control is perhaps one of the most important components of the entire system. It provides background or solar noise rejection by allowing the receiver to view only the spot that is to be illuminated by the laser. The background rejection is sufficient to allow operation of the lidar system during daylight hours. Depending on the application, FOV settings of 2–6 mrad are typically used during daylight hours even when high-gain photomultiplier tube detectors are employed in the system. FOV adjustments to 20 mrad are sometimes used during nighttime experiments. Subsequent to FOV control, collimation allows improved utilization of the received beam for (1) narrow-band interference filtering of extraneous, background noise prior to detection for temporal domain measurements and (2) spectral dispersion for fluorescence applications. Beam splitter 2 is the point at which the temporal and spectral measurement coordinates are established within the lidar hardware. (Refer also to Section 6.1.1 and Figure 6.1.)

The temporal domain measurements naturally divide into two separate segments.

1. The slant range timing, which represents the distance from the aircraft to the target, is obtained by ordinary time-of-flight measurement of the excitation laser pulse and its resulting echo or backscattered surface return (see Figure 6.1 also). For an aircraft platform operating at 150 m altitude the two-way elapsed time is $\sim 1\ \mu s$. Electronic counters are readily available that allow measurements of range with a precision of ~ 0.02 m.
2. The surface return, volume backscatter, and bottom reflection (if any) are typically obtained with a wide-bandwidth waveform recorder. For oceanic and terrestrial lidar applications these waveform recorders typically span a total extent of only ~ 100–200 ns with contiguous, individual time bins having widths of ~ 5 ns. The time delay shown in Figure 6.2 is used to allow the range timing module to initiate the waveform recording at the proper moment.

The spectral measurement portion of the lidar system receives collimated background, on-wavelength, as well as laser-induced water Raman and fluorescence backscatter signals. This multicolor radiation is dispersed by diffraction grating (or by prism) and focused on N detectors. The number of detectors is primarily dictated by the resolution and the sensitivity required by the experiments. As before, a time delay is used in series with the detectors to allow the high-speed digitizers to be activated in synchronism with the received fluorescence pulses. The *N* detection elements can be individual high-gain photomultiplier tubes, a silicon diode array, or an image dissector. These detectors may also be used in conjunction with image intensifiers for added sensitivity. The final choice depends on the sensitivity, spectral resolution, and spectral time decay measurement requirements. Lidar systems having many of the characteristics and features generally described here have been reported in detail in the literature (Bristow, 1978; Bristow et al., 1981; Hoge et al., 1980; Hoge and Swift, 1981b; O'Neil et al., 1980; Capelle et al., 1983; Franks et al., 1983). The status of airborne laser fluorosensing was recently discussed by O'Neil et al. (1981).

6.2 LIDAR MEASUREMENTS OF OCEAN SURFACE, VOLUME, AND BOTTOM

The airborne pulsed neon laser experiments by Hickman and Hogg (1969) amply demonstrated that the ocean surface, intervening volume, and the bottom could be probed remotely with airborne laser systems. Accordingly, the physical, chemical, and biological states of the surface, volume, and bottom regions were thus irreversibly declared fair game for the scientist interested in obtaining such knowledge in ways that only lidar remote sensing could uniquely provide.

While the ocean volume contains by far the majority of the important chemical and biological constituents, we will nevertheless begin the discussion of oceanic applications with the ocean surface. This is done for several reasons. First, historically some of the initial airborne lidar experiments were conducted to measure the sea surface wave profile or determine the on-wavelength backscattering cross section. Second, the airborne lidar measurement of fluorescence was first conducted on oil slicks residing on the sea surface. Third, the concept of using the volumetrically derived seawater Raman backscatter as the reference signal for lidar measurements can be introduced early in a formal, closed-form theoretical, and quantitative way for demonstrating airborne lidar metric measurements of oil films on the sea surface. In turn, this will (a) demonstrate the need for (and the importance of) the water Raman signal in most, if not all, oceanic lidar measurements; (b) lead naturally to its later use as a reference signal for measurement and/or correction of water column attenuation during the volumetric measurement of ocean constituents; and (c) encourage the reader to suggest and/or develop oceanic lidar measurement applications (and their attendant theories) based on this inherent oceanic lidar return signal. Thus, at the risk of excessive detail, the theory of oil fluorescence measurements will be fully discussed in Sections 6.2.1.4 and 6.2.1.5 in order to directly illustrate the quantitative potential that lidar measurements possess.

6.2.1 Measurements of Natural and Contaminated Ocean Surfaces

6.2.1.1 Lidar Backscatter

Outside of the initial bathymetric measurements discussed above, some of the first published accounts of lidar measurement of the uncontaminated ocean surface dealt with backscatter cross-sectional or signal return experiments. Indeed, it became necessary to first understand how the narrow coherent transmitted laser beam interacted with the ocean surface, and furthermore how the resulting small FOV telescopic receivers intercept and process such optical returns. One of the first published reports of airborne laser backscatter measurements from the sea surface was given by Jelalian (1968). While he (as well as Hickman and Hogg) made quantitative measurements, probably the most important finding of these original lidar investigators was that narrow-beam optical radar techniques from airborne platforms were feasible and presented no overriding physical complexity. Jelalian (1968) used a nadir-pointed 1.06-μm pulsed Nd–YAG laser at various aircraft altitudes to show that the surface return is range-square dependent and that the target cross section per unit area was larger for calm water than for seas with a measurable sea state. Petri (1977) made more extensive off-nadir measurements from a platform suspended from the southernmost Chesapeake Bay bridge span. He was also able to show that the return signal backscattered at normal incidence is larger for low sea states (corresponding to low wind speeds) and smaller for higher sea states. This is illustrated in Figure 6.3. Likewise, his data vividly showed that lidar systems that must operate off nadir should do so only if sufficient winds exist to disturb the sea surface and thereby produce sufficient surface facets to yield an adequate return signal to effect or at least initiate a lidar measurement. Anyone who attempts to operate a lidar system over calm water will readily observe that strong mirrorlike specular reflections occur near nadir and that significantly lower signals are received at 5°–10° from the nadir. Theoretical calculations by Swennen (1968) had predicted the decrease in signal with increasing wind speed for a normally incident beam. One can infer from his calculations that the wind-disturbed surface deflects the energy into the off-nadir wings (see Petri's data, Figure 6.3). Phillips (1979) formulated the general backscatter problem in terms of the wavenumber spectrum of the sea surface and showed that the amplitude (as well as the temporal) fluctuations depend on the wavenumber spectrum within the laser footprint. Unfortunately, his wavenumber development lacked sufficient data for the parameters to further verify the full applicability of the technique. While numerous airborne lidar experiments have been performed to date, a paucity of data has been published except as cited above. Bufton et al. (1983) has recently provided airborne scanning laser data at wavelengths of 337 nm, 532 nm, and 9.5 μm. He showed that the mean backscatter had a Gaussian-like shape with

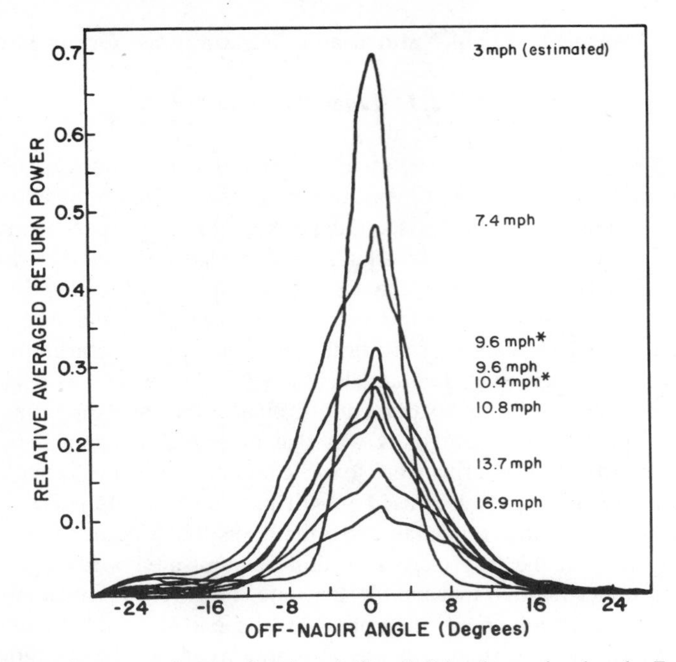

Figure 6.3. Average return signal as function of off-nadir lidar observational angle. Each transmit-power-normalized curve represents average of 18 azimuth positions (except single azimuth position averages labeled by asterisk). Data collected at 16 scans per second with receiver field of view of 24 mrad (Petri, 1977).

a FWHM ranging from 11° to 24° for wind speeds of 1.8–6.5 m/s. At or near nadir the normalized standard deviation ranged from 0.1 to 0.6. Furthermore, by airborne calibration against beach sand, an effective Lambertian reflectance of about 20% was found. This is of course about an order of magnitude higher than the theoretical Fresnel reflectance of an air–sea interface.

From a practical standpoint the laser backscatter measurements and theoretical developments performed to date allow one to only approximately ascertain the field conditions and lidar operating parameters under which satisfactory data can be obtained. As a specific example, consider the backscatter data adopted from Petri (1977) and given in Figure 6.3. From this data it is rather clear that a lidar with a scan angle of ±20° off nadir would have adequate surface return to initiate a measurement only within ±6° of nadir on days when the wind speed was ~3 mph. The data also indicate that operation beyond ±25°, even on the windiest days, would in all probability be marginal. Increasing the laser footprint size will probably increase the return signal. Likewise, opening the receiver FOV will no doubt improve the detection of the surface return pulse but will in turn lead to a compromise in SNR during daylight hours. As may be seen later, it is possible to have sufficient wind to produce a ruffled sea surface for adequate backscattered surface return and still have detection problems. These detection difficulties can be caused by the wave-damping effects of, for example, a target oil slick or monomolecular film or strong near surface wave–current interaction and/or strong air–sea surface temperature gradient (McClain et al., 1982b). These wave-damping mechanisms modify both the height and slope probability density of the surface waves within the footprint so that the surface return is reduced.

It should be noted in passing that a conically scanning lidar system (Hoge et al., 1980) can be used to ascertain the airborne platform attitude should such orientational information be required. Byrnes and Fagin (1978) applied optimal filtering analyses to scanning data from the NASA AOL. They showed that the roll and pitch of the aircraft can be obtained with an accuracy comparable to that obtained from the gyros of an onboard inertial navigation system. The effects of large scan angles, sea state and currents, laser pulse width and repetition rate, and so on, were not studied to indicate the range of practicality.

6.2.1.2 Airborne Laser Profilometry

Laser profilometry (Ross et al., 1968) of ocean surface waves was apparently the first application of a lidar to oceanic surface wave studies. The potential of high spatial (and metric elevational) resolution of laser profilometry was quickly recognized, and a short time later (Ross et al., 1970) it was applied to the scientific study of sea surface wave development under fetch-limited wind conditions. In fact, these studies also served to show the rapid acceptability of laser instrumentation by the scientific community since these airborne experiments were made in conjunction with microwave profilometry measurements. In spite of its continuing simplicity of airborne instrument application, laser sea surface wave profilometry data interpretation advances have probably been the important ingredient to its recent increased acceptability. Long (1979) and Kats and Spevak (1980) detailed the procedures for analyzing surface gravity wave spectra observed in a moving reference frame. Additonally, due to (1) the recognition that the frequency spectrum of a homogeneous wave field can be distorted by flight direction relative to the wave field (McClain et al., 1980) and (2) the utilization of recently developed analytical and theoretical techniques (McClain et al., 1982b), laser profilometry has been shown to be useful in the measurement of sea state variations across oceanic fronts. In fact, a profilometer can be used to determine the directional properties of a wave field with an accuracy of better than $\pm 10\%$, and the wavelength of each dominant component can also be determined to within 10%. For the verification of wave–wave and wave–current interactions, airborne laser profilometry may well be the only way to obtain the high-resolution, synoptic data required. Recently, laser profilometry has been used to simultaneously measure ocean wave elevation and return power to identify those portions of the ocean wave yielding the highest backscatter cross section (Hoge et al., 1984). Laser profilometry has similarly been used to assist in the study of the electromagnetic bias of microwave altimeter systems when applied to satellite ocean topography measurements (Walsh et al., 1984; Choy et al., 1984).

6.2.1.3 Oceanic Geoid Measurements by Laser Altimetry

Lidar altimeters operating from space platforms promise improved elevation measurement precision since laser pulse widths in the 70-fs domain have recently been demonstrated (Fork et al., 1982). No doubt, even shorter pulse width lasers will become available in this decade. The space-borne altimetric measurement of the oceanic geoid by lidar has of course not yet been attempted. The theoretical background for such measurements have been given by Gardner (1982) and by Tsai and Gardner (1982). Their developments have beginnings in satellite microwave radar altimetry, which has proven highly successful during the past decade (McGoogan and Walsh, 1979). High-precision laser altimetry would allow higher spatial resolution mapping of the geoid and potentially the inference of ocean surface currents through high-resolution ocean surface elevational measurement of geostrophic effects. These ocean currents are in part responsible for the transport of biological and chemical constituents in the ocean. The ocean current transport is rather easily seen in satellite ocean color and temperature images of the ocean surface (Gordon et al., 1983). There are several reasons for discussing laser altimetry of the ocean surface within the confines of a book that focuses on laser remote chemical analysis. As we shall see in Section 6.2.1.5, the transmitted and backscattered pulse time-of-flight measurement by such lidar systems yields sea surface elevation measurements but simultaneously can give surface roughness information via pulse energy amplitude (Hoge et al., 1984). The surface roughness in turn determines the coupling of wind energy into the upper surface layers of the ocean. In turn, these wind-generated waves and currents are in part responsible for the transport of the chemical and biological constituents within the water column. Given sufficient laser power and receiver aperture, the same lidar altimeter performing altimetry and/or surface wave elevational measurements could be used simultaneously to measure the laser-induced fluorescence of phytoplankton photopigments (chlorophyll *a* and phycoerythrin), dissolved organic material (DOM), surface slicks, and so on. Present-day airborne lidar fluorosensors are capable of simultaneous measurements of the sea surface elevation and spectral fluorescence (Hoge and Swift, 1983a), but it will probably be several decades before such multiuse space-borne lidars are operational. One would expect to first see existing passive ocean color scanners mated with microwave altimeters, wherein the altimeter is used to determine sea state and ocean currents for respective correction of ocean color images for glint and the determination (and ultimate understanding of) possible transport current mechanisms. Gradually, then, the microwave altimeter could be supplanted with the laser altimeter. Ultimately, given sufficient and expected technological progress, the laser altimeter could advance to simultaneous measurements of laser-induced fluorescence as well as geoid and surface wave determinations. Coupled with space-borne passive ocean color scanners similar to those in existence today, such lidar systems could yield data of significant use to the oceanographic community. The future challenge is to concurrently obtain sufficient laser energy and receiver aperture size to make oceanic and terrestrial application of such space-borne lidar color (fluorescence) sensors practical.

In the meantime, the study of the electromagnetic bias problem, which limits the ultimate accuracy of space-borne altimeters, must also be continued (Walsh et al., 1984; Hoge et al., 1984; Choy et al., 1984). Furthermore, the detailed satellite laser altimetry studies of Tsai and Gardner (1982) and the joint height-slope probability density function developments of Huang et al. (1983, 1984) should be combined with the space-borne laser studies to perhaps provide additional insight into the problems of space-borne laser altimetry.

6.2.1.4 Sea Surface Oil Slick Detection, Measurement, and Identification

Petroleum hydrocarbon inputs to the world's oceans are estimated to be about six million metric tons each year (Myers and Gunnerson, 1976). International interest in the study of spilled oil continues, as evidenced by the technical papers originating during the past decade from the United States (Kung and Itzkan, 1976; Hoge and Swift, 1980; Hoge and Swift, 1983a; Hoge, 1983), Canada (O'Neil et al., 1980; Rayner et al., 1978; Rayner and Szabo, 1978), The Netherlands (Visser, 1979), Italy (Burlamacchi et al., 1983), Japan (Sato et al., 1978), and the USSR (Abramov et al., 1977; Bogorodskiy et al., 1977).

Unlawful spillage cannot easily be deterred unless the oil can be classified according to some criteria. While numerous passive sensors can detect the oil slick and thus measure its areal extent, none has the capability to classify the oil. However, Fantasia et al. (1971) and also Measures and Bristow (1971) separately concluded that the laser-induced fluorescence spectrum (including fluorescence lifetime) showed significant promise. A helicopter-borne fluorosensor later demonstrated for the first time that the fluorescence spectrum could be detected (Fantasia and Ingrao, 1974) remotely. It has been shown that the shape of the fluorescence color spectrum alone is adequate to separate the oil into three classifications: (1) light-refined products such as diesel fuel, (2) crude oils, and (3) heavy oils such as Bunker C fuel oil (O'Neil et al., 1980). Given the large number of oils that produce different spectra (Fantasia and Ingrao, 1974), a classification into only three types would not be sufficient to allow deterrent policing for surreptitious spillage. However, Measures et al. (1974) have demonstrated that fluorescence decay time spectroscopy could be used to discriminate between similar hydrocarbons. Furthermore, since the fluorescence conversion efficiency is highly variable among the oils (Fantasia and Ingrao, 1974), the avenue of development seems clear: a classification scheme based on oil fluorescence decay time spectroscopy and conversion efficiency should be undertaken.

However, a thin portion of an oil slick will produce less fluorescence than a thick portion so that a different conversion efficiency would apparently be found throughout a slick, thus rendering the fluorescence conversion efficiency method questionable if not useless. What was needed, then, was a method of remotely sensing the fluorescence conversion efficiency of oil slicks independent of their thickness. The theory for such a remote oil fluorescence classification technique was published by Kung and Itzkan (1976), and airborne validation experiments were begun in 1978 (Hoge and Swift, 1980, 1983a; Hoge, 1982). The Kung and Itzkan method of calculating the characteristic infinite-thickness absolute spectral fluorescence conversion efficiency from oil films of finite thicknesses used the seawater Raman backscatter as the reference signal for calibration. The theory also accounted for the background fluorescence caused by dissolved organic material, or *Gelbstoff*. They showed that under most conditions the absolute oil fluorescence spectral conversion efficiency (AOFSCE) could be determined within a factor of 2 accuracy if the oil film is $\geqslant 0.5$ μm thick, the oil fluorescence is greater than the *Gelbstoff* fluorescence, and the water column beam attenuation at the laser wavelength and at the Raman backscatter wavelength is less than 1.2 m^{-1} somewhere along the flight path. (This latter requirement is to allow at least one measurement of the depth-resolved water Raman decay constant in reasonably clear water. Assuming long-term drift is not a problem, the integrated Raman and its corresponding depth-resolved decay constant are both constant multiples of each other at all other positions of the flight where a Raman signal can be obtained.) These criteria being satisfied, the theory allows a general classification of oils without a priori knowledge of the oil film thickness. Furthermore, once the oil is positively identified, we shall see later that the oil film thickness can in turn be measured. Of course, this latter measurement aids in the determination of the volume of oil spilled.

The AOFSCE theory has been field tested using the NASA AOL (Hoge and Swift, 1980, 1983a). The absolute oil fluorescence spectral conversion efficiency η_f is given by (Kung and Itzkan, 1976)

$$\eta_f = \frac{\sigma N_w T_e T_r}{\gamma_e + \gamma_r} \frac{\theta_f}{1 - \Delta_r^{1+\epsilon}} \left(1 - \frac{\Delta_r^{1+\epsilon}}{\Xi_f} \right) \tag{6.1}$$

where σ = Raman cross section of water

N_w = number density of water

T_e, T_r = transmissivities of air–sea interface at laser emission and Raman wavelengths, respectively

γ_e, γ_r = effective or beam attenuation coefficient of seawater at laser emission and Raman wavelengths, respectively

θ_f = oil fluorescence signal at wavelength λ_f as normalized by water Raman signal measured outside slick, K_f/R (see Figure 6.4*a, b*)

Δ_r = Raman signal amplitude measured while over oil slick (with fluorescence contributions from oil and *Gelbstoff* at λ_r removed by interpolation and subtraction) normalized by water Raman from clean sea region, R'/R (Figure 6.4*a*)

Ξ_f = oil fluorescence measured over oil slick and normalized by fluorescence signal at λ_f observed outside slick, K_f/J_f

Finally,

$$\epsilon = \frac{\kappa_f - \kappa_r}{\kappa_e + \kappa_r} \tag{6.2}$$

where κ_f, κ_r, and κ_e are the extinction coefficients of the oil at wavelengths λ_f, λ_r, and λ_e, respectively. Figure 6.4*a* contains the essential definitions of many of the above parameters defined for LIF of oil and water column constituents.

The first multiplicative term of Equation (6.1) is known as the seawater Raman conversion efficiency:

$$\psi = \frac{\sigma N_w T_e T_r}{\gamma_e + \gamma_r}. \tag{6.3}$$

Thus, a straightforward or simplified interpretation of Equation (6.1) is that the absolute oil fluorescence spectral conversion efficiency is the water column Raman conversion efficiency modified by two additional multiplicative terms whose values depend on the extinction and spectral fluorescence properties of the oil. Ideally, the oil fluorosensing lidar can be equipped to remotely measure the sum of the effective attenuation coefficients of water, $\gamma_e + \gamma_r$, by utilizing the temporal variation or depth-resolved Raman signal backscattered from the water column. Then

$$\gamma_e + \gamma_r = \frac{c\tau}{2n} \tag{6.4}$$

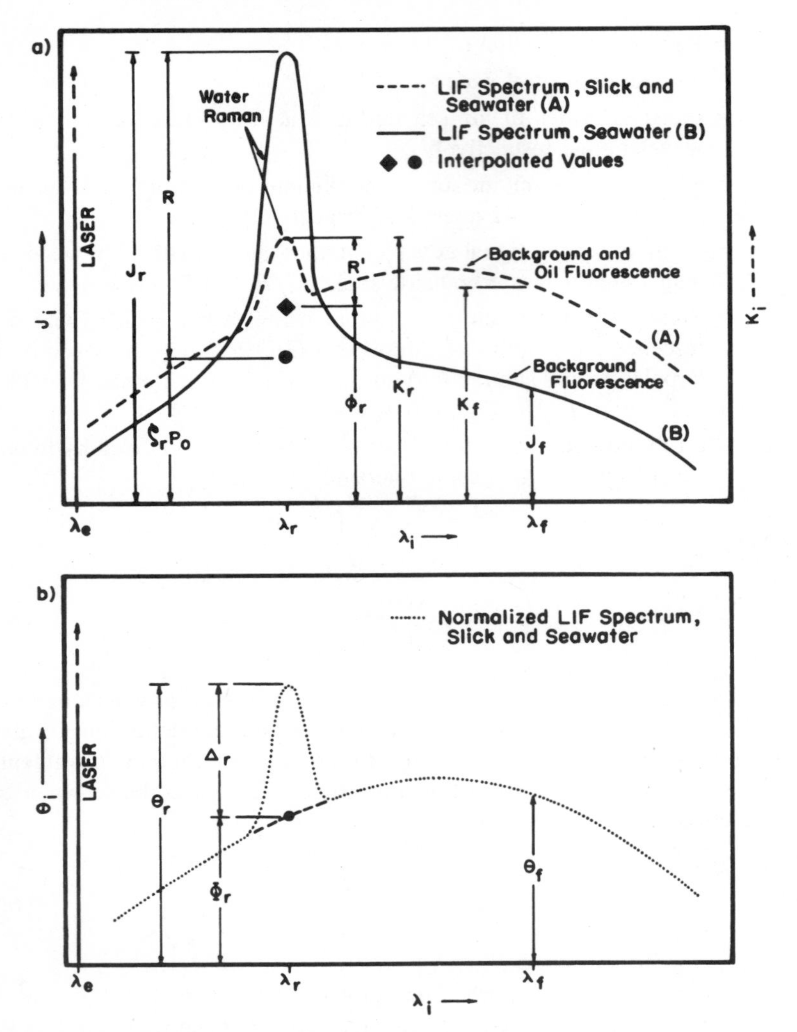

Figure 6.4. (*a*) Theoretical or expected laser-induced fluorescence spectral waveforms over (A) ocean water covered by optically thin oil film and (B) ocean water. (*b*) Spectral waveform of ocean water normalized by water Raman backscatter R (Hoge and Swift, 1980).

where c is the velocity of light, n is the index of refraction, and τ is the Raman signal decay time constant computed from total elasped, or two-way, time decay data. During our oil spill experiments, the NASA AOL could not be rapidly reconfigured from the spectral mode to the temporal or depth-resolved mode in order to measure τ. Instead, an estimate of ψ from the water optical transmission data taken by truth ships in the immediate vicinity of the spill was used. It is recognized that the ultimate utility of AOFSCE for the general typing of oil spills with an airborne laser fluorosensing system requires a measurement of ψ independent of any supporting surface vessels and during the same time frame as the airborne oil fluorescence data is acquired. Therefore, the latter sections of this chapter will describe progress to date in the measurement of depth-resolved water Raman during separate field experiments. The reader will no doubt recognize the general oceanographic usefulness of the remote measurement of ψ and its interpretation via Equation (6.3) of the effective extinction coefficient of the water column.

Field experiments to investigate the Kung–Itzkan theory have been performed with the NASA AOL. The AOL instrument was configured with a 337-nm nitrogen laser transmitter having an energy of 1 mJ in a 10-ns pulse. The fluorosensing mode was used for the airborne spectral measurements described herein. The range of AOL transmitter and receiver parameters used in the oil spectral measurements are essentially described in some of the previous publications referenced above. For convenience, the important parameters are summarized in the fluorosensor portion of Table 6.1.

During November 1978 the AOL participated (together with the Canada

Table 6.1. AOL Operating Parameters

	Oil Fluorosensing Mode, Nitrogen Laser
Transmitter	
Wavelength	337.1 nm
Bandwidth	0.1 nm
Pulse width	10 ns
Pulse rate	$\leqslant$100 Hz
Peak output power (max)	100 kW
Beam divergence	4 mrad
Receiver	
Bandwidth	3500–8000 Å
Spectral resolution (min)	11.25 nm
Field of view	1–20 mrad, variable, vertical, and horizontal
Temporal resolution	8–150 ns, variable
Experiment	
Aircraft altitude	150 m
Aircraft velocity	75 m/s
Measurement background	Day

Center for Remote Sensing; O'Neil et al., 1980) in a series of missions designed to measure the effects of a dispersant on the spreading rates of crude oils spilled on the ocean surface. Information regarding the spills is provided in Table 6.2. Additional physical characteristics of the two crude oils are given in Table 6.3. The data analyzed were from missions flown over the LaRosa slick on November 3, 1978, and the Murban slick on November 9, 1978.

Figure 6.5 shows the calculated AOFSCE at 500 nm for the Murban crude oil as a function of oil film thickness. The thickness could be obtained since (a) the oil type was known a priori and (2) the extinction coefficients had been measured in the laboratory (Hoge and Kincaid, 1980; Hoge, 1982). Although the theoretical model should recover the same conversion efficiency for all thicknesses, the fluorescence yield is apparently too low for films <2 μm thick. This is especially evident for the more fluorescent Murban crude. A minimum limit for the AOFSCE application was not totally unexpected since the calculations of Kung and Itzkan (1976) showed that the film should be >0.5 μm to attain a conversion efficiency within a factor of 2 accuracy using a lidar having instrument parameters similar to the AOL.

A single self-contained lidar for AOFSCE measurements must be capable of depth-resolved water Raman backscatter measurement at some position along the flight line in order to complete the calibration. Few fluorosensors are yet configured to perform this portion of the AOFSCE measurement during the same mission. However, such instrument configuration problems are tenable (and have actually been accomplished on the AOL). This was not the case during the oil spills discussed herein, so experiments to ascertain the feasibility of performing the requisite AOFSCE depth-resolved seawater Raman backscatter was performed separately (Hoge and Swift, 1983a,b). A mission was flown off the coast of Virginia in the Atlantic Ocean. It consisted of a single flight line originating about 30 km offshore and intersecting the beach. This gave a continually varying water column attenuation. As with the oil spill experiments, a 337-nm pulsed laser was used. With the AOL configured into the

Table 6.2. Experiment and Data Summary

Date	Crude Oil Type	Dispersant Applied	AOL Data Passes Before Dispersant		Total AOL Data Passes
			Nonscan	Scanning	
November 2, 1978	Murban	After 2 h	0	0	18
November 3, 1978	LaRosa	After 2 h	3	5	33
November 9, 1978	LaRosa	Immediately	0	0	37
	Murban	Immediately	0	1	17

Table 6.3. Spilled Oil Types and Selected Physical Characteristics

			Fluorescence		Extinction Coefficient (nm^{-1})			
Oil	Source	Type Crude	Emission	Peak Wavelength (nm)	337.1 nm	381 nm	500 nm	API°
LaRosa	Venezuela	Heavy	Low	490	896	434	189	23.3
Murban	Middle East	Light	High	505	265	95	26	38.1

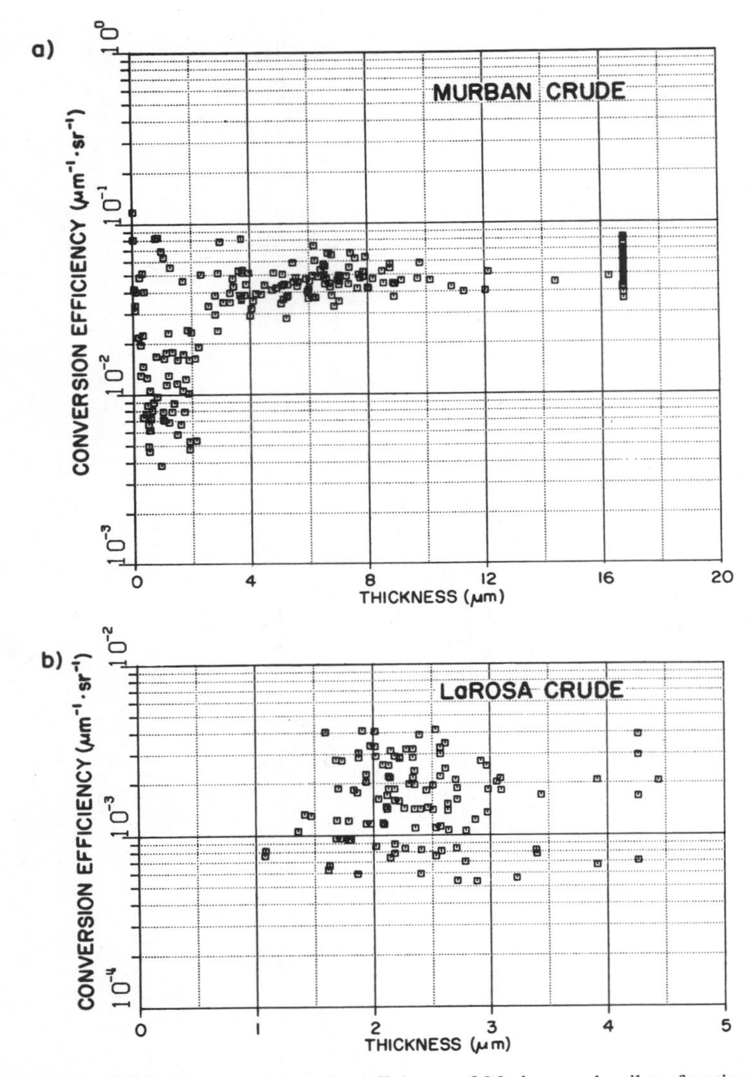

Figure 6.5. (*a*) Oil fluorescence conversion efficiency of Murban crude oil as function of film thickness. (*b*) Oil fluorescence conversion efficiency of LaRosa crude oil as function of film thickness (Hoge and Swift, 1980).

bathymetric mode (Hoge et al., 1980), the depth-resolved waveforms in Figure 6.6 were obtained.

Figures 6.6*a–c* show typical depth-resolved waveforms of the laser-induced water Raman backscatter. Note that the offshore waveforms (Figure 6.6*a*) have a larger amplitude and extend to greater depths. As one proceeds shoreward

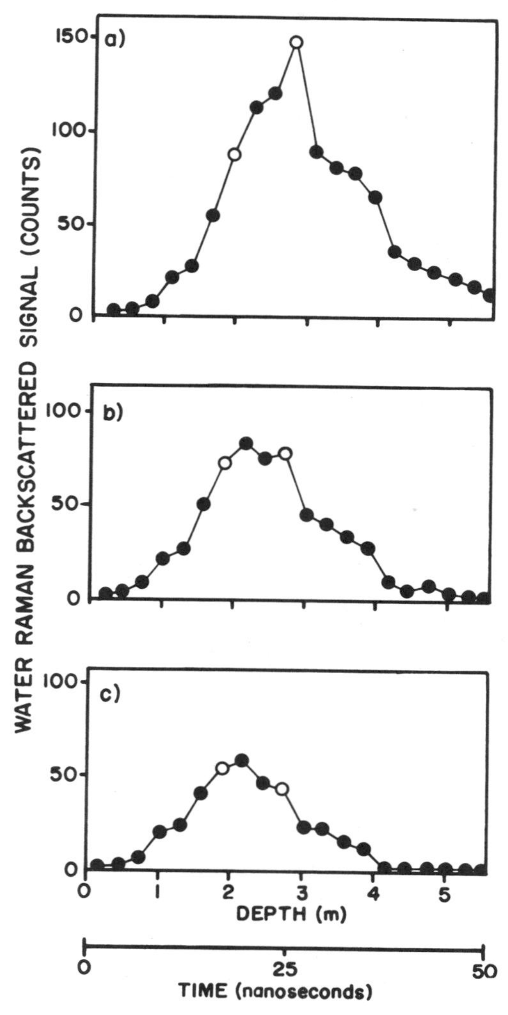

Figure 6.6. Depth-resolved, 337.1-nm laser-induced water Raman backscatter waveforms. Open circles at channels 7 and 10 are labels for subsequent discussion of turbidity cell structures in Section 6.2.2.4 (Hoge and Swift, 1983c).

(Figures 6.6*b*,*c*) into more turbid water, the backscatter from the deeper regions of the water column is progressively lost, as would be expected. At the same time, the peak amplitude of the backscattered Raman signal also declines. As a consequence, the instantaneous peak amplitude of the Raman waveform progressively shifts toward shallower depths as increasingly more turbid water is encountered. Later in this chapter we shall discuss from theoretical considerations the general shape of the backscattered Raman pulse and its relationship to the system response, sea surface, and water volume attenuation characteristics. The channel-to-channel sensitivity variations within each waveform were not correctable because of an inadvertent and unfortunate loss of calibration data in the time frame the experiment was conducted. The relative and qualitative features of the data together with the resulting conclusions are, however, unaffected. Recently developed calibration techniques together with new waveform digitizing equipment now permit significantly better quality data from the AOL.

A straightforward convolutional model accounts quite well for the observed pulse shape is increasingly turbid water. It can be shown (Hoge and Swift, 1983c) that the Raman pulse shape can be represented rather well by the multiple convolution of a Gaussian lidar system response, Gaussian sea surface height and slope probability density, and an exponential water column response,

$$R(t) = A \exp\left(\frac{-t}{\tau}\right) \exp\frac{\sigma^2}{2\tau^2}\left[1 + \operatorname{erf}\left(\frac{t}{\sigma} - \frac{\sigma}{\tau}\right)\right] \tag{6.5}$$

where

$$\sigma^2 = \sigma_g^2 + \sigma_t^2 \tag{6.6}$$

and σ_g denotes the rms width of the lidar system and σ_t is related to the width of the sea surface density $\sigma_\zeta = c\sigma_t/2$, where c is the velocity of light. From Equation (6.4) the parameter τ is related to γ_e and γ_r, the effective attenuation coefficients of the seawater at the laser and Raman wavelengths, by

$$\tau = n/2c(\gamma_e + \gamma_r) \tag{6.7}$$

where n is the index of refraction of seawater, and the factor of 2 accounts for the fact that the digitized waveform is recorded for the entire two-way propagation path. The constant A is

$$A = 1/c\pi S^2 \tag{6.8}$$

where S^2 is the mean square value of the total slope. For this simple convolution model, it is seen that the amplitude of the backscattered pulse is principally influenced by the mean square value of the total slope S^2. The remaining shape is, of course, contributed by the incident pulse width, sea state in the footprint, and the water attenuation function. For rapid computational purposes (1) a series expansion or (2) a rational approximation of the error function for argument z, erf(z), may be used (Abramowitz and Stegun, 1964; Hayne, 1980).

For initial estimates assume $\gamma_e = 2\gamma_r$ (Hoge and Swift, 1981a). To estimate the rms width of the laser system response, we note that our 337.1-nm nitrogen laser has an inherent pulse width of ~8 ns. We have measured a FWHM pulse width η of ~12 ns with the entire system including the waveform digitizer. But the FWHM pulse width is related to the rms width by $\eta = 2\sigma_g\sqrt{2 \ln 2}$. Thus, $\sigma_g = 5$ ns.

The mean square total slope S^2 may be obtained from the Cox and Munk (1954) relation

$$S^2 = 0.003 + 0.005W \tag{6.9}$$

For a wind speed W of 5 m s^{-1}, the rms wave height σ_ζ would for this wind speed be on the order of 0.7 m (Hoge et al., 1984). Then, $\sigma_t^2 = 4\sigma_\zeta^2 c^2$ yields a value of ~5 ns. However, the AOL radiated (and viewed) spot on the ocean surface is ~0.6-m diameter from our operating altitude of 150 m. Therefore, the instantaneous rms wave height within this spot for any single pulse is probably no greater than ~3 cm. This would yield a negligible contribution of the Raman pulse shape compared to an incident pulse width of ~8 ns.

Using the values of the above parameters, the Raman pulse given by Equation (6.5) is plotted in Figure 6.7 together with the Gaussian lidar system response and water column exponential decay function. Note that (1) the incremental entry of the laser pulse into the water and (2) the attenuation within the water column causes the Raman pulse to have a longer leading edge risetime. This results in a maximum or peak amplitude that is shifted several nanoseconds late with respect to the lidar (flat-surface) system response.

Furthermore, the tail of the Raman pulse is, of course, lengthened by the depth or range extent reached by the exciting laser pulse. Accordingly, as the water column attenuation increases, one would expect this trailing portion of the pulse to be quite sensitive to this attenuation because of the exponential behavior of the effect. This is indeed what happens in this computational model as well as the experimental field data (cf. Figure 6.6). In Figure 6.7 the uppermost water attenuation curve and resulting Raman pulse correspond to the most transmissive water we have observed in the coastal region near Wallops Island (Hoge et al., 1980). In this case we assumed $\gamma_e = 2\gamma_r = 0.1$ m^{-1}. The other two lowermost curves in Figure 6.7 correspond to succeedingly less transmissive water, having $\gamma_e = 0.6$ and 1.0, respectively. Note that progressively more water column attenuation yields a lower peak amplitude and a shift of this peak

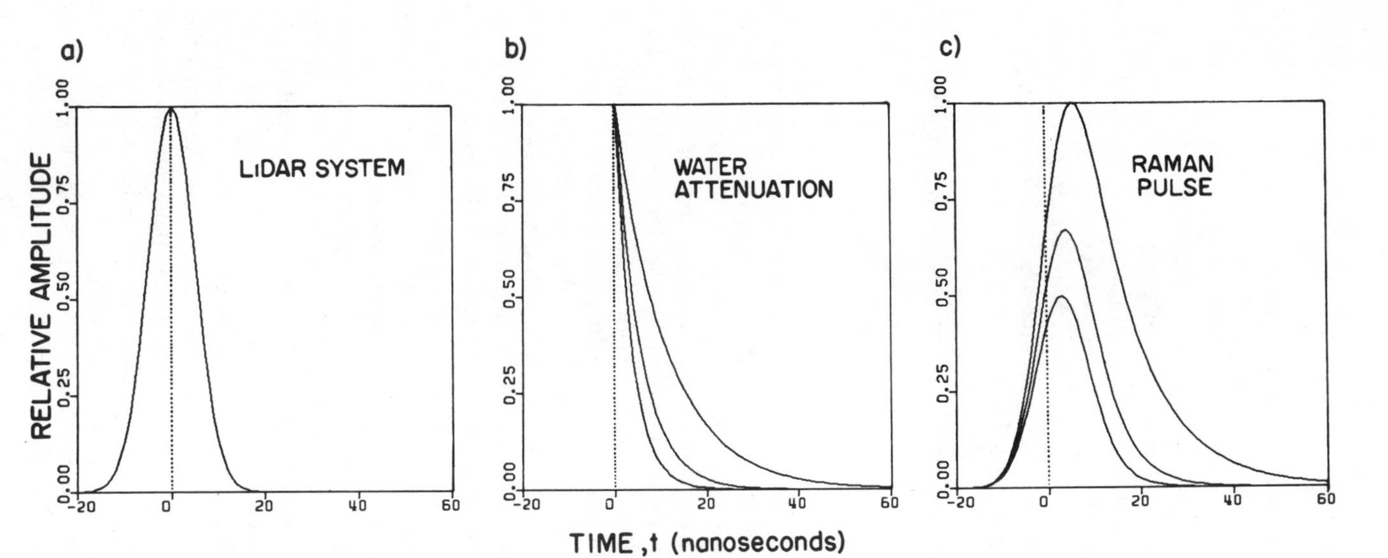

Figure 6.7. (*a*) Assumed Gaussian lidar system response function and (*b*) exponential water attenuation function. (*c*) Backscattered Raman pulse $R(t)$ calculated from convolution of Gaussian lidar system response, Gaussian sea surface height and slope probability density, and exponential water column response (Hoge and Swift, 1983c).

to shorter time and thus shallower water depths. This shift of the peak of the Raman pulse (as well as other attendant changes in pulse shape) would continue only until the Raman pulse becomes essentially equivalent to the lidar system response. At this point, of course, the amplitude of the Raman pulse (but not the on-wavelength Fresnel backscatter) tends to zero since no water molecules can be accessed due to the higher turbidity. One can easily see that the quantitative measurement of the effective attenuation coefficient is ultimately limited by the lidar system response. In our lidar system the principal contribution to the response is the transmitted pulse width. Therefore, if one wishes to conduct measurements in highly turbid water bodies, much shorter laser pulselengths than used in our initial equipment will be required. However, as Kung and Itzkan (1976) point out, the water column decay τ need only be measured at one position along the flight line where the water is sufficiently clear to yield a measurement for AOFSCE purposes. Then, if the system does not drift during the completion of the line, the τ at a more turbid region of the test site can be obtained by simple ratio with the total integrated Raman signal. Thus, it appears that oil spills can be classified according to their spectral conversion efficiency by using the water Raman signal for absolute calibration.

As was noted earlier, Equation (6.7) is obtained from Equation (6.4). The γ's are generally called the *effective attenuation coefficients* since it is not always known whether (a) the beam attenuation coefficient (frequently denoted by α or c, the latter not to be confused with the velocity of light) or (b) the diffuse attenuation coefficient is required. Confusion prevailed for several years until Gordon (1982) gave an interpretation for lidar elastic or on-wavelength backscattered signals. He used Monte Carlo techniques to solve the radiative transfer equation and found that the backscattered power is a decaying exponential function of time over the time interval required for the photons to travel four attenuation lengths through the water. Perhaps more importantly, it was also found that the effective attenuation coefficient of this exponential decay is strongly dependent on the parameters of the lidar system as well as on the optical properties of the water. The ratio of the radius of the spot on the sea surface viewed by the lidar receiver optics to the mean free path of photons in the water is the significant parameter. For values of this parameter near zero, the decay is determined by the beam attenuation coefficient, while for values greater than $\sim$5–6, the decay is given by the diffuse attenuation coefficient. Between these two extremes Gordon (1982) determined that the interpretation of the effective attenuation coefficient requires, essentially, complete knowledge of the inherent optical properties of the water: the beam attenuation coefficient and the volume scattering function.

As shall be seen later in the discussion of LIF of chlorophyll and its normalization by the water Raman signal, results essentially similar to Gordon's were found by Poole and Esaias (1982). Thus, it would seem that an oil classification theory is available and that no fundamental obstacles exist that would block its airborne field application. Of course, much more field work needs to be performed on a variety of oil types and under a variety of sea surface and water column conditions.

Once an oil type has been identified (or already known a priori), it is possible to measure the thickness of an oil film. From the results of Kung and Itzkan (1976) the thickness is given by

$$d = -\frac{1}{\kappa_e + \kappa_r} \ln\left(\frac{R'}{R}\right) \tag{6.10}$$

where K_e and K_r are the extinction coefficients of the oil at the laser and Raman backscatter wavelengths, respectively. Consistent with Figure 6.4, R' is the Raman signal obtained while over the oil whereas R is the Raman signal from an uncontaminated segment of the water column outside the slick region.

Figure 6.8 shows several channels of a typical laser fluorosensor output during overflight of an oil slick. These data were taken over a Murban-type oil by the NASA AOL in the scan mode (Hoge and Swift, 1980). Similar data were obtained during the same field experiments by the CCRS fluorosensor in a nonscanning mode (O'Neil et al., 1980). Notice that the Raman backscatter signal in channel 2 (and to a lesser degree, channel 1) depresses when the water column is increasingly obscured by the oil slick. Concurrently, the fluorescence emission in other channels (e.g., channels 3, 4, 5, and 12) increases. The 337-nm laser pulse is attenuated by the oil film immediately before entering the water column. Likewise, the isotropically emitted water Raman at 381 nm is attenuated upon its return passage through the slick. This attenuation is respectively accounted for in the model by the extinction coefficients κ_e and κ_r [Eq. (6.10)]. Upon complete transversal of the slick, the Raman signal that exceeds the *Gelbstoff* background returns to its prior value R. The modulation appearing on the fluorescence and the Raman signals in Figure 6.8 is caused by the AOL conical scanner sweeping the laser beam into and out of the oil slick. Knowing the off-nadir pointing angle of the scanner and the azimuth encoder angle, one can readily calculate the relative position of each LIF spectrum and resulting thickness. Thus, a three-dimensional map can be produced as the result of a single pass over the slick. A typical oil film thickness map is given in Figure 6.9. During this passage across the slick the scanning system mapped the rightmost edge of the slick. The thickness gradient as well as the volume is readily calculated from this data. The percentage of the thickness is given in the included table. Much more field work needs to be done to further validate the theoretical and field experimental techniques discussed here.

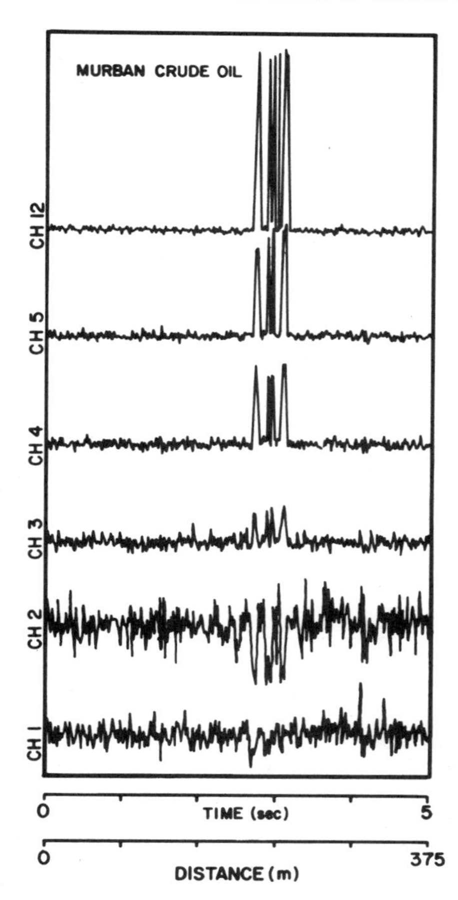

Figure 6.8. Return signal amplitudes recorded in fluorosensor spectral channels 1, 2, 3, 4, 5, and 12 during overflight along right edge of Murban crude oil slick. Conical scanner operation gives rise to 5-Hz amplitude modulation in all Raman and fluorescence (channels 3, 4, 5, 12) channels (Hoge and Swift, 1980).

In coastal and/or harbor regions, the water bodies are likely to be rather turbid. Then, the laser may not penetrate the water column to sufficient depth to produce a usable water Raman signal. Under these conditions one can resort to the oil fluorescence backscatter method of film thickness measurements. Hoge (1983) used the theoretical model developed by Horvath et al. (1971) and later modifed by Kung and Itzkan (1976) to calculate the thickness from the oil fluorescence backscatter. In Figure 6.4*a*, curve *A* shows the background and oil fluorescence, the seawater Raman, and some of the quantities to be discussed for the situation where the airborne lidar is positioned over the oil slick. Curve *B* of Figure 6.4*a* is an analogous situation for a lidar over seawater only. The notation used herein is primarily as found in Kung and Itzkan (1976) and Hoge

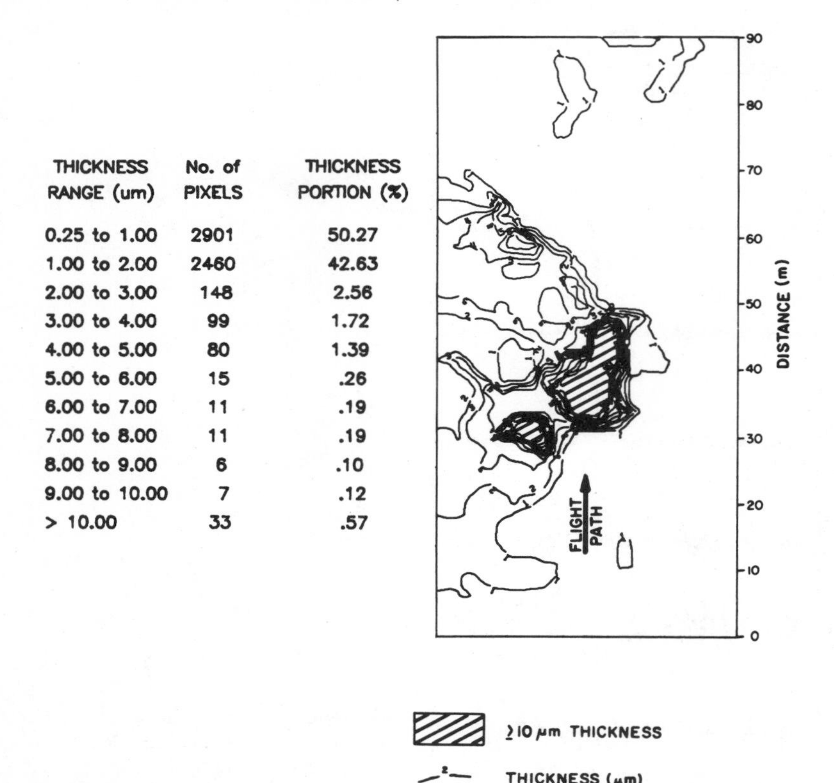

THICKNESS RANGE (um)	No. of PIXELS	THICKNESS PORTION (%)
0.25 to 1.00	2901	50.27
1.00 to 2.00	2460	42.63
2.00 to 3.00	148	2.56
3.00 to 4.00	99	1.72
4.00 to 5.00	80	1.39
5.00 to 6.00	15	.26
6.00 to 7.00	11	.19
7.00 to 8.00	11	.19
8.00 to 9.00	6	.10
9.00 to 10.00	7	.12
> 10.00	33	.57

Figure 6.9. Oil film thickness contour plot produced from water Raman backscatter data. Conical scanner azimuth angle, slant range, off-nadir angle, and aircraft velocity allowed determination of spatial position of thickness values (Hoge and Swift, 1980).

(1983). While the aircraft is over the slick, the return signal S_i in the ith channel of the fluorosensor can be expressed as

$$K_i = \eta_i P_0 \Big\{ 1 - \exp\big[-(\kappa_e + \kappa_i) d \big] \Big\} + \zeta_i P_0 \exp\big[-(\kappa_e + \kappa_i) d \big] + \delta_{ir} \psi P_0 \exp\big[-(\kappa_e + \kappa_i) d \big] \tag{6.11}$$

where η_i = fluorescence conversion efficiency in ith wavelength channel for optically thick oil film

P_0 = incident laser power

κ_e, κ_i = extinction coefficient of oil at laser excitation wavelength and at any ith wavelength channel, respectively

d = thickness of oil

ζ_i = fluorescent conversion efficiency of an optically thick seawater column exclusive of water Raman

ψ = Raman conversion efficiency of seawater

δ_{ir} = delta function to select Raman channel r

When the aircraft is outside or beyond the slick, $d = 0$, and the return signals are defined as J_i:

$$J_i = \zeta_i P_0 + \delta_{ir} \psi P_0 \tag{6.12}$$

Assume now that the water mass outside the slick is the same as present beneath the oil film. Evaluate Equations (6.11) and (6.12) at the same fluorescent wavelength $\lambda_i = \lambda_f \neq \lambda_r$ so that $i = f \neq r$. Combining the two resulting equations and solving for the thickness d yields

$$d = -\frac{1}{\kappa_e + \kappa_f} \ln \left(\frac{\kappa_{f,\infty} - \kappa_f}{\kappa_{f,\infty} - J_f} \right) \tag{6.13}$$

where

$$\mathrm{K}_{f,\infty} = \eta_f P_0 \tag{6.14}$$

Here $S_{f,\infty}$ is interpreted as the fluorescence signal intensity for an optically thick film at wavelength λ_f. The reason for this definition can be more clearly understood by evaluating Equation (6.11) at λ_f as d approaches large values. Thus, from Equation (6.13), to obtain the oil film thickness, one must have (a) the extinction coefficients of the oil at the laser excitation and fluorescence wavelengths, (b) the fluorescence level from an optically thick portion of the oil, (c) the fluorescence level of the film portion whose thickness is to be measured, and finally (d) the background fluorescence of naturally occurring materials J_f. The quantities described in (b), (c), and (d) can be obtained from the airborne lidar signals, as we shall see. The extinction coefficients κ_e and κ_f must be known a priori or obtained from a sample of oil immediately upon completion of the experiment. Equation (6.13) tells us that the background fluorescence J_f must only be subtracted from the infinite thickness fluorescence level before normalizing the residual fluorescence $\mathrm{K}_{f,\infty} - \mathrm{K}_f$. The $J_f = 0$ is a physical condition that occurs in deep-ocean areas where little organic material fluorescence is found (Hoge and Swift, 1981a; Leonard et al., 1979). In these types of water the fluorescence depth technique may offer little advantage over the water Raman suppression method. The reason for this is that the water Raman backscatter is rather easily obtained since the attenuation of open ocean water is minimum. Ocean regions would, however, offer an excellent location to compare the Raman suppression and fluorescence depth thickness measurement techniques.

The $K_{f,\infty}$, K_f, and J_f components are illustrated in Figure 6.10 for a hypothetical nonscanning flight line passage over a wind-driven oil slick having an optically thick nontransparent head and a thinner tail. Using the definitions in Figure 6.10, Equation (6.14) may be written as

$$d = -\frac{1}{\kappa_e + \kappa_f} \ln \frac{F'}{F} \tag{6.15}$$

This illustrates more clearly that the thickness is related to the residual fluorescent component while over the slick F', normalized by the peak or background corrected, saturated fluorescence component obtained from the optically thick portions of the slick, F. Equation (6.15) may be used to calculate the oil film thickness of the spills by using the ratio of the fluorescence from the thinner regions F' to that in the optically thick area F. The extinction coefficients κ_e and κ_f for each oil may be separately measured in the laboratory using fresh oil samples (Visser, 1979; Hoge and Kincaid, 1980; Hoge, 1982).Most probably, the main practical problem with this fluorescence depth technique is finding and/or defining the peak, "saturated" fluorescence level, F, in an optically thick region of slick, as shown in Figure 6.10.

Future oil spill experiment design should be concerned with utilizing an oil that (a) has minimum coagulation properties, (b) is sufficiently lightweight to

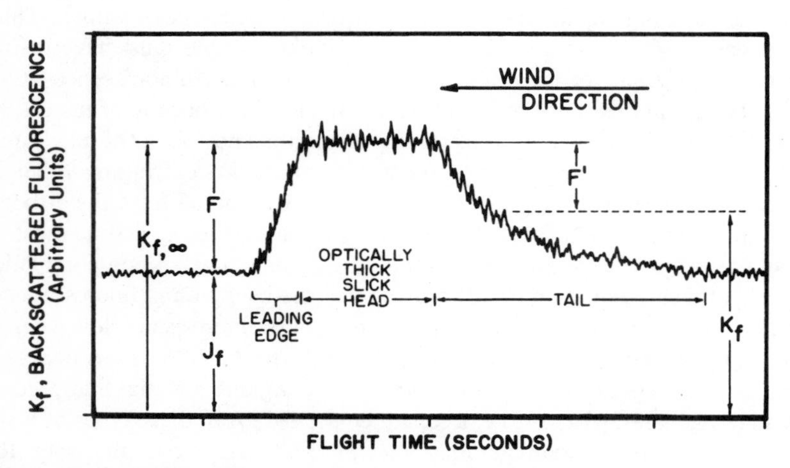

Figure 6.10. Hypothetical backscattered fluorescence signature expected during lidar overflight of wind-driven, optically thick oil slick (Hoge, 1983).

spread rapidly and be deployed in light wind conditions, and (c) contains a sufficiently large amount of oil to form a large optically thick target for airborne sensor investigation.

6.2.1.5 Natural and Man-Made Monomolecular Films

It should be pointed out that there are no doubt some physical and/or physico-chemical properties of oil spills on seawater that are not adequately modeled by the Kung and Itzkan theoretical development. For example, all the seawater Raman backscatter suppression modeled by R' in Figure 6.4 and experimentally demonstrated in Figure 6.8 is apparently not due to the κ_e and κ_r optical extinction of the oil film. There are experimental indications that nearly transparent, monomolecular films deployed on the sea surface can similarly suppress the Raman backscatter and the *Gelbstoff* fluorescence (Huhnerfuss et al., 1985). Thus, to some extent the wave-damping properties of oils can be expected to suppress the Raman and *Gelbstoff* fluorescence in addition to the $\kappa_e + \kappa_r$ optical attenuation normally observed during oil film overflights.

A typical oleyl monomolecular film (Huhnerfuss et al., 1978, 1981, 1983a,b) on the ocean will suppress the airborne laser-induced Raman (and dissolved organic fluorescence) signals, as shown in Figure 6.11 (Huhnerfuss et al., 1985). The oleyl alcohol film thickness is on the order of 3–30 nm, and for all practical purposes the laser-induced fluorescence and the optical absorption or extinction for such films is extraordinarily small and undetectable from aircraft altitude. Thus, the gravity waves and capillary wave structure are modified by the film. Such physical changes provide vivid contrast with respect to the surrounding unaltered sea surface and are therefore readily seen by the naked eye (Huhnerfuss and Garrett, 1981; Brockmann et al., 1982). Likewise, these physical surface changes give rise to a change in the wave height and slope probability densities, which determine the laser beam entry and ultimate exit of the Raman scatter and constituents fluorescence from the water column. Note that the oleyl films are only detectable in a differential or relative sense with respect to the surrounding water. If the entire ocean surface were covered with such a slick, it could only be detected in an absolute sense, a task not easily accomplished with the absolute calibration difficulty of today's lidar systems.

The modification of oceanic lidar signals by man-made monomolecular slicks serves as a warning that naturally occurring slicks (Garrett, 1983) can give a similar modification of lidar signals from the ocean surface and water column. Few locations in the world's oceans are devoid of these perturbing natural slicks. Accordingly, more effective means must be found to quantitatively study monomolecular slicks as well as deal with their effects on lidar measurements. The absolute calibration of oceanic lidars will probably help. The concurrent operation of other sensors measuring surface backscatter and emission (micro-

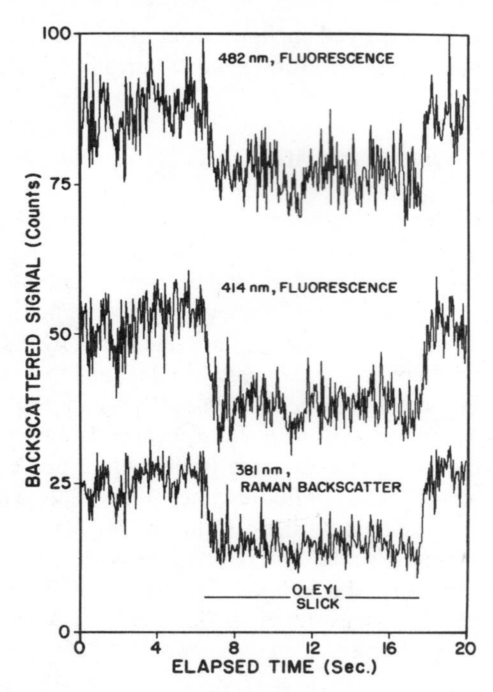

Figure 6.11. Profiles of 337.1-nm laser-induced fluorescence of dissolved organic material and water Raman backscatter obtained during lidar overflight of oleyl alcohol monomolecular film. For ease of comparison 30 and 60 count biases have been added to 414- and 482-nm profiles, respectively, in order to separate plots physically within single figure (Huhnerfuss et al., 1985).

wave scatterometers and radiometers) is recommended. Additionally, more man-made oleyl alcohol slick field experiments need to be executed both in the presence and absence of natural slicks.

The modification of lidar signals by monomolecular slicks prompts one to propose their use in the possible calibration or improvement of petroleum oil slick measurements. Under the singular assumption that a nonattenuating and nonfluorescent monomolecular film dampens ocean waves in the same way as the oil film, lidar data obtained over a monofilm could be used as the "neighboring sea" baseline against which the oil slick measurements are made. That is, instead of calibrating the Raman depression (and oil fluorescence) against the uncontaminated sea [such as done by the Kung and Itzkan (1976) technique], the comparison is performed against a monomolecular film deployed near the oil slick. Then, the Raman depression over the oil film will be attributed to the optical extinction $\kappa_e + \kappa_r$, as prescribed by the Kung–Itzkan model. Figure 6.12*a* shows a hypothetical or expected airborne lidar water Raman

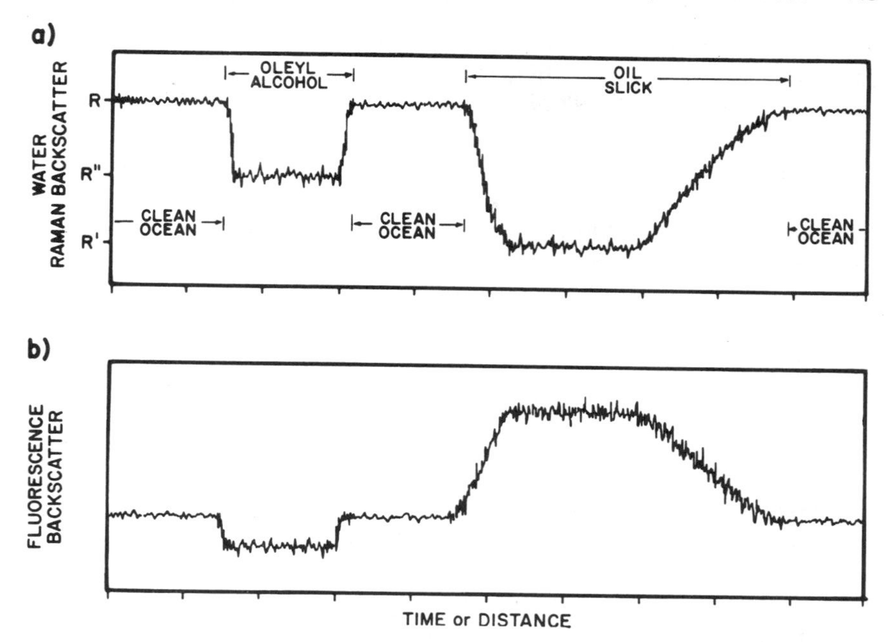

Figure 6.12. (*a*) Anticipated airborne laser-induced water Raman backscatter profile expected during sequential overflight of oleyl alcohol monomolecular film and oil slick both deployed on clean sea surface. (*b*) Airborne laser-induced fluorescence backscatter profile expected during overflight of oleyl alcohol film and oil slick.

backscatter profile that would be obtained over an oleyl alcohol monomolecular film deployed upon a "clean" sea surface near a target oil slick. This expected composite signature is assembled on the basis of separate airborne lidar observations of an oil slick (Hoge and Swift, 1980) and a monomolecular film (Huhnerfuss et al., 1985), examples of which were separately given in Figures 6.8 and 6.11. In the area covered by the monomolecular film, the seawater Raman backscatter decreases as a result of modifications of the laser beam and fluorescence backscatter spatial distribution caused by capillary and small gravity wave damping (further discussion is Section 6.2.3). After again crossing the clean sea, the oil slick is encountered, and the water Raman decreases because of both (a) the capillary and small gravity wave damping and (b) the attenuation of the laser and sea Raman backscatter by the slick material. Assuming the wave damping by the oil slick is the same as that caused by the monomolecular film, the effect described in (a) can be removed by using the monomolecular film measurements as the baseline against which the oil slick measurements are taken. Then, the optical attenuation in (b) as modeled by Kung and Itzkan remains.

Thus, rather than comparing the oil slick water Raman measurements to the surrounding clean, flat sea, as in the original Kung–Itzkan model, they are instead compared to the water Raman values obtained while over the monomolecular film. As a result of this experimental technique and data analysis procedure, the oil film thickness measurement, for example, would have less error contribution by wave-damping effects due to the oil.

Figure 6.12*b* shows a fluorescence backscatter profile that would be expected as the result of these same experimental conditions. Here, the *Gelbstoff* fluorescence is suppressed by the oleyl alcohol film, but this type of suppression is far exceeded by the oil fluorescence, as shown is the rightmost portion of Figure 6.12*b*.

The composite monomolecular film, oil slick, and seawater Raman signatures illustrated in Figure 6.12*a* are based on separately available airborne lidar data obtained to date (Figures 6.8 and 6.11). With this supporting experimental evidence, the Kung–Itzkan flat-sea model can be modified or redeveloped to incorporate the physical effects observed. Figure 6.13 (curve *C*) shows the expected suppression of the clean, open-sea lidar spectrum while over a monolayer. Also shown in Figure 6.13 is the expected spectrum (*B*) to be found over the clean sea. (Curve *B* is also given in Figure 6.4*a* for ease of compari-

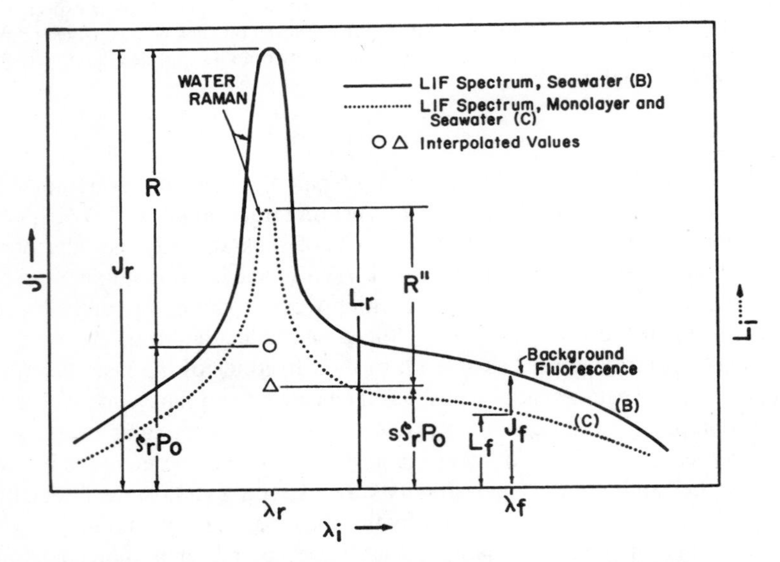

Figure 6.13. Theoretical airborne laser-induced emission spectrum expected while over a monomolecular film deployed on clean sea surface (curve *C*). Wave-damping effects coupled with lidar off-nadir observational angle can both lead to observed reduction of spectral backscatter emission relative to that found over clean sea (curve *B*).

son.) A suppression factor, $s = R''/R$, can be calculated from the data, such as given in Figure 6.11. This suppression factor is valid only for (a) sea surface conditions existing on a particular day and (b) lidar off-nadir viewing angle used (see Section 6.2.3 for further discussion). The channel-to-channel gains for the AOL were not well known during the time period of the monomolecular film experiment. For this reason also, the suppression factor shown herein may not be representative of values measured in figure experiments.

As before from the experimental data of Figure 6.11 (and as conceptually depicted in Figures 6.4*a*, 6.12, and 6.13), the integrated water column Raman found over the monomolecular film, R'', is related to the clean-sea Raman signal R by $R'' = sR$. Or, the airborne laser-induced water Raman signal obtained over the monomolecular film is some measurable fraction s of the water Raman signal obtained over the clean sea. Using procedures similar to those of Kung and Itzkan (1976), the flat-sea model can be renormalized against a monomolecular film (instead of the clean sea) to obtain

$$d = -\frac{1}{\kappa_e + \kappa_r} \ln \frac{R'}{sR} \tag{6.16}$$

which shows that the thickness is related to the Raman component while over the oil slick R' normalized by the suppressed Raman component over the monofilm, sR. The result in (6.16) is not totally unexpected. Furthermore, in the absence of a monomolecular film $s = 1$ and the expression reduces to the original Kung–Itzkan thickness Equation (6.10). Equation (6.16) may be used to calculate the film thickness of oil spills using the ratio of the water Raman over the oil to the water Raman signal observed over the monomolecular film.

Figure 6.14 illustrates the impact upon the lidar measurement and data system of including the wave-damping effects within the oil film thickness calculation [Eq. (6.16)]. For $s = 1$ no suppression has occurred, and the oil film thickness is as shown in the lowermost curve for a Murban-type oil. From Table 6.3, the extinction coefficients for Murban at the laser, Raman, and fluorescence wavelengths of 337.1 and 381 nm are 265 and 95 mm^{-1}, respectively (Hoge and Swift, 1980). Note that wave-damping effects immediately consume 40% of the dynamic range of the instrument for a suppression factor $s = 0.6$ (uppermost curve in Figure 6.14). Comparing Figure 6.11, also, one can see that the oil film optical extinction effects can now only start at ~600 counts for an assumed 10-bit (1024 digital count) lidar data system. Note further that the largest thickness error occurs at the thickest portion of the oil film and that this error can easily approach one order of magnitude. The smallest thickness errors due to wave-damping effects occur in the thinner portions of the slick. It thus appears that oil film thickness estimates could be significantly improved if

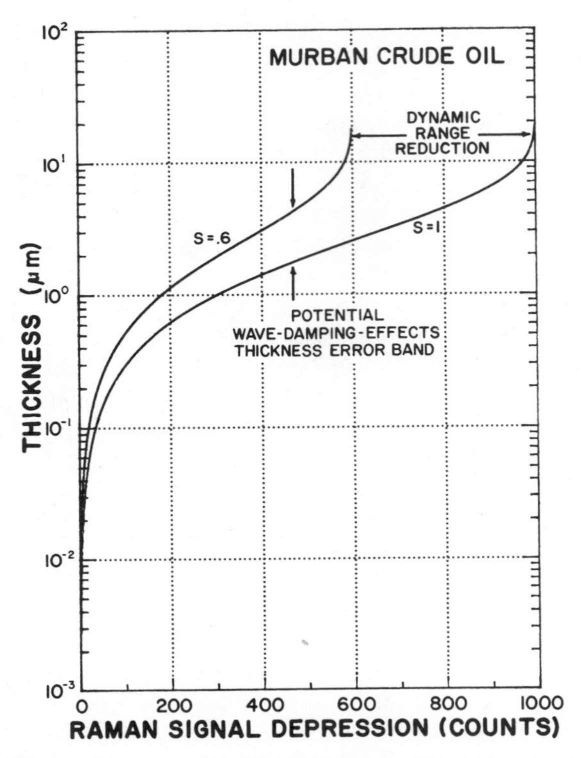

Figure 6.14. Dynamic range suppression of oceanic lidar data acquisition system caused by wave-damped signal reduction effects. Resulting potential errors in measurement of oil film thickness are nonlinear functions of water Raman signal depression.

monofilms are used for the baseline measurement to reduce wave-damping effects errors.

In Sections 6.2.1.4 and 6.2.1.5 we have devoted what would seem to be an inordinate amount of space to a subject that might by some be considered less important than the upcoming section on ocean volume measurements. However, an attempt has been made in these latter two sections to use the measurement of sea surface phenomena to illustrate the quantitative lengths to which lidar measurements may ultimately be extended. Lidar remote sensing is in its infancy. No doubt more quantitative models and field experiments can rightfully be expected in future years.

6.2.2 Measurements in Ocean Volume

This section will focus on the measurements in the volume below the ocean surface. While these volume measurements are affected by the state of the surface above it, we will forego the details of the effect of the sea surface waves on the volume measurements until Section 6.2.3.

6.2.2.1 Biological Measurements in Ocean Volume

The world's oceans comprise about three-quarters of Earth's surface and are responsible for approximately one-third of all primary production (Struum and Morgan, 1981). Primary production in marine waters is very closely correlated with chlorophyll *a* concentration (El-Sayed, 1970; Sorenzen, 1970). This is a key finding, not only because chlorophyll *a* is the principal ingredient in the photosynthetic process, but also because chlorophyll emits measurable fluorescence in virtually direct proportion to its concentration (Yentsch and Menzel, 1963). Thus, primary production can apparently be inferred by measurement of fluorescence. While this is a very simplified interpretation of a highly complex process, it forms the basis of remote sensing of chlorophyll and the ultimate goal of inferring primary productivity and total biomass in the ocean. However, the remote sensing of phytoplankton was assured only when it was found that the concentration could be evaluated from the in vivo chlorophyll fluorescence, that is, without extracting the pigment from the organism (Sorenzen, 1966).

The laboratory studies of Hickman and Moore (1970) and Friedman and Hickman (1971) strongly suggested the feasibility of chlorophyll measurements from remote platforms. Mumola and Kim (1972) sensed in vivo fluorescence of naturally occurring phytoplankton in the Chesapeake Bay from a pier-based platform. These experiments led to the airborne (helicopter) detection of chlorophyll (Kim, 1973). Mumola et al. (1973) suggested that more accurate chlorophyll measurements could be obtained if the four major color groups of algae (red, blue-green, green, and golden brown) were excited with a four-wavelength laser. The concept is discussed by Mumola et al. (1973) and by Browell (1977).

Without resorting to a maze of excitation and emission spectra, the reader can gain a general understanding of the laser-induced phytoplankton technique by referring to Figure 6.15. Adapted from Govindgee and Govindgee (1974), the lower portion of Figure 6.15 shows pigments important to the photosynthetic process. The arrows and annotation of the pigment blocks serves to label the dominant color of light absorbed as well as the energy transfer among (and spectral emission from) various major pigments. A specific and typical example of the fluorescence spectra observed during remote airborne excitation of phytoplankton with a high-power 532-nm blue-green pulsed laser (L) is shown in the upper portion of Figure 6.15. This spectrum was taken with the AOL during the Superflux experiments conducted in the Chesapeake Bay (Campbell and Thomas, 1981). The phycoerythrin and chlorophyll fluorescence emission band peaks are labeled as *P* and *C* and occur at ~580 and ~685, respectively. No carotenoid or phycocyanin fluorescence was observed. The water Raman back-

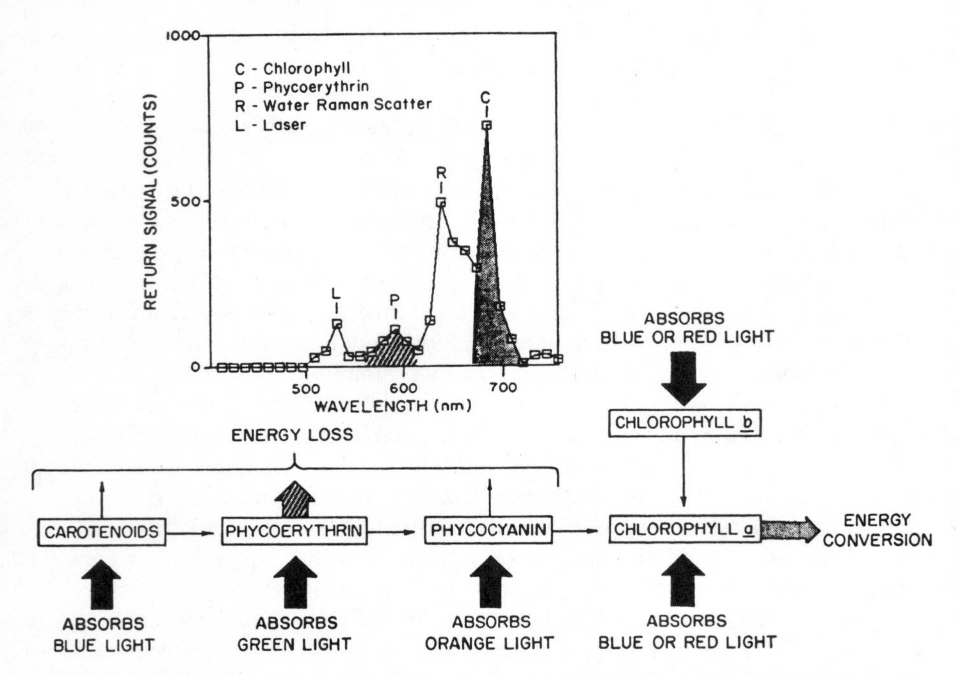

Figure 6.15. Model of energy absorption, transfer, and emission by and between principal pigments important to photosynthetic process (Govindgee and Govindgee, 1974; Browell, 1977). Inset: actual airborne 532-nm laser-induced spectrum of waterborne marine phytoplankton. This spectrum obtained in Atlantic Ocean near mouth of Chesapeake Bay (Hoge and Swift, 1981c).

scatter line is labeled *R*. The laser-induced Raman scatter is an inherent property of the water molecules in which the phytoplankton are growing and will always be produced even when no organisms are present. The water Raman can obscure (or cannot be distinguished from) phytoplankton fluorescence emissions that have narrow linewidths and occur at ~647 nm. The strongest water Raman emission is the OH^- stretch Stokes line and is situated at ~3350 cm^{-1} from the laser. Thus, the choice of laser wavelength is dictated not only by optimum excitation wavelengths of the target but also by the placement of the water Raman line to avoid obscuration of desired emission spectral bands from the waterborne target species. Selection of laser wavelength as well as the useful application of the water Raman emission for correction of spatial variability of water column attenuation properties is discussed subsequently in this section.

Concerning the excitation wavelength selection, the four previously mentioned wavelengths were chosen only for chlorophyll stimulation. It is now well known that the ~3350-cm^{-1} OH^- stretch water Raman backscatter can be easily detected remotely. Thus, an improper choice of the excitation wavelength could yield Raman interference in the 685-nm chlorophyll band. For example, one would not choose one of the excitation wavelengths to be ~557 nm since this would produce a Raman emission at 685 nm and would interfere simultaneously with the chlorophyll emission. As we shall discuss later, one would not choose an excitation wavelength at ~486 nm since this would similarly produce a Raman backscatter line at ~580 nm, the wavelength position of the phytoplankton phycoerythrin pigment fluorescence. Browell (1977) performed an analysis of lidar systems for phytoplankton studies and concluded among other things that the remote measurement of chlorophyll requires (a) optimum excitation wavelengths and (b) careful measurement of the water column attenuation coefficients. Regarding wavelength selection, a prototype laser was developed to stimulate chlorophyll at 454.4, 539.0, 598.7, and 617.8 nm and observe the resulting 685 nm fluorescence. This system was last operated in 1980 during the Superflux experiments in the Chesapeake Bay (Jarrett et al., 1981). The method used to handle the water column spectral attenuation coefficients will now be discussed in more detail.

The analysis of Browell (1977) pointed to the need for accurate knowledge of the water attenuation coefficients. Airborne field experiments subsequently demonstrated that the laser-induced water Raman backscatter could be used to effectively remove chlorophyll fluorescence signal variability due to horizontal spatial variations in water column attenuation. Bristow et al. (1981) reported the highly successful use of water Raman normalization of chlorophyll signals during flights over the fresh water of Lake Mead. Likewise, Hoge and Swift (1981b) reported similar successes during experiments in the marine waters of the North Sea, the Atlantic Ocean, and Chesapeake Bay estuary (Hoge and Swift, 1981c) and the northwestern Atlantic Ocean (Hoge and Swift, 1983b; Hoge and Swift, 1985). Figure 6.16*a* shows profiles of chlorophyll *a* fluorescence and water Raman backscatter along flight line 8 of a June 23, 1980, mission in the Chesapeake Bay. The effect of normalizing the chlorophyll fluorescence with the water Raman signal is seen by noting Figure 6.16*b*. The declining Raman signal gives a corresponding increase in the chlorophyll fluorescence. During four separate missions flown in June 1980 over the lower Chesapeake Bay and outflow into the Atlantic Ocean, the water-Raman-normalized chlorophyll *a* fluorescence yielded correlation coefficients of $r = 0.97$, 0.92, 0.81, and 0.82 in separate linear regressions against ship truth data. Without Raman normalization this level of agreement could not be obtained. The importance of applying Raman corrections to the various laser-induced fluorescence responses to correct for spatial variation in water column optical transmission properties cannot be overstressed since it is necessary for the precise recovery of even relative concentrations of various parameters.

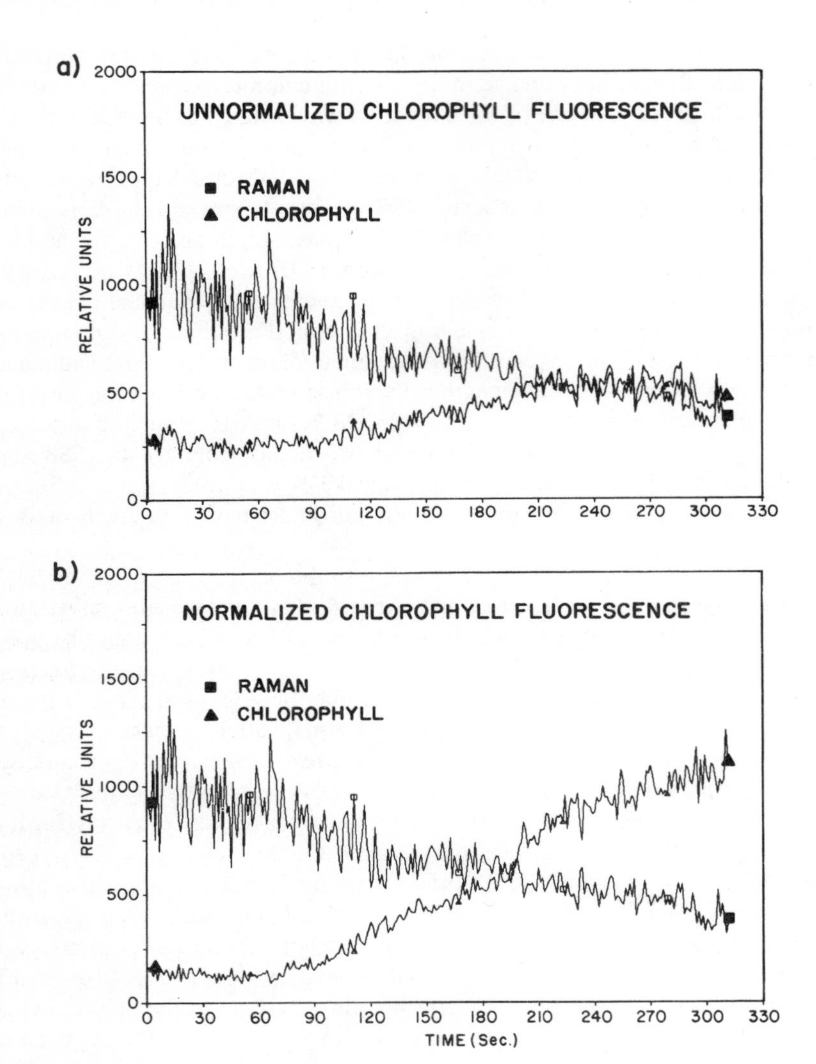

Figure 6.16. (*a*) Profiles of airborne laser-induced water Raman backscatter and unnormalized chlorophyll *a* fluorescence obtained during flight experiments in coastal waters of Atlantic Ocean. (*b*) Same profiles as in (*a*) except that chlorophyll *a* fluorescence corrected for spatial variability of water column attenuation by normalizing with water Raman backscatter signal strength (Hoge and Swift, 1981c).

The water Raman normalization technique has been succesfully demonstrated in lake, estuary, and marine bodies and is well established. The Monte Carlo simulation studies of Poole and Esaias (1982) showed that Raman normalization produces effectively linear response to chlorophyll concentration, particularly for excitation at 480 and 532 nm. Poole and Esaias (1983) suggested, however, that the Raman signal should be used cautiously, if at all, in linear algorithms to measure beam attenuation or irradiance (diffuse) attenuation coefficient in a quantitative sense. These latter results are in accord with those of Gordon (1982) and Poole and Esaias (1982).

Zimmerman et al. (1976) were apparently the first to observe the laser-induced fluorescence emission of in vivo phycoerythrin using a shipboard lidar remote sensing system. They conducted five stations in the marine waters of the Atlantic Ocean/Chesapeake Bay mouth. They found the 580-nm fluorescence peak and further discovered that its strength (as well as that of chlorophyll fluorescence) increased upon approaching the bay mouth. Celander et al. (1978) obtained the 337.1-nm laser-induced fluorescence emission spectra of six freshwater algae species. Using cultures of unknown concentration, they found spectral bands at 380 (water Raman), 460, 585, 665, and 725 nm. They did not identify the 460- and 585-nm peaks; however, the latter is probably phycoerythrin. Some of the species revealed bands centered near 515 nm and were also not identified. In general, the freshwater species displayed stronger blue-green fluorescence than marine algae.

The airborne LIF of phycoerythrin was observed in 1979 during experiments conducted in the North Sea and in the Chesapeake Bay (Hoge and Swift, 1981b). (A typical airborne LIF spectrum obtained by the AOL was previously given in the inset of Figure 6.15.) Airborne laser-induced phycoerythrin fluorescence was also observed in 1981 in the Nantucket Island–New York Bight region of the north Altantic Ocean (Hoge and Swift, 1982a). Furthermore, airborne laser observations of phycoerythrin have been reported during dedicated studies of Gulf Stream Warm Core Rings (Hoge and Swift, 1983b). Figure 6.17 shows the geographic location of WCR 82-B that was intensively studied in the spring of 1982. Figure 6.18*a* provides profiles of normalized chlorophyll and phycoerythrin fluorescence over this warm core ring (WCR) (Hoge and Swift, 1983b) during a mission flown on April 20, 1982. The water Raman shown in Figure 6.18*b* indicates that the optical transmission is higher in the inner portions of the ring. Likewise, a separate infrared sensor yielded data (Figure 6.18*c*) that verifies the core has a higher temperature than the boundaries. As before, a 532-nm pulsed, frequency-doubled Nd–YAG laser was used as the primary excitation source. The aircraft crossed the ring boundaries approximately 200 s from each end of this flight line. Elevated chlorophyll *a* and phycoerythrin fluorescence levels are found in this high-velocity boundary region that surrounds

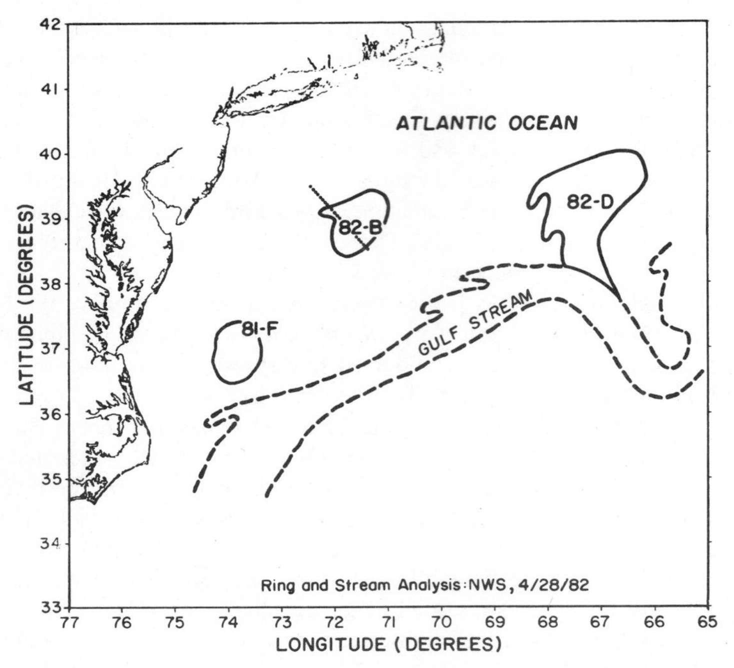

Figure 6.17. Location of Warm Core Ring 82-B at time of airborne experiment described in text. One of numerous airborne laser experiment flight lines shown across entire ring. Profiles of laser-induced chlorophyll *a* and phycoerythrin fluorescence as well as water Raman backscatter obtained in this particular flight line given in Figure 6.18 (Hoge and Swift, 1983b).

the ring. Figure 6.19 is a scatter plot of the chlorophyll and phycoerythrin segments labeled *A*, *B*, *C*, *D*, and *E* in Figure 6.18*a*. These scatter plots demonstrate the coherence of the chlorophyll and phycoerythrin fluorescence observed. The correlation coefficients between phycoerythrin and chlorophyll *a* fluorescence varied respectively from 0.85 to 0.653, as shown in the Figure 6.19. The proportion of phycoerythrin to chlorophyll *a* fluorescence is seen to be higher within the ring than in either of the boundary regions. The significance of these findings is at present not well understood. It has been suggested that these phycoerythrin–chlorophyll *a* ratio variations are caused by speciation changes. The ship data from these experiments has not yet allowed this to be proven conclusively. To illustrate the potential importance of airborne lidar fluorosensing to the study of mesoscale oceanographic features, Figure 6.20 shows a chlorophyll fluorescence contour map generated from data such as previously shown in Figure 6.18. Similar maps of phycoerythrin fluorescence and Raman scatter can also be generated. Such features can be mapped in 2–3 h depending

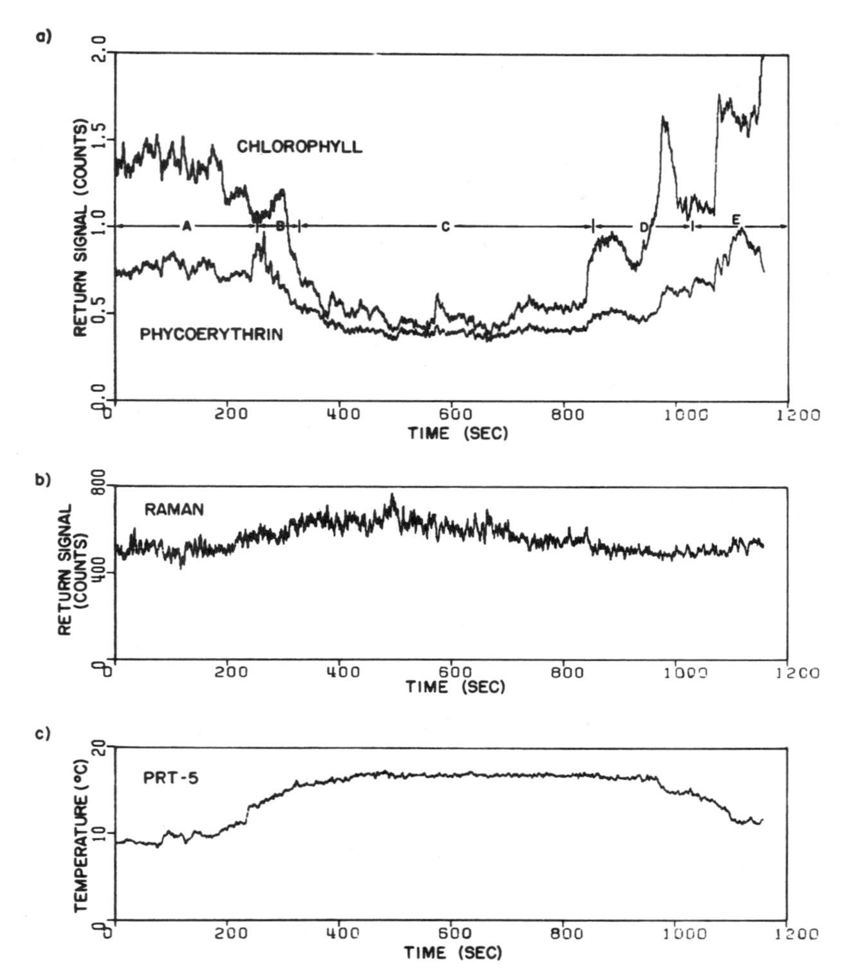

Figure 6.18. Profiles of airborne laser-induced (*a*) chlorophyll *a* and phycoerythrin fluorescence, (*b*) water Raman backscatter, and (*c*) infrared radiometer surface temperature obtained during flight across Warm Core Ring 82-B (Hoge and Swift, 1983b).

on the type of aircraft platform used and the flight pattern configuration. Such mapping can also be executed during overcast conditions or during other such times that a satellite coverage is not available (including darkness).

A shipboard laser system for in vivo stimulation of the water constituents is suggested to complement the usual chlorophyll extraction, flow-through fluorescence, and taxanomic observations. Such shipboard laser experiments should avoid water-pumping systems and use the forward end of the ship to reduce

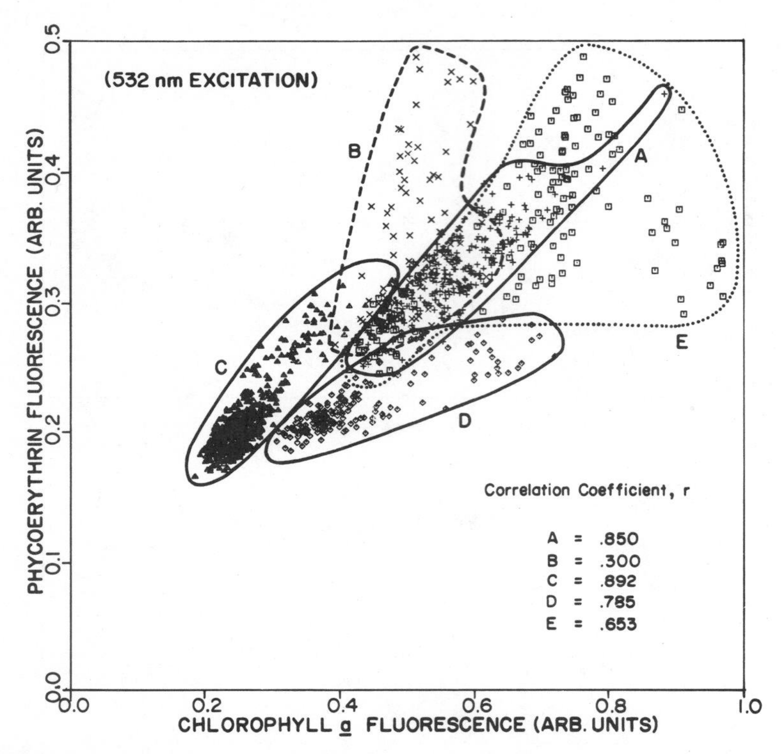

Figure 6.19. Scatter diagram of chlorophyll *a* and phycoerythrin fluorescence occurring within several discrete sections of flight line across Warm Core Ring 82-B. Portions *A–E* correspond to those regions of flight line in Figure 6.18 (Hoge and Swift, 1983b).

possible disturbances that may rupture fragile organisms. This is particularly true if the experiments are performed in estuarine regions such as the Chesapeake Bay. Exton et al. (1983) showed that Chesapeake Bay estuarine samples are dominated by cryptophytes whose phycoerythrin fluorescence was significantly modified by a pumping circulation system that apparently ruptured the cell walls of the organism. On the other hand, Exton found that coastal marine samples found near the bay mouth contained mostly cyanobacteria whose stronger cell walls are not easily disrupted. The phycoerythrin fluorescence from these plankton were not noticeably affected by pump agitation. Apparently, a shipboard laser system could assist scientists to some degree with speciation. Houghton et al. (1983) showed that a pier-based lidar system having 7.5 nm resolution was capable of detecting and distinguishing differences between the phycoerythrin spectral from cyanobacteria (peak at 576 nm) and cryptophytes

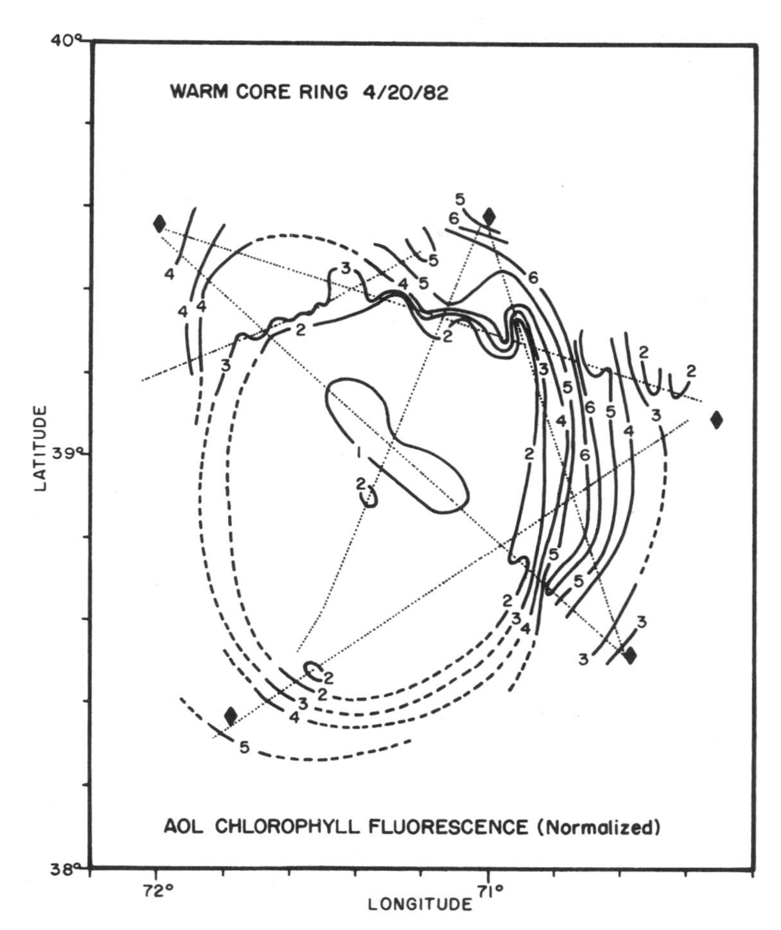

Figure 6.20. Contour plot of chlorophyll *a* fluorescence produced from data such as presented in Figure 6.18. Similar contour plots of phycoerythrin fluorescence can also be generated in similar fashion (Hoge and Swift, 1985).

(peak at 586 nm), both of which can be major components of nearshore/estuarine nanophytoplankton communities. Coupled with recent flow cytometers (Yentsch and Yentsch, 1984), advances such as these promise improved water surface truth for validating airborne laser results.

The reader should note that the phycoerythrin plots herein are not expressed in milligrams per cubic meter such as is normally done for chlorophyll. The reason is that phycoerythrin extraction and measurement techniques are not as well developed as for chlorophyll. A method for the extraction and quantification of phycoerythrin from algae has been recently described (Stewart, 1982). The effectiveness of this reported technique is not known.

Because of the specificity of the lidar data, it is very useful for validating passive sensors operating alongside the lidar. However, a single, combined active–passive sensor would allow still more applications than two separated instruments on the aircraft. A combined active–passive ocean color instrument allows reasonably rapid calibration of both modes, studies of the fluorescence-to-pigment ratio and its possible link to phytoplankton fluorescence yield, validation of passive ocean color in-water algorithms and instrumentation as discussed above, and studies of the effect of sea state on passive ocean color measurements. Such studies will probably be reported in future years since the NASA AOL fluorosensor has recently been modified to operate concurrently as a multichannel passive ocean color sensor.

Future space-borne fluorosensors should carry a comparison passive color sensor to provide for comparative data in the same footprint. Then global validation and comparison techniques can be rapidly assessed.

6.2.2.2 Chemical Constituents in Ocean Volume

Probably the first airborne lidar detection of chemicals in the environment was reported by O'Neil et al. (1973). They used a CW helium–cadmium laser at 442 nm to excite fluorescence of the effluent from a paper mill having high concentrations of lignin sulfonates. Bristow (1978) also reported the airborne laser-induced fluorescence of the outflow of this paper mill by using a 337-nm pulsed nitrogen laser. The main features of the work by Bristow (1978) has also been summarized by Measures (1984). The lignin sulfonate laser-induced emission spectrum is quite similar to that of naturally occuring dissolved organic carbon (DOC), and this could easily preclude the direct, unambiguous measurement of pulp mill effluent. It is well known that the fluorescence spectrum of natural waters is quite broad and typically devoid of identifiable spectral characteristics. However, the total signal under the fluorescence emission curve is more informative and is apparently proportional to organic loads in the water bodies (Measures et al., 1975). But emission amplitudes are difficult to calibrate absolutely, and Bristow et al. (1973) suggested that the OH^- stretch water Raman be used as a reference to calibrate the return signal.

Numerous investigators have found that the water-Raman-normalized fluorescence of natural water is to some degree correlated to the total organic content of the water (Measures et al., 1975; Zimmerman and Bandy, 1975; Bristow and Nielsen, 1981). These findings are in general agreement with earlier worker's conclusions (Kalle, 1966; Christman and Ghasseni, 1966; Sylvia et al., 1974). After a very detailed laboratory feasibility study Bristow and Nielson

(1981) concluded that the Raman-normalized fluorescence emission induced in surface waters by UV radiation could be used to provide a unique airborne remote sensing capability for monitoring the concentration of DOC. At present we know of no airborne lidar field experiments that have been conducted expressly to measure surface water DOC. However, Hoge and Swift (1982b) used the organic material to map fronts in the German Bight. Water Raman normalization was recommended by Bristow and Nielsen (1981), but for the high-DOC levels found in coastal regions, the UV laser-induced Raman may be difficult to obtain reliably.

While the laser-induced fluorescence emission signal of naturally occurring organic materials (*Gelbstoff*) is highly correlated to the amount of DOC (Bristow and Nielsen, 1981), the use of this signal in remote measurements is complicated by the fact that a major DOC constituent (fulvic acid) has been shown to be affected by both pH and by trace metals (Saar and Weber, 1980). The fluorescence of DOC is quenched, or reduced in intensity, by complexation with metals (Ryan and Weber, 1982). Furthermore, the quenching is wavelength dependent (Vodacek and Philpot, 1985), with the long-wavelength components in general being quenched more than the shorter wavelengths as the metal concentration increases. If one has thorough knowledge of the pH of the body of water (e.g., a lake), the concentration of metal may possibly be inferred from the wavelength-dependent quenching. Since iron causes more quenching than aluminum and copper (Vodacek and Philpot, 1985), the dominant metal present in the lake may also be required information. Knowledge of the amount of metals complexed, or associated with organics, is important since they are highly toxic in the free state. However, if the acidity of the lake increases, through acid rain or other causes, the complexing can be disrupted, the metal freed, and accordingly the toxicity will rise. Although low pH is also toxic, elevated levels of, for example, aluminum appear to have serious effects (Cronan and Schofield, 1979).

Vertucci (1985) found that the differences in the laser-induced fluorescence spectra of natural lake waters were sufficient to be used in regression models to predict sample DOC, pH, and aluminum. The prediction accuracy of DOC and aluminum was improved when the data were analyzed in separate pH classes. The studies were performed to determine the potential of airborne laser fluorosensing for remotely detecting and measuring lake acidification.

Yentsch (1973) was perhaps the first to suggest in a laser remote sensing forum that salinity might be inferred from the detection of the organic material or yellow substances or *Gelbstoff*. He showed that a high degree of correlation exists between yellow substances and salinity where the Merrimack River mixes with Gulf of Maine waters. Figure 6.21 shows a plot of laboratory laser-induced fluorescence of Savannah River/Atlantic Ocean water samples as a function of

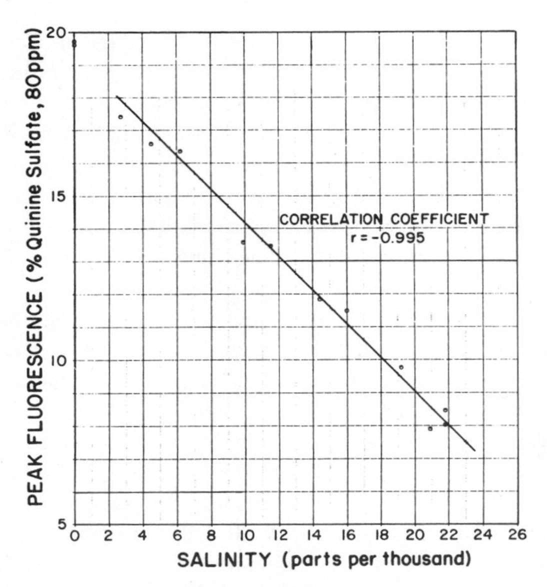

Figure 6.21. Variation of 337.1-nm laser-induced fluorescence of dissolved yellow substances and salinity. Samples gathered in vicinity of Savannah River mouth.

salinity. These latter data were provided by James Yungel of the AOL support laboratory. The reference signal for the organic fluorescence in the latter figure was an 80-ppm solution of quinine sulfate. To use this latter technique for remote measurement, the organic material must be conserved, that is, there must be no unknown sources or sinks. There are many conflicting opinions as to the source of the dissolved organic material and whether the sinks (if any) can be properly accounted for. Rather than discussing the merits in detail here, we suggest that the technique be tried in an airborne field environment selected to minimize the source/sink issues.

Airborne lidar observation of the sulfate ion Raman scatter may provide a salinity measurement method and a way to avoid the above problems. Houghton (1973) suggested that salinity might be measured remotely by detection of the Raman spectrum of the sulfate ion, SO_4^{2-}. The technique is based on the constancy of composition, which says that the ratio iof SO_4^{2-} to chlorine is a constant regardless of salinity. He recommended standardizing the SO_4^{2-} by using the OH^- stretch water Raman signal. Relative to remote sensing, several serious problems immediately present themselves. First, the SO_4^{2-} Raman is weaker than the OH^- stretch water Raman and only has a peak strength that is approximately equal to the H_2O bending mode Raman line. Second, the spectral width of the SO_4^{2-} Raman is significantly less than the OH^- stretch water Raman depending on the laser linewidth. This suggests the need for a high-resolution spectrofluorometer to allow quantitative utilization of the spectra. Third, the sulfate ion Raman rides upon, and can easily be obscured by, the naturally occurring dissolved and particulate organic fluorescence background. Ironically, the organic fluorescence interference is strongest in coastal regions where salinity maps would be most useful. The degree to which these problems affect the airborne remote detection and utilization can only be assessed by actual field experimentation. As Houghton (1973) points out, it will probably be necessary to utilize a multichannel fluorometer. This of course would allow the use of interpolation techniques (Kung and Itzkan, 1976) to accommodate or correct for the organic fluorescence.

Perhaps the first controlled chemicals to be remotely sensed by lidar were organic dyes dumped into the Atlantic Ocean for tracing purposes. Hoge and Swift (1981a) reported the absolute measurement of tracer dye concentration in the ocean using airborne laser-induced water Raman backscatter as the calibration reference signal. O'Neil et al. (1980) overflew the same dye spill and used the dye fluorescence data to conclude that such Rhodamine WT fluorescence could easily be distinguished from oil spill fluorescence. Hoge and Swift (1981a) used the scanning AOL fluorosensor mode to map the concentration down to about 2 ppb.

Figure 6.22 shows a typical lidar fluorosensor response to a dye spill in the open ocean. About 18.9 L of 20% solution of Rhodamine WT were used in these experiments. During this pass the AOL instrument was conically scanning with a full cone angle of 10°. The 5-Hz scan frequency can be seen in the fluorescent data. During this particular pass, the aircraft was principally flown over the leftmost edge of the irregularly shaped dye plume. Thus, the conical scanner during portions of the flight line caused the lidar system to view the plume only during approximately each half-cycle of the scan frequency of 5 Hz. Accordingly, the output response of each LIF channel is amplitude modulated by the successive views into (and away from) the dye plume. During this pass more than 20 incursions into and out of the dye were observed. Shown are channels 2, 11, 20, 22, and 24. Channel 11 is located between the Raman and dye emissions and is included to illustrate the typical fluorosensor response or behavior in spectral regions that have little or no physical participation. Channel 2 is rather closely centered upon the Raman band and contains the major portion of this spectral emission. Note that the Raman signal is approximately 10 times stronger than the dye emission. Also, no discernible Raman suppression is found in channel 2 as the dye is encountered and its fluorescence emission rises as shown in channels 20, 22, and 24. The absence of Raman suppression should be compared to the very noticeable Raman suppression over oil spills (Figure

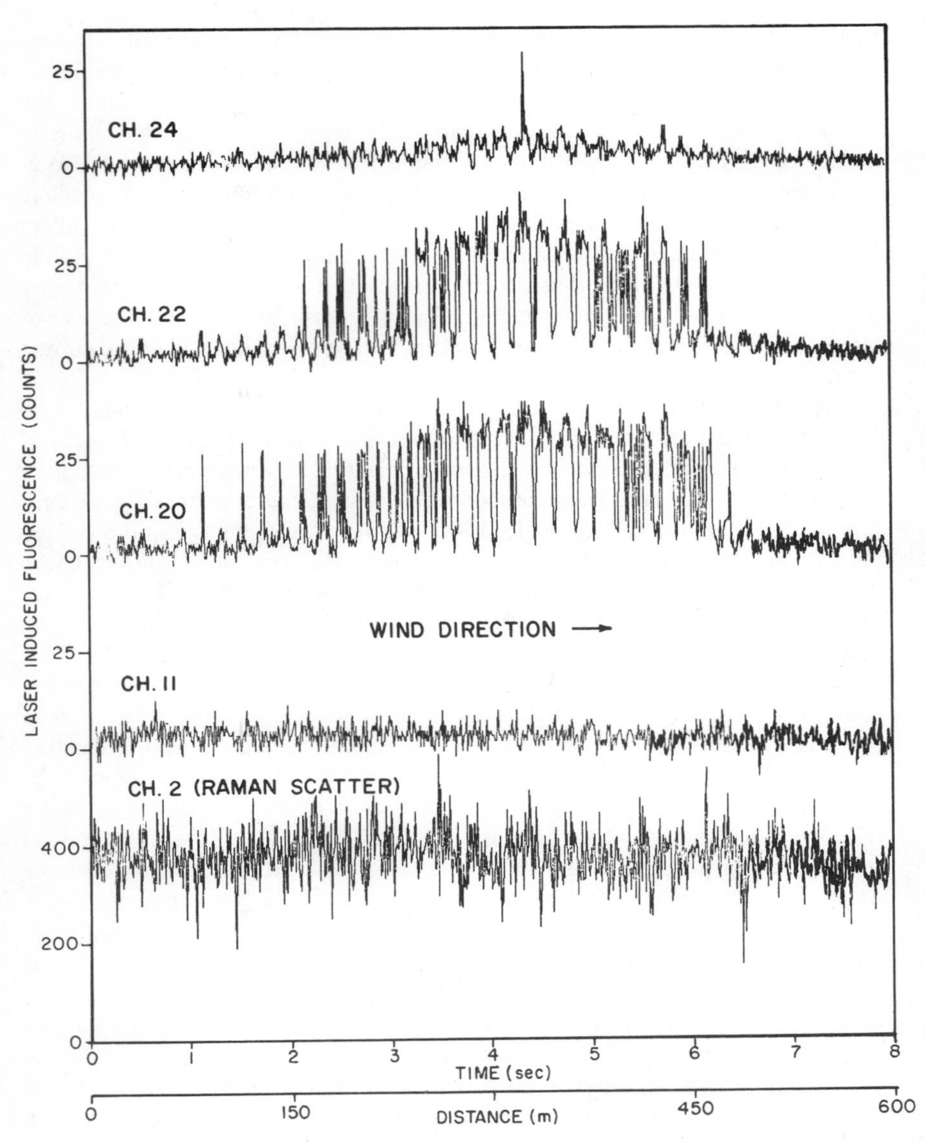

Figure 6.22. Airborne laser-induced spectral emissions from Rhodamine WT tracer dye dispersed in ocean. Channel 2 contained principal portion of water Raman backscatter centered at ~381 nm. Channel 11 included to show typical fluorosensor response in spectral regions not containing Raman and dye emissions. Absence of any Raman suppression in channel 2 shows that dye contributes little attenuation at 337.1 and 381.0 nm. The 5-Hz amplitude modulation of fluorescence is result of conical scanning of system during overflight. All channels contiguous and have bandwidths of 11.25 nm (Hoge and Swift, 1981a).

6.8). These experimentally observed, nondepressed Raman signal data completely agreed with the Canada Center for Remote Sensing laser fluorosensor results obtained over the same dye plume from another aircraft flying at 450 m altitude at the same time (O'Neil et al., 1980). Our interpretation of these channel 2 Raman data is that the dye produces no discernible change in the optical properties of the water column so that the extinction sum $\gamma_e + \gamma_r$ is essentially constant throughout the entire test region both inside and outside the plume. Or viewed another way, the absorption by the dye molecules contributes very little to the attenuation or extinction in the water column at the laser emission wavelength and the Raman wavelength (Hoge and Swift, 1981a). The nondepression of the water Raman signal over this chemical dye spill also suggests, but does not prove, that the Raman depressions found, for example, in deep-ocean phenomenon (such as the boundary of a warm core ring, Section 6.2.2.1) are due to scatter and not absorption.

Figure 6.23 is a dye concentration contour map produced from the corrected channel 20 data shown in Figure 6.22. Azimuthal scan angle data, slant range to the sea surface, as well as roll-and-pitch aircraft altitudes are recorded simultaneously with the spectral channel values to provide a mechanism for projecting the horizontal position of each laser pulse. The roll-and-pitch parameters are provided by a Litton LTN-51 inertial navigation system (INS). Note that the image is in three sections, which must be laid end to end to yield the entire dye plume edge. The conical scan pattern of small dots represents the actual spatial position of the radiated laser pulse intercepting the ocean surface. The contours were produced from the fluorescent intensity values observed at each dot position. Good agreement is found for spatial positions where the conical scan pattern intersects. The contours are labeled in arbitrary units of digital counts. The 10-digital-count contour actually corresponds to a concentration of 3.33 ppb, a 20-digital-count contour to 6.66 ppb, and so on.

A scanning airborne laser fluorosensor has the potential to provide relatively high density dye concentration maps without the need for extensive in situ sampling (Hoge and Swift, 1981a). The technique utilized here requires laboratory calibration with dilute dye concentrations that span the range of the ultimate field dye concentrations. The water Raman intensity must be obtained in both the laboratory and airborne data. The ratio of the fluorescence to Raman signals in the airborne data is compared to the same ratio found in the prepared laboratory dilutions to determine the field concentrations. One of the major errors in this technique involves the water column attenuation coefficient at the laser, Raman, and dye fluorescence emission wavelengths. It can be shown (Hoge and Swift, 1981a), however, that the errors produced by attenuation in the water column are quite manageable, particularly if the water mass can be accurately categorized by the Jerlov (1976) classifications.

Other dye-tracing experiments have been conducted (O'Neill et al., 1981)

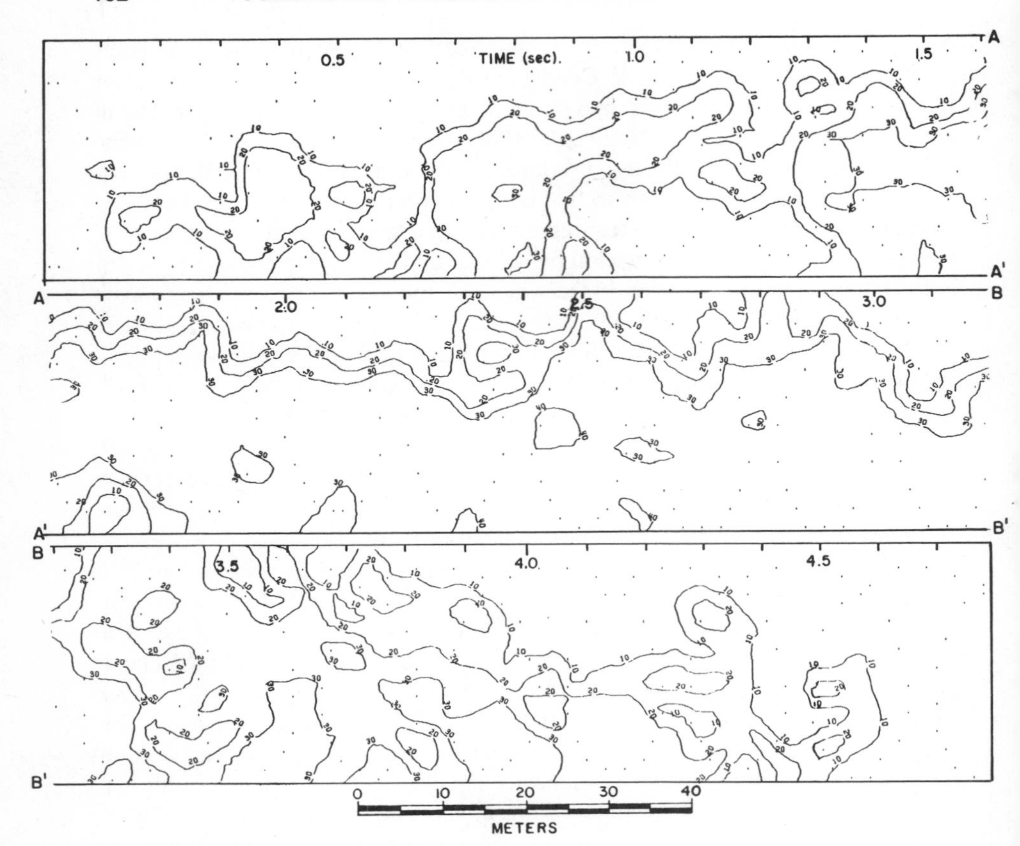

Figure 6.23. Dye concentration contour plot or image produced from channel 20 results in Figure 6.22 together with simultaneously recorded azimuthal scan angle data. Entire contiguous image may be seen in usual configuration by joining three segments of *AA'* and *BB'*, respectively. Roll and pitch of aircraft corrected using data from Litton LTN-51 inertial navigation system. Alternatively, roll and pitch may also be obtained directly from scanning lidar slant range data. Contours labeled as 10 digital counts correspond to 3.33 ppb dye concentration, 20 digital counts to 6.66 ppb, etc. (Hoge and Swift, 1981a).

with the AOL reconfigured to use a neon laser (λ_L = 540 nm). The fluorescence of Rhodamine 6G dye was still observable at a concentration of 4 ppb even though the pulse energy was about 3% of that available when N_2 was used in the same laser cavity. Using a wavelength in the green part of the spectrum, which is better able to penetrate the water than the UV wavelengths, allows the sensor to measure the dye concentration over a longer water column. With a time-resolved laser fluorosensor and a target substance with a sufficiently short fluorescence lifetime, the concentration as a function of depth could be measured, greatly enhancing the interpretation of such dispersion experiments. The above tracer dye remote measurement technique is of course applicable to the measurement of near-shore and estuarine calculation.

Another application of the tracer dye technique suggested by Hickman (1973) was the remote measurement of temperature. In this method a calibrated mixture of two dyes whose fluorescence emission intensities are temperature sensitive (such as Rhodamine B and Eosin Y) would be deployed together in the water column. The plume would then be excited by a single laser to produce an emission peak at two different wavelengths characteristic of each dye. The ratio of the strength of the two peaks would yield the relative temperature. The absolute temperature calibration would be obtained by in situ sampling. To improve the accuracy of the results, the data should be corrected for salinity effects. Conversely, if the water temperature is already known, salinity variability could be inferred.

6.2.2.3 Ocean and Estuarine Front Detection

When two or more different water masses meet, zones having a high degree of spatial and temporal variability are frequently formed. For example, when lower salinity estuaries and rivers empty into oceanic waters, sites of locally intense mixing and small-scale interaction occur at the boundary of such water masses. The interactions of the water masses may be further enhanced by inherent bathymetric features (shoals, banks, etc.), differing regional wave and current velocities, and temperature–density gradients (Bowman and Esaias, 1977). Conventional techniques cannot easily and economically provide the spatial resolution and wide-area coverage needed to adequately sample within the appropriate length and time scales present in these regions. Airborne lidars have the ability to map large areas over reasonably synoptic scales and with high spatial resolution. Since fronts or zones of convergence tend to concentrate biological and chemical waterborne constituents (Sick et al., 1978), the laser-induced spectral fluorescence characteristics of lidars appear particularly appealing. Furthermore, the on-wavelength backscatter from the surface and water column potentially allows measurement of (a) local surface wave structure modification or gradients and (b) entrained/resuspended sediments and/or particulates, respectively. This section will address the modest and limited application efforts to date.

Hoge and Swift (1983c) have shown that the depth-resolved water Raman backscatter waveform can reveal turbidity cell structure in the ocean. This cell structure is obtained by selecting and plotting two separate depth-resolved chan-

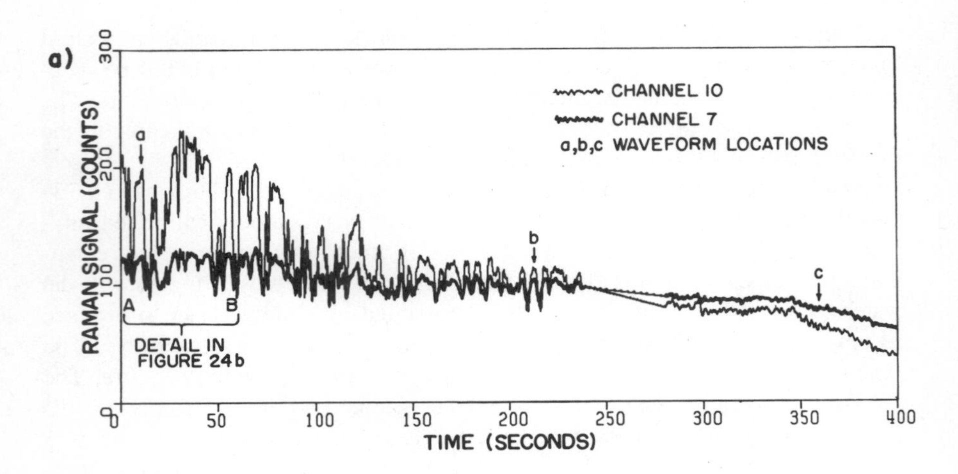

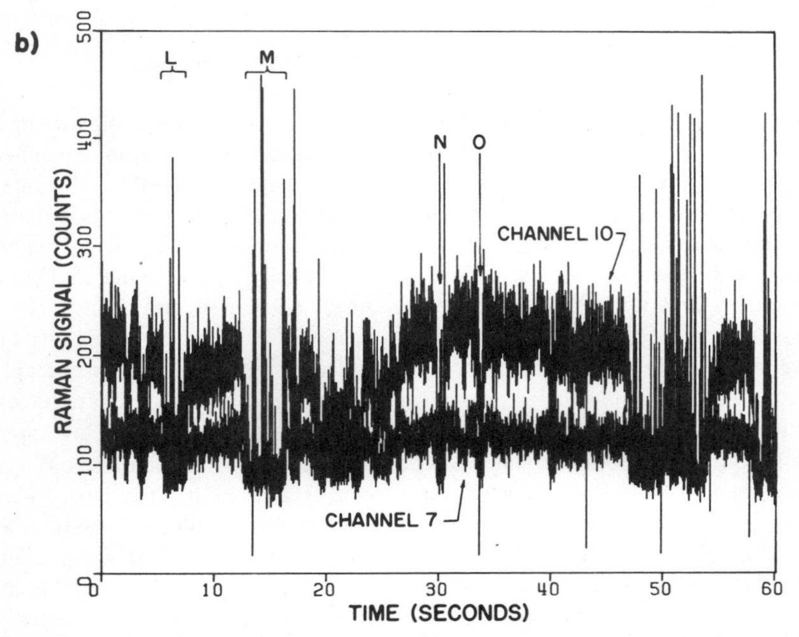

Figure 6.24. (*a*) Profile of channels 7 and 10 from airborne depth-resolved water Raman backscatter waveforms corresponding, respectively, to measurements made at depths of 1.8 and 2.5 m. Arrows show locations of depth-resolved waveforms of Figure 6.6. Data gap indicated by straight lines connecting profiles between 240 and 280 s (Hoge and Swift, 1983c). (*b*) Detailed plot of initial flight line portion labeled *A–B* in Figure 6.24. Presence of numerous clear water cells of various sizes readily apparent. Samples of sharp declines in water Raman backscatter can be found at positions *L*, *M*, *N*, and *O* (Hoge and Swift, 1983c).

nels from the 381-nm Raman backscatter waveform (channels 7 and 10 from Figure 6.6 of Section 6.2.1.4). Figure 6.24*a* vividly shows the gross cell structure in the coastal waters off the Virginia coast during a 40-km westward flight from clearer water to the more turbid surf water near the beach. Figure 6.24*b* shows the detailed cell structure observed in the clearer water portions of the flight line. The waveform shape is modeled rather well, as discussed in Section 6.2.1.4 and in Figure 6.7.

The water column or depth-integrated fluorescence and Raman signals may also be used in broader scale field experiments to map the location and relative strength of fronts (Hoge and Swift, 1982b). During the 1979 Maritime Remote Sensing Experiment (MARSEN) the AOL participated in seven flights in the North Sea. Several of these missions were used to map the German Bight where the Elbe River and Weser River deposit large quantities of fresh water laden with dissolved and particulate materials. The naturally occurring *Gelbstoff* or dissolved organic material was assumed to be a conservative tracer during the 3-h time period of the mapping exercise. The AOL instrumentation was configured as listed in Table 6.1. Figures 6.25*a*,*b* show the water Raman and unnormalized *Gelbstoff* fluorescence from a typical flight line. Notice, for the most part, that changes in the Raman profile are mirrored in the *Gelbstoff* profile. Although the sense of movement is generally the same in both profile traces, the degree of the corresponding movement is quite variable. This can be seen by comparing the segments designated 1, 2, 3, 4, and 5 in Figure 6.25. Normalization of the fluorescence signal with the Raman backscatter signal significantly reduces the effect of the changes in water optical transmission upon the final, corrected fluorescence signal. (Normalization is also discussed in Section 6.2.2.1.) Figure 6.25*c* shows the profile of the normalized *Gelbstoff* fluorescence. Notice that the amplitude remains relatively smooth over most of the flight line except at the points labeled 1–5. Both (a) the relative strength or localized slopes within the normalized *Gelbstoff* profile (in counts / km) and (b) the sign of slope were then used to delineate the fronts. Front location maps were generated from these slope/sign variations. It was determined that the spatial distribution of the fronts found by this airborne lidar technique was in very reasonable agreement with previously published oceanographic ship results.

Mesoscale eddy or warm core ring mapping by airborne lasers was discussed in Section 6.2.2.1. Essentially the same 532-nm laser and AOL configuration were used during the Superflux experiments (Campbell and Thomas, 1981) to map chlorophyll and phycoerythrin fluorescence as well as Raman backscatter (Hoge and Swift, 1981c). Sarabun (1981) used the AOL multichannel fluorosensing lidar data in a cluster analysis to define water mass boundaries and/or water types. He was able to show that the data allowed the generation of maps having physically plausible water mass delineations.

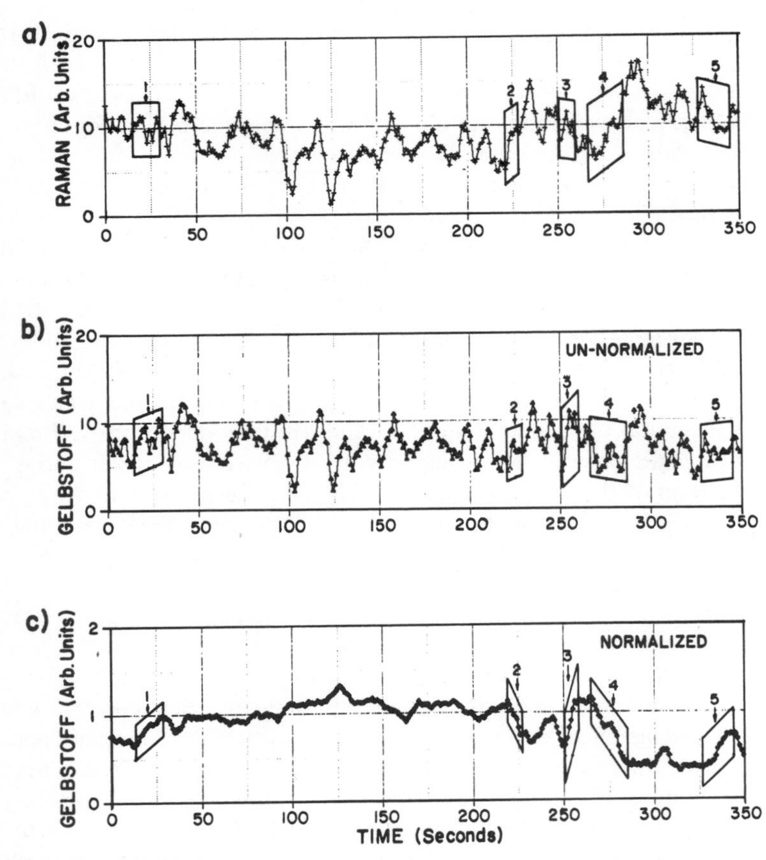

Figure 6.25. Profiles of (*a*) depth-integrated water Raman backscatter and (*b*) *Gelbstoff* fluorescence obtained in German Bight region of North Sea. (*c*) *Gelbstoff* fluorescence normalized by water Raman backscatter from (*a*). Normalization of *Gelbstoff* fluorescence with Raman backscatter significantly reduces effect of water optical transmission and allows easier identification of fronts whose locations can be found, e.g., at 1, 2, 3, etc. (Hoge and Swift, 1982b).

6.2.2.4 Optical Properties of Ocean Water Column

The world's oceans cover almost 75% of Earth's surface and are responsible for an estimated 30–45% of all primary production (Strumm and Morgan, 1981; Riley and Chester, 1971). It is a significant challenge to measure and monitor this vast and important surface area. Furthermore, it is an even bigger challenge to measure the upper layer and/or volume beneath this surface. Presently, measurements in the volume are being accomplished for the most part by airborne and satellite passive remote sensing. However, better physical, chemical, and biological measurements in the volume strongly depend on the models used to invert the data. Accordingly, these models can probably be improved by inclusion of accurate measurements of the optical properties of the water column by airborne lidars. Furthermore, a potentially important and concurrently related application of airborne oceanic lidars is the validation and/or calibration of passive, optical satellite measurements in a region of the ocean chosen to be generally representative of vast areas of the globe. Then, the lidar-calibrated satellite sensor can be used outside and beyond the validation area to obtain improved data to allow, for example, better global productivity estimates. Of course, even in the absence of passive ocean sensor requirements for lidar optical properties and validation/calibration data, airborne lidar measurements of the upper ocean are important in their own right since large areas of the ocean can be surveyed rapidly, even under cover of darkness if necessary.

As we have seen, the accuracy of biological, chemical, and physical water column measurements by lidar is strongly driven by the optical properties of the water column. The volumetrically derived water Raman backscatter has of course been used to significant advantage to correct for the attenuation properties of the water column. Beyond this, however, the quantitative measurement of the inherent optical properties of the ocean by remote sensing is difficult and still requires some in situ measurements.

As pointed out earlier in this chapter, Gordon (1982) investigated the effects of multiple scattering on the interpretation of oceanic lidar data. Using Monte Carlo techniques to solve the radiative transfer equation, he found that, after removal of geometric loss factors, the on-wavelength or elastic backscattered power is a decaying exponential function of time over the time interval required for photons to travel four attenuation lengths through the water. The effective attenuation coefficient of the exponential decay was found to be dependent on the ratio of the radius of the spot on the sea surface viewed by the lidar to the mean free path of the photons in water. When this parameter is nearly zero, the exponential decay is determined by the beam attenuation coefficient, whereas for values greater than about 6, the exponential decay asymptotically approaches, or is given by, the attenuation coefficient for downwelling irradiance (diffuse attenuation coefficient). For intermediate values of the parameter the interpretation of the effective attenuation coefficient apparently requires complete knowledge of the inherent optical properties of the water: the beam attenuation coefficient and the volume scattering function. Gordon's results were essentially corroborated in the Monte Carlo modeling of water Raman normalization by Poole and Esaias (1982).

Independently, Koerber and Phillips (1982) used Monte Carlo techniques to determine that accurate estimates of the absorption coefficient a can be obtained from measurements of the decay rate of the lidar backscatter envelope provided that an adequately large FOV is used. The technique uses an airborne lidar system to transmit a laser pulse vertically downward into the water column and subsequently records the temporal, on-wavelength or elastic backscattered signal. They also concluded that the backscatter signal can be used to estimate the scattering properties of seawater, for example, the volume scattering function at 180°, $\beta(\pi)$. However, this $\beta(\pi)$ must be converted to values of the scattering coefficient b by using a nonlinear empirical relationship that in turn depends on the in situ measured volume scattering functions of Petzold (1972). The obvious weakness of this method lies in the fact that the empirical relationship derived for a particular region may not be valid in all water masses. This, then, is roughly equivalent to the suggestion by Gordon (1982) that for intermediate values of the ratio of the lidar-viewed spot radius to the photon mean free path, a complete knowledge of the inherent optical properties is required.

Subsequently, Phillips et al. (1984) described an analytical technique for the independent measurement of two basic, intrinsic optical properties of seawater: the absorption and scattering coefficients. The technique relied on the previous Monte Carlo simulations that provided the necessary data to establish the feasibility of an airborne laser technique (Koerber and Phillips, 1982) and an earlier theoretical study (Phillips and Koerber, 1984) that utilized the Monte Carlo results to establish the limits of validity of the analytic model. Using a full Monte Carlo simulation (as opposed to semianalytic Monte Carlo techniques), Phillips et al. (1984) showed that in the asymptotic limit of large FOV (and corresponding sea surface spot diameter) the effective attenuation coefficient γ is the absorption coefficient a rather than the diffuse attenuation as found by Gordon (1982). They argue the finding that $\gamma \rightarrow a$ in the large-FOV limit becomes evident when the independent variable in the model is considered to be time rather than depth. Then, every photon received at a given time has obviously spent the same amount of time in the water and has also traveled the same total distance in the water, irrespective of how many times it has been scattered. Since the only process reducing the number of such photons is absorption, the effective attenuation coefficient equals the absorption coefficient. Phillips et al. (1984) further conclude that, apart from geometric factors, the amplitude of the backscatter signal at the water surface and its time decay in the water column allow both the absorption and scattering coefficients of water to be determined independently. In situ measurements were used to confirm the validity of the technique. Clearly, however, more modeling and field measurements were recommended to resolve the findings of Phillips et al. (1984) relative to those of Gordon (1982) and Poole and Esaias (1982).

In spite of efforts thus far it seems that (1) assumptions about the water mass being overflown must be invoked or, (2) as Gordon (1982) suggests, the complete knowledge of the inherent optical properties (beam attenuation coefficient and volume scattering function) of the water must be known if the ratio of the observed ocean spot to photon mean free path is from ~0.5 to 5 or 6.

6.2.3 Effect of Surface Waves on Measurements in Ocean Volume

To this point we have considered lidar measurements of the sea surface and materials deployed upon it (Section 6.2.1). Also, the measurement of constituents within the ocean volume has been discussed separately (Section 6.2.2). However, there is strong experimental evidence that variability in sea surface elevation modulates signals obtained from within the water column. The height and slope statistical distributions of the sea surface waves are known to significantly affect the in-water, spatial or volumetric distribution of an incident laser beam upon its initial, vertically downward passage through the surface (Bobb et al., 1978). Wave slope statistics have actually been obtained from optical radiance measurements made below the sea surface using the sun as the light source (Stotts and Karp, 1982). Furthermore, lasers immersed in the ocean have been used to measure wave slopes by directing the beam vertically upward into a four-quadrant detector (Schau, 1978; Tang, 1981; Tang and Shemdin, 1983). Thus, by induction, the uncollimated, spherically isotropic fluorescence and Raman backscatter from the water column must be spatially redistributed into the observational hemisphere above the sea surface. Clearly, then, there would be some manifestation of the distributed sea surface embedded within the volumetrically generated return signals. The sea surface perturbation of water-column-derived lidar backscatter signals was first reported by Hoge and Swift (1983a). This relationship is illustrated in Figure 6.26*a*, which shows profiles of the depth-integrated Raman signal and the sea surface elevation, both plotted as a function of time for an expanded segment of pass 2 over the LaRosa slick previously discussed in Section 6.2.1.4. The sea surface elevation was determined by subtracting the mean slant range measurement acquired by the lidar altimeter subsystem from individual slant range measurements and inverting the results such that the peaks represent wave crests and conversely the depressions represent wave troughs. [Low-frequency vertical aircraft motion (Krabill and Martin, 1983; Hoge et al., 1984) has not been removed from the data; however, this motion is of no consequence in the following discussion.] The correspondence (outside the oil slicks themselves) of higher Raman signal to wave crests and lower Raman signal to wave troughs persisted throughout the oil spill data sets. This correspondence or correlation also exists in depth-resolved water Raman backscatter signals, as seen in Figure 6.26*b*. Note particularly the apparent phase shift between the Raman signal peak and the wave crests. While there are a few exceptions, variations in Raman backscatter signal from the ocean volume appear to lag corresponding variations in the sea surface record. This

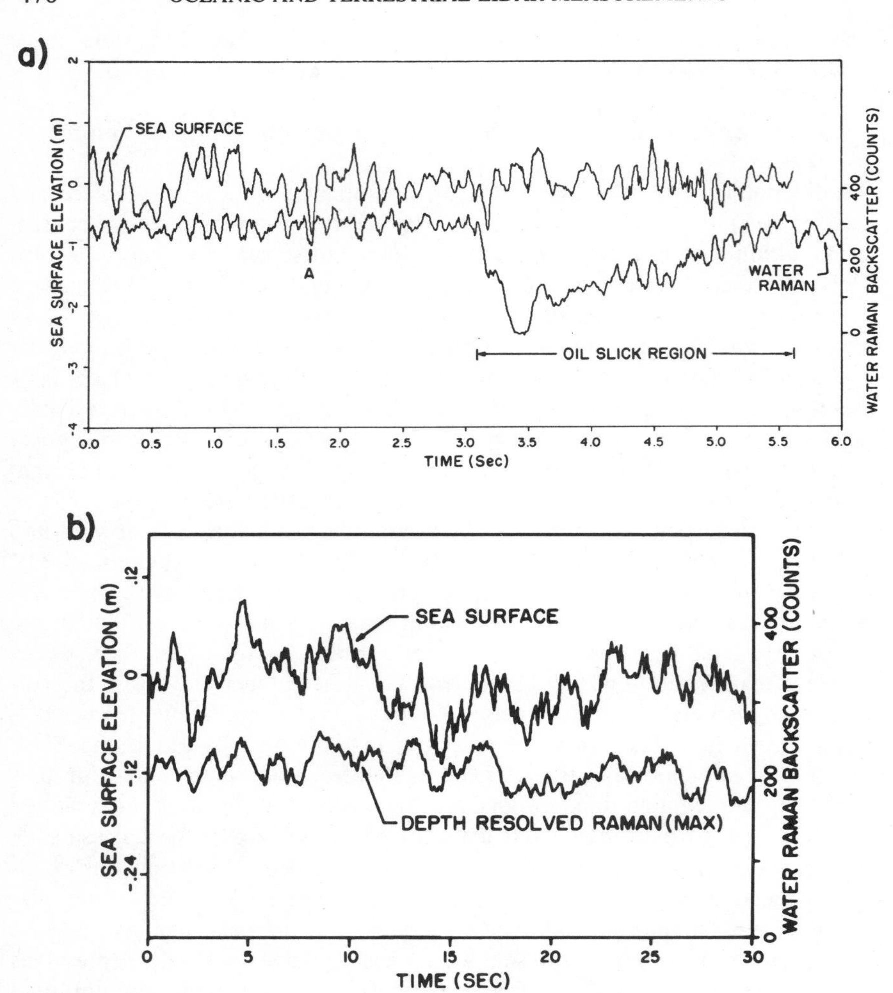

Figure 6.26. (*a*) High correlation of depth-integrated water Raman backscatter with sea surface elevation. Correlation persists over all but thickest portion of oil slick. Spatial phase lag between two measurements is particularly recognizable at position *A*. (*b*) Correlation of depth-resolved water Raman backscatter with sea surface elevation. Sea surface derived from range data obtained from leading edge of depth-resolved water Raman return pulse (Hoge and Swift, 1983a).

is quite evident in Figure 6.26*a* at the position labeled *A*, where the local Raman depression lags the wave trough. This apparent phase shift is caused by the off-nadir pointing angle of the lidar system relative to mean sea level (Hoge et al., 1984). It should also be noted that the detection of the ocean surface is dominated by Fresnel reflection with some Mie backscatter components from the ocean volume while the Raman backscattered signal is strictly volumetric.

The relationship between the sea surface wave structure and the resulting modulation of volumetrically derived fluorescence and Raman backscatter is not well understood and is not easily modeled in closed form. Gelhaar (1982) was the first to show by computer simulation and some limited closed-form analysis that lidar fluorescence signals from the volume are modulated by the sea surface. Gehlaar (1982) developed a three-dimensional computer simulation based on ray-tracing principles. He used this model to analyze the effect of sea surface structure, or elevation, on lidar fluorescence signals generated within the ocean volume. He found that the depth-resolved fluorescence return signals possessed a maximum that was positioned several meters below the surface. This signal maximum is attributed to allowable refractory exit channels leading to the detector. If one assumes that a monochromatic sine wave corrugation makes up the sea surface, the depth of the signal maximum can be calculated exactly and is found to be in good agreement with the computer model results. Gehlaar (1982) further validated the computer simulation model by calculations based on (1) the transformation of solid angles by the curved boundaries on the surface and (2) statistical methods. Generally, it was found throughout the calculations that the signal maxima found just below the surface is followed by a signal reduction or fall-off with depth that is steeper than would be attributed to the usual $\gamma_e + \gamma_f$ effective attenuation coefficient losses. His work is a good first-start effort and points to the need for adequate experimental verification and further modeling.

In Section 6.2.1.5 manmade monomolecular slicks on the ocean surface were discussed in the context of their use in the calibration of petroleum oil slick measurements. In truth, since these monomolecular slicks are, for all practical purposes, nonfluorescent and nonattenuating, they should be considered useful in the study of fluorescence/Raman signal modulation by surface waves. Huhnerfuss et al. (1985) have experimentally shown that airborne lidar fluorescence and Raman signals are significantly modified by monomolecular films. They suggest that one of the possible reasons for the observed lidar signal depressions is the alteration of the spatial distribution of the volumetrically derived backscatter signals by the monomolecular film. Figure 6.27*a* (upper) illustrates qualitatively that (because of wave damping) the monomolecular slick region has a wave slope distribution whose variance or width is smaller than found for the clean sea region. Thus, the volumetrically derived backscatter signals detected while overflying a monomolecular film will have a more sharply peaked spatial distribution in the upward or vertical direction normal to the mean sea surface (Figure 6.27*a*, lower). For the nonslick or clean sea, the wave slope distribution and corresponding lidar backscattered signal spatial distribution are given in Figures 6.27*b* (upper) and 6.27*b* (lower) respectively.

Then, if one is lidar-sensing the water column beneath the film, very near to the nadir direction, the backscattered signals would actually be enhanced relative to those found in the clean, non-film-covered sea nearby. However, if the

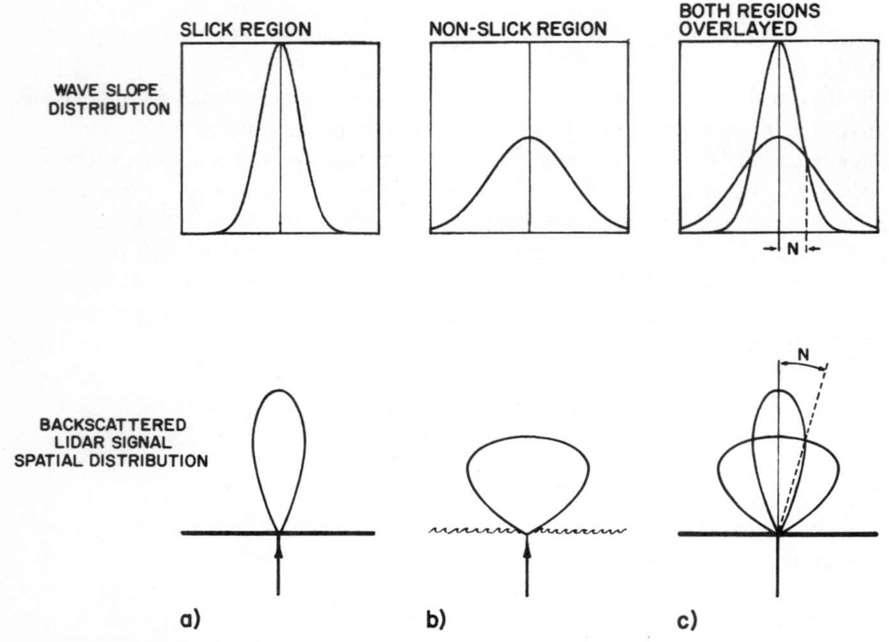

Figure 6.27. Wave slope distribution and backscatter lidar signal spatial distribution expected for (*a*) monomolecular slick and (*b*) nonslick region of ocean. (*c*) Overlay of respective distributions in (*a*) and (*b*) illustrating lidar observational angle at which slick could not be distinguished from open ocean.

airborne system is viewing the film-covered sea at a sufficiently large off-nadir angle relative to the vertical, the lidar signals while over the slick will be reduced (relative to the clean-sea region). This is illustrated in Figure 6.27*c*, where both the wave slope (and lidar backscatter spatial) distributions for slick and nonslick regions have been overlayed. A larger lidar backscatter signal is obtained while over the nonslick region if the observing angle is as depicted in distribution overlays of Figure 6.27*c*. The observations at large off-nadir angles ($>N$) are essentially being made in the higher power peripheral portion of the broadly peaked spatial distribution of the upwelling radiant flux in the nonslick region. Figure 6.27*c* further shows that the slick and nonslick regions cannot be easily discriminated on the basis of backscattered power if the lidar off-nadir observing angle is equal to N, as shown. The angle N would probably not be known a priori but could be measured with a scanning lidar by determining the angle at which the slick and nonslick backscattered signal strengths are equal. In Figure 6.11, discussed in Section 6.2.15, the AOL fluorescence and Raman backscatter data were obtained at an off-nadir angle of $\sim 7°$, resulting in signal depressions while over the monofilm.

In this qualitative explanation we have avoided discussing the fact that the narrow transmitted laser beam spatial distribution incident on the sea surface is also altered at the air–sea interface. However, based on recent studies it appears that the in-water, volumetric spatial redistribution of the incident laser beam upon its downward passage through the surface (Bobb et al., 1978) [and subsequent multiple scattering within the ocean volume (Lerner and Summers, 1982)] is small (and potentially ignorable) compared to the spherically isotropic spatial distribution of the fluorescence and Raman scatter that this incident beam generates.

Karp (1976) theoretically studied the transmission of a laser beam from above (as well as from below) the sea surface for down-link (and up-link) satellite optical communications feasibility studies. His work (and included references) and that of Bobb et al. (1978) could serve, in addition to Gehlaar's (1982) work as the research base from which monostatic airborne lidar sea surface modulation modeling studies could be initiated. The theoretical work of Luchinin (1979) and Luchinin and Sergiyevskaya (1982) should also be evaluated to determine if redistribution of the surface incident laser beam by surface waves could be described as a special case of their solar ray formalism. While there is strong experimental evidence for sea surface perturbation of volumetrically derived lidar signals, the degree of influence on the signal spatial distribution by Mie scattering within the medium should also be evaluated, quantitatively if possible. In this regard, the theoretical descriptions of the propagation and interaction of narrow light beams in seawater (Yura, 1973; Arnush, 1972; Makarevich et al., 1969; Romanova, 1968; Bravo-Zhivotovskiy et al., 1969) should also be evaluated.

6.2.4 Bathymetry and Other Water Column and Ocean Bottom Measurements

As discussed previously, Hickman and Hogg (1969) first demonstrated that an airborne laser system could be used to measure water depth. They showed that a single pulse, when transmitted vertically downward to the water's surface, is in turn partially reflected from that surface as well as the water–bottom interface. Thus, there is generated a pair of pulses whose separation is proportional to the water depth. The low-power 60-μJ neon laser pulse that they used was adequate to map depths to nearly 8 m from an airborne platform flown 150 m above the water's surface. Kim et al. (1975) and Kim (1977) used a 2-kW, 6-ns pulsed neon laser at 540 nm and a 2-MW, 8-ns, frequency-doubled Nd–YAG laser at 532 nm to obtain bathymetric profiles in the Chesapeake Bay and

Key West, Florida, regions. Depths to almost 10 m were obtained in waters having an effective attenuation coefficient of 0.175 m^{-1}. His work has been adequately reviewed by Measures (1984).

By far, the principal hinderance to the useful, routine application of airborne coastal bathymetry is the two-way exponential water column attenuation of the transmitted pulse and the reflected pulse from the bottom. Accordingly, one must virtually scale the laser power upward by an order of magnitude in order to achieve operation to twice the depth. Thus, with the inefficiency of lasers, one must carefully choose the season in order to avoid attenuation by plankton growth and further avoid months when winds will stir and entrain bottom sediments into the water column (Hoge et al., 1980). Finally, one must choose sites whose bottom can actually be reached by laser sounding. Some of the practical considerations for using an airborne laser system for routine survey have been reported by Enabnit et al. (1981). Furthermore, to be truly cost-effective in a high-density mapping mode, the laser bathymetry system must be able to scan the beam over a substantial swath. This requires a high-repetition-rate laser transmitter, receiver, and recording components. The principal cost factors have been studied by Enabnit et al. (1978), who reported that airborne laser hydrography was cost-effective relative to standard launch or ship techniques.

Hoge et al. (1980) demonstrated that conical scan methods were feasible for airborne laser hydrography. They used a 400-Hz, 7-ns neon laser having a 3-kW output at 540.1 nm to map portions of the Chesapeake Bay and the adjoining Atlantic Ocean. Depths to 10 m in the ocean and 4.6 m in the bay were measured. Water truth measurements were taken to validate the depth and beam attenuation coefficient, which allowed evaluation of the system's performance. Lidar bathymetry in the United States is presently being pursued on an active basis by the U.S. Navy. A modest effort continues to show good progress on their Hydrographic Airborne Laser Sounder (HALS). This 400-Hz, frequency-doubled Nd–YAG conical scan system is electronically configured to process analog bathymetry waveforms in real time and record the resulting depth. The system contains a time-gated or variable-gain PMT to reduce the strong surface return and enhance weak bottom returns. This PMT is subsequently followed by a logarithmic amplifier to further accommodate the wide dynamic range between the surface and bottom return signal strengths.

Interest in laser hydrography is not limited to the United States. The Australian coastal waters are so extensive that a laser hydrography program was begun there as an outgrowth of their airborne laser land profiler effort (Clegg and Penny, 1977). Their first bathymetric profiling tests in 1975 yielded 30–40-m depths using a 532-nm-wavelength laser. Another system (called WRELADS II) has full scanning capability, horizontal position fixing, and data-recording capabilities and is undergoing flight trials (Penny, 1982).

In Canada four different airborne techniques were evaluated for their coastal hydrography needs. Only lidar bathymetry was demonstrated to achieve the depth accuracy required by the Canadian Hydrographic Service Charting Standard (O'Neil, 1983).

Scientists with the Swedish government recently reported laser bathymetry efforts in the Baltic Sea (Steinvall et al., 1981). They used the Canadian lidar bathymeter (MK-II) to show that depths of 30 m could be mapped. The MK-II lidar delivered 5-MW pulses with 5-ns lengths and was flown at an altitude of 150 m.

In bathymetrically related efforts, researchers in Sweden had previously reported underwater laser–radar laboratory experiments in the areas of bathymetry and fish school detection (Fredriksson et al., 1979). The experiments were performed in a 27-m horizontal tank that had been fitted with mirrors to control beam direction. In these experiments, a nitrogen-laser-pumped dye laser was used to produce a 5-ns pulse in the green spectral region. The backscatter was collected by a receiver consisting of a Newtonian telescope, a PMT, a fast transient digitizer, and an on-line computer. Useful laser–radar echoes could be detected from the dark backs of herrings, sprats, and mackerels at "depths" to 15 m. Subsequently, some live fish were gently held a few centimeters underwater in small collars while the 337.1- and 532-nm laser-induced fluorescence was measured. Fluorescence was observed from both the backs and undersides of the fish with the latter signals being strongest. Unfortunately, the fluorescence (a) was quite weak and (b) had the same shape as the spectrum of naturally occurring DOC. They concluded that the strength was insufficient to be of practical use. Much earlier, laboratory and theoretical studies of fish school detection by airborne laser systems had been conducted in the United States (Murphee et al., 1973). They concluded that fish detection (using the on-wavelength backscatter signal) was feasible from a low-flying (150 m) aircraft using equipment available at that time.

In closing this section it should be mentioned that the operation in shallow water of a depth-integrating fluorescence lidar has not yet been reported. However, upon encountering shallow water, the received fluorescence signals are expected to increase for several reasons. First, the isotropically emitted fluorescence (stimulated at every depth location during downward propagation of the on-wavelength beam) will be reflected from the bottom to provide some increase in the backscattered signal. Second, there will be additional stimulation of the water column (and its constituents) by the on-wavelength radiation after reflection from the bottom. Thus, the on-wavelength reflection from the bottom effectively increases the apparent depth of the water column being sensed. Since the original transmitted beam remains spatially more compact (Lerner and Summers, 1982) than the isotropically emitted fluorescence it produced during vertical, downward traversal, the bottom reflection of this induced fluorescence is expected to be smaller than that produced by the vertically propagating on-wavelength reflection. Furthermore, the on-wavelength radiation is expected to dominate the bottom effects since it is reflected in a Lambertian manner from the bottom. This, in turn, improves its effectiveness in stimulation of the water column during its upward propagation. Note further that some of the fluorescence produced by the vertically propagating on-wavelength reflection also is ultimately reflected from the bottom. Finally, if the bottom contains fluorescent materials and/or vegetation, this fluorescence adds to that of the water column constituents. Of course, if one is observing the on-wavelength bottom return, the fluorosensor aperture gate can be manually or automatically adjusted in width and depth position to reject all bottom reflections.

6.3 TERRESTRIAL LIDAR MEASUREMENTS

6.3.1 Terrain and Vegetation Elevation Measurements

Fundamentally, terrain profiling is the simplest and perhaps the oldest application of an airborne laser (Jensen, 1967). Since the late 1960s and early 1970s, airborne lasers have been used for the metric measurement of terrain and ice surfaces. Noble et al. (1969) reported the airborne measurement of ice elevation with a commerically available modulated CW 632.8-nm helium–neon laser profiling system. This low-beam divergence lidar system was flown over ice in the Beaufort Sea to demonstrate the feasibility of using a laser system to accurately profile the sea ice surface and to measure the pressure ridge heights. The laser-derived sea ice surface profile clearly allowed the delineation of water, thin ice (first-year ice), and pack ice (multiyear ice). They found that for a small-spot-size nadir-pointed lidar such as this, the height variance (as well as the back-scattered signal strength) of the elevation profile data is higher over water than thin ice, and in turn, the thin-ice elevation variance (and associated backscatter signal strength) was higher than that of pack ice. Using essentially the same lidar, Ketchem (1971) and Tooma et al. (1976) obtained profiles of first-year and multiyear ice.

For land terrain profiling Penny (1972) used a doubled Nd–YAG laser with a 20-ns pulse width, an energy of 750 μJ/pulse, and a beam divergence of 0.3 mrad as the transmitter. The 10-cm receiver telescope and PMT detector/amplifier allowed operation to 5000 m above ground level. The laser system determined a profile of terrain heights above a reference surface that was 6000 m below an isobaric surface at the nominal aircraft altitude. The barometric height above the isobaric surface was measured by a sensitive differential pressure gauge. The horizontal position of the aircraft was determined through interpretation of photographs obtained during the laser profiling. Their system was not fully evaluated for mapping through tree-covered terrain, and no field trials were conducted over dense jungle terrain.

Mamon et al. (1978) reported the test flight results of a compact transceiver that was built by a commercial manufacturer. The transmitter used a GaAs solid-state laser, 15-cm-diameter collimating objective yielding a 1.5-mrad FOV. The laser pulse width was 10 ns and yielded a peak power of 25 W per pulse. The receiver consisted of a 20-cm-diameter collimating lens and an avalanche photodiode detector. Their system was designed for terrain elevation profile measurements even in the presence of trees. Since branches may give multiple target return signals, one or more return pulses might be received in response to one transmitted pulse. Thus, since the range to the ground is required, the last pulse from a series of multiple returns must be discriminated. Their system incorporated this so-called last-pulse selection. Data were taken over a deciduous forest at two different seasons of the year, yielding measurements during full-foliage and no-foliage conditions. From this standpoint alone, the GaAs lidar represented a slight improvement over the Australian system. However, no method was provided for removing the aircraft vertical motion as had the Australian system. As with the Australian system, the horizontal position control was provided by photographic data.

Krabill et al. (1980, 1984) presented the results of a series of joint NASA–U.S. Army (Corps of Engineers) terrain mapping experiments. They used the NASA AOL in the bathymetric mode with a laser repetition rate of 200 pps to obtain both profile and scan data over open and tree-covered terrain. The experiments were conducted to collect data similar to that a ground survey team would normally obtain for input to hydraulic–hydrologic models for simulating flow lines of streams. Both existing ground survey and independently obtained photogrammetry were used for ground truth. Figure 6.28 shows a comparison of the airborne laser data and a photogrammetrically derived profile. Notice that a 12-cm rms difference was obtained over terrain with no trees or bushes while a 50-cm rms difference was obtained over wooded terrain. No attempt was made to verify the accuracy of either the lidar or photogrammetric results. Some of the difference may have been due to errors in the photogrammetric profile. In fact, in at least one instance, the laser found verifiable terrain features that were completely missing in the final photogrammetry truth products. Note further in Figure 6.28 that the heights of the trees may be obtained quite easily. [Such vegetation features can also be found in the data of Penny (1972) and Mamon et al. (1978).] In lieu of a last-pulse tracker, the bathymetric mode of the AOL (Hoge et al., 1980) was used to advantage to find the forest floor terrain segments. Here, the depth-resolved or temporal-waveform-recording subsystem is used to obtain a "surface" return from the top of the forest canopy and the bottom return from the forest floor. Then the sum of the range to the "surface" (canopy) and the "depth" (tree height) yields the slant range from the aircraft

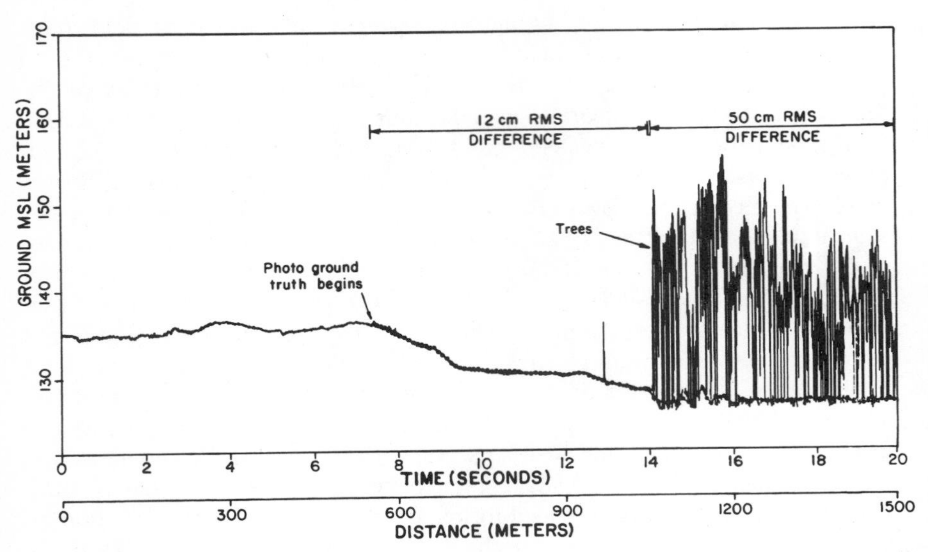

Figure 6.28. Comparison of airborne laser terrain survey and photogrammetrically derived profile. Bathymetric mode of AOL used to locate forest floor terrain segment (Krabill et al., 1980).

to the ground. Aircraft horizontal position for these tests was determined from a combination of inertial navigation system (INS) position, velocity, heading, track angle data, and photographic interpretation. The aircraft vertical position was determined from three (minimum) ground survey points, AOL range data, and a vertical accelerometer.

The lengths to which one must go to determine the aircraft position (and orientation) points to the fact that both navigation and positioning remain as problems for generalized application of an airborne lidar to terrain mapping and tree height measurement (Chapman, 1982). Vertical control can be satisfactorily accomplished over flight lines of several kilometers using an inexpensive vertical accelerometer along with several vertical reference points on each survey line. One can look with anticipation to the global positioning system (GPS) to fill the future positioning needs for these and similar airborne mapping applications.

With the AOL operating in the scan mode (Link et al., 1983a,b) it was shown that a three-dimensional terrain contour map could be obtained through forested areas. Figure 6.29*a* shows a composite of two separate passes of the AOL over the same forested ("leaves-off") target region. Again, the bathymetric mode of the AOL was used with a laser rate of 200 pulses/s. The laser data have been smoothed to remove the high-frequency contributions and thus

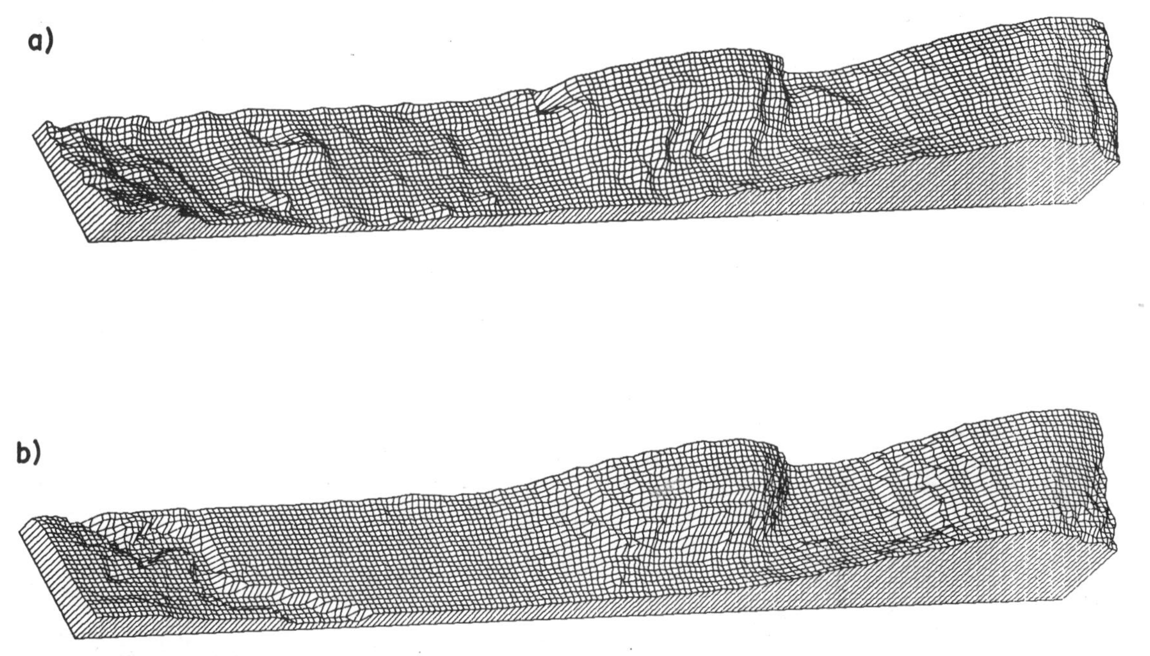

Figure 6.29. (*a*) Topography map of forested region composed of two passes of AOL (smoothed). Bathymetric mode of instrument used with 200-pps laser rate. (*b*) Photogrammetric truth map of same region as in (*a*) (Link et al., 1983a).

simplify comparison with the photogrammetry truth data (Figure 6.29*b*). An independent point-by-point comparison showed that almost 80% of the laser data were within 1 m of the photogrammetric data.

In summary, Krabill et al. (1980, 1984) showed that laser mapping of river valley cross sections in deciduous forest areas appears to be primarily restricted to winter foliage conditions to meet accuracy requirements. However, in areas covered predominantly by coniferous forests, useful ground surface elevations may be acquired at any time of the year. The use of an inexpensive accelerometer to effect vertical control over the distance of a normal flight line is an important advance in vertical position control (Krabill and Martin, 1983). This technique is now being used extensively in making airborne oceanographic laser (Hoge et al., 1984) and radar (Walsh et al., 1984) measurements.

Nelson et al. (1984) used a pulsed nitrogen laser in the AOL system (Hoge et al., 1980; Krabill et al., 1980) to determine the alteration of forest canopy characteristics as a result of insect infestation. Again, the bathymetric mode of the AOL instrument was used to obtain tree heights. They found that (a) the amount of healthy crown sensed over a flight line interval is most closely related to the number of transmitted laser pulses that hit the tree canopy but in which a ground return signal could not be found in the temporal waveform, (b) a comparison of laser profiles taken over healthy and over defoliated (by gypsy moth infestation) forest segments suggested that pulse penetration into the defoliated canopy is greater than the intrusion into a healthy canopy, (c) cross-sectional area (the canopy profile) increases as canopy density increases (this is apparently because denser canopies intercept the laser pulse higher in the canopy, hence yielding a cross-sectional area that is directly related to canopy closure), and (d) the numbers of ground returns increased significantly in defoliated canopies.

One of the most important practical uses of tree height information is the potential it has for yielding timber volume estimates. The relationship between canopy profile and timber volume is based on stratification of species (Maclean, 1982), and thus to obtain reliable estimates of volume, the tree species must be identified. Airborne tree species identification (simultaneously with tree height and canopy closure information) would perhaps allow high-speed timber volume estimates to be obtained. One possible technique that merits investigation is the use of laser-induced fluorescence of the tree's foliage at the same time as the metric height and canopy data are being obtained. Airborne lidars such as the AOL have the potential to gather the required LIF and metric data simultaneously. However, significant laboratory work must be accomplished before this technique can be expected to yield fruitful airborne results. Some of the most recent LIF of terrestrial plants are described in the next section.

6.3.2 Laser-Induced Fluorescence of Living Terrestrial Plants

Clearly, species identification is important insofar as it would allow the practical airborne measurement of timber volume and assessment of biomass. Species identification of all growing plants is also important particularly for foodcrop (and noxious weed infiltration) inventory reporting on a local, national, and international basis. Furthermore, the ability to discriminate among different varieties of the same species would allow regional identification of more naturally disease-resistant strains of different cultivated foodcrops. In total, the ability to identify (a) species, (b) varieties of the same species, (c) maturity within a variety or species, and (d) the previsual and contemporary stress upon the species would have far-reaching, economic and human resources development implications throughout the world. It is felt by many scientists that the multispectral, high-resolution capability of laser-induced fluorescence has enormous potential to address plant identification, maturity, and stress. This is especially evident when such laser techniques could be combined with existing satellite passive multispectral data that is now so easily obtained.

Apparently, the first reported utilization of laser-induced fluorescence in the study of intact plants was performed in Canada. Brach et al. (1977) used remote sensing laser spectroscopy techniques in a laboratory-greenhouse configuration. A pulsed nitrogen (337-nm) and a CW helium–cadmium laser (441 nm) were used to investigate the potential for recognition of both lettuce maturity and variety. While the techniques showed promise, no conclusions regarding their ultimate application were made. Subsequent LIF experiments by Brach et al. (1978) showed that the fluorescence induced by 410-nm laser excitation of a lettuce or grass crop increased as the lettuce or grass matured. However, the same experiments indicated that the laser fluorescence technique would not easily discriminate between different varieties of lettuce cultivars. Later experiments on grain crops (Brach et al., 1982) showed that the fluorescence quantum yield (and possibly the structure of the fluorescence spectra) could be used to differentiate between species and cultivars of a species. Additionally, these authors noted that it appeared likely that the pigment composition determines the fluorescence yield and the structure of the fluorescence spectral curve.

Chapelle et al. (1984a) used a 337-nm laser spectrofluorometer in laboratory studies to show that LIF measurements of plants offer potential for remotely detecting certain types of stress conditions and also for differentiating plant species. The 337.1-nm LIF spectrum of soybeans is shown in Figure 6.30. Typically, spectral peaks are found at 440, 525, 690, and 740 nm. Withholding water from the experimental plants caused dehydration and wilting and resulted in a general increase in fluorescence, as shown. Specifically, the increase in the

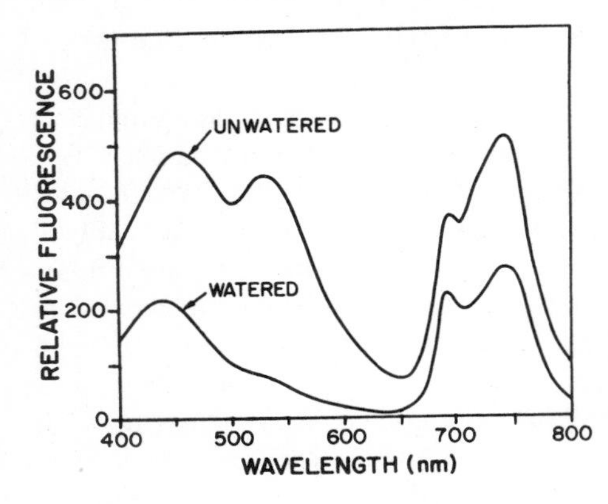

Figure 6.30. A 337.1-nm laser-induced fluorescence emission obtained in laboratory from soybean leaves. Most pronounced effect of dehydration is enhancement of 540-nm band (Chappelle et al., 1984a).

525-nm peak is most dramatic. A potassium-deficiency condition was found to result in a significant increase in the 690- and 740-nm spectral peaks of corn LIF spectra. Incubation of soybean leaves in 3-(3,4-dichlorophenyl)-1,1-dimethylurea (DCMU) led to a suppression of the 690- and 740-nm spectral peaks. Chapelle et al. (1984a) concluded that an impairment of the photosynthetic process leads to an increase in fluorescence at 690 and 740 nm, but a decrease in chlorophyll concentration such as by senescence leads to a decrease in 690- and 740-nm fluorescence with a corresponding increase in 440- and 525-nm fluorescence. The 690- and 740-nm bands are attributed to two different species of chlorophyll (Brown and Michel-Wolwertz, 1968). The spectral peaks at 440 and 525 nm have not yet been assigned to any molecule or plant pigment.

Chappelle et al. (1984b) later showed in the laboratory that deficiencies in phosphorus, nitrogen, and iron in intact corn plants are manifested primarily by a decrease in the fluorescence of the 690- and 740-nm bands. This is in contrast to the fluorescence increase in potassium-deficient corn. While the underlying reasons for the fluorescence changes are not actually known, these studies do suggest the potential usefulness of LIF measurements for the remote detection of nutrient deficiencies in plants.

Laboratory studies by Chappelle et al. (1985) also suggest that individual plant species may potentially be identifiable. They were able to show that the 337.1-nm LIF spectra of five major plant types were sufficiently different to allow their identification. The plant types were herbaceous dicots and monocots, conifers, hardwoods, and algae. Each would be identified by (a) the presence or absence of one or more of the 440-, 525-, 685-, and 740-nm peaks or (b) the strength of one or more of the spectral fluorescence peaks relative to

Figure 6.31. Laboratory 337.1-nm laser-induced fluorescence spectra of some selected conifers and hardwoods. Differences found in spectra suggests potential for discrimination of major plant types (Chappelle et al., 1985).

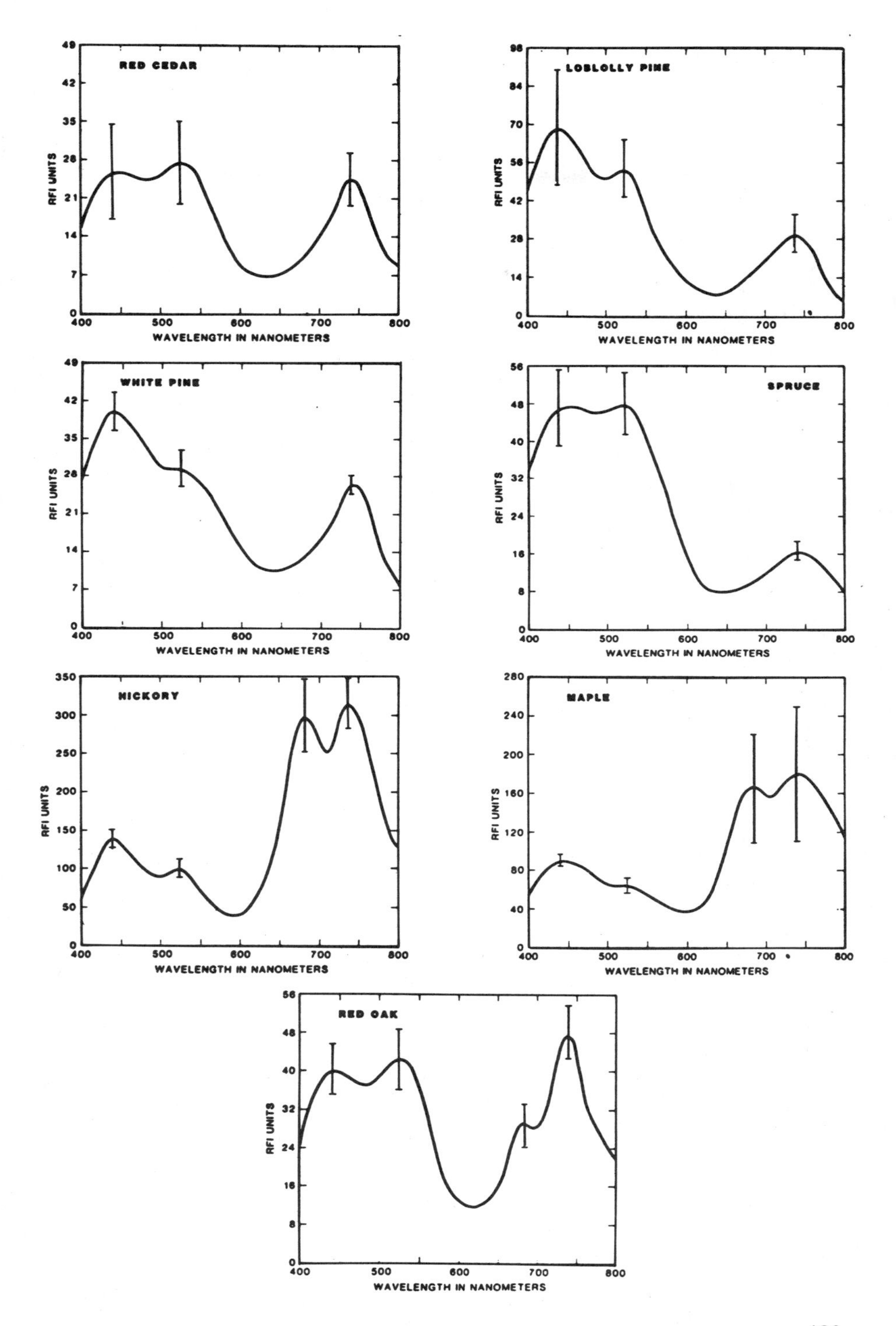
RED CEDAR
LOBLOLLY PINE
WHITE PINE
SPRUCE
HICKORY
MAPLE
RED OAK
RFI UNITS
WAVELENGTH IN NANOMETERS

another. Typical 337.1-nm LIF spectra of some conifers and hardwoods are shown in Figure 6.31.

Celander et al. (1978) reported the 337.1-nm laboratory LIF of the foliages of eight different trees including conifers and hardwoods. While some fluorescence emission peaks were always seen in the 400–600-nm region, the lack of observable fluorescence in the 680–750-nm region is not understandable. However, they did observe ~685- and 730-nm chlorophyll fluorescence in grass and in so-called ground vegetation such as peat moss and reindeer lichen. Since the spectra by Celander et al. (1978) were uncorrected, one can only conjecture that the instrument response or sensitivity in the red region was quite reduced in comparison to that of Chappelle et al. (1985), who (1) used a PMT detector having good sensitivity into the red region and (2) further corrected their spectra for the remaining roll-off in the red portion of the spectrum.

Hoge et al. (1983) showed that the LIF spectral emissions are of sufficient strength to be easily detectable from airborne altitudes of 150 m. They used a 532-nm frequency-doubled Nd–YAG laser and an excimer-pumped dye laser at 422 nm integrated into the AOL to conduct airborne experiments over natural

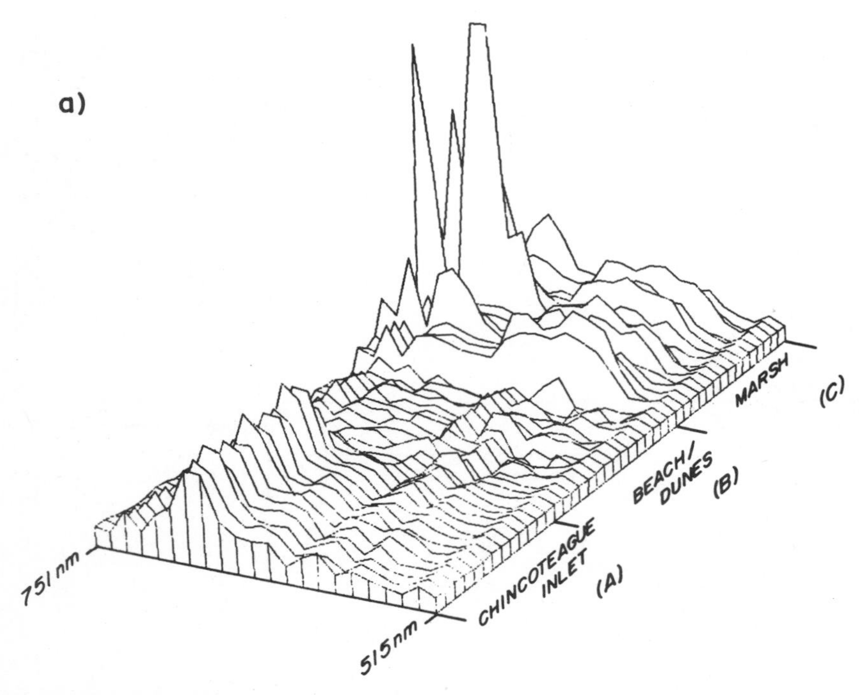

Figure 6.32. Individual, unaveraged laser-induced emission spectra obtained with AOL during overflight of (*a*) water, beach, and marsh portions of a flight line and (*b*) marsh, bush/tree, and marsh segments of same flight line. Pulse-to-pulse variability primarily induced by spatial variability of target and not system noise or solar background radiation (Hoge et al., 1983).

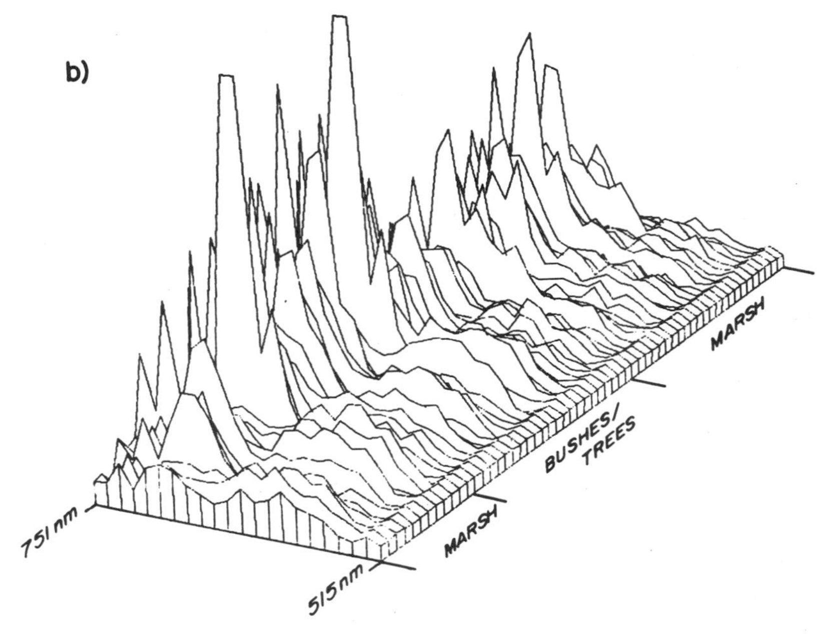

Figure 6.32. (*Continued*)

terrain. Grab samples were also collected by ground-truth teams to allow comparison of laboratory and airborne spectra. The dominant LIF peaks in the airborne and laboratory spectra were found at 690 and 740 nm and are due to chlorophyll. Figure 6.32*a* shows individual, or single-shot, 532-nm LIF spectral waveforms obtained during the course of a flight over marine water, beach, and marsh. Figure 6.32*b* shows a continuing portion of the flight line over marsh, bushes, and trees. Several important features of these spectra should be noted. The signal strength and the LIF spectral color are both highly variable over land targets (Figure 6.32*a*, section *C*) compared to marine targets (Figure 6.32*a*, section *A*). This pulse-to-pulse variability is caused by spatial variability of the target and is not a result of system noise or solar background radiation. The variability in the LIF chlorophyll fluorescence from marine targets is primarily caused by water surface gravity and capillary waves (once the water Raman has been used to correct for water column transmission variability).

The relative terrain and/or plant height were simultaneously measured by laser pulse time-of-flight measurements. These metric measurements offer further discrimination between trees, bushes, and grasses. Figure 6.33 shows the 685-nm chlorophyll fluorescence emission induced by the 422-nm, 100-kW, 5-ns dye laser pulses during the overflight. The on-wavelength backscattered return was utilized to obtain the plant and terrain elevation. Depth-resolved fluorescence techniques (similar to depth-resolved Raman backscatter, Section

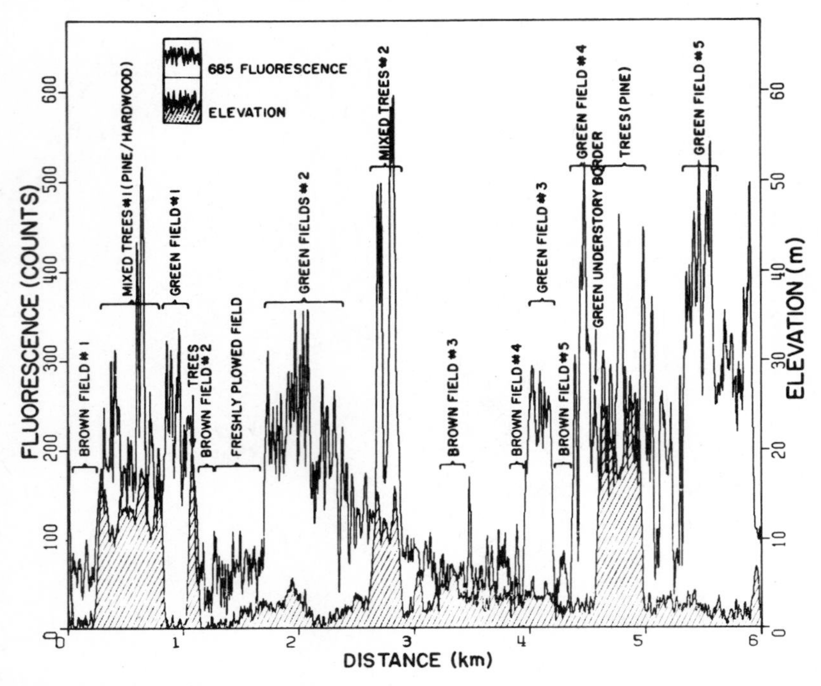

Figure 6.33. Profile of 422-nm LIF emission at 685 nm obtained across various targets of fields and trees. Terrain elevation shown by the cross-hatched profile. Note that some grass-covered fields yield fluorescence levels whose magnitude is approximately the same as that from stands of trees (Hoge et al., 1983).

6.2.2.4) can potentially allow the study of fluorescence emission as a function of leaf position on the plant if the laser pulse width and fluorescence emission response time are on the order of several nanoseconds or less.

We have shown that vegetation height and fluorescence are both measurable simultaneously. In Section 6.2.4 the success of bathymetry, and other water column measurements, was reviewed. The reader will quickly deduce that metric and fluorescence measurement (and wide-area mapping) feasibility studies for investigating submerged aquatic vegetation should be attempted. These measurements will, of course, be hampered by the attenuation of the water column but are nevertheless felt to be a natural extension of the previous bathymetry and plant terrain and fluorescence work.

Outside the United States and Canada, investigators in the USSR (Kanevskiy et al., 1983) have reported the laser irradiation (with a defocused beam) of leaves and subsequent image-intensifier recording of the 685-nm image from distances of 15–20 m. Such a system reportedly would not have adequate SNR for use at aircraft altitudes. They further reported use of a helicopter-borne lidar with a 440-nm, 15-mW laser (presumable He–Cd) to observe ~688 nm fluorescence of leaves in a 10-nm bandpass from altitudes of 15–20 m during cloudy days. It was further suggested that pulsed lasers and larger receiving telescopes could substantially increase the operating altitudes. Not surprisingly, their laboratory LIF spectra of beet leaves obtained during the course of the program displayed an elevated 735-nm peak relative to the 685-nm band.

6.3.3 Laser-Induced Fluorescence of Surface Mineral Deposits

DeNeufville et al. (1981) performed extensive work to evaluate the remote sensing of laser-induced fluorescence of the uranyl ion as a potential indicator of uranium occurring in geologic materials at Earth's surface. The most important characteristics of the uranyl ion fluorescence are illustrated by the laboratory excitation and fluorescence spectra of an efflorescent film from the Hale Quarry near Portland, Connecticut (Figure 6.34). Kasden et al. (1981) performed ground-based lidar fluorescence studies in a quarry known to have surface mineralization of the fluorescent uranyl ion UO_2^{2+} along its walls. They used a transmitter consisting of a flashlamp-pumped dye laser having a 35-mJ, 175-ns output at 425 nm in a 3-nm spectral bandwidth. The receiver consisted of an 18-cm reflecting telescope with a flexible filter/monochromator/PMT detector observing in the 490–600-nm spectral region. Efflorescent films and flakes of uranyl mineral were first located by visual fluorescence stimulated with a hand-held UV lamp. Then, the lidar was brought onto the target for remote measurements at distances of 32–40 m. The relative fluorescent brightness found was 330–1517 ppm A units, defined as 10^6 times the area concentration of meta-autunite that would have to be illuminated to observe the same brightness as measured from the target (deNeufville et al., 1981). During all the experiments, considerable temporal delay was inserted into the detection system since the fluorescence lifetimes varied from 154 to 398 μs. Based on the quarry measurements of fluorescence lifetime, the authors were then able to assign the shorter lifetime targets to the tetragonal (D_{4h}) uranyl minerals and the longer lived ones to D_{2h} or D_{3h} point symmetry such as possessed by liebigite, schroek-

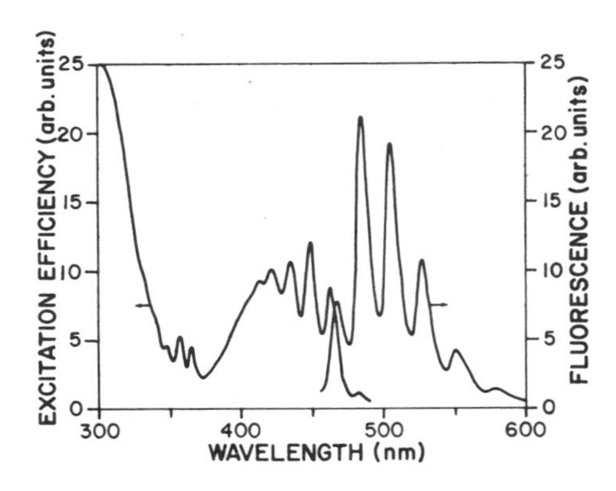

Figure 6.34. Excitation and fluorescence spectra of efflorescent film from Hale Quarry, Portland, CT. Emission at 5250 Å and excitation at 3300 Å, respectively. Chemical uranium 420–490 ppm (deNeufville et al., 1981).

ingerite, or andersonite (deNeufville et al., 1981). The fluorescence lifetime measurements were made with a filter detection system centered at 525 nm and having a 5-nm bandwidth.

Since a Fraunhofer line discriminator (FLD) (Watson, 1981) is able to detect luminescent rocks and minerals in daylight, it would seem natural to suggest that a lidar and a FLD be flown together on the same airborne platform to determine their complementarity (Chimenti, 1981).

Recently, laboratory evaluation of a laser fluorosensor was conducted to determine the feasibility of remotely detecting the fluorescent emission of organic effluents associated with coal processing (Capelle and Franks, 1979). This laser fluorosensor was able to separately accommodate N_2 or KrF excimer lasers and used a 20-cm modified Schmidt Cassegrain telescope and a filter–iris–PMT detector arrangement. For multichannel spectra, a PMT–monochromator and also an optical multichannel analyzer (OMA) were tested. The latter, of course, suffered from inadequate sensitivity. The N_2 lidar system was tested on a two-engine fixed-wing airborne platform at altitudes of 105–308 m over fabricated targets on the ground containing a coal solvent, quinine bisulfate, motor oil, and a detergent (Franks et al., 1983). The KrF implementation of the lidar was similarly flown over test targets at the Nevada Test Site, a naval petroleum reserve, and the Santa Barbara Channel oil field region. The 250-mJ, 249-nm KrF excimer produced a 16-ns pulse and was operated at 31 Hz. The detector consisted of a dichroic beam splitter to yield a visible channel (~360–460 FWHM) and a 290-nm UV channel, the latter using a solar blind CsTe PMT. Rather high SNRs were obtained, and petroleum-based by-products were detected in the visible spectral region from altitudes up to 610 m without PMT gating.

6.3.4 Terrain Imaging by Scanning Lidar Reflectance Measurements

During the course of airborne lidar experiments the on-wavelength pulse backscattered from the terrestrial target is most often used to (a) measure the aircraft slant range and/or tree height by temporal time-of-flight measurements or (b) synchronize, trigger, or otherwise gate a fluorosensor into active data acquisition at the proper time. The strength of this on-wavelength pulse is often recorded and infrequently used (Hoge et al., 1984), particularly in the mapping mode. This signal can, however, be used to advantage to image and delineate features on the ground and during postflight data analysis and in determining the position of the airborne platform upon completion of the mission. This mapping feature of the return pulse was investigated as a secondary objective of a recent geodynamic baseline monitoring experiment using aircraft laser ranging (Krabill et al., 1982). The NASA AOL was operated in the conical scan mode at 400 Hz using a 337.1-nm N_2 laser to interrogate a series of retroreflectors placed at known positions along the runways and taxiways of the Wallops Flight Facility (Figure 6.35*a*). The retroreflectors were easily discriminated on the basis of their specular, often-saturated return signal. However, since all return pulse amplitudes and their spatial positions had been recorded, a 337.1-nm reflectance map was generated. This lidar reflectance map is shown in Figure 6.35*b*. Notice that the concrete taxiways and leftmost portion of the runway yielded similar return power, while the black macadem portion gave a reduced signal. The marsh and creek regions (see upper left, Figure 6.35*a*) showed significant reductions in signal, which allowed their easy delineation. Note, also, that the grassy regions both within the interior of the runway–taxiway complex and outside of it produced the same return signal strength. While little

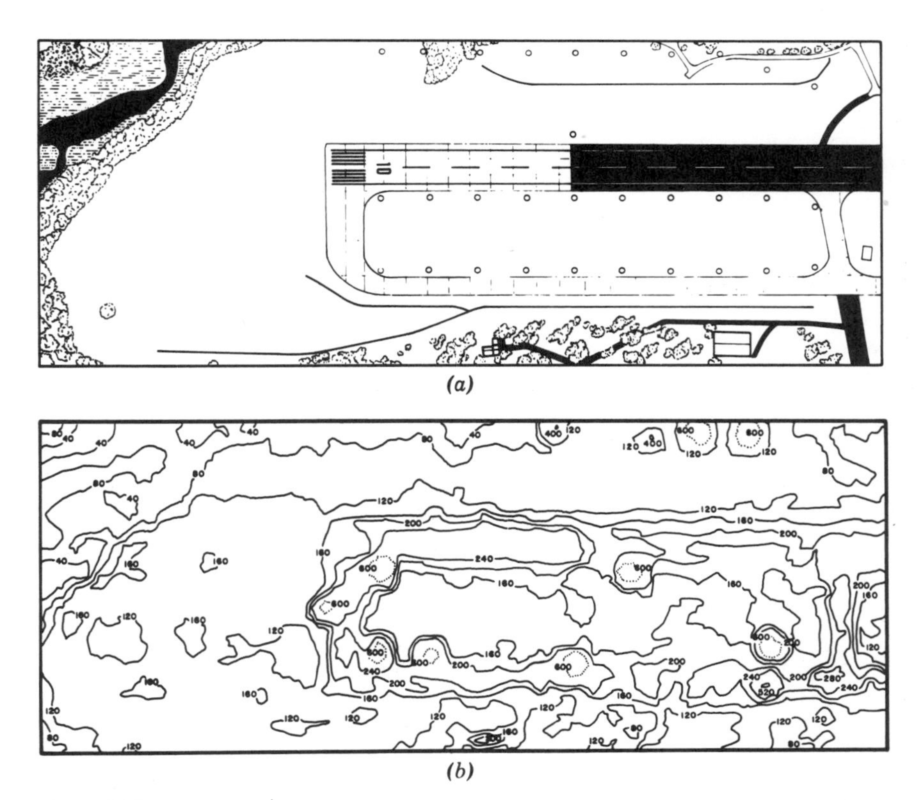

Figure 6.35. (*a*) Illustration of vegetation and terrain of selected test sites at Wallops Flight Facility. White, unaltered areas grass covered. Small circles denote location of retroreflectors deployed for baseline monitoring experiment. (*b*) Contour map of surface return signal intensity from target region in (*a*). Taxiways, runways, creek, and bushes as well as interrogated retroreflectors discriminated on basis of return signal strength.

analytical and algorithm development effort was expended in this modest effort, it shows the potential that exists, particularly when one notes that this lidar reflectance map was generated from data taken in total darkness. Considering the LIF of chlorophyll in trees, bushes, and grasses demonstrated recently (Hoge et al., 1983), numerous multiwavelength fluorescence images could also be generated with a single overflight. The high-repetition-rate metal vapor lasers operating in the visible would be a possible transmitter source. However, eye safety may then become more of a concern than when UV lasers are used. If military mapping research is the objective, this visible source may also be less desirable than UV laser transmitters.

6.4 FUTURE PROSPECTS FOR OCEANIC AND TERRESTRIAL LIDAR MEASUREMENTS

Oceanic and terrestrial lidar research and applications are both emerging fields that combine science, engineering, and technology. Exciting concepts have been proposed and investigated to date in ground-based laboratories and from shipboard and airborne platforms. Still newer and more sophisticated ideas will no doubt be proposed and tried. Hopefully, operation from Earth-orbiting satellite platforms will allow global oceanic and terrestrial lidar applications to be successfully pursued to operational status. The laser altimeter in the Apollo 15 orbital science payload certainly demonstrated that lidar operation from a satellite platform is feasible (Kaula et al., 1972).

For satellite applications, scaling the laser transmitters to higher powers (and receivers to larger apertures) will allow feasibility estimates to be quickly obtained in terms of prime power and sheer system weight and volume required for Earth orbit deployment. New problems and considerations such as eye safety and atmospheric two-way attenuation factors may require attention. As with present satellite sensors, the systems to handle the volume of data generated by global sensing will remain a challenge, especially if large numbers of fluorosensor bands are necessarily implemented.

Oceanic and terrestrial lidar systems will probably be erected (at least in part) upon the foundations of shuttle atmospheric lidar research (Browell, 1979) for winds (Huffaker, 1978), species and aerosol determination (Browell, 1983; Menzies and Shumate, 1978) as well as topography (Kobrick et al., 1981), and satellite Earth crustal motion experiments (Kahn et al., 1979).

Relative to the hydrosphere, many of the fundamental or basic problems have been listed and discussed, especially with regard to their possible solution by remote sensing (Goldberg, 1979). Certainly ship, aircraft, and satellite platforms carrying both active and passive sensors must all share a portion of the workload. Lidar sensors must continue to play a role in airborne sensing but must at the same time become established in the satellite platform domain. All scientists can look with fond hope to the use of satellite-based lidar systems to help provide answers to the many land, ocean, and atmospheric research questions. These three latter portions of Earth form a coupled, dynamic system, and the high temporal and spectral specificity of lidar data can, as well as any known sensor, help lace together the answers to many of the interdisciplinary questions of Earth science.

REFERENCES

Abramov, O. I., V. I. Yeremin, L. I. Lobov, and V. V. Polovinko (1977). Use of Lidar to Detect Oil Pollution on the Sea Surface. *Izv., Atmos. Oceanic Phys.* **13**(3), 232–234.

Abramowitz, M., and I. A. Stegun, eds. (1964). *Handbook of Mathematical Functions*, Dover, New York.

Arnush, D. (1972). Underwater Light-Beam Propagation in the Small-Angle-Scattering Approximation. *J. Opt. Soc. Am.* **62,** 1109–1111.

Bobb, L. C., G. Ferguson, and M. Rankin (1978). Laser Irradiance in the Sea: Theory and Experiment. Report No. NADC 78253-30, Naval Air Development Center, Warminster, Pennsylvania.

Bogorodskiy, V. V., M. A. Kropotkin, and T. Yu Sheveleva (1977). Determination of Oil Contamination of Water by a Scanning Optical Radar (Lidar). *Izv., Atmos. Oceanic Phys.* **13**(3), 914–917.

Bowman, M. J., and W. E. Esaias, eds., (1977). Oceanic Fronts in Coastal Processes. Proceedings of a Workshop held at the Marine Sciences Research Center, May 25–27, 1977, Springer-Verlag, New York.

Brach, E. J., J. M. Molnar, and J. T. Jasmin (1977). Detection of Lettuce Maturity and Variety by Remote Sensing Techniques. *J. Agric. Eng. Res.* **22,** 45–54.

Brach, E. J., M. A. Klyne, T. Phan, and J. J. Jasmin (1978). Use of Laser Fluorescence to Study Lettuce Growth and Development Under Controlled Environment. *Proc. Soc. Photo-Opt. Instrum. Eng.* **158,** 156–162.

Brach, E. J., B. Gasman, and L. J. Croix (1982). Development of a Laser Fluorosensor for Cultivars and Species Indentification of Grain Crops. *Z. Acker Pflanzenbau (J. Agron. Crop Sci.)* **151,** 6–16.

Bravo-Zhivotovskiy, D. M., L. S. Dolin, A. G. Luchinin, and V. A. Savel'yev (1969). Structure of a Narrow Light Beam in Sea Water. *Izv., Atmos. and Oceanic Phys.* **5,** 160–167.

Bristow, M. P. F. (1978). Airborne Mapping of Surface Water Pollutants by Fluorescence Spectroscopy. *Remote Sensing of Environment* **7,** 105–127.

Bristrow, M., and D. Nielsen (1981). Remote Monitoring of Organic Carbon in Surface Waters. Report No. EPA-80/4-81-001, U.S. Environmental Protection Agency, Las Vegas, Nevada.

Bristow, M., D. Nielsen, D. Bundy, and F. Furtek (1981). Use of Water Raman Emission to Correct Airborne Laser Fluorosensor Data for Effects of Water Optical Attenuation. *Appl. Optics* **20,** 2889–2906.

Bristow, M. P. F., W. R. Houston, and R. M. Measures (1973). Development of a Laser Fluorosensor for Airborne Surveying of the Aquatic Environment, in *The Use of Lasers for Hydrographic Studies*, H. H. Kim and P. T. Ryan, eds., NASA SP-375, Proceedings of a Symposium held at Wallops Island, Virginia, September 12, 1973, pp. 119–145, Washington, D.C.

Brockmann, U. H., H. Huhnerfuss, G. Kattner, H.-C. Broecker, and Gunter Hentzschel (1982). Artificial Surface Films in the Sea Area Near Sylt. *Limnology and Oceanography* **27,** 1050–1058.

Browell, E. V. (1977). Analysis of Laser Fluorosensor Systems for Remote Algae Detection and Quantification. NASA TN D-8447, Washington, D.C.

Browell, E. V., ed. (1979). Shuttle Atmospheric Lidar Research Program. NASA SP-433, Washington, D.C.

Browell, E. V. (1983). NASA Multi-Purpose Airborne DIAL System and Measurements of Ozone and Aerosol Profiles. *Appl. Optics* **22,** 522–534.

Brown, J. S., and M. R. Michel-Wolwertz (1968). Chlorophyll Fluorescence Near 720 μm in Englena Extracts. *Biomchem. Biophys. Acta* **155,** 288–290.

Bufton, J. L., F. E. Hoge, and R. N. Swift (1983). Airborne Measurements of Laser Backscatter from the Ocean Surface. *Appl. Optics* **22,** 2603–2618.

Burlamacchi, P., G. Cecchi, P. Mazzinghi, and L. Pantani (1983). Performance Evaluation of UV Source for Lidar Fluorosensing of Oil Films. *Appl. Optics* **22,** 48–53.

Byrnes, H. J., and S. L. Fagin (1978). Optimal Filtering and Analysis of Scanning Laser Data. Naval Ocean Research and Development Activity (NORDA) Technical Note No. 24, NSTL Station, MS.

Campbell, J. W., and W. E. Esaias (1983). Basis for Spectral Curvature Algorithms in Remote Sensing of Chlorophyll. *Appl. Optics* **22,** 1084–1093.

Campbell, J. W., and J. P. Thomas, eds. (1981). The Chesapeake Bay Plume Study, Superflux 1980. Proceedings of a Conference held in Williamsburg, VA, January 21–23, 1981, NASA Conference Proceedings Document CP-2188.

Capelle, G. A., and L. A. Franks (1979). Laboratory Evaluation of Two Laser Fluorosensor Systems. *Appl. Optics* **18,** 3579–3586.

Capelle, G. A., L. A. Franks, and D. A. Jessup (1983). Aerial Testing of a KrF Laser-Based Fluorosensor. *Appl. Optics* **22,** 3382–3387.

Celander, L., K. Fredriksson, B. Galle, and S. Svanberg (1978). Investigation of Laser-Induced Fluorescence with Applications to Remote Sensing of Environmental Parameters. Gotenburg Institute of Physics, Report No. GIPR-149, Gotenberg, Sweden.

Chapman, W. H. (1982). Aerial Profiling of Terrain System, in Proceedings of Surveying Requirements Meeting, February 2–5, 1982, U.S. Army Engineer Waterways Experiment Station, Vicksburg, MS, pp. 247–260.

Chappelle, E. W., F. W. Wood, Jr., J. E. McMurtrey III, and W. W. Newcomb (1984a).

Laser-Induced Fluorescence of Green Plants. 1: A Technique for the Remote Detection of Plant Stress and Species Differentiation. *Appl. Optics* **23,** 134–138.

Chappelle, E. W., J. E. McMurtrey III, F. M. Wood, Jr., and W. W. Newcomb (1984b). Laser-Induced Fluorescence of Green Plants. 2: LIF Caused by Nutrient Deficiencies in Corn. *Appl. Optics* **23,** 139–142.

Chappelle, E. W., F. M. Wood, Jr., W. W. Newcomb, and J. E. McMurtrey, III (1985). Laser-Induced Fluorescence of Green Plants. 3: LIF Spectral Signatures of Five Major Plant Types. *Appl. Optics* **24,** 74–80.

Chimenti, R. J. L. (1981). Optical Characteristics of Uranyl Geologic Targets, in *Workshop in Applications of Luminescence Techniques to Earth Resource Studies*, W. R. Hemphill and M. Settle, ed., LPI Technical Report 81-03, Lunar and Planetary Institute, Houston, pp. 69–71.

Choy, L. W., D. L. Hammond, and E. A. Uliana (1984). Electromagnetic Bias of 10 GHz Radar Altimeter Measurements of MSL. *Marine Geodesy* **8,** 297–312.

Christman, R. F., and M. Ghasseni (1966). Chemical Nature of Organic Color in Water. *J. AWWA* 723–731.

Clegg, J. E., and M. F. Penny (1977). Depth Sounding from the Air by Laser Beam. *Navigation (Australia)* **5** 541–547.

Cox, C., and W. Munk (1954). Measurement of the Roughness of the Sea Surface from Photographs of the Sun's Glitter. *J. Opt. Soc. Am.* **44,** 838–850.

Cronan, C. S., and C. L. Schofield (1979). Aluminum Leaching Response to Acid Precipitation: Effects of High Elevation Watershed in Northeast. *Science* **204,** 304–306.

deNeufville, J. P., A. Kasden, and R. J. L. Chimenti (1981). Selective Detection of Uranium By Laser-Induced Fluorescence: A Potential Remote-Sensing Technique. 1: Optical Characteristics of Uranyl Geological Targets. *Appl. Optics* **20,** 1279–1296.

El-Sayed, S. A. (1970). Phytoplankton Production of the South Pacific and the Pacific Sector of the Antarctic. *Scientific Exploration of the South Pacific*, National Academy of Sciences, Washington, D.C.

Enabnit, D. B., L. R. Goodman, G. K. Young, and W. J. Shaughnessy (1978). The Cost Effectiveness of Airborne Laser Hydrography. National Oceanic and Atmospheric Administration Technical Memorandum NOS-26, Rockville, MD.

Enabnit, D. B., G. C. Guenther, J. Williams, and F. Skove (1981). An Estimate of the Area Surveyable with an Airborne Laser Hydrography System at Ten U.S. Sites. National Oceanic and Atmospheric Administration Technical Report OTES-5, Rockville, MD.

Exton, R. J., W. M. Houghton, W. Esaias, L. W. Haas, and D. Hayward (1983). Spectral Differences and Temporal Stability of Phycoerythrin Fluorescence in Estuarine and Coastal Waters Due to the Domination of Labile Cryptophytes and Stabile Cyanobacteria. *Limnology and Oceanography* **28,** 1225–1231.

Fantasia, J. F., and H. C. Ingrao (1974). Development of Experimental Airborne Laser Remote Sensing System for the Detection and Classification of Oil Spills. *Proc. of the 9th Intern. Symp. on Remote Sensing of the Environment*, 15–19 April 1974, Paper 10700-1-X, 1711–1745.

Fantasia, J. F., T. M. Hard, and H. C. Ingrao (1971). An Investigation of Oil Fluorescence as a Technique for Remote Sensing of Oil Spills, Report No. DOT-TSC-USCG-71-7, Transportation Systems Center, Dept. of Transportation, Cambridge, MA.

Fork, R. L., C. V. Shank, and R. T. Yen (1982). Amplification of 70-fs Optical Pulses to Gigawatt Powers. *Appl. Phys. Lett.* **41,** 223–225.

Franks, L. A., G. A. Capelle, and D. A. Jessup (1983). Aerial Testing of an N_2 Laser Fluorosensor System. *Appl. Optics* **22,** 1717–1721.

Fredriksson, K., B. Galle, K. Nystrom, S. Svanberg, and B. Ostrom (1979). Marine Laser Probing: Results from a Field Test. Department of Physics, Chalmers University Report No. 245, Gotenberg, Sweden.

Friedman, E. J., and G. D. Hickman (1971). Laser-Induced Fluorescence in Algae: A New Technique for Remote Detection. Final Report, sponsored by NASA, Wallops Island, VA, under Contract NAS6-02081, Sparcom, Inc., Alexandria, VA.

Gardner, C. S. (1982). Target Signatures for Laser Altimeters: An Analysis. *Appl. Optics* **21,** 448–453.

Garrett, W. D. (1983). Frequency and Distribution of Natural and Pollutant Organic Sea Slicks. NRL Memorandum Report #5075, Naval Research Laboratory, Washington, D.C.

Gehlhaar, U. (1982). Computer Simulations and Theory of Oceanograpic Fluorescence Lidar Signals: Effect of Sea Surface Structure. *Appl. Optics* **21,** 3743–3755.

Goldberg, E. D., ed. (1979). Remote Sensing and Problems of the Hydrosphere. National Aeronautics and Space Administration (NASA) Conference Publication 2109, Washington, D.C.

Gordon, H. R. (1982). Interpretation of Airborne Oceanic Lidar: Effects of Multiple Scattering. *Appl. Optics* **21,** 2996–3001.

Gordon, H. R., D. K. Clark, J. W. Brown, R. H. Evans, and W. W. Broenkow (1983) Phytoplankton Pigment Concentrations in the Middle Atlantic Bight: Comparison of Ship Determinations and CZCS Estimates. *Appl. Optics* **22**, 20–36.

Govindjee, and R. Govindjee (1974). The Absorption of Light in Photosynthesis. *Sci. Am.* **231**(6), 68–82.

Grew, G. W., and L. S. Mayo (1983). Ocean Color Algorithm for Remote Sensing of Chlorophyll. NASA Technical Paper 2164, Langley Research Center, Hampton, VA.

Hayne, G. S. (1980). Radar Altimeter Mean Return Waveforms from Near-Normal-Incidence Ocean Surface Scattering. NASA Contractor Report 156864, prepared by Applied Science Associates under Contract NAS6-2810.

Hickman, G. D. (1973). Recent Advances in the Application of Pulsed Lasers to the Hydrosphere. NASA Conference on the Use of Lasers for Hydrographic Studies, NASA SP-375, 81–88.

Hickman, G. D., and J. E. Hogg (1969). Application of an Airborne Pulsed Laser for Near Shore Bathymetric Measurements. *Remote Sensing of Environment* **1,** 47–58.

Hickman, G. D., and R. B. Moore (1970). Laser Induced Fluorescence in Rhodamine and Algae. Proc. 13th Conf. Great Lakes Res., Int. Assoc. Great Lakes Res., 1–14.

Hoge, F. E. (1982). Laser Measurements of the Spectral Extinction Coefficients of Fluorescent, Highly Absorbing Liquids. *Appl. Optics* **21,** 1725–1729.

Hoge, F. E. (1983). Oil Film Thickness Using Airborne Laser-Induced Oil Fluorescence Backscatter. *Appl. Optics* **22,** 3316–3318.

Hoge, F. E., and J. S. Kincaid (1980). Laser Measurement of Extinction Coefficients of Highly Absorbing Liquids. *Appl. Optics* **19,** 1143–1150.

Hoge, F. E., and R. N. Swift (1980). Oil Film Thickness Measurement Using Airborne Laser-Induced Water Raman Backscatter. *Appl. Optics* **19,** 3269–3281.

Hoge, F. E., and R. N. Swift (1981a). Absolute Tracer Dye Concentration Using Airborne Laser-Induced Water Raman Backscatter. *Appl. Optics* **20,** 1191–1202.

Hoge, F. E., and R. N. Swift (1981b). Airborne Simultaneous Spectroscopic Detection of Laser Induced Water Raman Backscatter and Fluorescence from Chlorophyll-a, and other Naturally Occurring Pigments. *Appl. Optics* **20,** 3197–3205.

Hoge, F. E., and R. N. Swift (1981c). Application of the NASA Airborne Oceanographic Lidar to the Mapping of Chlorophyll and Other Organic Pigments. The Chesapeake Bay Plume Study, Superflux 1980, J. W. Campbell and J. P. Thomas, eds., NASA Conference Proceedings Document CP2188.

Hoge, F. E., and R. N. Swift (1982a). Airborne Laser-Induced Fluorescence of Upper Water Column Constituents of a Warm Core Gulf Stream Ring. *Trans. Amer. Geophys. Union* **63**(3), 60.

Hoge, F. E., and R. N. Swift (1982b). Delineation of Estuarine Fronts in the German Bight Using Airborne Laser-Induced Water Raman Backscatter and Fluorescence of Water Column Constituents. *Int. J. Remote Sening* **3**(4), 475–495.

Hoge, F. E., and R. N. Swift (1983a). Experimental Feasibility of the Airborne Measurement of Absolute Oil Fluorescence Spectral Conversion Efficiency. *Appl. Optics* **22,** 37–47.

Hoge, F. E., and R. N. Swift (1983b). Airborne Dual-Laser Excitation and Mapping of Phytoplankton Photopigments in a Gulf Stream Warm Core Ring. *Appl. Optics* **22,** 2272–2281.

Hoge, F. E., and R. N. Swift (1983c). Airborne Detection of Oceanic Turbidity Cell Structure Using Depth-Resolved Laser-Induced Water Raman Backscatter. *Appl. Optics* **22,** 3778–3786.

Hoge, F. E., and R. N. Swift (1985). Airborne Mapping of Laser-Induced Fluorescence of Chlorophyll *a* and Phycoerythrin in a Gulf Stream Warm Core Ring. Paper No. 18 in *Mapping Strategies in Chemical Oceanography*, A. Zirino, ed., *Advances in Chemistry Series*, No. 209, American Chemical Society, Washington, D.C., pp. 353–372.

Hoge, F. E., R. N. Swift, and E. B. Frederick (1980). Water Depth Measurement Using an Airborne Pulsed Neon Laser System. *Appl. Optics* **20,** 1191–1202.

Hoge, F. E., R. N. Swift, and J. K. Yungel (1983). Feasibility of Airborne Detection of Laser-Induced Fluorescence of Green Terrestrial Plants. *Appl. Optics* **22,** 2991–3000.

Hoge, F. E., W. B. Krabill, and R. N. Swift (1984). The Reflection of Airborne UV Laser Pulses from the Ocean. *Marine Geodesy* **8,** 313–344.

Horvath, R., W. L. Morgan, and S. R. Stewart (1971). Optical Remote Sensing of Oil Slicks: Signature Analysis and Systems Evaluation. Project 724104.2/1, Willow Run Laboratories, University of Michigan, Ann Arbor, Michigan.

Houghton, W. M. (1973). A Remote Sensing Laser Fluorometer, in *The Use of Lasers for Hydrographic Studies*, H. H. Kim and P. T. Ryan, eds., NASA SP-375, Proceedings of a Symposium held at Wallops Island, Virginia, September 12, 1973. NASA, Washington, D.C.

Houghton, W. M., R. J. Exton, and R. W. Gregory (1983). Field Investigation of Technique for Remote Laser Sensing of Oceanographic Parameters. *Remote Sensing of Environment* **13,** 17–32.

Huang, N. E., S. R. Long, C. C. Tung, Y. Yuan, and L. F. Bliven (1983). A Non-Gaussian Statistical Model for Surface Elevation of Nonlinear Random Wave Fields. *J. Geophys. Res.* **88,** 7597–7606.

Huang, N. E., S. R. Long, L. F. Bliven, and C. C. Tung (1984). The Non-Gaussian Joint Probability Density Function of Slope and Elevation for a Nonlinear Gravity Wave Field. *J. Geophys. Res.* **89,** 1961–1972.

Huffaker, R. M. (1978). Feasibility Study of Satellite-Borne Lidar Global Wind Monitoring System. National Aeronautics and Space Administration, Document TM-ERL-WPL-37.

Huhnerfuss, H., and W. D. Garrett (1981). Experimental Sea Slicks: Their Practical Applications and Utilization for Basic Studies of Air-Sea Interactions. *J. Geophys. Res.* **86,** 439–447.

Huhnerfuss, H., W. Alpers, and W. L. Jones (1978). Measurements at 13.9 GHz of Radar Backscattering Cross-Section of the North Sea Covered with an Artificial Surface Film. *Radio Science* **13,** 979–983.

Huhnerfuss, H., W. Alpers, W. L. Jones, P. A. Lange, and K. Richter (1981). The Damping of Ocean Surface Waves by a Monomolecular Film Measured by Wave Staffs and Microwave Radars. *J. Geophys. Res.* **86,** 429–438.

Huhnerfuss, H., W. Alpers, W. D. Garrett, P. A. Lange, and S. Stolte (1983a). Attenuation of Capillary and Gravity Waves at Sea by Monomolecular Organic Surface Films. *J. Geophys. Res.* **88,** 9809–9816.

Huhnerfuss, H., W. Alpers, A. Cross, W. D. Garrett, W. C. Keller, P. A. Lange, W. J. Plant, F. Schlude, D. L. Schuler (1983b). The Modification of *X* and *L* Band Radar Signals by Monomolecular Sea Slicks. *J. Geophys. Res.* **88,** 9817–9822.

Huhnerfuss, H., W. Garrett, and F. E. Hoge (1986). The Discrimination Between Crude Oil Spills and Monomolecular Sea Slicks by Airborne Lidar. submitted to *International Journal of Remote Sensing*, **7**(1), 137–150.

Jarrett, Olin, Jr., W. E. Esaias, C. A. Brown, Jr., and E. B. Pritchard (1981). Analysis of ALOPE Data from Superflux, in *Chesapeake Bay Plume Study, Superflux 1980*,

J. W. Campbell and J. P. Thomas, eds., NASA Conf. Pub. 2188, NOAA/NEMP III 81 ABCDFG 0042, NASA Langley Research Center, Hampton, VA, pp. 405–411.

Jelalian, A. V. (1968). Sea Echo at Laser Wavelengths. *Proc. IEEE* **56,** 828–835.

Jensen, H. (1967). Performance of an Airborne Laser Profiler, in Airborne Photo-Optical Instrumentation; Proceedings of a Seminar In-Depth, Society of Photo-Optical Instrumentation Engineers, pp. VIII-1–VIII-6.

Jensen, H., and K. A. Ruddock (1965). Applications of a Laser Profiler to Photogrammetric Problems. Paper presented to Amer. Soc. of Photogrammetry, Washington, D.C.

Jerlov, N. G. (1976). *Marine Optics*, Oceanography Series 14, Elsevier, Amsterdam.

Kahn, W. D., F. O. Vonbun, D. E. Smith, T. S. Englar, and B. P. Gibbs (1979). Performance Analysis of the Spaceborne Laser Ranging System. NASA Technical Memorandum 80330, Goddard Space Flight Center, Greenbelt, MD.

Kalle, K. (1966). The Problem of the Gelbstoff in the Sea. *Oceanogr. Mag. Biol. Ann. Rev.* **4,** 91–104.

Kanevskiy, V. A., V. S. Fedak, F. v. Ryazantsev, V. P. Il'in, Yu. R. Shelyag-Sosonko, S. M. Kochubey, and M. G. Sosonkin (1983). Opyt Primeneniye Electronno-Opticheskikh Sistem Dlya Distantsionnogo Issledovaniya Lyuminetsentsii Rastency. *Issledovaniye Zemli iz Kosmosa* **6,** 87–90.

Karp, S. (1976). Optical Communication between Underwater and above Surface (Satellite) Terminals. *IEEE Trans. Comm.* **COM-24,** 66–81.

Kasden, H. R., J. L. Chimenti, and J. P. deNeufville (1981). Selective Detection of Uranium by Laser-Induced Fluorescence: A Potential Remote-Sensing Technique. 2: Experimental Assessment of the Remote Sensing of Uranyl Geological Targets. *Appl. Optics* **30,** 1297–1307.

Kats, A. V., and I. S. Spevak (1980). Reconstruction of the Seawave Spectra from the Measurements of Moving Sensors. *Izv. Atmos. Ocean. Phys.* **16,** 194–200.

Kaula, W. M., G. Schubert, R. E. Lingenfelter, W. L. Sjogren, and W. R. Wollenhaupt (1972). Analysis and Interpretation of Lunar Laser Altimetry. Proceedings of Third Lunar Science Conference (Supplement 3, *Geochemica et Cosmochimica Acta*) **3,** 2189–2204.

Ketchem, R. D., Jr. (1971). Airborne Laser Profiling of the Arctic Ice Pack. *Remote Sensing of Environment* **2,** 41–52.

Kim, H. H. (1973). New Algae Mapping Technique by the Use of an Airborne Laser Fluorosensor. *Appl. Optics* **12,** 1454–1459.

Kim, H. H. (1977). Airborne Bathymetric Charting Using Pulsed Blue-Green Lasers. *Appl. Optics* **16,** 46–56.

Kim, H. H., P. O. Cervenka, and C. B. Lankford (1975). Development of an Airborne Laser Bathymeter. NASA TN-D-8079, October.

Kobrick, M., and C. Elachi (1981). A Shuttle Scanning Laser Altimeter for Topographic Mapping, in *Proceedings International Symposium on Remote Sensing of Environ-*

ment, Vol. 2, Environmental Research Institute of Michigan, Ann Arbor, MI, pp. 711–714.

Koerber, R. W., and D. M. Phillips (1982). Monte Carlo Simulation of Laser Backscatter from Sea Water. Tech. Report ERL-0224-TR, Department of Defense, Defense Science and Technology Organization, Electronics Research Laboratory, Adelaide, South Australia.

Krabill, W. B., and C. F. Martin (1983). Analysis of Aircraft Vertical Positioning Accuracy Using a Single Accelerometer. NASA Technical Memorandum 84417, Goddard Space Flight Center, Wallops Flight Facility, Wallops Island, VA.

Krabill, W. B., and R. N. Swift (1982). Preliminary Results of Shoreline Mapping Investigations Conducted At Wrightsville Beach, NC, in Proceedings of Surveying Requirements Meeting, February 2–5, 1982, U.S. Army Engineer Waterways Experiment Station, Vicksburg, MS, pp. 261–280.

Krabill, W. B., J. G. Collins, R. N. Swift, and M. L. Butler (1980). Airborne Laser Topographic Mapping Results from Initial Joint NASA/U.S. Army Corps of Engineers Experiment. NASA Technical Memorandum No. 73287, Wallops Island, VA.

Krabill, W. B., F. E. Hoge, and C. F. Martin (1982). Baseline Monitoring Using Aircraft Laser Ranging. NASA Technical Memorandum No. 73298, Wallops Island, VA.

Krabill, W. B., J. G. Collins, L. E. Link, R. N. Swift, and M. L. Butler (1984). Airborne Laser Topography Mapping Results. *Photogrammetric Engineering and Remote Sensing* **50,** 685–694.

Kung, R. T. V., and I. Itzkan (1976). Absolute Oil Fluorescence Conversion Efficiency. *Appl. Optics* **15,** 409–415.

Leonard, D. A., B. Caputo, and F. E. Hoge (1979). Remote Sensing of Subsurface Water Temperature by Raman Scattering. *Appl. Optics* **13,** 1732–1745.

Lerner, R. M., and J. D. Summers (1982). Monte Carlo Description of Time- and Space-Resolved Multiple Scatter in Natural Water. *Appl. Optics* **21,** 861–869.

Link, L. E., W. B. Krabill, and R. N. Swift (1983a). Prospectus on Airborne Laser Mapping System, in *Highway Geometries, Interactive Graphics, and Laser Mapping*, Transportation Research Record No. 923, National Research Council, National Academy of Sciences, Washington, D.C., pp. 1–7.

Link, L. E., W. B. Krabill, and R. N. Swift (1983b). A Prospectus on Airborne Laser Mapping Systems. Proceedings of Symposium and Workshop on Remote Sensing and Mineral Exploration, Ottawa, Canada, May 16–June 2, 1982, *Advances in Space Research*, Vol. 3, 309–322.

Long, R. B. (1979). On Surface Gravity Wave Spectra Observed in a Moving Frame of Reference. National Oceanic and Atmospheric Administration Technical Memorandum ERL AOML-38.

Luchinin, A. G. (1979). Influence of Wind Waves on the Characteristics of the Light Field Backscattered by the Bottom and the Intervening Water. *Izv., Atmos. Oceanic Phys.* **15,** 53–533.

Luchinin, A. G., and I. A. Sergiyevskaya (1982). On the Light Field Fluctuations Beneath a Wavy Sea Surface. *Izv. Atmos. and Oceanic Phys.* **18,** 656–661.

Maclean, G. A. (1982). Timber Volume Estimation Using Cross-Sectional Photogrammetric and Densitometric Methods, Masters Thesis, University of Wisconsin, p. 227.

Makarevich, S. A., A. P. Ivanov, and G. K. Il'ich (1969). The Structure of a Narrow Light Beam Emerging from a Layer of a Scattering Medium. *Izv., Atmos. Oceanic Phys.* **5,** 77–83.

Mamon, G., D. G. Youmans, Z. G. Sztankay, and C. E. Mongan (1978). Pulsed GaAs Laser Terrain Profiler. *Appl. Optics* **17,** 868–877.

McClain, C. R., D. T. Chen, and D. L. Hammond (1980). Gulf Stream Ground Truth Project: Results of the NRL Airborne Sensors. *Ocean Eng.* **7,** 55–97.

McClain, C. R., D. T. Chen, and W. Hart (1982a). On the Use of Laser Profilometry for Ocean Wave Studies. *J. Geophys. Res.* **87,** 9509–9515.

McClain, C. R., N. E. Huang, and P. E. LaViolette (1982b). Measurements of Sea-State Variations Across Oceanic Fronts Using Laser Profilometry. *J. Phys. Ocean* **12,** 1228–1244.

McGoogan, J. T., and E. J. Walsh (1979). Real-Time Determination of Geophysical Parameters from a Multibeam Altimeter, in *Remote Sensing of Earth from Space: Role of Smart Sensors*, Roger A. Breckenridge, ed., Vol. 67, *Progress in Astronautics and Aeronautics*, Paper No. 78-1735 of AIAA/NASA Conference on Smart Sensors Nov. 14–16, 1978, Hampton, VA.

Measures, R. M. (1984). *Laser Remote Sensing*, Wiley, New York.

Measures, R. M., and M. Bristow (1971). The Development of a Laser Fluorosensor for Remote Environmental Probing. Joint Conference on Sensing of Environmental Pollutants, Palo Alto, Nov. 1971, AIAA Paper 71-112; *Can. Aeron. Space J.* **17,** 421–422.

Measures, R. M., W. R. Houston, and D. G. Stephenson (1974). Laser Induced Fluorescence Decay Spectra—A New Form of Environmental Signature. *Opt. Eng.* **13,** 494–450.

Measures, R. M., J. Garlick, W. R. Houston, and D. G. Stephenson (1975). Laser Induced Spectral Signatures of Relevance to Environmental Sensing. *Can. J. Remote Sensing* **1,** 95–102.

Menzies, R. T., and M. S. Shumate (1978). Tropospheric Ozone Distribution Measured with an Airborne Laser Absorption Spectrometer. *Geophys. Res.* **83,** 4039–4043.

Mumola, P. B., and H. H. Kim (1972). Remote Sensing of Marine Plankton by Dye Laser Induced Fluorescence. Proc. of Conference, Engineering in the Ocean Environment. *IEEE* **3,** 204–207.

Mumola, P. B., O. Jarrett, Jr., and C. A. Brown, Jr. (1973). Multiwavelength Lidar for Remote Sensing of Chlorophyll *a* in Algae and Phytoplankton. NASA Conference on the Use of Lasers for Hydrographic Studies, NASA SP-375, 137–145.

Murphee, D. L., C. D. Taylor, J. K. Owens, H. R. Ebersole, and R. W. McClendon (1973). Mathematical Modeling and System Recommendations for the Detection of

Fish by an Airborne Remote Sensing Laser. Report No. EIRS-ASE-74-2, Mississippi State University, Engineering and Industrial Research Station, Contract 03-3-042-27 for the National Marine Fisheries Service, Bay St. Louis, MI.

Myers, E. P., and C. G. Gunnerson (1976). Hydrocarbons in the Oceans. NOAA Marine Ecosystems (MESA) Analysis Program Special Report, Boulder, CO.

Nelson, R., W. B. Krabill, and G. Maclean (1984). Determining Forest Canopy Characteristics Using Airborne Laser Data. *Remote Sensing of Environment* **15,** 201–212.

Noble, V. E., R. D. Ketchum, and D. B. Ross (1969). Some Aspects of Remote Sensing as Applied to Oceanography. *Proc. IEEE* **57,** 495–604.

O'Neil, R. A. (1983). Coastal Hydrography Using the CCRS Lidar Bathymeter. Paper presented at Conference on Ocean Environment and Its Interaction with Off Shore Structures, The Swedish Trade Fair Foundation, Goteborg, Sweden, March 1–4, 1983.

O'Neil, R. A., A. R. Davis, H. G. Gross, and J. Kraus (1973). A Remote Sensing Laser Fluorometer, in *The Use of Lasers for Hydrographic Studies*, H. H. Kim and P. T. Ryan, eds., NASA SP-375, Proceedings of a Symposium held at Wallops Island, Virginia, September 12, 1973.

O'Neil, R. A., L. Buja-Bijunas, and D. M. Rayner (1980). Field Performance of a Laser Fluorosensor for the Detection of Oil Spills. *Appl. Optics* **19,** 863–870.

O'Neil, R. A., F. E. Hoge, and M. P. F. Bristow (1981). The Current Status of Airborne Laser Fluorosensing. Proc. of Fifteenth Int. Sym. On Remote Sensing of Environment, Ann Arbor, Michigan, pp. 379–398.

Penny, M. F. (1972). WREMAPS: System Evaluation. WRE-Technical Note-811(AP), Dept. of Supply, Australian Defence Scientific Service, Weapons Research Establishment, Adelaide, Australia.

Penny, M. F. (1972). Laser Hydrography in Australia. Technical Report ERL-0229-TR, Department of Defence, Electronics Research Laboratory, Adelaide, South Australia.

Petri, K. I. (1977). Laser Radar Reflectance of Chesapeake Bay Waters as a Function of Wind Speed. *IEEE Trans. Geosci. Electron.* **GE-15,** 87–97.

Petzold, T. J. (1972). Volume Scattering Functions for Selected Ocean Waters. Final Report Ref. 73-8, Scripps Institution of Oceanography, La Jolla, California.

Phillips, D. M. (1979). Effects of the Wavenumber Spectrum of a Sea Surface on Laser Beam Reflection. *Aust. J. Phys.* **32,** 469–89.

Phillips, D. M., and B. W. Koerber (1984). A Theoretical Study of an Airborne Laser Technique for Determining Sea Water Turbidity. *Australian J. Phys.* **37,** 75–90.

Phillips, D. M., R. H. Abbot, and M. F. Penny (1984). Remote Sensing of Sea Water Turbidity with an Airborne Laser System. *J. Phys. D. Appl. Physics, Great Britain* **17,** 1749–1758.

Poole, L. R., and W. E. Esaias (1982). Water Raman Normalization of Airborne Laser Fluorosensor Measurements: A Computer Study. *Appl. Optics* **21,** 3756–3761.

Poole, L. R., and W. E. Esaias (1983). Influence of Suspended Inorganic Sediment on Airborne Fluorosensor Measurements. *Appl. Optics* **22,** 380–381.

Rayner, D. M., and A. G. Szabo (1978). Time-Resolved Laser Fluorosensors: A Laboratory Study of their Potential in the Remote Characterization of Oil. *Appl. Optics* **17,** 1624–1630.

Rayner, D. M., M. Lee, and A. G. Szabo (1978). Effect of Sea-State on the Performance of Laser Fluorosensors. *Appl. Optics* **17,** 2730–2733.

Remsberg, E. E., and L. L. Gordley (1978). Analysis of Differential Absorption Lidar from the Space Shuttle. *Appl. Optics* **17,** 624–630.

Riley, J. P., and R. Chester (1971). *Introduction to Marine Chemistry*, Academic Press, New York.

Romanova, L. M. (1968). Light Field in the Boundary Layer of a Turbid Medium with Strongly Anisotropic Scattering Illuminated by a Narrow Beam. *Izv., Atmos. Oceanic Phys.* **4,** 1185–1196.

Ross, D. B., R. A. Peloquin, and R. J. Sheil (1968). Observing Ocean Surface Waves with a Helium-Neon Laser. *Proc. 5th Symp. on Military Oceanography*, U.S. Navy Mine Defense Laboratory, Panama City, Florida.

Ross, D. B., V. J. Cardone, and J. W. Conaway, Jr. (1970). Laser and Microwave Observations of Sea-Surface Conditions for Fetch-Limited 17-to-25 m/sec Winds. *IEEE Transactions Geoscience Electronics* **GE-8,** 326–336.

Ryan, D. K., and J. H. Weber (1982). Copper (II) Complexing Capacities of Natural Waters by Fluorescence Quenching. *Environ. Sci. Tech.* **16,** 866–872.

Saar, R. A., and J. H. Weber (1980). Comparison of Spectrofluorometry and Ion-Selective Electrode Potentiometry for Determination of Complexes Between Fulvic and Heavy-Metal Ions. *Anal. Chem.* **52,** 2095–2099.

Sarabun, C. C., Jr. (1981). Mapping Watermass Boundaries Using Fluorosensing Lidar. The Chesapeake Bay Plume Study, Superflux 1980, J. W. Campbell and J. P. Thomas, eds., NASA Conference Proceedings Document CP2188.

Sato, T., Y. Suzuki, H. Kashiwagi, M. Nanjo, and Y. Kasui (1978). Laser Radar for Remote Detection of Oil Spills. *Appl. Optics* **17,** 3798–3803; A Method for Remote Detection of Oil Spills Using Laser-Excited Raman Backscattering and Backscattered Fluorescence. *IEEE J. Oceanic Eng.* **OE-E**(1), 1–4.

Schau, H. C. (1978). Measurement of Capillary Wave Slopes on the Ocean. *Appl. Optics* **17,** 15–17.

Sick, L. V., C. C. Johnson, and A. Engel (1978). Trace Metal Enhancement in the Biotic and Abiotic Components of an Estuarine Tidal Front. *J. Geophys. Res.* **83,** 4659–4667.

Sorenzen, C. J. (1966). A Method for the Continuous Measurement of "In Vivo" Chlorophyll Concentration. *Deep Sea Res.* **13,** 223–227.

Sorenzen, C. J. (1970). The Biological Significance of Surface Chlorophyll Measurements. *Limnology and Oceanography* **15,** 479–480.

Steinvall, O., H. Klevebrant, J. Leander, and A. Widen (1981). Laser Depth Sounding in the Baltic Sea. *Appl. Optics* **20,** 3484–3286.

Stewart, D. E. (1982). A Method for the Extraction and Quantitation of Phycoerythrin from Algae. NASA Contractor Report 165996 by Bionetics Corporation under Contract NAS1-16978, Hampton, VA.

Stotts, L. B., and S. Karp (1982). Wave Slope Statistics Derived from Optical Radiance Measurements Below the Sea Surface. *Appl. Optics* **21,** 978–981.

Struum, W., and J. Morgan (1981). *Aquatic Chemistry*, Wiley, New York.

Swennen, J. P. J. W. (1968). Time-Average Surface-Reflected Energy Received from a Collimated Beam of Radiant Energy Normally Incident on the Ocean Surface. *J. Opt. Soc. Am.* **58,** 47–51.

Sylvia, A. E., D. A. Bancroft, and J. D. Miller (1974). Detection and Measurements of Microorganics in Drinking Water by Fluorescence. *Am. Water Works Assoc. Tech. Conf. Proc.* Dec. 2–3, 1974, Dallas, TX.

Tang, S. (1981). Field Investigation of the Short Wave Modulation by Long Waves. Dissertation, University of Florida, Gainesville, Florida.

Tang, S., and O. H. Shemdin (1983). Measurement of High Frequency Waves Using a Wave Follower. *J. Geophy. Res* **88,** 9832–9840.

Tooma, S. G., R. D. Ketchum, Jr., R. A. Mennella, and J. P. Hollinger (1976). Comparison of Sea-Ice Type Identification Between Airborne Dual-Frequency Passive Microwaver Radiometry and Standard Laser/Infrared Techniques. *J. Glaciology* **15,** 225–238.

Tsai, B. M., and C. S. Gardner (1982). Remote Sensing of Sea State Using Laser Altimeters. *Appl. Optics* **21,** 3932–3940.

Vertucci, F. A. and A. Vodacek (1985). The Remote Sensing of Lake Acidification Using Laser Fluorosensing. Tech. Papers of the American Society Photogrammetry, 51st Annual Meeting, 10–15 March, Vol. 2, 793–802. Washington, D.C.

Visser, H. (1979). Teledetection of the Thickness of Oil Films on Polluted Water Based on the Oil Fluorescence Properties. *Appl. Optics* **18,** 1746–1749.

Vodacek, A., and W. D. Philpot (1985). Use of Induced Fluorescence Measurements to Assess Aluminum-Organic Interactions in Acidified Lakes. Technical Papers of the American Society of Photogrammetry, Washington, D.C., 51st Annual Meeting, vol. 2, March 11–15, 460–469.

Walsh, E. J., D. W. Hancock III, D. E. Hines, and J. E. Kinney (1984). Electromagnetic Bias of 36 GHz Radar Altimeter Measurements of MSL. *Marine Geodesy* **8,** 265–296.

Watson, R. D. (1981). Airborne Fraunhofer Line Discriminator Surveys in Southern California, Nevada, and Central New Mexico, in *Workshop on Applications of Luminescence Techniques to Earth Resource Studies*, W. R. Hemphill and M. Settle, eds., LPI Technical Report 81-03, Lunar and Planetary Institute, Houston, pp. 19–35.

Yentsch, C. S. (1973). The Fluorescence of Chlorophyll and Yellow Substances in Natural Water: A Note on the Problems of Measurements and the Importance of Their Remote Sensing, in *The Use of Lasers for Hydrographic Studies*, H. H. Kim and P. T. Ryan, eds., NASA SP-375, Proceedings of a Symposium held at Wallops Island, Virginia, September 12, 1973.

Yentsch, C. S. and D. W. Menzel (1963). A Method for the Determination of Phytoplankton Chlorophyll and Phaeophytin by Fluorescence. *Deep Sea Research* **10,** 221–231.

Yentsch, C. M., and C. S. Yentsch (1984). Emergence of Optical Instrumentation for Measuring Biological Properties. *Oceanogr. Mar. Biol. Ann. Rev.* **22,** 55–98.

Yura, H. T. (1973). Propagation of Finite Cross-Section Laser Beams in Sea Water. *Appl. Optics* **12,** 108–115.

Zimmerman, A., and A. R. Bandy (1975). The Application of Laser Raman Scattering to Remote Sensing of Salinity and Turbidity. Final Technical Report, Old Dominion University Research Foundation for NASA Langley Research Center under Master Contact Agreement NAS1-11707, Task Authorization No. 33.

Zimmerman, A. V., F. W. Paul, and R. J. Exton (1976). Research and Investigation of the Radiation Induced by a Laser Beam Incident on Sea Water. Contractor Final Report CR-14519, NASA Grant NSG-1096, Chesapeake College, Wye Mills, MD.

CHAPTER

7

FIBER-OPTIC REMOTE CHEMICAL SENSING

ULRICH J. KRULL and R. STEPHEN BROWN

Chemistry Department
University of Toronto
Ontario, Canada

7.1 INTRODUCTION

7.1.1 Conventional Remote Sensing for Chemical Analysis: Lidar

For many years, scientists have been probing the atmosphere with numerous instruments to determine various species concentrations, interactions, and motions. Knowledge of atmospheric conditions has been the fundamental incentive for development of remote sensing systems in situations where the targeted species may be hundreds of meters from the measurement device. However, such studies have been difficult since most commonly the instrumentation had to be elevated to the atmospheric level containing the species of interest. Such experiments are necessarily limited with regard to observation duration and sampling domain. A significant advance in remote sensing was achieved by development of ground-based passive sensing systems that could observe solar electromagnetic radiation that was scattered, emitted, or absorbed by atmospheric constituents. Severe limitations of this strategy still exist since the source of sample excitation and its atmospheric penetration depth cannot be controlled. The introduction of active sensing systems based on laser excitation sources has largely eliminated these problems, allowing remote sensing of the atmosphere at all times and at all levels (Measures, 1984).

Instrumentation is designed to be either monostatic or bistatic in operation and consists of a powerful UV, optical, or IR laser, an optical system for beam collimation as well as temporal and wavelength filtering, and a second optical receiver system for light collection, intensity measurement, and signal processing. The physical processes involved in signal generation provide an indication of the broad diversity of analytical parameters available for observation by remote optical techniques. Such processes include reflection, scatter, fluorescence, and differential absorption, which can be observed with respect to temporal features, wavelength shifts, and optical polarization.

The advent of active sensing systems for atmospheric analysis was in large part due to the evolution of laser technology, which provided the necessary wavelengths and excitation intensities. These latter parameters indicate major deficiences that exist in the experiment due to sample matrix limitations. Many samples do not provide sufficient molecular molar extinction coefficients or quantum yields to make them readily detectable at standard ambient concentrations, necessitating employment of very high intensity excitation. Limitations in available excitation wavelengths have been largely overcome by laser dye tuning, but often at the expense of intensity. The intensity problem is further compounded by the $1/r^2$ radiation power reduction with distance r and by interference from other airborne species. Dust, particulate matter, aerosol, and chemical interferents in the beam path are uncontrollable aspects of the laser remote sensing experiment that make replication and quantitative analysis extremely difficult. The physical limitations responsible for weak analytical signal evolution and substantial signal attenuation with distance have confined direct remote laser sensing to expensive, cumbersome instrumental systems with an intense laser, large high-quality optics, and very sensitive detection equipment usually capable of only line-of-site sampling and only suitable over relatively short distances.

Identification of the distance and interferent parameters as those of greatest impact on signal limitations clearly exemplifies the enormous advantages that could be achieved by use of guided optics. The development of fiber-optic technology for long-distance communications has provided the basis for laser remote sensing experiments that can avoid the most significant limitations encountered in conventional direct laser probing situations. Furthermore, an exciting vista of new opportunities for remote distributed sensing of solutions and gases exists by combining specialized selective chemistry with fiber-optic guides to produce active dedicated chemical sensors.

Fiber-optic sensors have been developed as a response to increasing demands placed on sensors, including greater sensitivity, selectivity, versatility, remote capability, and freedom from interferences. In many situations, an optical sensing strategy fulfills these demands.

Optical sensing probes were first developed as alternative sensors to physical or electrical probes. The latter devices measure parameters such as temperature and pressure. Systems designed as chemical sensors naturally followed, mainly using spectroscopic transduction mechanisms. The progress in the area of fiber optics has caused an increased interest in sensor development for two reasons. First, remote operation of established selective, spectroscopically based devices became possible, making them more versatile, and replacing nonoptical remote sensors. Second, new optical sensing strategies intrinsic to the optical fiber have resulted in additional unique sensors.

7.1.2 Chemical Sensor Categories and Remote Sensing Opportunities

Chemical sensors are devices that can quantitatively detect a specific analyte by means of continuous transduction of a dedicated localized selective chemical interaction. The four major categories of chemical sensors share in common a basic sensing philosophy where a chemically selective reagent phase is physically bound to a device structure suitable for probing any chemical interactions (Alberry et al., 1986; Schuetzle and Hammerle, 1986; Seiyama, 1983; Thompson and Krull, 1984). Chemical sensors generally strive to achieve operation analogous to the widely used physical sensor systems, which can monitor parameters such as temperature or pressure. Table 7.1 summarizes the ideal attributes of chemical sensors and biosensors. The main problems facing the chemical sensor field originate with the unique chemistry required in selective matrices and the physical marriage of the selective chemistry to suitable device structures. The chemical sensor categories are largely defined by the device structures and are introduced in the following sections.

7.1.2.1 Electrochemical Sensors

The field of electrochemical sensors covers a broad range of topics that can be divided into the two main categories of ion-selective electrodes and electrolytic

Table 7.1. Characteristics of "Ideal" Chemical Sensors

Attribute	Comments
Selectivity	Desire specific interaction with analyte, no signal from matrix
Sensitivity	Provides highest possible response to change in concentration of analyte while maintaining linear concentration–response curve
Detection limit	As low as possible, to 10^{-15} M for some toxins and environmental contaminants
Response speed	Full-signal development within 1 s
Reversibility	Regeneration of analytical response within 1 s without loss of binding affinity
Size	Small sizes desirable for implantation into most sample matrices
Dependability/simplicity	Reliable operation by nonexpert personnel, also physically rugged, operational for years
Biocompatibility	For interfacing with living systems as monitor/controller
Cost	Low cost for disposible or widespread continuous use

systems. Ion-selective electrodes have been available for decades in the form of the glass or pH electrode, which is one of the most common analytical measurement instruments. The ion selectivity of these systems originates from a ubiquitous solid or liquid membrane containing chemical binding sites for certain species based on ionic charge, size, and mobility. The development of localized concentrations of ions can be monitored as a logarithmic variation of the electrode potential relative to that of a standard voltage provided by a reference electrode. This provides for a sensor that can respond to a very wide range of concentrations but at the same time is sensitive to large concentration errors with minor variations of measured millivolt potentials. These systems require an indicator and stable reference electrode pair, with electrical noise filters even for measurements performed in close proximity to the millivolt meter. The cumbersome sensor configuration, sensitivity to error based on minor potential variations, and susceptibility to external interference limit the applicability of ion-selective electrodes in remote sensing configurations. This is true for all forms of ion-selective electrodes, including those that have been modified for biochemical analysis as in the case of enzyme electrodes.

The second form of electrochemical transducer is based on amperometric or coulometric analysis where an analyte or reaction product is selectively oxidized or reduced, and the resulting electronic current is measured. This strategy affords some improvement in the sense that a variable potential is not measured, and the logarithmic concentration–signal relationship is avoided. However, the electronic currents that must be quantitatively measured can be quite small, and these systems are therefore again susceptible to extraneous electrical noise effects in remote sensing configurations.

7.1.2.2 Chemically Sensitive Field-Effect Transistors (CHEMFET)

Much research has been directed toward miniaturizing ion-selective electrodes and eliminating their electronic noise problems. One approach that has achieved widespread interest involves the interfacing of conventional potentiometric selective membrane chemistry to the gate region of a field-effect transistor (FET). Modulation of the gate voltage can control an electronic current within the transistor in a configuration that may occupy an area less than 1 mm^2. The transistor can also include further microelectronic hardware such as amplifiers and electrical noise filters. The close proximity of the device and the selective chemistry responsible for voltage development allows chemical control of an underlying measurable electronic current. Since voltages are not measured over significant distances, this should ideally provide systems of greatly improved signal stability. Unfortunately, the chemically sensitive FETs are not physically stable and suffer from electronic noise. Furthermore, transmission of the small transistor currents over large distances may again prove to be a major limitation in remote sensing applications.

7.1.2.3 Piezoelectric Sensors

Piezoelectric devices have found use as frequency controllers in many microelectronics applications. These systems are usually based on specially cut quartz crystals that can undergo lattice oscillation when placed in an alternating electric field. The mechanical oscillation of the crystal is very regular and can be accurately measured with a frequency counter and associated electronics. It has been established that minor mass changes or microviscosity changes caused by selective chemical interactions (Thompson et al., 1986) on the surface of such crystals can lead to substantial alterations of the fundamental oscillation frequency. The analytical signal is measured as a change in the frequency of an alternating electronic signal that can be amplified to substantial magnitude. The piezoelectric system may be suitable for remote sensing since absolute electronic signals need not be quantitatively evaluated. Significant limitations presently exist in deriving substantial chemical selectivity since nonselective surface adsorption can dominate the evolution of frequency changes.

7.1.2.4 Optical Sensors

Spectrophotometric techniques are among the oldest and best established methods in the arsenal of the analyst. The most common analysis for chemical speciation employs electromagnetic radiation at visible or near-visible wavelengths to interact with specific chromophores by electronic or vibrational excitation. The absorption of radiation by the sample can be quantitatively related to species concentration by the Beer–Lambert law. In a small sensor configuration the sample pathlength is often physically limited, thereby reducing analytical sensitivity. This can be particularly difficult if the chromophore concentration remains limited and the chromophore absorbs relatively weakly. The use of inherently more sensitive fluorescence modes and intense optical sources can overcome analytical signal limitations in some instances. A number of miniature cuvette systems for optical analysis have been prepared as dedicated chemical sensors. They generally operate by interfacing a compartmentalized selective reagent solution to the sample by means of a semipermeable membrane such as dialysis tubing. Such systems are in themselves limited in remote sensing applications since they operate by means of external energy input in contrast to the passive electrochemical systems. The optical radiation required for excitation and observation must travel a well-defined pathway without other physically or chemically induced intensity reduction or wavelength shift. These restrictions have led to the development of an optical train based on fiber-optic technology, producing a selective sensor known as an optrode (i.e. optical analogy to an electrode).

7.1.3 Light-Guided Chemical Sensor Strategy

The interfacing of optical fibers with selective wet chemical cells provides a flexible sensing arrangement that can be very small and may be operated without pathlength interference in a remote sensing capacity (Hirschfeld, 1986; Milanovich and Hirschfeld, 1984). Fiber-optic technology is capable of transmitting radiation by total internal reflection over distances of hundreds of meters without major intensity losses and eliminates much of the matrix interference found in conventional lidar experiments. The optical environment is well defined and easily predicted, and high-intensity radiation can be delivered to the target over large distances. In contrast to all other major sensor types, which rely on information transfer via electrical signals, the fiber-optic sensors are the only systems that can guarantee delivery and return of analytical information over extended distances in a form unaltered by electronic perturbations in the external environment. Further attractive features of fiber-optic systems with respect to the other major sensor types can be summarized as follows (Seitz, 1984; Wolfbeis, 1985):

1. Optical sensors are not restricted by the need for a reference signal as required in potentiometric systems, thereby reducing the complexity and potential for error.

2. Miniaturization is relatively simple and has provided sensor heads significantly smaller than achieved for electrochemical cells and chemically sensitive FET systems.
3. Selective chemistry unavailable to electrodes can open new horizons for chemical analysis and particularly biochemical assay.
4. Optical systems provide opportunities for multiwavelength analysis, which could permit simultaneous analysis of a number of different chemical species. This could be employed for actual quantitative multispecies analysis or, perhaps more significantly, for monitoring of the analytical reagent phase stability including criteria for temperature, ionic strength, or pH corrections.
5. Other optical parameters can also increase analytical flexibility relative to devices that rely on electrical transmission. Optical radiation can vary in polarization, phase, and temporal characteristics as well as wavelength and intensity to carry multidimensional information in a single fiber.

The disadvantages of fiber-optic sensors are not uncharacteristic of all the other major sensor types (Thompson and Krull, 1984). The selective chemistry of these systems all employ immobilized indicator phases, which tend to degrade with time. This process can be enhanced in optical systems by photobleaching, particularly when high-intensity laser sources are employed. The optical response thermodynamics do not obey Nernstian characteristics, and the dynamic concentration range may cover only three decades of concentration, thereby reducing flexibility of measurement of samples of widely differing concentration (Wolfbeis, 1985). Beer–Lambert law failure, concentration quenching, and inner filter effects are also possible. Careful shielding from ambient light, employment of only the minimum required radiation intensity, and design of the selective chemistry for the required concentration range are characteristics that are readily achievable for optrode development.

7.2 INSTRUMENTAL CONFIGURATIONS OF FIBER-OPTIC SENSING SYSTEMS

The components of a fiber-optic sensor are similar in principle to those required in any optical excitation experiment. The complexity of a dedicated sensing device is determined by the desired performance and application (Seitz, 1984). The source can consist of an inexpensive tungsten lamp or light-emitting diode combined with appropriate filters, though laser excitation provides an ideal source for long-distance transmission. The detector can consist of a phototube or photodiode and in the case of fluorescence is usually preceded by a decoupler to spectrally separate excitation radiation from the analytical signal. It is possible to employ a conventional spectrophotometer as the source and detector for observation of remote chemistry. One possible fluorometric configuration makes use of optical analysis with a double-grating Raman-type spectrometer that can separate the fluorescence signal from source radiation geometrically. Such a system would contain quantum counting systems and would in effect be a high-quantum-efficiency spectrofluorimeter (Hirschfeld, 1986).

The optical fiber and its marriage to the selective chemistry constitute the most analytically significant components of the sensor. Optical communication technology provides inexpensive fibers capable of optical transmission suitable for conventional laboratory instrumentation over distances of 1 km. The fiber characteristics are determined by their chemical composition, cross-sectional thickness and regularity, gradient indexing, cladding, and length. Accessible wavelength range is partially controlled by choice of fiber material; fused silica for measurements from 220 nm, glass for standard optical ranges, and plastic for wavelengths greater than 450 nm. Fiber composition can also be more complex and can include liquid-core fibers and highly birefringent polarization fibers (Payne, 1984). In the case of liquid-core systems, a composite structure of capillary glass with a liquid filler provides optical properties that vary with respect to temperature and pressure to a much larger extent than glass. Sensor operation can be based on changes in refractive index or Rayleigh scattering, which can be induced by chemical means in combination with intrinsic fiber sensors (see Section 7.2.1.1). Polarization-maintaining fibers employ optical birefringence originating in the fiber from anisotropic thermal stress induced by two areas of high-expansion glass placed on either side of the core. Such fibers can transmit linearly polarized radiation with little effect from environmental factors and are not very sensitive to physical twists or bends as experienced in coiled configurations.

The experimental flexibility achieved by manipulating fiber core thickness and refractive index distribution provides waveguides that can act by multiple total internal reflection with many simultaneous optical modes or single-mode systems that are inherently small and can make use of nonlinear optical effects. The diameter of the optical fiber can be reduced to the order of the wavelength of the transmitted radiation. Practically, chemical sensors in the submillimeter size range have been reported.

The fibers can serve to deliver and return optical radiation from a sample and must be efficiently interfaced to the chemistry (Milanovich and Hirschfeld, 1984). A sensor configuration that employs a separate wet chemical solution cell as a miniature reaction flask can be probed by fibers that are terminated by polished planar surfaces normal to the fiber length. Alternatively, the radiation can be focused onto the fiber by means of a spherical lens mounted at the end of the polished planar termination to improve reliability and performance. Such a lens must possess characteristics of chemical inertness, low optical absorptivity, and small size. Sapphire beads have been deemed appropriate for this purpose and have been used to concentrate excitation radiation onto smaller portions of reagent solution while collecting larger fractions of the available analytical signal. A sample cell enclosed in a silvered capillary tube can further assist the collection and return of optical radiation.

7.2.1 Single-Fiber Systems

7.2.1.1 Extrinsic and Intrinsic Sampling Strategies

Optical fibers can be employed in two distinct configurations for probing selective chemical interactions. The reagent phase can be compartmentalized and mounted at the end of a fiber. This is termed an extrinsic sensor and has the advantage that the reagent phase may be simply altered or replaced on a disposable basis. Furthermore, the chemistry can be located in an environment different from the fiber, providing flexibility for interfacing to difficult sample matrices such as high-pressure vessels or corrosive solutions. The second fiber-based sampling strategy involves manipulation of the cladding that usually surrounds the optically conductive core. When light is transmitted through the core, a significant component of the electromagnetic radiation will propagate beyond the physical boundary of the core to distances of a fraction of the wavelength. This evanescent wave will interact with the environment external to the core and can be the major feature responsible for loss of intensity as the radiation moves along the length of a fiber. Appropriate manipulation of the refractive index of the external core environment by enveloping the core in a cladding greatly reduces such intensity loss in fibers designed for long-range transmission. However, it is also possible to replace the cladding with the selective reagent chemistry, providing a situation where the evanescent wave is employed to optically interact with the analytical chemistry (Burghardt and Thompson, 1984; Place et al., 1985).

An optical fiber carries radiation by total internal reflection when an angle of reflection θ at an interface between two transparent media of different refractive indices is larger than some critical angle θ_c. The relationship between the critical angle and the refractive indices can be written as

$$\theta_c = \sin^{-1}(n_2/n_1)$$

where the light impinges the interface between the two transparent media from the optically denser material, implying that $n_1 > n_2$. The intensity propagation I of the evanescent wave into the medium of lower refractive index (n_2) is dependent on the reflection angle θ, wavelength, and a Fresnel transmission factor T:

$$I = T(\theta)\exp(-2Z/d_p)$$

where Z represents distance, and d_p is given by

$$d_p = \frac{\lambda/n_1}{2\pi\left[\sin^2\theta - (n_2/n_1)^2\right]^{1/2}}$$

The value of d_p is therefore defined as the distance where the electric field amplitude has decayed to $1/e$ of the value directly at the surface. The evanescent wave propagates with a d_p value in the range of a few hundred nanometers for optical wavelengths. Therefore, optical excitation by the evanescent wave will only occur close to the waveguide surface, eliminating bulk sample matrix effects and providing an optical technique well suited for surface spectral analysis. The multiple internal reflection excitation obtained from evanescent wave propagation provides for greatly improved relative sample pathlengths and signal-to-noise ratios in situations where the selective chemistry consists of a thin film. Such systems have been reported for both absorption and fluorescence modes and are considered as intrinsic sensors since the selective chemistry actually forms part of the conventional optical-fiber structure.

7.2.1.2 Transmission-Based Sensors

Transmission systems can operate in both extrinsic and intrinsic modes and are characterized by a source at one end and a detector system at the other. This configuration is relatively awkward for remote sensing using a spectrophotometer since the fiber would extend to the selective chemistry and then return to the analytical instrument, effectively doubling the required fiber length. Optical coupling in an extrinsic mode to a remote chemical cell is more complicated since the fiber must enter the sample zone, and the resulting interaction may be monitored by a second fiber leading back to the spectrophotometer. The intrinsic operational mode is much better suited for transmission operation due to the simplicity of instrumental coupling, the ease of chemical interfacing and availability of suitable optical probes, and the analytical sensitivity manipulation achieved by coating different lengths of fiber.

7.2.1.3 Reflection-Based Sensors

A preferred fiber configuration for optrode design makes use of a single fiber that acts to conduct excitation radiation to the selective chemistry and return the analytical signal via the same route. This can be accomplished by a reflective coating mounted at the end of the sample housing, which can direct radiation back into the fiber. Similar results can be obtained with reagent phases deposited on microscopic polymer beads causing radiation scatter from a sample reservoir or from trapped fluorescent reagent solutions that can act as independent optical sources. This latter technology can also be employed for intrinsic sensors where a fluorescent reagent is covalently immobilized around a fiber core at the sensing tip of the fiber.

7.2.2 Bifurcated Fibers

Occasionally, it is advantageous to use two fibers to couple with a chemical sensor head in a reflection or fluorescent experiment. One fiber can be used to guide excitation radiation to the sample while the other would guide the analytical radiation to a photodetector. Such systems offer the advantages of observing only the common zone of the reagent phase illuminated by the cone of probe radiation and sampled by the cone of accepted analytical radiation, improvements in the capability for separation and filtering of different wavelengths, and simplified placement of optics in the spectrophotometer. In the case of a single-fiber probe, it is evident that reflected or fluorescent radiation must pass through a beam splitter to select the analytical information from the probe radiation and to direct the desired radiation into a photodetector. Bifurcated fibers eliminate the need for beam splitters.

7.2.3 Interferometry

Interferometric sensors rely on the change in optical pathlength and/or index of refraction produced by the measured parameter (11). Optically matched fibers, or composite fibers, which consist of two conducting pathways in a single fiber housing, provide the basis for an interferometry cell. A selective chemical cladding on the surface of an indicator arm of a matched fiber set can induce refractive index or mechanical changes through the chemical interaction process to change the optical path relative to the reference arm. Little work has been reported with respect to chemical sensors based on interferometry, though numerous fiber devices for position sensing have been described. Reversible polymerization reactions such as those found in natural antibody–antigen interactions should provide suitable chemistry for changing the physical properties of an indicator fiber.

7.2.4 Multiplexed Distributed Systems

One of the significant advantages of fiber-optic sensors for remote analysis is based on the fact that a conventional laboratory instrument in a quality control laboratory can remotely sample in (almost) real time a number of different points in a distributed array, such as may be applicable in a large processing plant (Milanovich and Hirschfeld, 1984). The communications industry has developed the capacity to selectively and efficiently switch and sample many different fibers at high speed. The multiple measurements of a distributed system require similar optical multiplexing or switching to accommodate large numbers of fibers. The required signal-to-noise ratio would determine the length of sampling time for each input channel. A remote sensing configuration over distances of several hundred meters and sampling 120 points at concentrations of parts per billion has been reported. Such distributed networks greatly reduce sensor costs since the number of individual instruments is no longer of the same order as the number of points to be sampled. This substantially increases analytical flexibility and applicability for controlling on-line systems in situ for feedback control and optimization by virtue of the implementation of the large number of sensors required for most realistic operations.

7.3 CONTEMPORARY FIBER-OPTIC SENSOR DEVELOPMENT

From the perspective of this chapter, fiber-optic sensors may be divided into three generations.

The first generation includes sensors that do not involve a chemical process. These rely on fiber transmission characteristics themselves as indicators of external parameters. The first generation is not of great relevance in this chapter, as no chemical sensing is involved. These sensors will be briefly reviewed as an introduction to the general configuration of fiber-optic sensors.

The second generation is composed of extrinsic sensors that are capable of remote adaptation of classical spectroscopic instruments. Here the fiber optic is simply used as a lightguide, bringing incident and resultant light to and from a sensing area. While most remote fiber-optic chemical sensors are currently of the second generation, the number of classical optrodes that may benefit from remote capability is nearing saturation. Problems encountered with these devices include slow sensor response due to the need for diffusion of chemical species into a sensing environment, difficulty in miniaturization of the sensor area, and coupling of a fiber to the sensing component.

In the third generation, the fiber optic is intrinsically a component of the sensing area. This generally involves binding a spectroscopically active region to the surface or distal end of a fiber. A few sensors of this type have been developed, with many other possibilities proposed. In addition to avoiding many of the problems of the second generation, these have also introduced the potential for evolution of new types of sensors. Where spectroscopic interaction with the fiber is via the evanescent component of the light in the fiber, these sensors are surface sensitive and independent of bulk sample interferences. In addition to working sensors, new data pertaining to the long neglected field of total internal reflection spectroscopy is also a result of this technology.

A potential fourth generation, where the sensing region is used also as a portion of the lightguide, will be discussed in a later section.

7.3.1 First-Generation Sensors

The principal characteristic of an optical fiber that may be altered by its external environment (often used in first-generation sensors) is its index of refraction, n. Changes in n are most easily measured by interacting light from two fiber sections, one used as a reference, and observing the interference pattern that develops as one value of n changes with respect to the other. Factors affecting n include external pressure and stress upon the fiber. Such parameters, and others that affect them, may be measured as changes in transmitted intensity due to interference. Fiber-optic interferometers have been developed for detection of sound and water pressure (Cole et al., 1977). Single-fiber interferometers have been devised using phenomena such as the Doppler effect (Durst and Krebs, 1984), as an anemometer, and by the transmission of circularly polarized light through a birefringent medium (Pratt et al., 1984), as a magnetometer.

7.3.2 Second-Generation Sensors

Second-generation sensors have been designed to use many spectroscopic principles, with varying degrees of success. A cross section of optrode designs are reviewed here, with a more complete listing to be found in Table 7.2. Most may be divided into categories of absorbance, reflectance or scattering, and fluorescence.

Brenci et al. (1984) used the changing absorbance spectrum of a cobalt–chloride solution as a probe for temperature. Absorption at 655 nm was measured from 25 to 50 °C as a temperature-dependent signal. Comparison with absorbance at 800 nm eliminated light source drift problems, and a resolution of 0.4 °C or better was achieved with commercial silica/silicone fibers, a tungsten lamp, and interference light filters. The optrode was isolated from the environment by encasing the solution in a closed, opaque container, which should eliminate chemical and stray-light interferences. Response time, though not stated, should be as fast as the conduction of heat into the optrode. The complete lack of metallic components in this sensor makes it ideal for environments where metals may not be introduced, such as the microwave and high-intensity radio fields used in some clinical procedures.

A remote absorbance cell was devised by Freeman et al. (1985) by simply encasing a fiber in a vespel housing into which a slot had been cut (Figure 7.1).

Table 7.2. Fiber-Optic Sensors

Analyte	Probe	Mechanism	Reference
Sound	Fiber	Interferrometric signal	Cole et al., 1977
Magnetism	Fiber	Interferrometric signal	Pratt et al., 1984
Methane	—	Infrared absorbance	Stueflotten et al., 1984
Radiation	Semiconductor	Scintillation	Batchellor and Carless, 1984
Temperature	Cobalt solution	Absorbance spectrum	Brenci et al., 1984
Copper II	—	Visible absorbance	Freeman et al., 1985
Absorbers	—	In vivo absorbance	Coleman et al., 1984
p-Nitrophenyl phosphate	Alkaline phosphatase	Reflectance from membrane with immobilized enzyme	Arnold, 1985
Algae	—	Intrinsic fluorescence	Lund, 1984
pH	Bromothymol blue	Reflectance from indicator immobilized on external surface	Kirkbright et al., 1984*a*
pH	Bromothymol blue	Reflectance from indicator immobilized on fiber end	Kirkbright et al., 1984*b*
Al (III)	Morin	Fluorescence on formation of complex with immobilized morin	Saari and Seitz, 1983
pH	HOPSA	Fluorescence of dye	Zhujun and Seitz, 1984a
CO_2	HOPSA	CO_2-permeable membrane over pH sensor	Zhujun and Seitz, 1984b
Halothane/O_2	Decacyclene	Fluorescence quenching	Wolfbeis et al., 1985
O_2	Perylene dibutyrate	Fluorescence quenching	Peterson et al., 1984
In vivo pH	Phenol red	Absorbance of light by dye	Abraham et al., 1985a
In vivo pCO_2	Phenol red	CO_2-permeable membrane over pH sensor	Abraham et al., 1985b

The portion of the fiber in the slot was removed, forming a light path of known length. Radiation at 820 nm was absorbed by conventional Beer–Lambert principles as an indicator of copper sulfate concentration in an electroplating solution. In addition to the remote capability of this "spectrophotometer," the problems associated with handling and cleaning absorbance cells were eliminated.

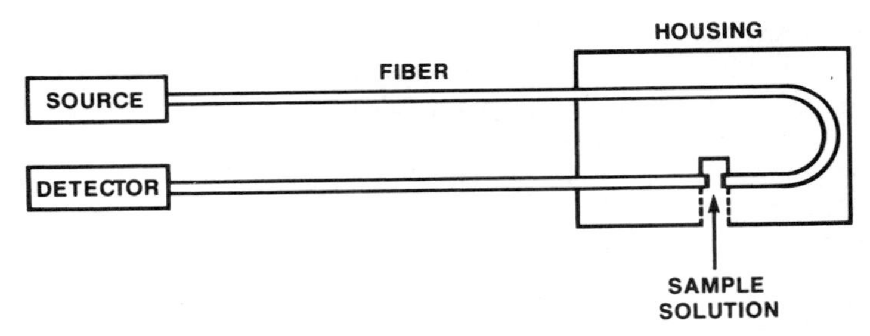

Figure 7.1. Fiber-optic-based absorption cell indicating simplicity of construction of probe head and great potential for probe miniaturization. Fiber alignment and finish of fibers at sample solution well are important considerations in optimization of optical radiation transfer (Freeman et al., 1985).

The response time of this sensor should be practically instantaneous. The vulnerability to interference of this type of optrode makes it useful only in reasonably predictable environments.

A catheter-style probe for making absorbance measurements in vivo has been built by Coleman et al. (1984; Figure 7.2). A single fiber was used for both

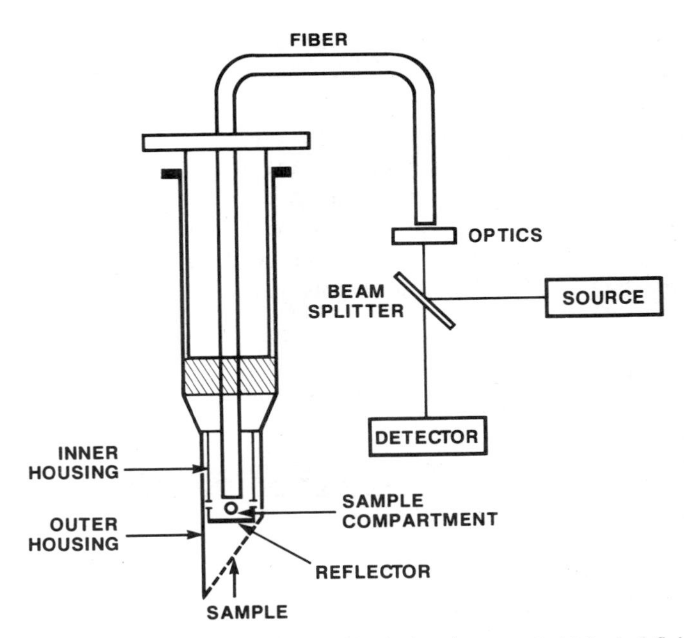

Figure 7.2. Hypodermic-mounted fiber-optic probe designed to sample biological fluids as they enter miniature solution sample well housed within syringelike needle. Reflector placed behind solution compartment ensures efficient capture of radiation back into fiber. This system represents extrinsic sampling mode based on single-fiber transmission (Coleman et al., 1984).

incident and resultant radiation, allowing for further miniaturization. The probe was essentially a hypodermic with a fiber inserted down into the needle, making it a probe. A hole was drilled through the side of the steel probe, and a reflector was placed below this hole across from the fiber end. This formed a small, mobile absorbance chamber for spectrophotometric measurements. Although the lack of selectivity prohibits any significant use of this sensor, it is a notable design from which biomedical fiber-optic sensors may be fabricated.

Selectivity was introduced in a fiber-optic sensor by Kirkbright et al. (1984b). A reflectance-type sensor featured a pH indicator dye embedded into a support polymer that was coated onto the end of a fiber (Figure 7.3). Bromothymol blue gave pH indication as a function of reflectance over the range 6.5–9.5. It was presumed that other pH ranges may be selected by choosing other dyes. Selectivity was introduced by coating the reflective membrane with polytetrafluoroethylene (PTFE), which is permeable to protons but to very few other solution components. The sensor showed no response to solution turbidity. However, significant signal changes were observed due to changes in ionic strength and solution temperature. Response time was given as 65 s for 63% of final value.

Fluorescence was used by Zhujun and Seitz (1984a, 1984b) in developing optrodes for pH and CO_2 (Figure 7.4). The trisodium salt of 8-hydroxy-1,3,6-pyrenetrisulfonic acid (HOPSA) was immobilized on an anion exchange membrane at the end of a fiber. In contact with solution, there was an equilibrium between the species HOPSA and OPSA—that depends on the pH of the solution. The fluorescence signal at different wavelengths specific to each of these species was used as an indicator of pH, with the greatest sensitivity occurring around the pK_a value of 7.3. Alternatively, a ratio of two peaks, at wavelengths that corresponded to each species, was used with this sensor. Use of the ratio allowed the device to be self-correcting in terms of source intensity, temperature fluctuations, fluorescence quenching, and loss of HOPSA from the membrane. This optrode thus exhibited good selectivity and was essentially self-calibrating

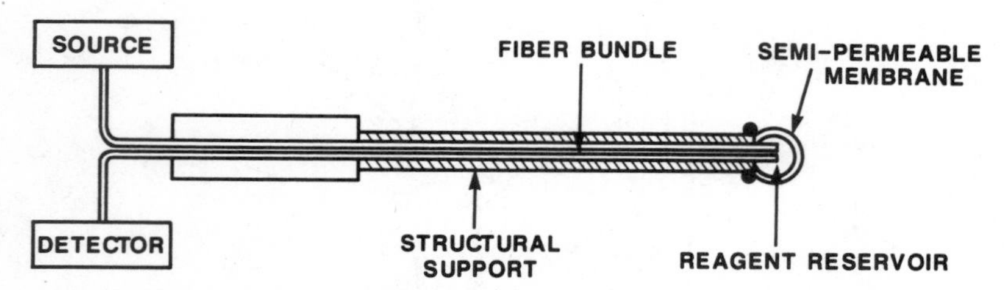

Figure 7.3. Extrinsic mode bifurcated fiber-optic probe design where optical radiation may be carried by single fibers or fiber bundles to and from replaceable sampling compartment. Sample cell consists of fluorescent or light-scattering reagent phase separated from external environment by selectively permeable membrane (Kirkbright et al., 1984b).

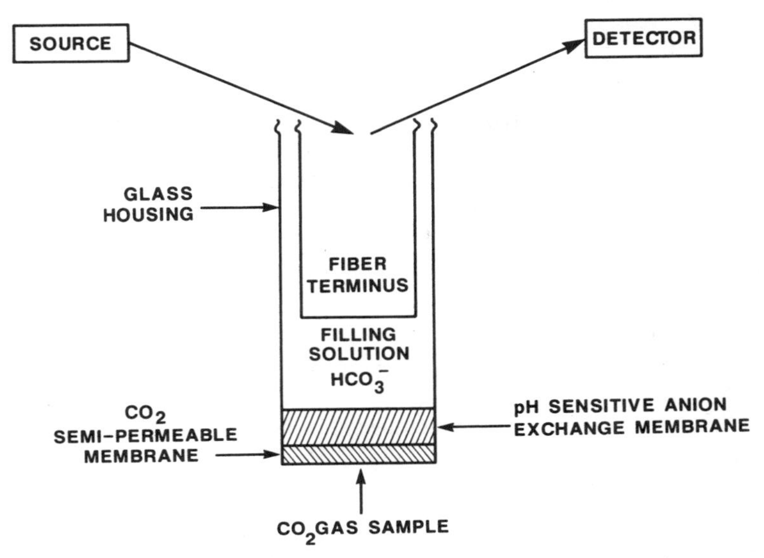

Figure 7.4. Example of selective chemical cell proven successful in extrinsic mode bifurcated fiber probe design. CO_2 gas selectively passed through semipermeable membrane into reagent layer consisting of pH-sensitive anion exchange membrane containing pH-sensitive immobilized fluorophore (Zhujun and Seitz, 1984b).

under a wide variety of circumstances. The pH sensor was applied as a CO_2 sensor by placing it in a solution behind a membrane permeable only to CO_2 and measuring the pH of the resultant carbonic acid solution. Standard deviations of 0.02–0.08 were reported for pH near 7.3. Temperature and ionic strength changes were observed to equally affect both emission wavelengths, eliminating these prime sources of interference. Response times were 70–120 s for a pH change of 6–8 and approximately 5 min for a drastic CO_2 concentration change from 2×10^{-3} to 4×10^{-3} *M*. Optrodes very similar to these, using phenol red as an indicator, were used in vivo in dogs (Abraham et al., 1985a and 1985b), demonstrating their functionality in real systems.

A similarly configured optrode was utilized by Wolfbeis et al. (1985) in constructing a sensor for halothane (2-bromo-2-chloro-1,1,1-trifluoroethane, an anaesthetic) and oxygen. A sensor based on the quenching of fluorescence of various polycyclic aromatic hydrocarbons (PAH) was found to respond to both halothane and oxygen. By coating the original sensor with PTFE, which is permeable to O_2 but not halothane, a halothane-independent signal resulted. This oxygen sensor was used to subtract the oxygen component from the signal of the original dual sensor, enabling measurement of both halothane and oxygen concentrations. Precision for this optrode was 0.2% and was superior to a Clark oxygen electrode at higher oxygen concentrations. Response time (90% final value) was given as 15–20 s for halothane and 10–15 s for oxygen.

An in vivo oxygen probe based on fluorescence quenching was constructed by Peterson et al. (1984) (e.g., see Figure 7.3). Perylene dibutyrate was immobilized onto small silica beads that were then encased in a porous polypropylene envelope. This envelope formed the probe, with two fibers coupled into one end. A deuterium light source furnished blue excitation light, which was focused into one of the fibers. Through the other fiber, emitted green fluorescent light as well as scattered incident blue light were directed to a photodetector. The analytical signal was a function of the ratio of these two light signals, automatically correcting for changes in source intensity and in recovery and transmission of fluorescent light. Comparison with an electrode gave a maximum error of 1 Torr over the range 0–150 Torr, with a typical error of a few tenths of a torr. Analysis of interference showed a 0.6% decrease in fluorescence per degree Celsius increase in temperature and significant quenching by halothane as the only serious factors. Response time was reported as 30 s for 63% of the final value.

The above examples are representative of the current state of the second generation of remote fiber-optic chemical sensors. Several reviews have been written that include some comment on this topic (Hirschfield, 1986; Place et al., 1985; Seitz, 1984; Wolfbeis, 1985). A summary of the achievements and problems of chemical optrodes may be found in Table 7.3. It is apparent that for fiber-optic-based sensors to realize the potential ascribed to them, some major obstacles must be overcome. For true integration into real-time feedback devices, response time must be lowered. For the sensors to be of significant value, remote operation will not be enough. Selectivity and sensitivity rivaling the best instrumental capabilities must be achieved, and true compositional analysis must be made possible. This will require sophisticated optrodes capable of monitoring complex analytes in complicated environments at trace levels. Fortunately, there is every indication that this will be possible.

7.3.3 Third-Generation Sensors

The in vivo capability, potential low cost and disposibility, nonmetallic construction, and nonelectrical operation of fiber-optic sensors makes them most attractive to biomedical analytical applications. This is evident in the third-generation prototype sensors that have been developed. Thus, the end target for the sensor may be stated as systems having the selectivity and sensitivity of a fluorescent binding biochemical assay, the interferences of a chromatographically purified sample, the ability to function in a living biological environment (biocompatible), the ability to respond quickly (in a few seconds), and high reversibility to analyte so as to rapidly follow environmental changes.

Table 7.3. Chemical Optrodes

Advantages	Disadvantages
No equivalent of reference or junction potential	Sensitive to ambient light
Easily miniaturized	General lack of long-term stability
Free from electrical, magnetic and static interference	Not "log sensitive," usually limited dynamic range
Capable of more than one signal on each fibre (multiplexing)	Only applicable to spectroscopically active sensing strategies
Easily made remote, workable in many environmental situations	

The first true third-generation optrode was a sensor used by Newby et al. (1984) to detect adsorption of protein onto a glass surface. The cladding was stripped from the end section of a fiber, a reflective surface was coated onto the distal end, and adsorption of protein onto the fiber was monitored by evanescent excitation of fluorescence. The fluorescent dye Rhodamine 6G was covalently labeled onto IgG protein, and adsorbed protein was detected independently of bulk solution protein.

Munkholm et al. (1986) covalently bound a polyacrylamide layer to the distal end of a fiber. A pH sensor was made by incorporating fluorescein into the polyacrylamide. However, the authors were more interested in proof of concept than in another pH optrode. It was stated that sensors of the evanescent type are not practical because of the small surface area and signal available (in spite of the working prototypes that exist). The advantage of avoiding a membrane between the sensor and the sample is illustrated by the Munkholm optrode, as response time was reported as 9 s.

One of the fundamental uses of fluorescence in biochemistry is in fluoroimmunoassay. A fiber-optic approach to assay techniques was proposed by Andrade et al. (1985). In this proposal, analysis for an antigen (Ag) is based on reaction with a targeting antibody (Ab). For fluoroassay, one or both of these must be fluorescent. When intrinsic fluorescence does not occur, one of the components must be labeled without losing the specificity of binding between them. As a typical assay strategy, Andrade described first covalently attaching Ab to the fiber surface and then reacting the surface with Ag. An increase in fluorescence during this binding reaction would be proportional to Ag concentration in the analyte solution. If Ag was not fluorescent and could not be labeled, a further reaction with Ab would result in a secondary fluorescent signal, still proportional to the concentration of Ag in the original analyte solution. Kinetic studies of the binding process would also be possible, as the evanescent excitation method allows detection of the bound molecules independent of molecules in the bulk sample.

The Andrade proposal is demonstrated by a fiber-optic sensor fabricated by Dahne et al. (1984) to assay for human IgG. To determine IgG concentration, antibody anti-human IgG was first immobilized onto a fiber surface. Antigen IgG was then allowed to bind onto the antibody from the solution, the extent of binding being proportional to the concentration of the IgG. The actual extent of binding was determined by further reaction of the IgG with fluorescently labeled antibody (FITC-rabbit anti-human IgG). This reaction was monitored through evanescently induced fluorescence. The fluorescence intensity observed was significant at an antigen concentration of 1 μg/mL IgG.

Deposition of a fluorescently labeled bilayer lipid membrane onto a fiber was proposed by Krull et al. (1986) as the basis of a sensor. It was demonstrated that the membrane probes phloretin and valinomycin caused significant changes in the fluorescence signal observed when they were allowed to perturb a fluorescently labeled monolayer. This was proposed as the fundamental mechanism of a fiber-optic sensor in which a protein receptor is incorporated into a membrane, and fluorescence is monitored as a function of binding to the receptor (Figure 7.5). The function of proteins in biological systems is in part a result

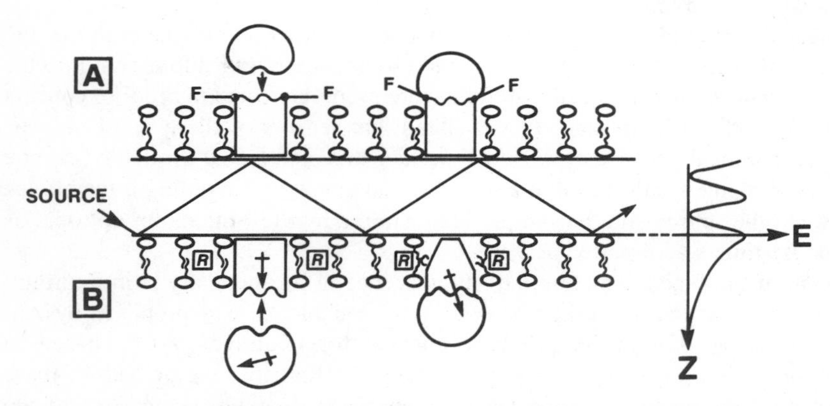

Figure 7.5. Intrinsic mode fiber-based sensor employs chemically selective lipid-membrane-based cladding and optical excitation by evanescent wave strategy. (*a*) Receptor covalently conjugated with fluorophore (F) selectively binds with stimulant, resulting in alteration of emission characteristics of fluorophore. (*b*) Fluorophore reporter molecule (R) sensitive to membrane structure and/or electrostatic fields resides permanently in chemically selective lipid membrane matrix. Receptor can selectively bind to stimulant, changing membrane structure and/or local electrostatic fields, thereby altering emission characteristics of reporter fluorophore. Evanescent excitation schematically represented as exponential decay of radiation into environment immediately adjacent to surface of optical fiber. Energy of standing wave represented as E, and distance axis from fiber surface is z (see Section 7.2.1.1).

of their ability to perturb the surrounding cell membrane, by affecting membrane structure or electrostatic fields, when a binding event occurs at the protein. This should cause a change in the fluorescence of fluorophores in the membrane, as was observed with the perturbation caused by the membrane probes. The result would be a generic optrode series, where each sensor was of the same basic structure, except for the protein chosen according to the analyte. Some advantages of this strategy are: the use of molecules such as large transmembrane proteins, which are known to be degenerated and to lose their activity outside of a membrane environment; the natural selectivity and sensitivity demonstrated by some biological membrane systems; the potential of affecting many fluorescent molecules with a single binding event or chemical signal amplification; and perhaps the opportunity to further understand the functioning of biological membranes.

7.4 CURRENT DEVELOPMENTS

7.4.1 Intrinsic Sensor

The immediate development of third-generation intrinsic sensors will likely center on the fiber–optic adaptation of current total internal reflection (TIR) technology. This implies that sensor transduction systems will be based on polymer or biochemical fiber coatings, as these are currently the materials being thus studied. This is partly the basis for emphasis on the biochemical assay for current and proposed sensors.

While evanescent excitation of fluorescence is well characterized, a second concern is the subsequent conduction of fluorescence in the desired light guide. It would at first appear that only a very small portion of emitted fluorescence would be at an angle that would be conducted down the fiber. However, work by Lee et al. (1979) indicates that a reasonable signal may be expected. Using a semicircular prism, fluorescence signal versus angle of transmittance into the prism was measured. The signal was observed to peak at the critical angle between the prism and the fluorescent coating, independent of the angle of the exciting radiation. An explanation of this phenomenon will not be pursued here. However, this indicates that a peak fluorescence signal occurs at exactly the angle where the signal begins to be transmitted down the fiber. The observation of fluorescence reported in the literature confirms this prediction. Another implication of this result is that the fluorescent coating must have a higher index of refraction than the surrounding medium or the resultant signal would be lost to that medium once the fiber extended beyond the coated region and the critical angle then increased.

Spectroscopic study and subsequent manipulation of organized biological depositions on glass surfaces has been known for some time (see, e.g., Sagiv, 1980). The evanescent excitation of fluorescently labeled monolayers on glass surfaces is gaining popularity. Weis et al. (1982) used TIR fluorescence to study the interaction of rat basophil leukemia cells with monolayers of phospholipid deposited onto glass surfaces. The surface sensitivity of the evanescent excitation allowed for separation of adsorbed fluorescent species from those remaining in solution and also prevented interference due to other fluorescing species that may have been present. The use of intersecting lasers resulted in an interference pattern in the observed fluorescence, which may open new methods of studying localized phenomena on fiber surfaces like those using microscopy on flat surfaces, if intersecting laser beams may be directed together through a fiber. Axelrod et al. (1983) utilized TIRF microscopy in various such studies of interaction at surfaces. These included studying binding rates and surface diffusion through photobleaching of fluorescent molecules and then monitoring recovery of fluorescence and using correlational spectroscopy. Thompson et al. (1984) observed the dipole orientation of adsorbed fluorescence emitters by rotating the plane of a polarized laser excitation source, thus rotating the subsequent evanescent field.

Such TIR experiments suggest the feasibility for development of a series of sensors incorporating the same biological systems. The most obvious is an adaptation of classical fluoroimmunoassay techniques (see, e.g., Petrossian and Owicki, 1984) that have measured hapten concentrations in the nanomolar range using the quenching of fluorescence as the measured signal. The surface sensitivity of the evanescence suggests a potentially more sensitive method of detecting fluorescence of analyte bound to an optrode, preventing interferences from the bulk sample such as other fluorescing species and absorption of the signal by colored species. Photobleaching and recovery could be used where there is an exchange between bound analyte and that in solution, to study the binding reaction, as well as to determine concentration. Phase effects, such as interferometric patterns, may be used as additional dimensions in sensors, providing an even greater range of measured properties. These will rely on the increasing use of single-mode fibers, which maintain phase information over long fiber transmission distances.

7.4.2 Chemically Selective Light Guide

An intrinsic fiber sensor is usually designed as a fiber core coated with a chemically selective cladding that is probed by the salient evanescent wave. These devices can be quite sensitive if selective coatings are relatively thick and highly optically absorptive. However, the evanescent wave intensity and penetration depth is limited, and capture of fluorescent radiation by the fiber from the cladding is limited, reducing the analytical sensitivity of the device. An exciting direction in chemically selective fiber-optic development is based on an intrinsic fiber where the analyte directly affects the optical properties of the fiber core (see Figure 7.6). The evanescent wave would not participate as the source of excitation in this scenario since the selective chemistry would be located within the light guide. This maximizes physical parameters such as excitation radiation intensity as well as optimal conditions for radiation capture or transmission and advantageously integrates the chemistry and transducer into a single entity.

The major problem with such a sensor is determined by the limited experimental experience with chemically selective matrices that are known to act as efficient light guides. Work has been reported for thin-film light guides prepared by depositing high-refractive-index light guides on low-refractive-index substrates (Ulrich and Martin, 1971). Studies have further included liquid solution deposition onto glass of materials such as epoxy, polyurethane, photoresist, multilayer lipid matrices, and dye-doped polymers such as polyurethane–Rhodamine 6G mixtures (Ulrich and Weber, 1972). Little work has been reported regarding chemically selective optical fiber cores, but experiments with lipid matrices (Pitt and Walpita, 1980) and dye-doped polymers provide strong support for development potential. Considering the optical conductivity of lipid multilayers and their inherent advantages for chemical receptor incorporation and activation, it would seem that a structurally supported lipid matrix free from structural defects could serve as a highly refined remote chemical sensor (see Figure 7.6).

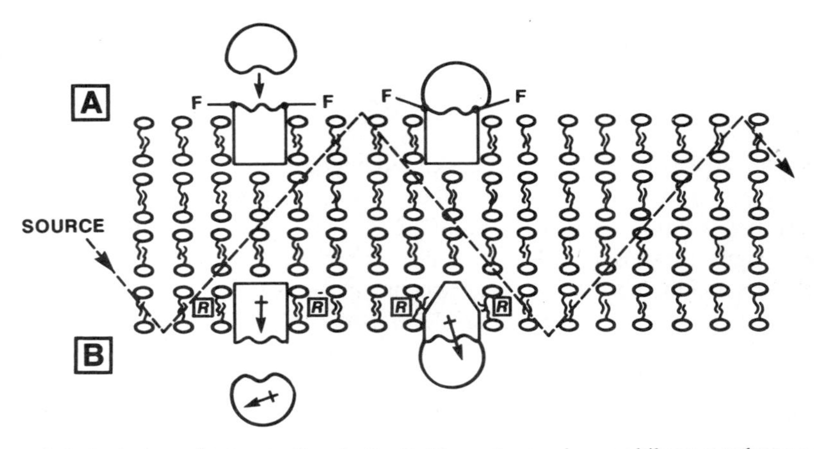

Figure 7.6. Intrinsic mode chemically selective lipid membrane where multilayer membrane acts as waveguide. This configuration maximizes excitation radiation intensity as well as fluorescence radiation capture and transmission in chemical scheme based on processes described in Figures 7.5*a* and *b*.

7.5 FUTURE DEVELOPMENTS

The development of remote fiber-optic sensors is continuing at a rapid pace. Presently, a number of clearly defined limitations exist that can be classified as either instrumental or chemical problems.

7.5.1 Instrumentation

Research in the area of fiber-optic sensors will continue to stress multiplexing development for multiple distributed real-time analysis. The main instrumental impediment to development is, surprisingly, that the majority of fiber-optic research is becoming specialized for the communications industry. This involves devices that operate in the IR region. Noted areas are the development of new fibers and of semiconductor lasers, for which the state of the art is exclusively IR. Essentially all spectroscopic processes of a quantitative nature involve visible–near UV radiation. Development in this region will occur eventually, not only as a result of the demand for sensors, but also as a result of the relatively new field of optoelectronics. This should eventually result in integrated sensor devices capable of environmental, chemical, and instrumental drift corrections as well as signal amplification and chemometrics, all via optical components. Greater emphasis will be placed on the multidimensional information-carrying capacity of fibers, though current work continues with simpler one- or two-wavelength systems concentrating on improvements in distance, signal quality, and signal quantity. Further optrode designs stressing miniaturization, improved sample interfacing, and multiple-pass configurations for signal-to-noise ratio improvements can be expected.

7.5.2 Chemistry

The chemical components in fiber-optic devices include both the fiber and the selective reagent chemistry. New fiber materials based on composites will be developed for improvements in reproducibility, long-term stability, and shorter wavelength transmission. This may also improve the surface characteristics of the fibers and the potential to bind reagents to the surface. The possible sensing phases are limited only by the possible spectroscopic transducers available. Emphasis must be placed on improvements in the marriage of the selective chemistry to the optics, reducing leaching loss, and degradation of the interface. Developments in the selective chemistry will include a much wider range of optically active probes for a much larger number of analytes. Directions for improvement will be governed by requirements for sensitivity, stability, reproducibility, and resistance to interference or contamination. Further work will also be directed toward the evolution of chemical monitoring of optrode performance characteristics and their relative temporal variations.

These points are illustrated by the biological interest noted previously in this chapter. Current usage of the more significant systems has been restricted to controlled laboratory conditions, but the marriage of these biosensing layers to fibers will have to be more regulated and more robust than simple deposition if an optrode that will operate in a real environment is desired. At the same time, these are model systems in terms of such characteristics as sensitivity and selectivity, often far outperforming conventional chemical analysis, especially on a real-time basis.

The importance of the development of remote fiber-optic sensors lies mainly in distributed in situ industrial and in vivo applications of a generic sensor based on this strategy. Such applications lie mainly in the biomedical field, owing to both clinical need and commercial potential, but include everything from compositional process control to environmental monitoring.

REFERENCES

Abraham, E., D. R. Markle, S. Fink, H. Ehrlich, M. Tsang, M. Smith, and A. Meyer (1985a) Continuous Measurement of Intravascular pH with a Fiber Optic Sensor. *Crit. Care Med.* April, 354.

Abraham, E., D. R. Markle, G. Pinholster, and S. Fink (1985b). Non-invasive Measurement of Conjunctival pCO2 with a Fiber Optic Sensor. *Crit. Care Med.* April, 352.

Alberry, J., B. Haggett, and D. Snook (1986). You Know It Makes Sensors, *New Scientist* **109**(1495), 38.

Andrade, J. D., R. A. van Wagenen, D. E. Gregonis, K. Newby, and J.-N. Lin (1985). Remote Fiber-optic Biosensors Based on Evanescent-excited Fluoroimmunoassay: Concept and Progress. *IEEE Trans. Elect. Dev.* **ED-32**(7), 1175.

Arnold, M. A. (1985). Enzyme-based Fiber Optic Sensor. *Anal. Chem.* **57,** 565.

Axelrod, D., N. L. Thompson, and T. P. Burghardt (1983). Total Internal Reflection Fluorescence. *J. Microscopy* **129,** 19.

Batchellor, C. R., and D. Carless (1984). Fiber Optic Sensor for High Energy Radiation, in R. Th. Kersten and R. Kist, eds., *International Conference on Optical Fiber Sensors*, VDE-Verlag GmbH, Berlin, Offenbach, p. 143.

Brenci, M., G. Conforti, R. Falciai, A. G. Mignani, and A. M. Scheggi. (1984). Thermochromic Transducer Optical Fiber Temperature Sensor, in R. Th. Kersten and R. Kist, eds., *International Conference on Optical Fiber Sensors*, VDE-Verlag GmbH, Berlin, Offenbach, p. 155.

Burghardt, T. P., and N. L. Thompson (1984). Effect of Planar Dielectric Interfaces on

Fluorescence Emission and Detection. Evanescent Excitation with High-Aperture Collection. *Biophys. J.* **46,** 729.

Cole, J. H., R. L. Johnson, and P. G. Bhuta (1977). Fiber-optic Detection of Sound. *J. Acoust. Soc. Am.* **62,** 1136.

Coleman, J. T., J. F. Eastham, and M. J. Sepaniak (1984). Fiber Optic Based Sensor for Bioanalytical Absorbance Meausrements. *Anal. Chem.* **56,** 2247.

Dahne, C., R. M. Sutherland, J. F. Place, and A. S. Ringrose (1984). Detection of Antibody-antigen Reactions at a Glass-liquid Interface: A Novel Fiber-optic Sensor Concept, in R. Th. Kersten and R. Kist, eds., *International Conference on Optical Fiber Sensors*, VDE-Verlag GmbH, Berlin, Offenbach, p. 75.

Durst, F., and H. Krebs (1984). Glassfiber LDA-systems and Their Application to Internal Combustion Engines, in R. Th. Kersten and R. Kist, eds., *International Conference on Optical Fiber Sensors*, VDE-Verlag GmbH, Berlin, Offenbach, p. 750.

Freeman, J. E., A. G. Childers, A. W. Steele, and G. M. Hieftje (1985). A Fiber-optic Absorption Cell for Remote Determination of Copper in Industrial Electroplating Baths. *Anal. Chim. Acta* **177,** 121.

Hirschfeld, T. (1986). Remote and In-situ Composition Analysis: Have Your Cake, Eat It Too. *InTech* **33**(2), 45.

Kirkbright, G. F., R. Narayanaswamy, and N. A. Welti (1984a) Studies with Immobilised Chemical Reagents Using a Flow-cell for the Development of Chemically Sensitive Fibre-optic Devices. *Analyst* **109,** 15.

Kirkbright, G. F., R. Narayanaswamy, and N. A. Welti (1984b) Fibre-optic pH Probe Based on the Use of an Immobilized Colorimetric Indicator. *Analyst*, **109,** 1025.

Krull, U. J., C. Bloore, and G. Gumbs (1986). Supported Chemoreceptive Lipid Membrane Transduction by Fluorescence Modulation: The Basis of an Intrinsic Fibre-optic Biosensor. *Analyst* **111,** 259.

Lee, E-H., R. E. Benner, J. B. Fenn, and R. K. Chang (1979). Angular Distribution of Fluorescence from Liquids and Mono-dispersed Spheres by Evanescent Wave Excitation. *Appl. Opt.* **18,** 862.

Lund, T. (1984). Fluorescence Sensor Based on Fiber Optics, *IEE Proceedings* **131(H),** 49.

Measures, R. M. 1984. *Laser Remote Sensing*, Wiley, New York.

Milanovich, F. P., and T. Hirschfeld (1984). Remote Fibre Fluorimetry: Using Optics for On-stream Analysis. *InTech* **31**(3), 33.

Munkholm, C., D. R. Walt, F. P. Milanovich, and S. M. Klainer (1986). Polymer Modification of Fiber Optic Chemical Sensors as a Method of Enhancing Fluorescence Signal for pH Measurement. *Anal. Chem.* **58,** 1427.

Newby, K., W. M. Reichert, J. D. Andrade, and R. E. Benner (1984). Remote Spectroscopic Sensing of Chemical Adsorption Using a Single Multimode Optical Fiber. *Appl. Opt.* **23,** 1812.

Payne, D. N. (1984). Optical Fibres for Sensors, in R. Th. Kersten and R. Kist, eds.,

International Conference on Optical Fiber Sensors, VDE-Verlag GmbH, Berlin, Offenbach, p. 353.

Peterson, J. I., R. V. Fitzgerald, and D. K. Buckhold (1984). Fiber-optic Probe for In-vivo Measurement of Oxygen Partial Pressure. *Anal. Chem.* **56,** 62.

Petrossian, A., and J. C. Owicki (1984). Interaction of Antibodies with Liposomes Bearing Fluorescent Haptens. *Biochim. Biophys. Acta* **776,** 217.

Pitt, C. W., and L. M. Walpita (1980). Light Guiding in Langmuir-Blodgett Films. *Thin Sol. Films* **68,** 101.

Place, J. F., R. M. Sutherland, and C. Dahne (1985). Opto-electronic Immunosensors: A Review of Optical Immunoassay at Continuous Surfaces. *Biosensors* **1,** 321.

Pratt, R. H., R. E. Jones, P. Extanca, G. D. Pitt, and K. W. Foulds (1984). Optical Fiber Magnetometer Using a Stabilized Semiconductor Laser Source, in R. Th. Kersten and R. Kist, eds., *International Conference on Optical Fiber Sensors*, VDE-Verlag GmbH, Berlin, Offenbach, p. 45.

Saari, L. A., and W. R. Seitz (1983). Immobilized Morin as a Fluorescence Sensor for Determination of Aluminum (III). *Anal. Chem.* **55,** 667.

Sagiv, J. (1980). Organized Monolayers by Adsorption. I. Formation and Structure of Oleophobic Mixed Monolayers on Solid Surfaces. *J. Am. Chem. Soc.* **102,** 92.

Schuetzle, D., and R. Hammerle, eds. (1986). *Fundamentals and Applications of Chemical Sensors*, Vol. 309, American Chemical Society Symposium Series, Washington, D.C.

Seitz, W. R. (1984). Chemical Sensors Based on Fiber Optics. *Anal. Chem.* **56,** 16A.

Seiyama, T., K. Fueki, J. Shiokawa, and S. Suzuki, eds. (1983). *Proceedings of the International Meeting on Chemical Sensors*, Elsevier-Kodansha, Tokyo.

Stueflotten, S., T. Christensen, S. Iversen, J. O. Hellvik, K. Almas, T. Wien, and A. Graav (1984). An Infrared Fiber Optic Gas Detection System, in R. Th. Kersten and R. Kist, eds., *International Conference on Optical Fiber Sensors*, VDE-Verlag GmbH, Berlin, Offenbach, p. 87.

Thompson, M., C. L. Arthur, and G. K. Dhaliwal (1986). Liquid-phase Piezoelectric and Acoustic Transmission Studies of Interfacial Immunochemistry. *Anal. Chem.* **58,** 1206.

Thompson, M., and U. J. Krull (1984). Biosensors and Bioprobes. *Trends Anal. Chem.* **3,** 173.

Thompson, N. L., H. M. McConnell, and T. P. Burghardt (1984). Order in Supported Phospholipid Monolayers Detected by the Dichroism of Fluorescence Excited with Polarized Evanescent Illumination. *Biophys. J.* **46,** 739.

Ulrich, R., and R. J. Martin (1971). Geometrical Optics in Thin Film Light Guides. *Appl. Optics* **10,** 2077.

Ulrich, R., and H. P. Weber (1972). Solution-deposited Thin Films as Passive and Active Light-guides. *Appl. Optics* **11,** 428.

Weis, R. M., K. Balakrishnan, B. A. Smith, and H. M. McConnell (1982). Stimulation of Fluorescence in a Small Contact Region Between Rat Basophil Leukemia Cells and Planar Lipid Membrane Targets by Coherent Evanescent Radiation. *J. Biol. Chem.* **257,** 6440.

Wolfbeis, O. S. (1985). Fluorescence Optical Sensors in Analytical Chemistry. *Trends Anal. Chem.* **4,** 184.

Wolfbeis, O. S., H. E. Posch, and H. W. Kroneis (1985). Fiber Optical Fluorosensor for Determination of Halothane and/or Oxygen. *Anal. Chem.* **57,** 2556.

Zhujun, Z., and W. R. Seitz (1984a). A Fluorescence Sensor for Quantifying pH in the Range from 6.5 to 8.5. *Anal. Chim. Acta* **160,** 47.

Zhujun, Z., and W. R. Seitz (1984b). A Carbon Dioxide Sensor Based on Fluorescence. *Anal. Chim. Acta* **160,** 305.

INDEX

(*continued from front*)

Vol. 66. **Solid Phase Biochemistry: Analytical and Synthetic Aspects.** Edited by William H. Scouten

Vol. 67. **An Introduction to Photoelectron Spectroscopy.** By Pradip K. Ghosh

Vol. 68. **Room Temperature Phosphorimetry for Chemical Analysis.** By Tuan Vo-Dinh

Vol. 69. **Potentiometry and Potentiometric Titrations.** By E. P. Serjeant

Vol. 70. **Design and Application of Process Analyzer Systems.** By Paul E. Mix

Vol. 71. **Analysis of Organic and Biological Surfaces.** Edited by Patrick Echlin

Vol. 72. **Small Bore Liquid Chromatography Columns: Their Properties and Uses.** Edited by Raymond P. W. Scott

Vol. 73. **Modern Methods of Particle Size Analysis.** Edited by Howard G. Barth

Vol. 74. **Auger Electron Spectroscopy.** By Michael Thompson, M. D. Baker, Alec Christie, and J. F. Tyson

Vol. 75. **Spot Test Analysis: Clinical, Environmental, Forensic and Geochemical Applications.** By Ervin Jungreis

Vol. 76. **Receptor Modeling in Environmental Chemistry.** By Philip K. Hopke

Vol. 77. **Molecular Luminescence Spectroscopy—Parts I and II: Methods and Applications.** Edited by Stephen G. Schulman

Vol. 78. **Inorganic Chromatographic Analysis.** Edited by John C. MacDonald

Vol. 79. **Analytical Solution Calorimetry.** Edited by J. K. Grime

Vol. 80. **Selected Methods of Trace Metal Analysis: Biological and Environmental Samples.** By Jon C. VanLoon

Vol. 81. **The Analysis of Extraterrestrial Materials.** By Isidore Adler

Vol. 82. **Chemometrics.** By Muhammad A. Sharaf, Deborah L. Illman, and Bruce R. Kowalski

Vol. 83. **Fourier Transform Infrared Spectrometry.** By Peter R. Griffiths and James A. de Haseth

Vol. 84. **Trace Analysis: Spectroscopic Methods for Molecules.** Edited by Gary Christian and James B. Callis

Vol. 85. **Ultratrace Analysis of Pharmaceuticals and Other Compounds of Interest.** Edited by S. Ahuja

Vol. 86. **Secondary Ion Mass Spectrometry: Basic Concepts, Instrumental Aspects, Applications and Trends.** By A. Benninghoven, F. G. Rüdenauer, and H. W. Werner

Vol. 87. **Analytical Applications of Lasers.** Edited by Edward H. Piepmeier

Vol. 88. **Applied Geochemical Analysis.** by C. O. Ingamells and F. F. Pitard

Vol. 89. **Detectors for Liquid Chromatography.** Edited by Edward S. Yeung

Vol. 90. **Inductively Coupled Plasma Emission Spectroscopy: Part I: Methodology, Instrumentation, and Performance; Part II: Applications and Fundamentals.** Edited by J. M. Boumans

Vol. 91. **Applications of New Mass Spectrometry Techniques in Pesticide Chemistry.** Edited by Joseph Rosen

Vol. 92. **X-Ray Absorption: Principles, Applications, Techniques of EXAFS, SEXAFS, and XANES.** Edited by D. C. Konnigsberger

Vol. 93. **Quantitative Structure–Chromatographic Retention Relationships.** By Roman Kaliszan

Vol. 94. **Laser Remote Chemical Analysis.** Edited by Raymond M. Measures